MECHANICS OF SOLIDS AND SHELLS

Theories and Approximations

MECHANICS OF SOLIDS AND SHELLS

Theories and Approximations

Gerald Wempner
Professor emeritus
Georgia Institute of Technology
Atlanta, Georgia
U.S.A.

Demosthenes Talaslidis
Professor
Aristotle University Thessaloniki
Thessaloniki
Greece

CRC Press
Taylor & Francis Group
Boca Raton London New York

CRC Press is an imprint of the
Taylor & Francis Group, an informa business

CRC Press
Taylor & Francis Group
6000 Broken Sound Parkway NW, Suite 300
Boca Raton, FL 33487-2742

First issued in paperback 2019

ISBN-13: 978-0-8493-9654-0 (hbk)
ISBN-13: 978-0-367-39569-8 (pbk)

Library of Congress Cataloging-in-Publication Data

Wempner, Gerald, 1928-
Mechanics of solids and shells : theories and approximations / Gerald Wempner, G. Talaslidis Demosthenes.
p. cm. — (Mechanical engineering series)
Includes bibliographical references and index.
ISBN 0-8493-9654-9 (alk. paper)
1. Mechanics, Applied. 2. Solids. 3. Shells (Engineering) I. Demosthenes, G. Talaslidis. II. Title. III. Mechanical engineering series (Boca Raton, Fla.)

TA350 .W525 2002
620.1′.05—dc21 2002073733

Library of Congress Card Number 2002073733

Visit the Taylor & Francis Web site at
http://www.taylorandfrancis.com

and the CRC Press Web site at
http://www.crcpress.com

Preface

The mechanics of solids and the mechanics of shells have long histories; countless articles and books have been written by eminent scholars. Most books address specific aspects which are necessarily restricted. Many are limited by kinematical assumptions, e.g., small strains and/or small rotations. Others are confined to specific behaviors, e.g., elasticity. This book is also necessarily restricted but in a different way:

This book is intended as a reference for scholars, researchers, *and* practitioners, to provide a reliable source for the mathematical tools of analysis *and* approximation. As such, those aspects which are common to all continuous solids are presented in generality. These are the kinematics (Chapter 3), the kinetics (Chapter 4), and the energetics (Chapter 6); the only common limitations are the continuity and cohesion of the medium. The basic quantities, e.g., strains, stresses, energies, and the mathematical relations are developed precisely. To achieve the requisite precision and to reveal the invariant properties of physical entities, the foundations are expressed via the language of vectorial and tensorial analysis (Chapter 2). Additionally, geometrical and physical interpretations are emphasized throughout. Since large rotations play a central role in structural responses, e.g., instabilities, the decomposition of rotations and strains is given special attention and unique geometrical interpretations. Where special restrictive conditions are invoked, they are clearly noted.

With a view toward practical applications, the authors have noted the physical implications of various approximations. In the same spirit, only a few specific results are presented; these are the "exact" solutions of much significance in engineering practice (simple bending and torsion, and examples of actual stress concentrations).

The utility of this book is enhanced by the unification of the various topics: The theories of shells (Chapters 9 and 10) and finite elements (Chapter 11) are couched in the general concepts of the three-dimensional continuous solid. Each of the variational principles and theorems of three dimensions (Chapter 6) has the analogous counterpart in the two dimensions of the shell (Chapter 9).

This book does provide original presentations and interpretations: Principles of work and energy (Chapter 6) include the basic concepts of Koiter's monumental work on stability at the critical load. Additionally, the various complementary functionals are expressed in terms of the alternative strains and stresses; the various versions are fully correlated and applicable to finite deformations. The presentation of the Kirchhoff-Love shell (Chapter 9) includes original treatments for the plastic behavior of shells.

A final chapter views the finite element as a device for the approximation of the continuous solution. The mathematical and physical attributes are described from that viewpoint. As such, the presentation provides a meaningful bridge between the continuum and the discrete assembly.

Our rationale for the content and the structure of this book is best exhibited by the following sequential decomposition:

- The foundations of all theories of continuous cohesive solids are set forth in generality in the initial Chapters 3 and 4.
- The established theories of elasticity, plasticity, and linear viscoelasticity are presented in the subsequent Chapter 5. These are cast in the context of classical thermodynamics. The specific mathematical descriptions of materials are limited to those which have proven effective in practice and are supported by physical evidence.
- The principles of work and energy are presented in the next Chapter 6; the basic forms are given without kinematical limitations. Only the Castigliano theorem is restricted to small deformations.
- The formulations of linear elasticity and viscoelasticity are contained in Chapter 7. This facet of our subject has evolved to the extent that it is vital to a complete understanding of the mechanics of solids. Here, the basic formulations are couched in the broader context of the general theory. These linear theories are included so that our book provides a more complete source of reference.
- The differential geometry of surfaces is essential to the general theories of shells. Chapter 8 presents the geometric quantities and notations which are employed in the subsequent treatment of shells.
- The mechanics of shells is presented in two parts: Chapter 9 sets forth a theory which is limited only by one kinematic hypothesis; a normal is presumed to remain straight. No restrictions are imposed upon the magnitude of deformations. As such, the mechanics of Chapter 9 encompasses the Kirchhoff-Love theory of the subsequent Chapter 10. The latter includes the traditional formulations of elastic shells, *but* also presents original theories of elastic-plastic shells.

- The final Chapter 11 places the basic notions of finite elements in the context of the mechanics of solids and shells. The most fundamental aspects are described from the mathematical and mechanical perspectives.

We trust that the preceding preview serves to reveal our intended *unification* of the general foundations, the various theories and approximations.

The authors' perspectives have been influenced by many experiences, by interactions with colleagues and by the efforts of many predecessors. A few works are most notable: The classic of A. E. H. Love, the insightful works of S. P. Timoshenko, the lucid monograph by V. V. Novozhilov, and the text by A. E. Green and W. Zerna are but a few which have shaped our views. Our grasp of energetic formulations and instability criteria are traceable to the important contributions by E. Trefftz, B. Fraeijs de Veubeke, and W. T. Koiter. Throughout this text, the authors have endeavored to acknowledge the origins of concepts and advances. Inevitably, some are overlooked; others are flawed by historical accounts. The authors apologize to persons who were inadvertently slighted by such mistakes.

On a personal note, Gerald Wempner must acknowledge his most influential teacher: His father, Paul Wempner, was a person with little *formal* education, but one who demonstrated the value of keen observation and concerted intellectual effort. Demosthenes Talaslidis would like to express his gratitude to his wife Vasso for her tolerance and support throughout preparation of this book. Both authors owe a debt of gratitude to Vasso Talaslidis and Euthalia Papademetriou; their forbearance and hospitality have enabled our collaboration. The authors are also indebted to Professor Walter Wunderlich who encouraged their earlier research. Finally, the authors are obliged to Mrs. Feye Kazantzidou for her careful attention to the illustrations.

Gerald Wempner and Demosthenes Talaslidis
Atlanta and Thessaloniki, July 2002

Contents

Chapter 1

Introduction

Much of engineering is concerned with solid bodies in various forms: Components of machines, instruments, structures, and the vessels that contain the fluids of chemical and biological processes are commonplace. In every application, the body performs some function which requires that it sustain loadings under given environmental conditions. The body fails mechanically if the material fractures or deforms excessively.

In a study of the mechanical behavior, two classifications of solid bodies are evident: First, bulky bodies which undergo imperceptible changes of shape, such as the thick walls of a pressure vessel; second, thin bodies which are often quite flexible, such as the skin of an aircraft. However, localized behavior of a small element is similar in most bodies, and consequently, certain fundamentals apply to both categories. These fundamentals constitute the initial chapters of this book.

An element of a structure, a rod, plate or shell, can experience large deflections and rotations, though strains and stresses may remain small enough that neither yielding nor fracture is imminent. In such flexible bodies, excessive deflections, vibrations or buckling are often the criteria of failure. Then any description or approximation of the deformation must accommodate moderate rotations, but admit simplifications for small strains. These are essential ingredients in the theories of shells which are explored in Chapters 9 and 10.

Existing capabilities for digital computation provide powerful means to obtain accurate approximations to problems of solids and shells. The concept of the finite element is a most effective and rational basis for such approximations. Our final chapter casts the formulation of finite elements within the rigorous foundations of the preceding mechanics of the continuum.

1.1 Purpose and Scope

As the need arises, the engineer or scientist attacks each new problem by means available at the time. As new theories, methods, and solutions emerge for different classes of engineering problems, so has the mechanics of solid bodies evolved and matured, but has often been fragmented. The various aspects have been presented in books devoted to the strength of materials, the theory of elasticity, the theory of plasticity, the theory of plates and shells, etc. At intervals during the evolution of a discipline, the pieces must be unified to provide greater perspective toward the discipline as a whole. Such unification has been done, most admirably, within the discipline of elasticity; in particular we note the classic of A. E. H. Love [1], the monograph by V. V. Novozhilov [2], and the outstanding book by A. E. Green and W. Zerna [3].

In recent years, various scholars have developed a unified treatment of the mechanics of continuous media, both fluid and solid. Notable examples are the treatises by C. Truesdell and R. A. Toupin [4] and by C. Truesdell and W. Noll [5]. From a theoretical viewpoint, this unification has been eminently successful. From the practical viewpoint of the engineer, who may be less mathematically inclined, such generality imposes an unwarranted burden: In particular, he must employ the spatial (Eulerian) view of the fluid and the material (Lagrangean) view of the solid. Consequently, he finds himself encumbered by two distinctly different reference systems. On the one hand, he has the fixed system of coordinates which proves useful in the analysis of fluids; that is the Eulerian viewpoint. On the other hand, he is compelled to identify a reference state of the solid, to follow the particles of the medium, and, therefore, to use the *initial* coordinates which label the particles during all subsequent motions; that is the Lagrangean viewpoint. Stated otherwise, we must recognize an essential difference between the solid and the fluid; particles, lines, and surfaces of the solid *must* remain *contiguous* (barring fracture or slippage) while the particles of fluids are free to separate and intermingle.

This book is intended for *engineers* interested in the *applied* mechanics of *solids*. Therefore, the treatment is facilitated by a single system of coordinates, those which locate the particles of a reference, or initial state. Arbitrary curvilinear coordinates identify particles, lines, and surfaces in the reference state and in any subsequent state. The *identity* of a line or surface is unchanged from our viewpoint, but its geometrical properties, lengths, and curvatures are altered by deformation. Indeed, such geometrical properties provide the mathematical description of the deformation. To facilitate understanding and subsequent application, physical and/or geometrical interpretations are ascribed to the various mathematical entities and/or operations. To aid the uninitiated reader, occasional examples are cast in initial rectangular coordinates.

The aim of this book is a unified presentation of the *foundations* of the mechanics of solids and the *practical theories* which characterize the behavior of deformable bodies. To facilitate the practical application of the theory, final chapters present the fundamentals of approximations which describe the behavior of thin shells and the properties of finite elements. These are not approached here as mathematical exercises based upon prescribed axioms, nor as unfounded computational schemes. On the contrary, our attention is directed to the underlying *assumptions* and the *approximations* which lead to various mathematical formulations. Consequently, our approach to each topic is seldom the most expedient one; instead, we strive for a constructive presentation which reveals the origins, approximations and limitations, thereby providing the basis for effective usage.

The mechanics of solids has evolved throughout the centuries and encompasses many, often diverse, viewpoints. To be useful, our treatment is necessarily confined to established theories of continuous solids and to practical means of approximation. References are also limited to original works and sources which augment our presentations.

Because many current and future problems may require more precise formulations, the kinematic foundations are general and include an accurate description of finite strains and displacements. The nonlinearities are specifically noted and special attention is given to moderately large rotations which play an important role in analyzing flexible bodies.

The initial chapters are devoted to the fundamentals: kinematics, dynamics, energetics, and behavior of materials. A chapter on the linear theories of elasticity and viscoelasticity draws upon the preceding foundations to exemplify the concepts and simultaneously to provide a ready reference for the practitioner. A chapter on energy principles provides the basis for the approximations and the theories of shells and finite elements, which are formulated in the final chapters. Each of the specific topics or theories is couched in the general theory of the initial chapters and, consequently, each is made a part of the general theory of solid bodies.

Like the warp and woof of a fabric, a practical theory of a solid body is built upon the underlying concepts of the general theory, interlaced with assumptions which are bred of experience. Such theories are seldom entirely satisfactory, but serve our immediate needs until such time as new insights, or circumstances, lead to advancements and alternatives. To achieve a unified perspective and a basis for subsequent study, we lay down the general concepts and then introduce the assumptions which support the contemporary theories of solids and structures.

1.2 Mechanical Concepts and Mathematical Representations

From the works of Newton and Lagrange, we have gained our knowledge and understanding of theoretical mechanics. Their alternative but complementary viewpoints are rendered in sharp focus via the mathematics of vectors and tensors. The power of vectorial representation, algebra and calculus, provides the insights to the actions (force, moment, stress, etc.) of the Newtonian mechanics. The beauty of Lagrange's description rests in large measure upon the invariance of work and energy. In the mechanics of continuous media, the essential variables are precisely identified and interrelated by the alternative forms of tensorial representation:
Alternative vectorial bases, covariant and contravariant tensorial components, provide the means to interpret the physical components of the Newtonian mechanics, and to associate those with the invariants of Lagrangean mechanics. The alternative representations of covariance and contravariance are indispensable tools. From a mathematical standpoint, they establish invariance. As practical devices, they provide physical interpretations, e.g., the directions of actions are identified with the alternative vectorial bases. Finally, it must be noted that the tensorial notation and invariance require the dual forms of indicial notation (superscripts and subscripts), as described in Chapter 2.

1.3 Index Notation

Index notation is a means of labeling the elements in a collection of quantities. An index is a letter (usually a lowercase) which appears as a subscript or superscript. In the present work, the Latin index represents any of the numbers 1, 2, or 3, because the system is used here to identify elements in a three-dimensional space.

We illustrate this notation as follows: Three coordinates of a rectangular system are denoted by x_1, x_2, x_3. A typical coordinate is denoted by x_i with the understanding that i has the range 1, 2, 3. The unit vector tangent to the line of coordinate x_i is denoted by $\hat{\boldsymbol{\imath}}_i$; the three vectors comprise a triad $(\hat{\boldsymbol{\imath}}_1, \hat{\boldsymbol{\imath}}_2, \hat{\boldsymbol{\imath}}_3)$. In certain two-dimensional problems, a Greek index is used to represent either of the numbers 1 or 2 only. As an example, we mention the curvilinear coordinates θ^α $(\alpha = 1, 2)$, which describe the position of a particle on a surface. The vector tangent to the coordinate line θ^α is denoted by $\boldsymbol{a}_\alpha$.

Index notations, together with their appropriate positioning, play an important role in mechanics: By the conventions of tensor analysis (Chapter 2) these notations serve to identify physical attributes which are independent of a coordinate system; in particular, those properties that are invariant (unchanged by a transformation of coordinates). This is essential to any basic theory, since coordinates are always a matter of convenience as dictated by the geometrical form, the analytical or computational capabilities.

1.4 Systems

One advantage of index notation is that a system comprising many elements or components can be represented by one. For example, any of the three components V_1, V_2, V_3 is represented by V_i; collectively, the three components constitute a system of the first order. A second-order system is represented by a typical element A_{ij} with two indices. The indices i and j can independently assume the values 1, 2, 3; they are termed *free* indices. The second-order system has nine components. A system of nth order has n free indices and contains 3^n components. Finally, an element without indices (a system of order zero) is an invariant (see Section 2.7).

1.5 Summation Convention

Any term in which a Latin (or Greek) index is repeated denotes a sum of all terms obtained by assigning numbers 1, 2, 3 (or 1, 2), in turn, to the repeated index; for example,

$$a_i b_i \equiv \sum_{i=1}^{3} a_i b_i = a_1 b_1 + a_2 b_2 + a_3 b_3,$$

$$a_{AM} b_{AN} \equiv \sum_{A=1}^{3} a_{AM} b_{AN} = a_{1M} b_{1N} + a_{2M} b_{2N} + a_{3M} b_{3N},$$

$$a_\alpha b_\alpha \equiv \sum_{\alpha=1}^{2} a_\alpha b_\alpha = a_1 b_1 + a_2 b_2.$$

If several indices are repeated in a term, each repetition signifies a summation; for example:

$$
\begin{aligned}
a_{ji}b_{ji} &\equiv \sum_{j=1}^{3}\sum_{i=1}^{3} a_{ji}b_{ji} \\
&= \sum_{i=1}^{3}(a_{1i}b_{1i} + a_{2i}b_{2i} + a_{3i}b_{3i}) \\
&= a_{11}b_{11} + a_{21}b_{21} + a_{31}b_{31} + a_{12}b_{12} + a_{22}b_{22} + a_{32}b_{32} \\
&\quad + a_{13}b_{13} + a_{23}b_{23} + a_{33}b_{33}.
\end{aligned}
$$

A repeated index is sometimes referred to as *dummy* index, since the meaning is not changed if the index is replaced by any other. For example:

$$a_{ii} = a_{jj}.$$

At times we may wish to signify only that element in which certain free indices are equal. Then such indices are represented by the same minuscule (lowercase letter) but the repeated index is underlined. Such repeated but underlined indices do not indicate a summation. For example, $A_{i\underline{i}}$ or $A_{\underline{ii}}$ represents any of the three elements A_{11}, A_{22}, A_{33}; recall that A_{ii} denotes the sum:

$$A_{ii} = A_{11} + A_{22} + A_{33}.$$

Caution: The same index cannot appear more than twice in a term; such a repetition would be meaningless.

It is understood that a free index which appears on both sides of an equation is assigned the same number on both sides. For example, the equation

$$a_j = b_j + c_j$$

represents any of the three equations,

$$a_1 = b_1 + c_1,$$

$$a_2 = b_2 + c_2,$$

$$a_3 = b_3 + c_3.$$

The index j is a free index since it represents any of the numbers 1, 2, or 3. Both free and dummy indices are contained in the following:

$$a_{ij}b_j = a_{i1}b_1 + a_{i2}b_2 + a_{i3}b_3.$$

The free index i is yet to be assigned a number 1, 2, or 3; the dummy index j is repeated and indicates the summation.

A system can be *contracted* to obtain another of lower order by a summation. For example, a_{ijj} represents a component of a first-order system, since there is one free index i; it is obtained from a system of third order, which has components a_{ijk}, by a summation with respect to the second and third indices.

Since the indicial notation is used throughout the text, an uninitiated reader is urged to examine each expression and to expand sums whenever the meaning is not apparent. As one learns a foreign language, conscious translation is necessary until the individual comes to think in that language. In our study, the language of indicial notation is well worth learning because it is both mathematically convenient and physically meaningful.

1.6 Position of Indices

An index may appear as a subscript or as a superscript. In general, the position of an index is a distinguishing feature of an element; for example, a component A_i is distinct from a component A^i. Very few exceptions arise and these are specifically noted.

The positioning of indices is a device of tensor analysis and serves to establish the invariance of quantities under transformation of coordinates. Since physical entities must be independent of our choice of coordinates, such invariance is essential; to that end, positioning of indices serves a useful purpose. Specific examples abound in our subsequent mathematical descriptions of deformable bodies.

1.7 Vector Notation

In the customary way, a vector is denoted by boldface type. A unit vector is signified by a caret ($\hat{}$) placed over the symbol; for example, $\boldsymbol{n}$ denotes a vector, but $\hat{\boldsymbol{n}}$ denotes a unit vector.

1.8 Kronecker Delta

The symbol δ_{ij} is defined as follows:

$$\delta_{ij} \equiv \begin{Bmatrix} 1 & \text{if } i = j \\ 0 & \text{if } i \neq j \end{Bmatrix}. \tag{1.1}$$

To illustrate its utility, consider the scalar product of unit vectors $(\hat{\boldsymbol{\imath}}_1, \hat{\boldsymbol{\imath}}_2, \hat{\boldsymbol{\imath}}_3)$, associated with a rectangular coordinate system (x_1, x_2, x_3). All the products are compactly given by

$$\hat{\boldsymbol{\imath}}_i \cdot \hat{\boldsymbol{\imath}}_j \equiv \delta_{ij}. \tag{1.2}$$

Consider a sum in which one of the repeated indices is on the Kronecker delta, for example,

$$a_{ij}\delta_{ik} = a_{1j}\delta_{1k} + a_{2j}\delta_{2k} + a_{3j}\delta_{3k}.$$

Only one term of the sum does not vanish, the term in which i equals k. Consequently, the sum reduces to

$$a_{ij}\delta_{ik} = a_{kj}.$$

Notice that the summation, involving one index i of the Kronecker delta and one of another factor, has the effect of substituting the free index k of the Kronecker delta for the repeated index i of the other factor.

Observe that the partial derivatives of independent variables x_i are expressed by the Kronecker delta as follows:

$$\frac{\partial x_i}{\partial x_j} = \delta_{ij}. \tag{1.3}$$

Note that the Kronecker delta is an exceptional symbol which may carry the indices in any position:

$$\delta_{ij} = \delta_{ji} = \delta^i_j = \delta^{ij} = \delta^{ji}. \tag{1.4}$$

Finally, according to the "summation convention" (Section 1.5):

$$\delta_{ii} = \delta^i_i = \delta^{ii} = 3,$$

$$\delta_{ik}\delta_{ik} = \delta_i^k\delta_k^i = \delta^{ik}\delta^{ik} = 3.$$

1.9 Permutation Symbol

The permutation symbol ϵ_{ijk} is defined as follows:

$$\epsilon_{ijk} \equiv \left\{ \begin{array}{ll} +1 & \text{if } i,\ j,\ k \text{ are an even permutation of } 1,\ 2,\ 3 \\ -1 & \text{if } i,\ j,\ k \text{ are an odd permutation of } 1,\ 2,\ 3 \\ 0 & \text{if any two indices are the same number} \end{array} \right\}. \tag{1.5}$$

To demonstrate the utility of this system, we note that the vector product of orthogonal unit vectors ($\hat{\boldsymbol{\imath}}_1$, $\hat{\boldsymbol{\imath}}_2$, $\hat{\boldsymbol{\imath}}_3$) is given by

$$\hat{\boldsymbol{\imath}}_i \times \hat{\boldsymbol{\imath}}_j = \epsilon_{ijk}\hat{\boldsymbol{\imath}}_k. \tag{1.6}$$

By interchanging two indices, the result is:

$$\epsilon_{ijk} = -\epsilon_{jik} = -\epsilon_{ikj} = -\epsilon_{kji}; \tag{1.7}$$

whereas, by cyclic interchanging of all indices, the sign of the permutation symbol remains the same:

$$\epsilon_{ijk} = \epsilon_{jki} = \epsilon_{kij}. \tag{1.8}$$

Also, we observe that the value of a 3×3 determinant with elements a_{ij} is expressed in the form:

$$|a_{ij}| = \tfrac{1}{6}a_{ir}a_{js}a_{kt}\epsilon_{ijk}\epsilon_{rst} \tag{1.9}$$

$$= \tfrac{1}{6}(a_{ii}a_{jj}a_{kk} - 3a_{ii}a_{jk}a_{kj} + a_{ij}a_{jk}a_{ki} + a_{ji}a_{kj}a_{ik}) \tag{1.10}$$

$$= a_{i1}a_{j2}a_{k3}\epsilon_{ijk}. \tag{1.11}$$

A few useful contractions are:

$$\epsilon_{ijk}\epsilon_{lmk} = \delta_{il}\delta_{jm} - \delta_{im}\delta_{jl}, \tag{1.12a}$$

$$\epsilon_{ijk}\epsilon_{ljk} = 2\delta_{il}, \tag{1.12b}$$

$$\epsilon_{ijk}\epsilon_{ijk} = 6. \tag{1.12c}$$

In subsequent chapters, we require the expanded form of a determinant, such as,

$$|e_{ij} - \lambda\delta_{ij}|.$$

Here, e_{ij} signifies any second order system. Using (1.7), (1.9) to (1.11), and (1.12a–c), this determinant assumes the following form:

$$|e_{ij} - \lambda\delta_{ij}| = -\lambda^3 + I_1\lambda^2 - I_2\lambda + I_3. \tag{1.13}$$

The expressions for the coefficients I_1, I_2, and I_3 are given by:

$$I_1 = e_{ii}, \tag{1.14a}$$

$$I_2 = \tfrac{1}{2}(e_{ii}e_{jj} - e_{ij}e_{ji}), \tag{1.14b}$$

$$\begin{aligned} I_3 &= \tfrac{1}{6}(e_{ii}e_{jj}e_{kk} - 3e_{ii}e_{jk}e_{kj} + e_{ij}e_{jk}e_{ki} + e_{ji}e_{kj}e_{ik}) \\ &= e_{i1}e_{j2}e_{k3}\epsilon_{ijk} = |e_{ij}|. \end{aligned} \tag{1.14c}$$

Finally, note that the permutation symbol is another exception which may carry the indices as subscripts or superscripts, that is,

$$\epsilon_{ijk} = \epsilon^{ijk}. \tag{1.15}$$

1.10 Symmetrical and Antisymmetrical Systems

If the elements of a system are unchanged when two indices are interchanged, then the system is *symmetrical* with respect to those two indices. For example, a third-order system has an element A_{ijk}; it is symmetrical in the first and second indices if

$$A_{ijk} = A_{jik}.$$

If the sign of the element is reversed when two indices are interchanged, then the system is *antisymmetrical* (or skewsymmetrical) with respect to the two indices. For example, a system of third order has an element B_{ijk}; it is *antisymmetrical* in the first and second indices if

$$B_{ijk} = -B_{jik}.$$

A system may exhibit both symmetries and antisymmetries. We encounter such example in Section 2.9:

$$R_{ijkl} = -R_{jikl}, \qquad R_{ijkl} = -R_{ijlk}, \qquad R_{ijkl} = R_{klij}.$$

The Kronecker delta δ_{ij} and the permutation symbol ϵ_{ijk} are notable examples of completely symmetrical and antisymmetrical systems, respectively.

Any system can be decomposed into one part which is symmetrical in two indices and another part which is antisymmetrical. For example, if A_{ij} denotes an element of a system, then an element of a symmetrical system is

$$A_{(ij)} \equiv \tfrac{1}{2}(A_{ij} + A_{ji}). \tag{1.16}$$

An element of an antisymmetrical system is

$$A_{[ij]} \equiv \tfrac{1}{2}(A_{ij} - A_{ji}). \tag{1.17}$$

It follows that

$$A_{ij} = A_{(ij)} + A_{[ij]}. \tag{1.18}$$

If A_{ij} and B_{ij} are symmetrical and antisymmetrical, respectively, then the sum $A_{ij}B_{ij}$ vanishes. The proof follows:

$$\begin{aligned} A_{ij}B_{ij} &= \tfrac{1}{2}(A_{ij} + A_{ji})B_{ij} \\ &= \tfrac{1}{2}A_{ij}B_{ij} + \tfrac{1}{2}A_{ij}B_{ji} \\ &= \tfrac{1}{2}A_{ij}(B_{ij} + B_{ji}) \\ &= 0. \end{aligned}$$

Conversely, if B_{ij} is antisymmetrical and if $A_{ij}B_{ij} = 0$, then A_{ij} is symmetrical. A notable example is a sum in which one factor is the permutation symbol; for example, if

$$A_{ij}\epsilon_{ijk} = 0, \tag{1.19}$$

then

$$A_{ij} = A_{ji}. \tag{1.20}$$

1.11 Abbreviation for Partial Derivatives

It is convenient to denote partial differentiation with a comma followed by the index of the independent variable, for example:

$$\frac{\partial A}{\partial x_i} \equiv A_{,i}. \tag{1.21}$$

1.12 Terminology

For brevity we sometimes refer to a collection of quantities by a reference to the typical quantity. Thus, we refer to "the coordinates x_i," meaning the coordinates x_1, x_2, x_3, or "vector a_k," meaning the vector with components a_1, a_2, a_3. In such terms of reference a symbol with free indices is used to denote the entire system. However, mathematically the symbol denotes a single element of the group.

1.13 Specific Notations

For the convenience of the reader, the following list includes the definitions of the most prevalent and the most important notations:

General Notations

$\tilde{(\,.\,)}$, $\bar{(\,.\,)}$	a tilde or bar over a symbol signifies a specific reference system
$\dot{(\;)}$	a dot over a symbol signifies time derivative (rate) or increment
$\delta(\,.\,)$	signifies an increment or a first-order variation
$\equiv$	signifies an identity or a definition
$\doteq$	signifies an approximation
$(\,.\,)\vert$	a vertical bar signifies covariant differentiation with respect to the undeformed system; see (2.63) to (2.65)

$(\,.\,)\vert^*$	an asterisk next to the vertical bar signifies covariant differentiation with respect to the deformed system
$(\,.\,)\Vert$	a double bar signifies covariant differentiation with respect to the undeformed surface; see (8.77a, b)
$(\,.\,)\Vert^*$	an asterisk appended to a double bar signifies covariant differentiation with respect to the deformed surface
$[\ldots]_P$, $(\,.\,)]_P$	the suffix (e.g., *P*) appended to a bracket signifies evaluation at that point (*P*)
$(\,.\,)_i$	a Latin index represents any of the numbers 1, 2, or 3
$(\,.\,)_\beta$	a Greek index represents any of the numbers 1 or 2
$(\,.\,)_{i\underline{i}}$ or $(\,.\,)_{\underline{ii}}$	repeated but underlined indices negate summation
$A_{(ij)}$, $A_{[ij]}$	element of a symmetrical, antisymmetrical system
${}^*T^{ij}$	component of a tensor associated with the deformed system
$\vert A_{ij}\vert$	determinant of a matrix with elements A_{ij}
x_i or x^i	coordinates of a rectangular/Cartesian system
θ^i	coordinates of an arbitrary curvilinear system
δ^{ij}, δ_{ij}, δ^i_j	Kronecker delta of (1.1) and (1.2)
ϵ^{ijk}, ϵ_{ijk}	permutation symbol of (1.5) and (1.6)
$(\,.\,)_{,i}$	a comma signifies partial differentiation with respect to the coordinate θ^i
D, *P*, *E*, *R*	uppercase prefixes or suffixes signify *D*issipation, *P*lastic (*P*ermanent), *E*lastic, *R*eversible

Geometrical Quantities/Deformation

$\boldsymbol{r}$, $\boldsymbol{R}$	position vector before (2.1) and after (3.2) deformation
$\hat{\boldsymbol{\imath}}_i$	unit vector tangent to the x_i line of a Cartesian coordinate system (Figure 2.1)
$\boldsymbol{g}_i$, $\boldsymbol{g}^i$	tangent, normal base vector of a curvilinear system of coordinates θ^i; see (2.5), (2.6)
$\hat{\mathbf{e}}_i$, $\hat{\mathbf{e}}^i$	unit vector tangent to the θ^i line, normal to the θ^i surface (undeformed state); see (2.14a, b)
$\acute{\boldsymbol{g}}_i$, $\acute{\boldsymbol{g}}^i$	rigidly rotated versions of the initial base vectors $\boldsymbol{g}_i$ and $\boldsymbol{g}^i$, respectively; see (3.87a, b)

$\hat{e}'_i$	unit vector in direction $\acute{\boldsymbol{g}}_i$
$\boldsymbol{G}_i$, $\boldsymbol{G}^i$	tangent, normal base vector of a curvilinear system of coordinates θ^i (deformed configuration); see (3.3), (3.4)
g_{ij}, g^{ij}	components of associated metric tensors of the undeformed system (2.9), (2.10)
G_{ij}, G^{ij}	components of associated metric tensors of the deformed system (3.6a, b)
$\sqrt{g}$	$\equiv \boldsymbol{g}_1 \cdot (\boldsymbol{g}_2 \times \boldsymbol{g}_3)$ (2.23a); volume metric (measure) of the undeformed system
$\sqrt{G}$	$\equiv \boldsymbol{G}_1 \cdot (\boldsymbol{G}_2 \times \boldsymbol{G}_3)$ (3.11a); volume metric (measure) of the deformed system
e_{ijk}, e^{ijk}	$\equiv \sqrt{g}\,\epsilon_{ijk}$, $\epsilon^{ijk}/\sqrt{g}$ (2.24a, b)
E_{ijk}, E^{ijk}	$\equiv \sqrt{G}\,\epsilon_{ijk}$, $\epsilon^{ijk}/\sqrt{G}$ (3.17a, b)
ℓ, L	length of undeformed, deformed line
s, S	area of undeformed, deformed surface
s_t, s_v	undeformed bounding surface where tractions, displacements are prescribed
v, V	volume of undeformed, deformed medium
D	dilation; ratio of elemental deformed to undeformed volume (3.12)
Γ_{ijk}, Γ^k_{ij}	Christoffel symbols of the first and second kind, respectively, for the undeformed system (2.46a, b), (2.47)
$R^n_{\cdot\,ijk}$	component of the Riemann-Christoffel tensor associated with the undeformed system; see (2.67)
$\boldsymbol{V}$, $\dot{\boldsymbol{V}}$, $\ddot{\boldsymbol{V}}$	displacement vector of Figure 3.2 (3.18a, b), velocity and acceleration vector, respectively
γ_{ij}	component of Cauchy-Green strain tensor (3.31)
ϵ_{ij}	physical component of strain (3.35a, b)
γ^N_{ij}	$(N = 1, \ldots, M)$ component of inelastic strain (5.40)
$d\epsilon^P_{ij}$	plastic-strain increment
$\lambda_{(i)}$	principal value of strain tensor γ_{ij} (3.52a, b)
h_{ij}	component of engineering strain tensor (3.96)
η_{ij}	deviatoric strain of (3.150)

I_i $(i = 1, 2, 3)$; invariant of the Cauchy-Green strain tensor γ^i_j (3.72a–c)

$\bar{I}_i$ $(i = 1, 2, 3)$; invariant of the engineering strain tensor h^i_j (3.75a–c)

$'e$, $'h$ alternative measures of volumetric strain; see (3.73a–c), (3.74a–c)

C_{ij} stretch tensor of (3.78)

$r_k^{\bullet n}$ rotation tensor in an arbitrary system of coordinates θ^i (3.87a, b)

$\dot{\omega}_{ij}$ rate of rotation tensor; see (3.138a, b) and (3.141a, b)

$\dot{\boldsymbol{\omega}}$ rate of "spin" vector $\dot{\boldsymbol{\omega}} = \dot{\Omega}^i \boldsymbol{g}_i = \dot{\omega}^i \boldsymbol{G}_i = \dot{\bar{\omega}}^i \acute{\boldsymbol{g}}_i$; see (3.140)

Stress

$\boldsymbol{\tau}^i$ stress vector upon θ^i surface, per unit of deformed area (4.1)

$\boldsymbol{\sigma}^i$ stress vector upon θ^i surface, per unit of undeformed area

$\boldsymbol{f}$, $\tilde{\boldsymbol{f}}$ body force per unit (initial) volume, body force per unit mass

$\boldsymbol{t}^i$ tensorial stress vector associated with deformed area; see (4.19a)

$\boldsymbol{s}^i$ tensorial stress vector associated with undeformed area; see (4.19c)

σ^{ij} physical component of stress vector $\boldsymbol{\tau}^i$ per unit area of deformed body; see (4.15a)

$\overset{*}{\sigma}{}^{ij}$, $\bar{\sigma}^{ij}$ physical components of stress vector $\boldsymbol{\sigma}^i$ per unit area of undeformed body; see (4.15b, c)

τ^{ij} tensorial component of stress vector $\boldsymbol{t}^i$; $\tau^{ij} = \boldsymbol{G}^j \cdot \boldsymbol{t}^i$ (4.20a)

s^{ij} tensorial component of stress vector $\boldsymbol{s}^i$; $s^{ij} = \boldsymbol{G}^j \cdot \boldsymbol{s}^i$ (4.20c)

t^{ij} tensorial component of stress vector $\boldsymbol{s}^i$; $t^{ij} = \acute{\boldsymbol{g}}^j \cdot \boldsymbol{s}^i$ (4.20e)

$\tilde{s}^i$ principal stresses (4.35)–(4.37)

$'\tau^{ij}$, $'s^{ij}$, $'t^{ij}$ components of stress deviators; see (4.43a) and (4.47a, b)

$\boldsymbol{T}$ surface traction per unit initial area

$\overline{\boldsymbol{T}}$ surface traction which acts upon the part s_t of the boundary

Properties of Materials

ρ_0, ρ	mass density of undeformed, deformed medium
E^{ijkl}	Hooke's stiffness tensor of (5.49), (5.51a)
D_{ijkl}	Hooke's flexibility tensor of (5.55a, b)
E_T^{ijkl}	Tangent modulus of elastoplastic deformations (5.135)
α^{ij}	coefficient of thermal expansion; see (5.49) and (5.51b)
E	elastic modulus (Young's modulus) (5.78)
ν	Poisson's ratio of (5.79)
G	shear modulus of elasticity (5.87)
λ	Lamé coefficient of (5.85)
K	bulk modulus of (5.84) and (5.86)
Y	yield stress in simple tension
$\mathcal{Y}(s^{ij}) = \bar{\sigma}^2$	yield condition of (5.119)
G^P	strain hardening parameter of (5.131)
μ	coefficient of viscosity
$k(t)$	creep compliance
$m(t)$	relaxation modulus

Thermal Quantities

T	absolute temperature
Q	thermal energy, heat supplied (per unit mass)
S	entropy density (per unit mass) defined by (5.5a)
$\dot{\boldsymbol{q}}$	heat flux (per unit deformed area)

Work and Energy Applied to a Continuous Body

w_s, ${}_\rho w_s$	work expended by the stresses per unit volume, per unit mass
w_t	work of tractions $\boldsymbol{T}$ (per unit area) of the bounding surface
w_f	work of body forces per unit volume
W_s	work of all internal forces (stresses)

W_f, W_t	work of all body forces, tractions upon the entire body
W	total work upon the entire body (see Section 6.14)
π_f, π_t	potential of body forces and surface tractions, respectively (per unit initial volume)
u	internal-energy potential per unit deformed volume
$u_0(\gamma_{ij})$	potential (internal energy) per unit undeformed volume
$\bar{u}_0(h_{ij})$	potential (internal energy) per unit undeformed volume
$u_{c0}(s^{ij})$	complementary energy per unit undeformed volume
$\bar{u}_{c0}(t^{ij})$	complementary energy per unit undeformed volume
U	internal-energy potential of (6.117)
Π_f, Π_t	potential of all body forces (6.118) and surface tractions (6.119)
$\mathcal{U}(\gamma_{ij}, S)$	internal-energy density (per unit mass) (5.23)
$\mathcal{G}(s^{ij}, T)$	Gibbs potential (5.27)
$\mathcal{F}(\gamma_{ij}, T)$	free-energy potential per unit mass (5.30)
$\mathcal{V}$, $\overline{\mathcal{V}}$	total potentials; see Sections 6.15, 6.17, and 9.8
$\mathcal{V}_c$, $\overline{\mathcal{V}}_c$	complementary potentials; see Sections 6.17 to 6.19 and 9.8
$\mathcal{V}^*$, $\overline{\mathcal{V}}^*$	modified potentials; see Sections 6.16 and 9.8
$\mathcal{V}_c^*$, $\overline{\mathcal{V}}_c^*$	modified complementary potentials; see Sections 6.20 and 9.8

Notations for Shells

${}_0\boldsymbol{r}$, ${}_0\boldsymbol{R}$	position vector to a point on the undeformed (9.4), deformed (9.5) reference surface
s_0, S_0	area of undeformed, deformed reference surface
θ^α	coordinates on the reference surface ($\alpha = 1, 2$)
θ^3	coordinate, length along the normal to the undeformed reference surface
$\boldsymbol{a}_i$, $\boldsymbol{a}^i$	tangent, normal (reciprocal) base vectors of the undeformed system θ^i; see (8.6a, b) and (8.7)
$\boldsymbol{A}_i$, $\boldsymbol{A}^i$	tangent, normal base vectors of the deformed system θ^i (9.3)

$\hat{\boldsymbol{a}}_3$, $\hat{\boldsymbol{N}}$	unit vectors normal to the undeformed (8.9a, b), deformed (10.2b) surface
$a_{\alpha\beta}$, $a^{\alpha\beta}$	covariant, contravariant metric tensors of the coordinates θ^α in the undeformed surface; see (8.11a, b) and (8.40)
$A_{\alpha\beta}$, $A^{\alpha\beta}$	covariant, contravariant metric tensors of the coordinates θ^α in the deformed surface
a	$\equiv g(\theta^1, \theta^2, 0) = \lvert a_{\alpha\beta} \rvert$; see (8.14a, b)
A	$\equiv \lvert A_{\alpha\beta} \rvert$
$\overline{e}_{\alpha\beta}$, $\overline{e}^{\alpha\beta}$	components of permutation tensor in the initial system $\boldsymbol{a}_\alpha$ (8.17a–d)
$\overline{E}_{\alpha\beta}$, $\overline{E}^{\alpha\beta}$	components of permutation tensor in the deformed system $\boldsymbol{A}_\alpha$
$\overline{\Gamma}_{\alpha\beta\gamma}$, $\overline{\Gamma}^{\gamma}_{\alpha\beta}$	Christoffel symbols of the undeformed system θ^α
${}^*\overline{\Gamma}_{\alpha\beta\gamma}$, ${}^*\overline{\Gamma}^{\gamma}_{\alpha\beta}$	Christoffel symbols of the deformed system θ^α
$b_{\alpha\beta}$, $B_{\alpha\beta}$	components of curvature tensors associated with the undeformed (8.27a, b), deformed (9.18a) reference surface
$\widetilde{k}$, $\widetilde{K}$	Gaussian curvature of the undeformed (8.59), deformed reference surface
$\widetilde{h}$	mean curvature associated with the undeformed reference surface (8.60)
$\boldsymbol{b}_i$	rotated base triad of (9.8) according to (9.9) and (9.11)
$C_{\alpha\beta}$, C_{3i}	covariant components of the stretch of the reference surface (9.9) and the stretch of the θ^3 line (9.10), respectively
$h_{\alpha\beta}$, $h_{3\alpha}$	covariant components of the engineering strain of the reference surface (9.42a) and transverse shear strain (9.42c), respectively
h_{33}	components of the transverse extensional strain (9.42d)
$\kappa_{\alpha\beta}$, $\kappa_{\alpha 3}$	flexural strain (9.42b) and gradient of transverse shear strain (9.42e)
${}_0\gamma_{\alpha\beta}$	components of Cauchy-Green strain tensor at the reference surface (10.13a)
$\rho_{\alpha\beta}$	alternative flexural strain of (9.77) or change-of-curvature tensor of (10.13b)
$\boldsymbol{N}^\alpha$, $\boldsymbol{T}$, $\boldsymbol{M}^\alpha$	resultant force ($\boldsymbol{N}^\alpha$, $\boldsymbol{T}$) and moment ($\boldsymbol{M}^\alpha$) vectors of [(9.36a), (9.37)] and (9.36b), respectively

$N^{\alpha i}$, T^i tensorial components of tensorial force vectors ($\boldsymbol{N}^\alpha$, $\boldsymbol{T}$); see (9.38a, b), (9.40a, b), and (10.9a–c)

$M^{\alpha i}$ tensorial components of tensorial moment vector $\boldsymbol{M}^\alpha$; see (9.39a, b), and (10.9d, e)

$\boldsymbol{\mathcal{N}}^\alpha$, $\boldsymbol{\mathcal{M}}^\alpha$ force and moment per unit length of an edge (9.61a, b)

$\boldsymbol{F}$, $\boldsymbol{C}$ resultants (force and couple) of external forces upon a shell; see (9.46a–d), (9.49a, b)

Chapter 2

Vectors, Tensors, and Curvilinear Coordinates

2.1 Introduction

Many machine parts and structures have curved surfaces which dictate the use of curvilinear coordinate systems. In particular, it is difficult to formulate a general theory of shells without an adequate knowledge about the coordinates of curved surfaces.

Although a rectangular system may be inscribed conveniently in certain bodies, deformation carries the straight lines and planes to curved lines and surfaces; rectangular coordinates are deformed into arbitrary curvilinear coordinates. Consequently, we are unavoidably led to explore the differential geometry of curved lines and surfaces using curvilinear coordinates, if only in the deformed body.

In general, coordinates need not measure length directly, as Cartesian coordinates. The curvilinear coordinates need only define the position of points in space, but they must do so uniquely and continuously if they are to serve our purpose in the mechanics of continuous media. A familiar example is the spherical system of coordinates consisting of the radial distance and the angles of latitude and longitude: Spherical coordinates have the requisite continuity except along the polar axis; one is a length and two are angles, dimensionless in radians.

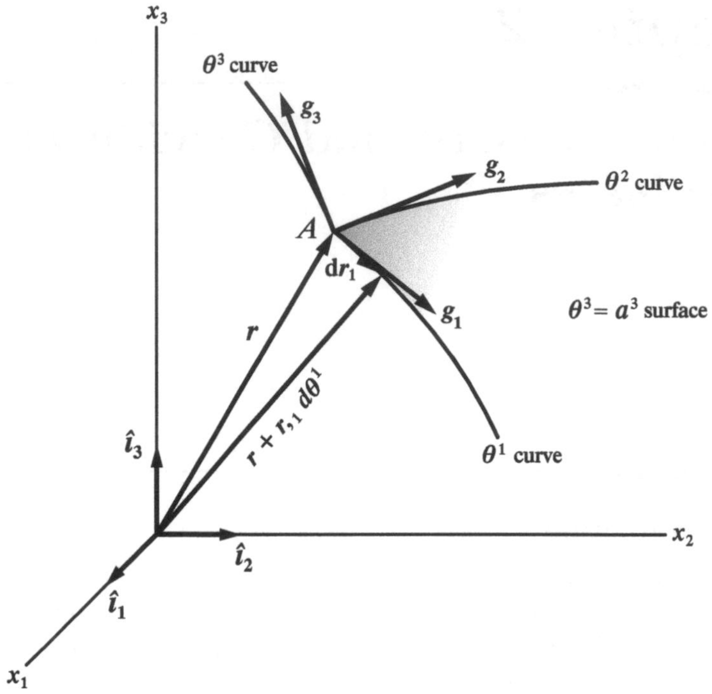

Figure 2.1 Curvilinear coordinate lines and surfaces

2.2 Curvilinear Coordinates, Base Vectors, and Metric Tensor

The position of point A in Figure 2.1 is given by the vector

$$\boldsymbol{r} = x_i \hat{\boldsymbol{\imath}}_i, \tag{2.1}$$

where x_i are rectangular coordinates and $\hat{\boldsymbol{\imath}}_i$ are unit vectors as shown.

Let θ^i denote arbitrary curvilinear coordinates. We assume the existence of equations which express the variables x_i in terms of θ^i and vice versa; that is,

$$x_i = x_i(\theta^1, \theta^2, \theta^3), \qquad \theta^i = \theta^i(x_1, x_2, x_3). \qquad (2.2), (2.3)$$

Also, we assume that these have derivatives of any order required in the subsequent analysis.

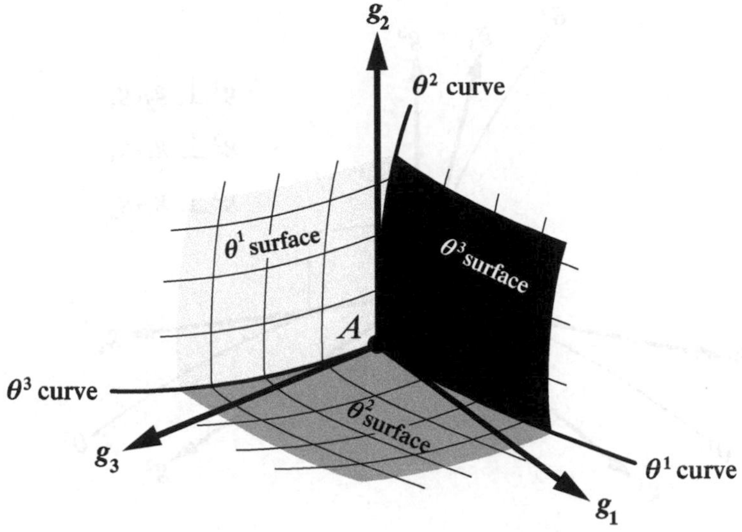

Figure 2.2 Network of coordinate curves and surfaces

Suppose that $x_i = x_i(a^1, a^2, a^3)$ are the rectangular coordinates of point A in Figure 2.1. Then $x_i(\theta^1, a^2, a^3)$ are the parametric equations of a curve through A; it is the θ^1 curve of Figure 2.1. Likewise, the θ^2 and θ^3 curves through point A correspond to fixed values of the other two variables. The equations $x_i = x_i(\theta^1, \theta^2, a^3)$ are parametric equations of a surface, the θ^3 surface shown shaded in Figure 2.1. Similarly, the θ^1 and θ^2 surfaces correspond to $\theta^1 = a^1$ and $\theta^2 = a^2$. At each point of the space there is a network of curves and surfaces (see Figure 2.2) corresponding to the transformation of equations (2.2) and (2.3).

By means of (2.2) and (2.3), the position vector $\boldsymbol{r}$ can be expressed in alternative forms:

$$\boldsymbol{r} = \boldsymbol{r}(x_1, x_2, x_3) = \boldsymbol{r}(\theta^1, \theta^2, \theta^3). \tag{2.4}$$

A differential change $d\theta^i$ is accompanied by a change $d\boldsymbol{r}_i$ tangent to the θ^i line; a change in θ^1 only causes the increment $d\boldsymbol{r}_1$ illustrated in Figure 2.1. It follows that the vector

$$\boldsymbol{g}_i \equiv \frac{\partial \boldsymbol{r}}{\partial \theta^i} = \frac{\partial x_j}{\partial \theta^i} \hat{\boldsymbol{\imath}}_j \tag{2.5}$$

is tangent to the θ^i curve. The tangent vector $\boldsymbol{g}_i$ is sometimes called a *base vector.*

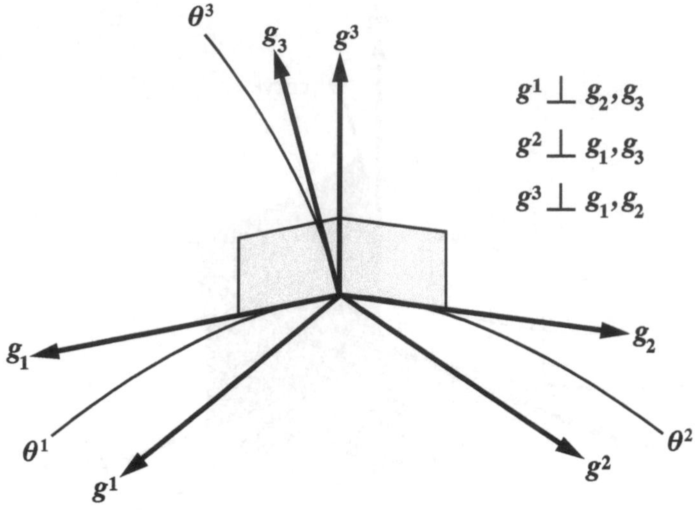

Figure 2.3 Tangent and normal base vectors

Let us define another triad of vectors $\boldsymbol{g}^i$ such that

$$\boldsymbol{g}^i \cdot \boldsymbol{g}_j \equiv \delta^i_j. \tag{2.6}$$

The vector $\boldsymbol{g}^i$ is often called a *reciprocal* base vector. Since the vectors $\boldsymbol{g}_i$ are *tangent* to the *coordinate curves*, equation (2.6) means that the vectors $\boldsymbol{g}^i$ are *normal* to the *coordinate surfaces*. This is illustrated in Figure 2.3. We will call the triad $\boldsymbol{g}_i$ *tangent base vectors* and the triad $\boldsymbol{g}^i$ *normal base vectors*. In general they are not unit vectors.

The triad $\boldsymbol{g}^i$ can be expressed as a linear combination of the triad $\boldsymbol{g}_i$, and vice versa. To this end we define coefficients g^{ij} and g_{ij} such that

$$\boldsymbol{g}_i \equiv g_{ij}\boldsymbol{g}^j, \qquad \boldsymbol{g}^i \equiv g^{ij}\boldsymbol{g}_j. \qquad (2.7), (2.8)$$

From equations (2.6) to (2.8), it follows that

$$g_{ij} = g_{ji} = \boldsymbol{g}_i \cdot \boldsymbol{g}_j, \qquad g^{ij} = g^{ji} = \boldsymbol{g}^i \cdot \boldsymbol{g}^j, \qquad (2.9), (2.10)$$

$$g^{im} g_{jm} = \delta^i_j. \tag{2.11}$$

The linear equations (2.11) can be solved to express g^{ij} in terms of g_{ij}, as follows:

$$g^{ij} = \frac{\text{cofactor of element } g_{ij} \text{ in matrix } [g_{ij}]}{|g_{ij}|}. \tag{2.12}$$

The first-order differential of $\boldsymbol{r}$ is $d\boldsymbol{r} = \boldsymbol{r}_{,i}\, d\theta^i \equiv \boldsymbol{g}_i\, d\theta^i$, and the corresponding differential length $d\ell$ is given by

$$d\ell^2 = d\boldsymbol{r} \cdot d\boldsymbol{r} = g_{ij}\, d\theta^i\, d\theta^j. \tag{2.13}$$

Equation (2.13) is fundamental in differential geometry, and the coefficients g_{ij} play a paramount role; they are components of the *metric tensor*. Literally, a component $g_{\underline{ii}}$ provides the measure of length (per unit of θ^i) along the θ^i line; a component g_{ij} $(i \neq j)$ determines the angle between θ^i and θ^j lines.

For practical purposes, we may employ unit vectors $\hat{\mathbf{e}}_i$ and $\hat{\mathbf{e}}^i$, tangent to the θ^i line and normal to the θ^i surface, respectively:

$$\hat{\mathbf{e}}_i = \frac{\boldsymbol{g}_i}{\sqrt{g_{\underline{ii}}}}, \qquad \hat{\mathbf{e}}^i = \frac{\boldsymbol{g}^i}{\sqrt{g^{\underline{ii}}}}. \tag{2.14a, b}$$

If the coordinate system is orthogonal, then

$$g^{\underline{ii}} = \frac{1}{g_{\underline{ii}}}, \qquad g^{ij} = g_{ij} = 0 \qquad (i \neq j).$$

In the rectangular Cartesian system of coordinates,

$$g^{ij} = g_{ij} = \delta^{ij} \equiv \delta_{ij}.$$

In accordance with (2.5) and (2.9),

$$g_{ij} = \frac{\partial x_k}{\partial \theta^i} \frac{\partial x_k}{\partial \theta^j}. \tag{2.15}$$

By the rules for multiplying determinants

$$|g_{ij}| = \left| \frac{\partial x_i}{\partial \theta^j} \right| \left| \frac{\partial x_j}{\partial \theta^i} \right| = \left| \frac{\partial x_i}{\partial \theta^j} \right|^2.$$

Since this quantity plays a central role in the differential geometry, we employ the symbol

$$g \equiv |g_{ij}| = \left| \frac{\partial x_i}{\partial \theta^j} \right|^2. \tag{2.16a, b}$$

Also, from (2.11), it follows by the rules for multiplying determinants and by using (2.16a) that

$$|g^{im}g_{jm}| = |g^{ij}||g_{ij}| = |g^{ij}|g = |\delta^i_j| = 1,$$

and therefore

$$|g^{ij}| = \frac{1}{g}. \tag{2.16c}$$

2.3 Products of Base Vectors

The vector product of $\boldsymbol{g}_i$ and $\boldsymbol{g}_j$ is normal to their plane but zero if i equals j. In general, the vector product has the form

$$\boldsymbol{g}_i \times \boldsymbol{g}_j = M\epsilon_{ijk}\boldsymbol{g}^k, \tag{2.17a}$$

where M is a positive scalar and ϵ_{ijk} is the permutation symbol defined by (1.5) and (1.6). Conversely,

$$\boldsymbol{g}^i \times \boldsymbol{g}^j = N\epsilon^{ijk}\boldsymbol{g}_k, \tag{2.17b}$$

where N is another positive scalar. To determine M and N from (2.17a) and (2.17b), we form the scalar (dot) products of these equations with $\boldsymbol{g}_m\epsilon^{ijm}$ and $\boldsymbol{g}^m\epsilon_{ijm}$, respectively, and recall (1.12c) to obtain

$$M = \tfrac{1}{6}\epsilon^{ijk}\boldsymbol{g}_k \cdot (\boldsymbol{g}_i \times \boldsymbol{g}_j), \tag{2.18a}$$

$$N = \tfrac{1}{6}\epsilon_{ijk}\,\boldsymbol{g}^k \cdot (\boldsymbol{g}^i \times \boldsymbol{g}^j). \tag{2.18b}$$

Now, the vectors $\boldsymbol{g}_i$ of (2.18a) can be expressed in terms of the reciprocal vectors $\boldsymbol{g}^i$ according to (2.7); then

$$M = \tfrac{1}{6}\epsilon^{ijk}g_{im}g_{jn}g_{kl}\,\boldsymbol{g}^l \cdot (\boldsymbol{g}^m \times \boldsymbol{g}^n). \tag{2.19a}$$

It follows from (2.19a), (2.17b), and (1.9) that

$$M = \tfrac{1}{6}\epsilon^{ijk}\epsilon^{mnl}g_{im}g_{jn}g_{kl}N = |g_{ij}|N. \tag{2.19b}$$

In like manner,

$$N = |g^{ij}|M. \tag{2.20}$$

Recall (2.5) and (1.6):

$$\boldsymbol{g}_i \equiv \boldsymbol{r}_{,i} = \frac{\partial x_k}{\partial \theta^i}\,\hat{\boldsymbol{\imath}}_k, \qquad \hat{\boldsymbol{\imath}}_i \times \hat{\boldsymbol{\imath}}_j = \epsilon^{ijk}\,\hat{\boldsymbol{\imath}}_k.$$

Then (2.18a) assumes the forms:

$$M = \tfrac{1}{6}\epsilon^{ijk}\epsilon^{lmn}\frac{\partial x_l}{\partial \theta^i}\frac{\partial x_m}{\partial \theta^j}\frac{\partial x_n}{\partial \theta^k} = \left|\frac{\partial x_i}{\partial \theta^j}\right|.$$

By the notation (2.16a)

$$M = \sqrt{g}. \tag{2.21a}$$

In accordance with (2.16a), (2.19b), and (2.21a),

$$N = \frac{1}{\sqrt{g}}. \tag{2.21b}$$

With the notations of (2.21a, b), the equations (2.17a, b) take the forms:

$$\boldsymbol{g}_i \times \boldsymbol{g}_j = \sqrt{g}\,\epsilon_{ijk}\,\boldsymbol{g}^k, \qquad \boldsymbol{g}^i \times \boldsymbol{g}^j = \frac{\epsilon^{ijk}}{\sqrt{g}}\,\boldsymbol{g}_k. \tag{2.22a, b}$$

It follows that

$$\sqrt{g}\,\epsilon_{ijk} = \boldsymbol{g}_k \cdot (\boldsymbol{g}_i \times \boldsymbol{g}_j), \qquad \frac{\epsilon^{ijk}}{\sqrt{g}} = \boldsymbol{g}^k \cdot (\boldsymbol{g}^i \times \boldsymbol{g}^j). \tag{2.23a, b}$$

Let us define

$$e_{ijk} \equiv \sqrt{g}\,\epsilon_{ijk}, \qquad e^{ijk} \equiv \frac{1}{\sqrt{g}}\,\epsilon^{ijk}. \tag{2.24a, b}$$

Then, according to (2.22a, b) and (2.23a, b)

$$e_{ijk} = \boldsymbol{g}_k \cdot (\boldsymbol{g}_i \times \boldsymbol{g}_j), \qquad e^{ijk} = \boldsymbol{g}^k \cdot (\boldsymbol{g}^i \times \boldsymbol{g}^j), \tag{2.25a, b}$$

$$\boldsymbol{g}_i \times \boldsymbol{g}_j = e_{ijk}\,\boldsymbol{g}^k, \qquad \boldsymbol{g}^i \times \boldsymbol{g}^j = e^{ijk}\boldsymbol{g}_k. \tag{2.26a, b}$$

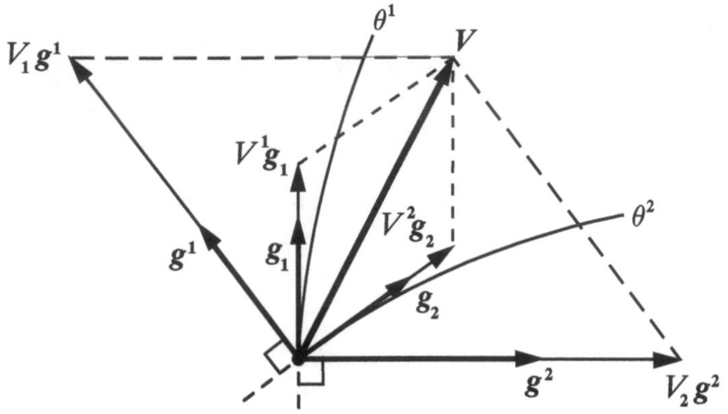

Figure 2.4 Contravariant and covariant components of vectors

2.4 Components of Vectors

Any vector can be expressed as a linear combination of the base vectors $\boldsymbol{g}_i$ or $\boldsymbol{g}^i$. Thus, the vector $\boldsymbol{V}$ has alternative forms (see Figure 2.4):

$$\boldsymbol{V} = V^i \boldsymbol{g}_i = V_i \boldsymbol{g}^i. \tag{2.27a, b}$$

The alternative components follow by means of (2.6) to (2.8)

$$V^i = \boldsymbol{g}^i \cdot \boldsymbol{V}, \qquad V_i = \boldsymbol{g}_i \cdot \boldsymbol{V}, \tag{2.28a, b}$$

$$V^i = g^{ij} V_j, \qquad V_i = g_{ij} V^j. \tag{2.29a, b}$$

The components V^i and V_i are called *contravariant* and *covariant components*, respectively. The full significance of these adjectives is discussed in Section 2.7.

In general, the base vectors $(\boldsymbol{g}_i, \boldsymbol{g}^i)$ are not unit vectors. The vector $\boldsymbol{V}$ can also be expressed with the aid of the unit base vectors $(\hat{\mathbf{e}}_i, \hat{\mathbf{e}}^i)$ defined by (2.14a, b). By (2.14a, b) and (2.27a, b), we have

$$\boldsymbol{V} = V^i \sqrt{g_{\underline{ii}}}\, \hat{\mathbf{e}}_i = V_i \sqrt{g^{\underline{ii}}}\, \hat{\mathbf{e}}^i. \tag{2.30a, b}$$

The coefficients $(V^i \sqrt{g_{\underline{ii}}})$ and $(V_i \sqrt{g^{\underline{ii}}})$ are known as the *physical compo-*

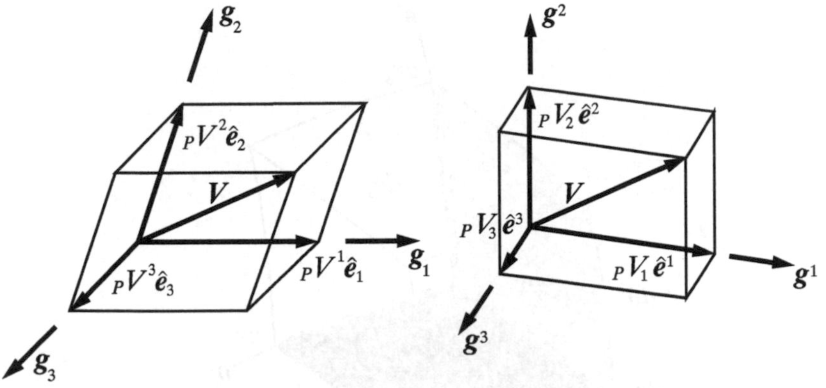

Figure 2.5 Physical components

nents of the vector $\boldsymbol{V}$:

$$_PV^i \equiv V^i\sqrt{g_{\underline{ii}}}, \qquad _PV_i = V_i\sqrt{g^{\underline{ii}}}. \tag{2.31a, b}$$

Notice that the physical components $_PV^i$ and $_PV_i$ are in directions tangent to the θ^i line and normal to the θ^i surface, respectively. The physical components (2.31a, b) represent the lengths of the sides of the parallelograms formed by the vectors $(V^i\boldsymbol{g}_i)$ and $(V_i\,\boldsymbol{g}^i)$ of Figure 2.4. In the three-dimensional case, these components are the lengths of the edges of the parallelepipeds constructed on the vectors $(V^i\boldsymbol{g}_i)$ or $(_PV^i\hat{\boldsymbol{e}}_i)$ and $(V_i\,\boldsymbol{g}^i)$ or $(_PV_i\hat{\boldsymbol{e}}^i)$ (see Figure 2.5). Furthermore, by forming the scalar product of the vector $\boldsymbol{V}$ of equation (2.27b) with $\hat{\boldsymbol{e}}_j$, by considering (2.14a), and the orthogonality condition (2.6), it follows that the term $V_i/\sqrt{g_{\underline{ii}}}$ is equal to the length of the orthogonal projection of $\boldsymbol{V}$ on the tangent to the θ_i line. In a similar way, by multiplying (2.27a) with $\hat{\boldsymbol{e}}^j$ and in view of (2.14b) and (2.6), it can be shown that the term $V^i/\sqrt{g^{\underline{ii}}}$ is equal to the length of the orthogonal projection of $\boldsymbol{V}$ on the normal to the θ_i surface.

The products of vectors assume alternative forms in terms of the contravariant and covariant components. For example,

$$\boldsymbol{V}\cdot\boldsymbol{U} = V^iU_i = V_iU^i = V_iU_jg^{ij} = V^iU^jg_{ij}, \tag{2.32a–d}$$

$$\boldsymbol{V}\times\boldsymbol{U} = e_{ijk}V^iU^j\boldsymbol{g}^k = e^{ijk}V_iU_j\boldsymbol{g}_k, \tag{2.33a, b}$$

$$\boldsymbol{U}\cdot(\boldsymbol{V}\times\boldsymbol{W}) = e_{ijk}U^iV^jW^k = e^{ijk}U_iV_jW_k, \tag{2.34a, b}$$

$$\boldsymbol{U}\times(\boldsymbol{V}\times\boldsymbol{W}) = (U^iW_i)(V^j\boldsymbol{g}_j) - (U^iV_i)(W^j\boldsymbol{g}_j). \tag{2.35}$$

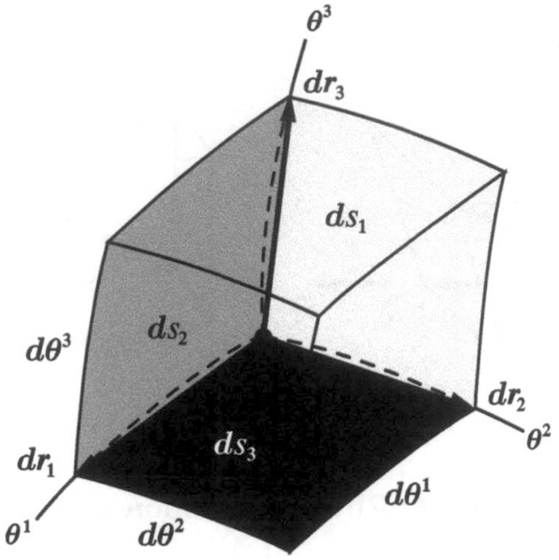

Figure 2.6 Elemental parallelepiped

Recall that the result of the vector product $\boldsymbol{V} \times \boldsymbol{U}$ is a vector normal to the plane formed by the vectors $\boldsymbol{V}$ and $\boldsymbol{U}$. The magnitude of this vector is equal to the area of the parallelogram with sides $\boldsymbol{V}$ and $\boldsymbol{U}$. From (2.33a) it is apparent that the vector product satisfies the anticommutative law:

$$\boldsymbol{V} \times \boldsymbol{U} = -\boldsymbol{U} \times \boldsymbol{V}. \tag{2.36}$$

Recall also that the value of the scalar triple product $\boldsymbol{U} \cdot (\boldsymbol{V} \times \boldsymbol{W})$ represents the volume of the parallelepiped whose edges are the vectors $\boldsymbol{U}$, $\boldsymbol{V}$ and $\boldsymbol{W}$. It follows from (2.34a), the definitions (2.24a), and the properties of the permutation symbol ϵ_{ijk} (1.7) and (1.8) that

$$\boldsymbol{U} \cdot (\boldsymbol{V} \times \boldsymbol{W}) = \boldsymbol{V} \cdot (\boldsymbol{W} \times \boldsymbol{U}) = \boldsymbol{W} \cdot (\boldsymbol{U} \times \boldsymbol{V}) \qquad (2.37\text{a, b})$$

and

$$\boldsymbol{U} \cdot (\boldsymbol{V} \times \boldsymbol{W}) = -\boldsymbol{U} \cdot (\boldsymbol{W} \times \boldsymbol{V}), \tag{2.38a}$$

$$= -\boldsymbol{V} \cdot (\boldsymbol{U} \times \boldsymbol{W}), \tag{2.38b}$$

$$= -\boldsymbol{W} \cdot (\boldsymbol{V} \times \boldsymbol{U}). \tag{2.38c}$$

2.5 Surface and Volume Elements

An elemental volume dv bounded by the coordinate surfaces through the points $(\theta^1, \theta^2, \theta^3)$ and $(\theta^1 + d\theta^1, \theta^2 + d\theta^2, \theta^3 + d\theta^3)$ is shown in Figure 2.6. In the limit, the volume element approaches

$$\begin{aligned} dv &= d\boldsymbol{r}_1 \cdot (d\boldsymbol{r}_2 \times d\boldsymbol{r}_3) \\ &= \boldsymbol{r}_{,1} \cdot (\boldsymbol{r}_{,2} \times \boldsymbol{r}_{,3})\, d\theta^1\, d\theta^2\, d\theta^3 \\ &= \boldsymbol{g}_1 \cdot (\boldsymbol{g}_2 \times \boldsymbol{g}_3)\, d\theta^1\, d\theta^2\, d\theta^3 \\ &= \sqrt{g}\, d\theta^1\, d\theta^2\, d\theta^3. \end{aligned} \tag{2.39}$$

The elemental parallelepiped of Figure 2.6 is bounded by the coordinate surfaces with areas ds_i and unit normals

$$\hat{\boldsymbol{e}}^i = \frac{\boldsymbol{g}^i}{\sqrt{g^{\underline{ii}}}}.$$

The area ds_1 is

$$ds_1 = \hat{\mathbf{e}}^1 \cdot (\boldsymbol{g}_2 \times \boldsymbol{g}_3)\, d\theta^2\, d\theta^3.$$

By (2.22a) we have

$$\begin{aligned} ds_1 &= \hat{\mathbf{e}}^1 \cdot \boldsymbol{g}^1 \sqrt{g}\, d\theta^2\, d\theta^3 \\ &= \sqrt{g^{11}}\sqrt{g}\, d\theta^2\, d\theta^3. \end{aligned} \tag{2.40a}$$

Likewise,

$$ds_2 = \sqrt{g^{22}}\,\sqrt{g}\, d\theta^3\, d\theta^1, \tag{2.40b}$$

$$ds_3 = \sqrt{g^{33}}\,\sqrt{g}\, d\theta^1\, d\theta^2. \tag{2.40c}$$

In subsequent developments (see, e.g., Sections 2.11 and 3.3), we enclose a region by an arbitrary surface s. Then, we require a relation which expresses an element ds of that surface in terms of the adjacent elements

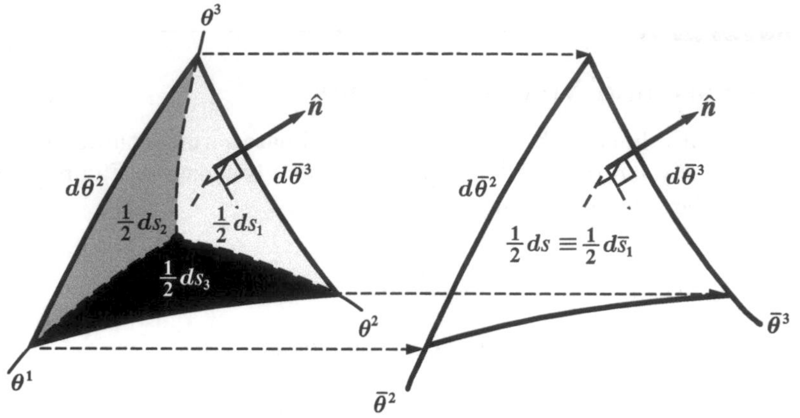

Figure 2.7 Elemental tetrahedron

ds_i of the coordinate surfaces. To that end, we examine the tetrahedron of Figure 2.7 which is bounded by triangular elements of the coordinate surfaces s_i and the triangular element of the inclined surface s. The areas of these elements are $(1/2)ds_i$ and $(1/2)ds$, where the ds_i correspond to the areas of the quadrilateral elements of Figure 2.6 and equations (2.40a–c). In keeping with the preceding development, one can view the inclined face as an element of another coordinate surface, viz., the $\bar{\theta}^1$ surface (lines $\bar{\theta}^2$ and $\bar{\theta}^3$ lie in surface $s \equiv \bar{s}_1$). The edges of the inclined triangular face have length $\sqrt{\bar{g}_{22}}\, d\bar{\theta}^{\,2}$ and $\sqrt{\bar{g}_{33}}\, d\bar{\theta}^{\,3}$; these are first-order approximations (valid in the limit: $d\theta^i$, $d\bar{\theta}^i \to 0$). The corresponding area of the inclined surface is [see (2.40a)]

$$\tfrac{1}{2}\, ds \equiv \tfrac{1}{2}\, d\bar{s}_1 = \tfrac{1}{2}\sqrt{\bar{g}^{11}}\,\sqrt{\bar{g}}\, d\bar{\theta}^{\,2}\, d\bar{\theta}^{\,3}.$$

For such enclosed tetrahedron

$$\tfrac{1}{2}\, ds_i\, \hat{\boldsymbol{e}}^i = \tfrac{1}{2}\, ds\, \hat{\boldsymbol{n}}, \tag{2.41}$$

wherein $\hat{\boldsymbol{n}}$ is the unit normal to s as $\hat{\boldsymbol{e}}^i$ is the unit normal to the surface s_i. Our reference to a second coordinate system $\bar{\theta}^i$ is merely an artifice to identify the area $d\bar{s}_1$ with the previous formulas for the area ds_i. Equation (2.41) holds for any surface s with unit normal $\hat{\boldsymbol{n}}$ and any regular coordinate system θ^i at a point of the surface. If $\hat{\boldsymbol{n}} = n_i \boldsymbol{g}^i$, then in accordance with (2.40a–c), equation (2.41) is expressed by three scalar equations:

$$\boldsymbol{g}_1 \cdot (ds_i\, \hat{\boldsymbol{e}}^i) = \sqrt{g}\, d\theta^2\, d\theta^3 = ds\, n_1, \tag{2.42a}$$

$$\boldsymbol{g}_2 \cdot (ds_i\, \hat{\boldsymbol{e}}^i) = \sqrt{g}\, d\theta^3\, d\theta^1 = ds\, n_2, \tag{2.42b}$$

$$\boldsymbol{g}_3 \cdot (ds_i\, \hat{\boldsymbol{e}}^i) = \sqrt{g}\, d\theta^1\, d\theta^2 = ds\, n_3. \tag{2.42c}$$

2.6 Derivatives of Vectors

In accordance with (2.5)

$$\boldsymbol{g}_{i,j} \equiv \frac{\partial \boldsymbol{g}_i}{\partial \theta^j} = \frac{\partial^2 \boldsymbol{r}}{\partial \theta^i\, \partial \theta^j} = \frac{\partial^2 x_p}{\partial \theta^i\, \partial \theta^j} \hat{\boldsymbol{\imath}}_p. \tag{2.43a–c}$$

Since the derivatives are supposed to be continuous,

$$\boldsymbol{g}_{i,j} = \boldsymbol{g}_{j,i}. \tag{2.44}$$

The vector $\boldsymbol{g}_{i,j}$ can be expressed as a linear combination of the base vectors $\boldsymbol{g}_i$ or $\boldsymbol{g}^i$:

$$\boldsymbol{g}_{i,j} = \Gamma_{ijk} \boldsymbol{g}^k = \Gamma^k_{ij} \boldsymbol{g}_k, \tag{2.45a, b}$$

where the coefficients are defined by

$$\Gamma_{ijk} \equiv \boldsymbol{g}_k \cdot \boldsymbol{g}_{i,j}, \qquad \Gamma^k_{ij} \equiv \boldsymbol{g}^k \cdot \boldsymbol{g}_{i,j}. \tag{2.46a, b}$$

The coefficients Γ_{ijk} and Γ^k_{ij} are known as the *Christoffel symbols of the first and second kind*, respectively. According to (2.44) and (2.46a, b), the symbols are symmetric in two lower indices; that is,

$$\Gamma_{ijk} = \Gamma_{jik}, \qquad \Gamma^k_{ij} = \Gamma^k_{ji}.$$

From (2.9), (2.44), and (2.46a) it follows that

$$\Gamma_{ijk} = \tfrac{1}{2}(g_{ik,j} + g_{jk,i} - g_{ij,k}). \tag{2.47}$$

Using (2.7), (2.8), and (2.46a, b), we obtain the following relations:

$$\Gamma_{ijk} = g_{kl} \Gamma^l_{ij}, \qquad \Gamma^k_{ij} = g^{kl} \Gamma_{ijl}. \tag{2.48a, b}$$

Differentiating (2.6) and employing (2.45a), we obtain

$$\boldsymbol{g}^i{}_{,k} \cdot \boldsymbol{g}_j = -\boldsymbol{g}^i \cdot \boldsymbol{g}_{j,k} = -\Gamma^i_{jk}.$$

It follows that

$$\boldsymbol{g}^{i}{}_{,k} = -\Gamma^{i}_{jk}\boldsymbol{g}^{j}. \tag{2.49}$$

The partial derivative of an arbitrary vector $\boldsymbol{V}$ (2.27a, b) has the alternative forms

$$\boldsymbol{V}_{,i} = V^{j}{}_{,i}\boldsymbol{g}_{j} + V^{j}\boldsymbol{g}_{j,i} = V_{j,i}\boldsymbol{g}^{j} + V_{j}\boldsymbol{g}^{j}{}_{,i}.$$

In accordance with (2.45b) and (2.49),

$$\boldsymbol{V}_{,i} = (V^{j}{}_{,i} + V^{k}\Gamma^{j}_{ki})\boldsymbol{g}_{j} \equiv V^{j}|_{i}\,\boldsymbol{g}_{j}, \tag{2.50a, b}$$

but also

$$\boldsymbol{V}_{,i} = (V_{j,i} - V_{k}\Gamma^{k}_{ji})\boldsymbol{g}^{j} \equiv V_{j}|_{i}\,\boldsymbol{g}^{j}. \tag{2.51a, b}$$

Equations (2.50b) and (2.51b) serve to define the *covariant derivatives* of the contravariant (V^i) and covariant (V_i) components of a vector.

Observe that the covariant derivative ($V^j|_i$ or $V_j|_i$) plays the same role as the partial derivative ($V_{j,i}$) plays in the Cartesian coordinate system, that the base vector ($\boldsymbol{g}_i$ or $\boldsymbol{g}^i$) plays the same role as the unit vector $\hat{\boldsymbol{\imath}}_i$ in the Cartesian system, and that the metric tensor (g_{ij} or g^{ij}) reduces to the Kronecker delta δ_{ij} in the Cartesian coordinates.

From the definition (2.46a), the Christoffel symbol $\overline{\Gamma}_{ijk}$ in one coordinate system $\bar{\theta}^i$ is expressed in terms of the symbols in another system θ^i by the formula:

$$\overline{\Gamma}_{ijk} = \Gamma_{lmn}\frac{\partial\theta^{l}}{\partial\bar{\theta}^{i}}\frac{\partial\theta^{m}}{\partial\bar{\theta}^{j}}\frac{\partial\theta^{n}}{\partial\bar{\theta}^{k}} + g_{lm}\frac{\partial\theta^{l}}{\partial\bar{\theta}^{k}}\frac{\partial^{2}\theta^{m}}{\partial\bar{\theta}^{i}\partial\bar{\theta}^{j}}. \tag{2.52}$$

Equation (2.52) shows that the Christoffel symbols *are not* components of a tensor (see Section 2.7).

From the definitions (2.16a), from (2.9), (2.12), and (2.49), it follows that

$$\frac{\partial g}{\partial g_{ij}} = g g^{ij}$$

and

$$\frac{1}{\sqrt{g}}\frac{\partial\sqrt{g}}{\partial\theta^{i}} = \Gamma^{j}_{ji}. \tag{2.53}$$

2.7 Tensors and Invariance

In the preceding developments we have introduced indicial notations, which facilitate the analyses of tensors and invariance. Also, certain terminology has been introduced, e.g., covariant, contravariant and invariant. We have yet to define tensors, tensorial transformations, and to demonstrate their roles in our analyses, most specifically the invariance of those quantities which are independent of coordinates. These specifics are set forth in the present section.

Recall that a system of nth order has n free indices and contains 3^n components in our three-dimensional space. If the components of a system, which are expressed with respect to a coordinate system θ^i, are transformed to another coordinate system $\bar{\theta}^i$ according to certain transformation laws, then the system of nth order is termed a tensor of nth order. The important attributes of tensors (e.g., invariant properties) are a consequence of the transformations which define tensorial components. The *explicit* expressions of tensorial transformations follow:

Consider a transformation from one coordinate system θ^i to another $\bar{\theta}^i$, that is to say,

$$\bar{\theta}^i = \bar{\theta}^i(\theta^1, \theta^2, \theta^3), \qquad \theta^i = \theta^i(\bar{\theta}^1, \bar{\theta}^2, \bar{\theta}^3).$$

The differentials of the variables $\bar{\theta}^i$ transform as follows:

$$d\bar{\theta}^i = \frac{\partial \bar{\theta}^i}{\partial \theta^j}\, d\theta^j .$$

This linear transformation is a prototype for the transformation of the components of a *contravariant tensor.*

In general, $T^i(\theta^1, \theta^2, \theta^3)$ and $\overline{T}^i(\bar{\theta}^1, \bar{\theta}^2, \bar{\theta}^3)$ are components of a *first-order* contravariant tensor in their respective systems, if

$$\overline{T}^i = \frac{\partial \bar{\theta}^i}{\partial \theta^j} T^j, \qquad T^i = \frac{\partial \theta^i}{\partial \bar{\theta}^j} \overline{T}^j . \tag{2.54a, b}$$

The component of a first-order tensor is distinguished by one free index and a contravariant tensor by the index appearing as a superscript.

The functions $F^{ij\cdots}(\theta^1, \theta^2, \theta^3)$ with n *superscripts* are the components of an nth-order contravariant tensor, if the components $\overline{F}^{ij\cdots}(\bar{\theta}^1, \bar{\theta}^2, \bar{\theta}^3)$ are given by the transformation:

$$\underbrace{\overline{F}^{ij\cdots}}_{n \text{ superscripts}} = \underbrace{\frac{\partial\bar{\theta}^i}{\partial\theta^p}\frac{\partial\bar{\theta}^j}{\partial\theta^q}\cdots}_{n \text{ partial derivatives}} \underbrace{F^{pq\cdots}}_{n \text{ superscripts}}. \tag{2.55}$$

Consider the transformation of the tangent base vectors $\boldsymbol{g}_i(\theta^i)$ to another coordinate system $\bar{\boldsymbol{g}}_i(\bar{\theta}^i)$; by the chain rule for partial derivatives, we have

$$\boldsymbol{g}_i \equiv \frac{\partial \boldsymbol{r}}{\partial\theta^i} = \frac{\partial \boldsymbol{r}}{\partial\bar{\theta}^j}\frac{\partial\bar{\theta}^j}{\partial\theta^i} \equiv \frac{\partial\bar{\theta}^j}{\partial\theta^i}\bar{\boldsymbol{g}}_j. \tag{2.56}$$

This linear transformation is the prototype for the transformation of the components of a *covariant tensor.*

The functions $P_{ij\cdots}(\theta^1,\theta^2,\theta^3)$ with n *subscripts* are components of an nth-order *covariant tensor*, if the components $\overline{P}_{ij\cdots}(\bar{\theta}^1,\bar{\theta}^2,\bar{\theta}^3)$ are given by the transformation:

$$\underbrace{\overline{P}_{ij\cdots}}_{n \text{ subscripts}} = \underbrace{\frac{\partial\theta^p}{\partial\bar{\theta}^i}\frac{\partial\theta^q}{\partial\bar{\theta}^j}\cdots}_{n \text{ partial derivatives}} \underbrace{P_{pq\cdots}}_{n \text{ subscripts}}. \tag{2.57}$$

A tensor may have mixed character, partly contravariant and partly covariant. The order of contravariance is given by the number of superscripts and the order of covariance by the number of subscripts. The components of a mixed tensor of order $(m+n)$, contravariant of order m and covariant of order n, transform as follows:

$$\underbrace{\overbrace{\overline{T}^{ij\cdots}_{kl\cdots}}^{m \text{ superscripts}}}_{n \text{ subscripts}} = \overbrace{\frac{\partial\bar{\theta}^i}{\partial\theta^p}\frac{\partial\bar{\theta}^j}{\partial\theta^q}\cdots}^{m \text{ partial derivatives}} \underbrace{\frac{\partial\theta^r}{\partial\bar{\theta}^k}\frac{\partial\theta^s}{\partial\bar{\theta}^l}\cdots}_{n \text{ partial derivatives}} \underbrace{\overbrace{T^{pq\cdots}_{rs\cdots}}^{m \text{ superscripts}}}_{n \text{ subscripts}}. \tag{2.58}$$

Observe that the transformations (2.55), (2.57), and (2.58) express the tensorial components in one system $\bar{\theta}^i$ as a *linear* combination of the components in another θ^i. The distinction between the contravariance (superscripts) and covariance (subscripts) is crucial from the mathematical and physical viewpoints. First, we observe that addition of tensorial components is meaningful if, and only if, they are of the same form, the same order of contravariance and covariance; then the sum is also the component of a tensor of that same order: e.g., if A_k^{ij} and B_k^{ij} are components of tensors in a system θ^i, then ($C_k^{ij} = A_k^{ij} + B_k^{ij}$) is also the component of a tensor in that system. The proof follows from (2.58). Second, the product of tensorial components is also the component of a tensor: e.g., if T^{ij} and S_{mn} are components of tensors in a system θ^i, then the product $Q_{mn}^{ij} \equiv T^{ij} S_{mn}$ is also the component of a tensor in that system. Note that the latter is a tensor of fourth order. Again, the proof follows directly from the transformation (2.58).

Consider a summation of the form Q_{mi}^{ij}; this might be termed a "contraction," wherein a system of second order is contracted from one of fourth order Q_{mn}^{ij} by the repetition of the index i and the implied summation. According to (2.58)

$$\overline{Q}_{pq}^{kl} = \frac{\partial\bar{\theta}^k}{\partial\theta^i}\frac{\partial\bar{\theta}^l}{\partial\theta^j}\frac{\partial\theta^m}{\partial\bar{\theta}^p}\frac{\partial\theta^n}{\partial\bar{\theta}^q}Q_{mn}^{ij}.$$

The "contracted" system has the components

$$\overline{Q}_{pk}^{kl} = \frac{\partial\bar{\theta}^k}{\partial\theta^i}\frac{\partial\theta^n}{\partial\bar{\theta}^k}\frac{\partial\bar{\theta}^l}{\partial\theta^j}\frac{\partial\theta^m}{\partial\bar{\theta}^p}Q_{mn}^{ij}.$$

Note that

$$\frac{\partial\bar{\theta}^k}{\partial\theta^i}\frac{\partial\theta^n}{\partial\bar{\theta}^k} = \frac{\partial\theta^n}{\partial\theta^i} = \delta_i^n.$$

Therefore,

$$\overline{Q}_{pk}^{kl} = \frac{\partial\bar{\theta}^l}{\partial\theta^j}\frac{\partial\theta^m}{\partial\bar{\theta}^p}Q_{mi}^{ij}.$$

In words, the latter components are the components of a second-order tensor; it is a mixed tensor of first-order contravariant (one free superscript) and first-order covariant (one free subscript). It is especially *important* to observe that one repeated index is a superscript and one is a subscript; one indicating the contravariant character, the other indicating the covariant character. Such repetition of indices and implied summation (one superscript–one subscript) is an inviolate rule to retain the tensorial character.

Let us now consider the further "contraction" $Q_{ij}^{ij} = T^{ij} S_{ij}$. Again, T^{ij} and S_{ij}, hence Q_{ij}^{ij} are tensorial components. In the light of (2.55) and (2.57)

$$\overline{T}^{ij}\overline{S}_{ij} = \frac{\partial\bar{\theta}^i}{\partial\theta^m}\frac{\partial\bar{\theta}^j}{\partial\theta^n}\,T^{mn}\,\frac{\partial\theta^p}{\partial\bar{\theta}^i}\frac{\partial\theta^q}{\partial\bar{\theta}^j}\,S_{pq}.$$

According to the chain rule for partial differentiation

$$\overline{T}^{ij}\overline{S}_{ij} = \frac{\partial \theta^p}{\partial \theta^m}\frac{\partial \theta^q}{\partial \theta^n} T^{mn} S_{pq} = \delta^p_m \delta^q_n T^{mn} S_{pq} = T^{mn} S_{mn}. \tag{2.59}$$

In words, this quantity is unchanged by a coordinate transformation. Such quantities are *invariants*; they have the same value independently of the coordinate system. The invariance hinges on the notions of covariance and contravariance. An invariant function of curvilinear coordinates is obtained by summations involving repeated indices which appear once as a superscript and once as a subscript. Invariants have special physical meaning because they are not dependent on the choice of coordinates. They are easily recognized as zero-order tensors (no free indices). However, care must be taken that repeated indices appear once as a superscript and once as a subscript, for otherwise the sum is not invariant. Cartesian coordinates are the exception because the covariant and contravariant transformations are then identical. For example, the Kronecker delta δ^i_j is a tensor in the Cartesian system x_i. It can be written with indices up or down, that is, $\delta^i_j = \delta^{ij} = \delta_{ij}$. Note that g^{ij} and g_{ij} are the contravariant and covariant components obtained by the appropriate transformations of δ^{ij} from the rectangular to the curvilinear coordinate system:

$$g_{ij} = \frac{\partial x_k}{\partial \theta^i}\frac{\partial x_l}{\partial \theta^j}\delta_{kl} = \frac{\partial x_k}{\partial \theta^i}\frac{\partial x_k}{\partial \theta^j}, \qquad g^{ij} = \frac{\partial \theta^i}{\partial x_k}\frac{\partial \theta^j}{\partial x_l}\delta^{kl} = \frac{\partial \theta^i}{\partial x_k}\frac{\partial \theta^j}{\partial x_k};$$

δ^{ij} are components of the metric tensor in a Cartesian coordinate system.

Recall the definition (2.6) of the reciprocal vector $\boldsymbol{g}^i$ and also (2.56). Then

$$\bar{\boldsymbol{g}}^i \cdot \bar{\boldsymbol{g}}_j = \delta^i_j = \bar{\boldsymbol{g}}^i \cdot \boldsymbol{g}_k \frac{\partial \theta^k}{\partial \bar{\theta}^j}.$$

The latter holds generally if, and only if,

$$\bar{\boldsymbol{g}}^i = \boldsymbol{g}^m \frac{\partial \bar{\theta}^i}{\partial \theta^m}. \tag{2.60}$$

It follows from (2.56) and (2.60) that the components g_{ij} and g^{ij} transform according to the rules for covariant and contravariant components, respectively. Similarly, from equations (2.25a, b), the components e_{ijk} and e^{ijk} are, respectively, covariant and contravariant:

$$e_{ijk} = \frac{\partial x_l}{\partial \theta^i}\frac{\partial x_m}{\partial \theta^j}\frac{\partial x_n}{\partial \theta^k}\epsilon_{lmn}, \qquad e^{ijk} = \frac{\partial \theta^i}{\partial x_l}\frac{\partial \theta^j}{\partial x_m}\frac{\partial \theta^k}{\partial x_n}\epsilon^{lmn}.$$

Note: The Christoffel symbols do *not* transform as components of a tensor; see equation (2.52).

A mathematical theorem—known as the *quotient law*—proves to be useful in establishing the tensor character of a system without recourse to the transformation [i.e., (2.55), (2.57), and (2.58)]: If the product of a system $S^{pr\cdots}_{ij\ldots}$ with an *arbitrary* tensor is itself a tensor, then $S^{pr\cdots}_{ij\ldots}$ is also a tensor (for proof see, e.g., J. C. H. Gerretsen [6], Section 4.2.3, p. 46).

Note the significant features of tensors, the significance of the *notations* (superscripts signify *contravariance*, subscripts *covariance*), the *summation convention*, the *linear transformations* and the identification of *invariance*. Invariants are especially important to describe physical attributes which are independent of coordinates; the conventions enable one to establish such quantities. The linear transformation of tensorial components is also very useful: If all components vanish in one system, then all vanish in every other system. This means that any equation in tensorial form holds in every coordinate system.

2.8 Associated Tensors

Let T_{ij} denote the component of a covariant tensor. The component of an associated contravariant tensor is

$$T^{pq} = g^{pi} g^{qj} T_{ij}. \tag{2.61}$$

Because g^{ij} and T_{ij} are components of contravariant and covariant tensors, respectively, the reader can show that T^{pq} is the component of a contravariant tensor. Moreover,

$$T_{ij} = g_{pi} g_{qj} T^{pq}. \tag{2.62}$$

The components of the covariant, contravariant, or mixed associated tensors of any order are formed by raising or lowering indices as in (2.61) and (2.62), that is, by multiplying and summing with the appropriate versions of the metric tensor, g^{ij} or g_{ij}. Observe that the base vectors conform to this rule according to (2.7) and (2.8).

If the tensors are not symmetric in two indices, it is essential that the proper position of the indices is preserved when raising or lowering indices. For example, this can be accomplished by placing a dot ($\cdot$) in the vacant position:

$$T_i^{\cdot j} = g_{ip} T^{pj}, \qquad T^i_{\cdot j} = g^{ip} T_{pj},$$

$$T_{ij} = g_{ip} T^{p}_{\bullet j} = g_{jp} T_{i}^{\bullet p}, \qquad T^{ij} = g^{jp} T^{i}_{\bullet p} = g^{ip} T_{p}^{\bullet j}.$$

If T^{ij} is symmetric, that is, $T^{ij} = T^{ji}$, then there is no need to mark the position:

$$T_{i}^{\bullet j} = T^{j}_{\bullet i} = T^{j}_{i}.$$

2.9 Covariant Derivative

The essential feature of the covariant derivative is its tensor character. For example, $V^j|_i$ of (2.50b): Since $V^j|_i = \boldsymbol{g}^j \cdot \boldsymbol{V}_{,i}$, it follows from the tensorial transformation (Section 2.7) that $V^j|_i$ transform as components of a mixed tensor.

Because of the invariance property, it is useful to define covariant differentiation for tensors of higher order. Equation (2.50b) is the prototype for the covariant derivative of a contravariant tensor. The general form for the covariant derivative of a contravariant tensor of order m follows:

$$F^{ij\cdots m}|_p \equiv \underline{F^{ij\cdots m}{}_{,p}} + F^{qj\cdots m}\Gamma^{i}_{qp} + F^{iq\cdots m}\Gamma^{j}_{qp} + \cdots + F^{ij\cdots q}\Gamma^{m}_{qp}. \quad (2.63)$$

The covariant derivative of (2.63) includes the partial derivative (underlined) but augmented by m additional terms, each is a sum of products in which a component $F^{qj\cdots m}$ is multiplied by a Christoffel symbol Γ^{i}_{qp}; in each of the latter, the index p of the independent variable of differentiation and the dummy index q appear as subscripts, and a different index of the component $F^{ij\cdots}$ is the superscript; that superscript on $F^{ij\cdots}$, which is replaced by the dummy index q.

Equation (2.51b) is the prototype for the covariant derivative of a covariant component. The general form for a tensor of order n follows:

$$P_{ij\cdots n}|_p \equiv \underline{P_{ij\cdots n,p}} - P_{rj\cdots n}\Gamma^{r}_{ip} - P_{ir\cdots n}\Gamma^{r}_{jp} \cdots - P_{ij\cdots r}\Gamma^{r}_{np}. \quad (2.64)$$

The covariant derivative again includes the partial derivative (underlined) but again it is augmented by n additional terms, each is a sum of products in which a component $P_{rj\cdots n}$ is multiplied by a Christoffel symbol Γ^{r}_{ip}; here each symbol has the index p of the variable of differentiation as a subscript, the dummy index r as a superscript and successive subscripts of the component $P_{ij\cdots}$ appear as the second subscript on the Christoffel symbol Γ^{r}_{ip}; that subscript on $P_{ij\cdots}$, which is replaced by the dummy index r.

The covariant derivative of a mixed component follows the rules for the contravariant and covariant components: The partial derivative is augmented by terms formed according to (2.63) or (2.64) as the component has contravariant or covariant character, respectively:

$$T^{ij\cdots m}_{kl\cdots n}\big|_p = \underline{T^{ij\cdots m}_{kl\cdots n,p}} + T^{qj\cdots m}_{kl\cdots n}\Gamma^i_{qp} + \cdots + T^{ij\cdots q}_{kl\cdots n}\Gamma^m_{qp}$$

$$- T^{ij\cdots m}_{rl\cdots n}\Gamma^r_{kp} - \cdots - T^{ij\cdots m}_{kl\cdots r}\Gamma^r_{np}. \qquad (2.65)$$

In general, the order of covariant differentiation is not permutable. If A_j denotes a covariant component of a tensor, then generally

$$A_j|_{kl} \neq A_j|_{lk}.$$

According to (2.64)

$$A_j|_{kl} - A_j|_{lk} = R^{i}_{\cdot jkl}A_i, \qquad (2.66)$$

where

$$R^{i}_{\cdot jkl} \equiv \Gamma^i_{jl,k} - \Gamma^i_{jk,l} + \Gamma^m_{jl}\Gamma^i_{mk} - \Gamma^m_{jk}\Gamma^i_{ml}. \qquad (2.67)$$

These comprise the components of the *Riemann-Christoffel tensor* or the so-called *mixed-curvature tensor*. The significance of the latter term is apparent when we observe that all components vanish in a system of Cartesian coordinates; then $A_j|_{kl} = A_j|_{lk} = A_{j,kl}$. The components in any coordinates of an Euclidean space can be obtained by a linear transformation, i.e., the appropriate tensorial transformation from Cartesian components. It follows that the Riemann-Christoffel tensor vanishes in Euclidean space. This fact imposes geometrical constraints upon the deformation of continuous bodies, as described in the subsequent chapters.

The associated curvature tensor has the components

$$R_{ijkl} = g_{im}R^{m}_{\cdot jkl}, \qquad (2.68a)$$

$$= \Gamma_{lji,k} - \Gamma_{kji,l} + \Gamma^m_{jk}\Gamma_{ilm} - \Gamma^m_{jl}\Gamma_{ikm}, \qquad (2.68b)$$

$$= \tfrac{1}{2}(g_{jk,il} + g_{il,jk} - g_{ik,jl} - g_{jl,ik})$$

$$+ g^{mn}(\Gamma_{jkm}\Gamma_{iln} - \Gamma_{jlm}\Gamma_{ikn}). \qquad (2.68c)$$

From the definition (2.68c) and the symmetries of the components of the metric tensor ($g_{ij} = g_{ji}$) and of the Christoffel symbol ($\Gamma_{ijk} = \Gamma_{jik}$) it follows that

$$R_{ijkl} = -R_{jikl}, \tag{2.69a}$$

$$R_{ijkl} = -R_{ijlk}, \tag{2.69b}$$

$$R_{ijkl} = R_{klij}. \tag{2.69c}$$

2.10 Transformation from Cartesian to Curvilinear Coordinates

According to (2.55), (2.57) and (2.58), if all components of a tensor vanish in one coordinate system, then they vanish in every other. This means that an equation (or equations) expressed in tensorial form holds in every system. This is especially important in any treatment of a physical problem, since any coordinate system is merely an artifice, which is introduced for mathematical purposes: Often a Cartesian system is the simplest; an expression in the Cartesian system may be a special form of a tensorial component. Provided that the tensorial character is fully established, the generalization to another system is readily accomplished.

The Cartesian coordinates have several distinct features: The coordinate *is* length or, stated mathematically, the component of the metric tensor is the Kronecker delta:

$$g^{ij} = g_{ij} = \delta_{ij} \equiv \delta^{ij} \equiv \delta^i_j.$$

Also, the position of the suffix (superscript or subscript) is irrelevant. It follows too that the Christoffel symbols vanish; hence, the covariant derivative reduces to the partial derivative, e.g.,

$$V^i|_j = V_i|_j = V^i{}_{,j} = V_{i,j}.$$

Additionally, the components of vectors or second-order systems in orthogonal directions have simpler interpretations, geometrically and physically.

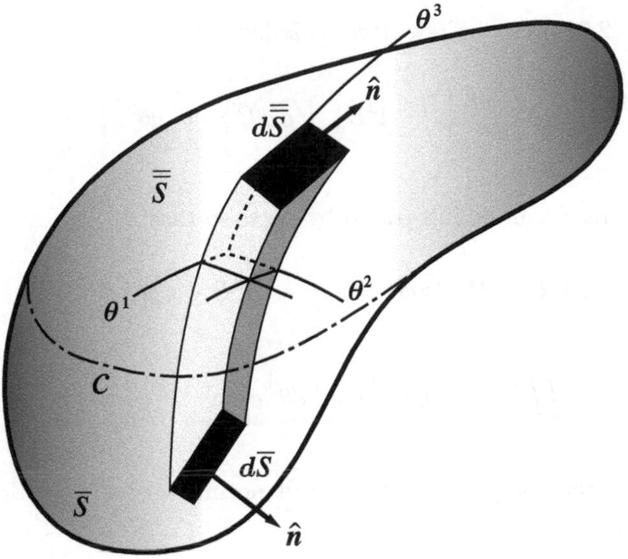

Figure 2.8 Region of integration

2.11 Integral Transformations

In the analysis of continuous bodies, especially by energy principles, integrals arise in the form

$$I \equiv \iiint_v A^i B_{,i}\, dv = \iiint_v A^i B_{,i} \sqrt{g}\, d\theta^1\, d\theta^2\, d\theta^3, \tag{2.70}$$

where the integration extends through a volume v. A^i and B are assumed continuous with continuous derivatives. The region of integration is bounded by the surface s in Figure 2.8. The entire bounding surface s is divided into surfaces $\bar{s}$ and $\bar{\bar{s}}$ by the curve c such that θ^3 lines are tangent to s along c. If the surface s is irregular, e.g., possesses concave portions, the derivation must be amended; the surface s must be subjected to additional subdivision. Still the final result (2.72) holds.

One term of (2.70) is

$$I_3 \equiv \iiint_v A^3 B_{,3} \sqrt{g}\, d\theta^3\, d\theta^1\, d\theta^2. \tag{2.71a}$$

The integral (2.71a) can be rewritten as follows:

$$I_3 \equiv \iint \left\{ \int_{\bar{s}}^{\bar{\bar{s}}} A^3 B_{,3}\,\sqrt{g}\,d\theta^3 \right\} d\theta^1\,d\theta^2.$$

The term in braces can be integrated by parts so that

$$I_3 = \iint \left\{ \left[A^3 B\,\sqrt{g}\,d\theta^1\,d\theta^2\right]_{\bar{\bar{s}}} - \left[A^3 B\,\sqrt{g}\,d\theta^1\,d\theta^2\right]_{\bar{s}} \right\}$$

$$- \iiint \left(A^3\sqrt{g}\right)_{,3} B\,d\theta^1\,d\theta^2\,d\theta^3.$$

If n_i represent the components of the unit normal vector $\hat{\boldsymbol{n}} = n_i \boldsymbol{g}^i$, then according to (2.42c)

$$\left[\sqrt{g}\,d\theta^1\,d\theta^2\right]_{\bar{\bar{s}}} = n_3\,d\bar{\bar{s}}, \qquad \left[\sqrt{g}\,d\theta^1\,d\theta^2\right]_{\bar{s}} = -n_3\,d\bar{s}.$$

The negative sign in the second equation arises because $\hat{\boldsymbol{n}} \cdot \hat{\boldsymbol{e}}^3 = -1$ on $\bar{s}$. The integral I_3 follows:

$$I_3 = \iint_s A^3 B\,n_3\,ds - \iiint_v \frac{1}{\sqrt{g}} \left(A^3\sqrt{g}\right)_{,3} B\,dv. \qquad (2.71b)$$

The other terms of the integral I are similar. Consequently,

$$I \equiv \iiint_v A^i B_{,i}\,dv = \iint_s A^i B\,n_i\,ds - \iiint_v \frac{1}{\sqrt{g}} \left(A^i\sqrt{g}\right)_{,i} B\,dv. \quad (2.72)$$

Equation (2.72) is one version of Green's theorem, one which is particularly useful in applications of the energy principles to continuous bodies. From (2.72) a series of useful formulations can be derived.

In a subsequent application of (2.72) we encounter vectors $\boldsymbol{s}^i$, which transform as contravariant tensorial components, and a vector $\delta\boldsymbol{V}$. Then from (2.72) it follows that

$$\iiint_v \boldsymbol{s}^i \cdot \delta\boldsymbol{V}_{,i}\,dv = \iint_s (\boldsymbol{s}^i\,n_i) \cdot \delta\boldsymbol{V}\,ds - \iiint_v \frac{1}{\sqrt{g}} \left(\sqrt{g}\,\boldsymbol{s}^i\right)_{,i} \cdot \delta\boldsymbol{V}\,dv. \qquad (2.73)$$

Integral transformations, which express certain surface integrals in terms of line integrals are presented in Section 8.11.

Chapter 3

Deformation

3.1 Concept of a Continuous Medium

All materials are composed of discrete constituents, crystals or molecules which, in turn, have various atomic structures. Additionally, voids may occur in real materials. Usually such discrete and distinctive parts are microscopic and randomly distributed. When one considers adjoining elemental, but macroscopic portions, the properties are often similar; variations are usually quite gradual. Indeed, our experiences tell us that most analyses of structural elements and mechanical components can be based on the notion of continuity. Cautiously, we assume that contiguous parts are cohesive and that properties vary continuously. These concepts of a continuous medium provide powerful means for the analyses of solid, but deformable, bodies. Specific discontinuities can be accommodated at prescribed sites or interfaces. In practice, the engineer must bear in mind that certain phenomena have their origins in the microstructure of the materials and, therefore, their treatment falls beyond the theories of continuous cohesive media.

Our mathematical treatment of the continuous medium is intentionally limited to means most suited to the mechanics of solids. Specifically, we recognize that our knowledge of the behavior and properties must be traced from an initial state. Accordingly, we identify each element of material with that reference state and follow the element throughout its deformational history. This so-called Lagrangean viewpoint enables us to adopt a single system of coordinates and obviates any need for alternatives with attendant and burdensome transformations.

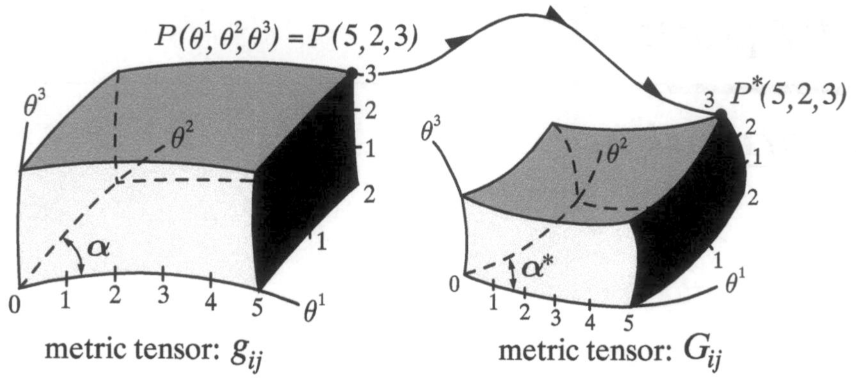

Figure 3.1 Convected coordinate system

3.2 Geometry of the Deformed Medium

In the mechanics of solids, it is necessary that we establish a reference state. We suppose that the reference state exists at time $t = 0$, and we tentatively view it as the undeformed state. In other words, we measure deformation from the initial configuration. In most instances, only the initial configuration is known a priori and our viewpoint is the only one practicable. Accordingly, we employ the initial coordinates of particles as independent variables. It may be helpful to imagine that an identifying label is affixed to each particle of the body; the same particle bears the same label at any subsequent time. Thus, if the particle P of Figure 3.1 has the initial coordinates ($\theta^1 = 5$, $\theta^2 = 2$, $\theta^3 = 3$), it retains the identifying numbers $(5, 2, 3)$ throughout the deformation, *but* the distances between the particles are, in general, changed. Coordinate systems which follow the deformation of the solid are called *convected* coordinate systems. The measure of distances between particles and the angle between θ^i and θ^j lines ($i \neq j$) in the initial and convected systems differ: In the initial system, distances and angles are determined by the metric tensor g_{ij} of equation (2.13); in the deformed system by the metric tensor G_{ij} of equation (3.8). When possible, quantities of the initial and subsequent states are distinguished by minuscules and majuscules, e.g., ℓ and L denote lengths in the initial and subsequent states, respectively.

The deformation of a medium is determined by the displacements of particles and lines of particles. The position of the particle P of Figure 3.2 before deformation is specified by the vector $\boldsymbol{r}$:

$$\boldsymbol{r} = \boldsymbol{r}(\theta^1, \theta^2, \theta^3) = x_i(\theta^1, \theta^2, \theta^3)\hat{\boldsymbol{\imath}}_i, \tag{3.1}$$

where θ^i are arbitrary curvilinear coordinates, x_i are the rectangular Carte-

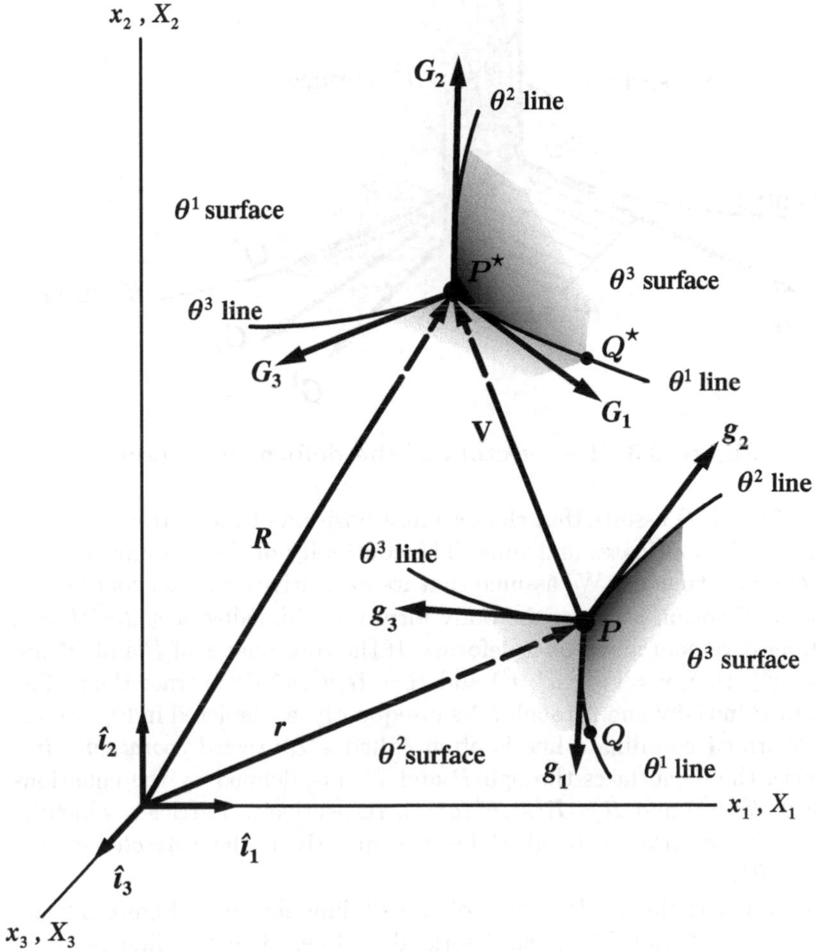

Figure 3.2 Displacement and deformation

sian coordinates of the particle, and $\hat{\boldsymbol{\imath}}_i$ are unit vectors parallel to the axes as shown in Figure 3.2. After deformation, the particle is at P^* with the position vector

$$\boldsymbol{R} = \boldsymbol{R}(\theta^1, \theta^2, \theta^3; t) = X_i(\theta^1, \theta^2, \theta^3; t)\hat{\boldsymbol{\imath}}_i. \tag{3.2}$$

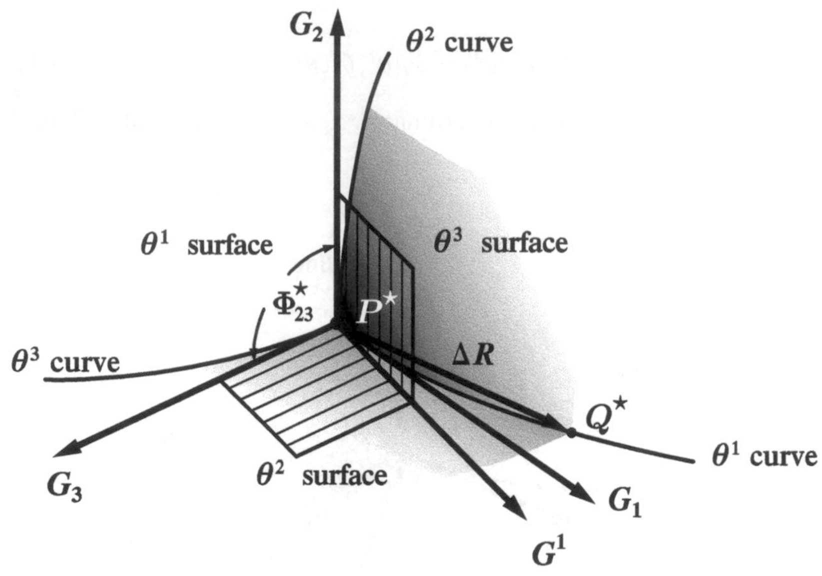

Figure 3.3 Base vectors of the deformed system

Equation (3.2) asserts that the terminal position of the particle depends on its initial coordinates and time. The continuity of the medium requires that $\boldsymbol{R}$ be continuous. We assume that its derivatives are also continuous.

The particles on a line of the body form a chain, called a *material line*, which must remain intact as it deforms. If the coordinates of P and P^* are (a^1, a^2, a^3), then $\boldsymbol{r} = \boldsymbol{r}(a^1, a^2, \theta^3)$ and $\boldsymbol{R} = \boldsymbol{R}(a^1, a^2, \theta^3; t)$ trace the θ^3 line through P initially and through P^* subsequently, as depicted in Figure 3.2. The deformed coordinate line is often called a *convected coordinate line*. Likewise, the θ^3 surfaces through P and P^* are defined by the equations $\boldsymbol{r} = \boldsymbol{r}(\theta^1, \theta^2, a^3)$ and $\boldsymbol{R} = \boldsymbol{R}(\theta^1, \theta^2, a^3; t)$, respectively. Particles, which lie initially on the surface through P, lie subsequently on the *convected surface* through P^*.

The two particles at P and Q of the θ^1 line shown in Figure 3.2 are displaced to P^* and Q^* of the deformed θ^1 line. The coordinates of P^* and Q^* differ by the increment $\Delta\theta^1$, just as the coordinates of P and Q in the undeformed configuration. The position of particle Q^* relative to particle P^* is given by the vector $\Delta\boldsymbol{R}$ originating at P^* and terminating at Q^* (see Figure 3.3). In the limit, as Q^* approaches P^*, the vector $\Delta\boldsymbol{R}$ is tangent to the θ^1 line at P^*. It follows that the derivative $\partial\boldsymbol{R}/\partial\theta^1$ is tangent to the θ^1 line at particle P^*. Collectively, the partial derivatives form the triad of *tangent base vectors*:

$$\boldsymbol{G}_i \equiv \frac{\partial \boldsymbol{R}}{\partial \theta^i} = \boldsymbol{R}_{,i}. \tag{3.3}$$

Since it is physically impossible for two coordinate surfaces to deform to coincidence, the tangent base vectors are always noncoplanar and any vector can be expressed as a linear combination of these base vectors. The tangent vector $\boldsymbol{G}_i$ is the deformed counterpart of the vector $\boldsymbol{g}_i$ which is tangent to the undeformed θ^i line.

An alternative base triad is formed of vectors $\boldsymbol{G}^i$ which are normal to the coordinates surfaces. The vector $\boldsymbol{G}^i$ is defined by the equations

$$\boldsymbol{G}^i \cdot \boldsymbol{G}_j \equiv \delta^i_j. \tag{3.4}$$

The normal vector $\boldsymbol{G}^1$, shown in Figure 3.3, is normal to the tangent base vectors ($\boldsymbol{G}_2$, $\boldsymbol{G}_3$) and, therefore, normal to the θ^1 surface. The vectors $\boldsymbol{G}^i$ are often termed reciprocal base vectors; here we call them by the descriptive phrase *normal base vectors*. Since the vector $\boldsymbol{G}^i$ is normal to the θ^i surface, the triad $\boldsymbol{G}^i$ must be noncoplanar; it follows that any vector can be expressed as a linear combination of the normal base vectors.

The vector $\boldsymbol{G}_i$ can be expressed as a linear combination of the normal base vectors and vice versa:

$$\boldsymbol{G}_i = G_{ij}\boldsymbol{G}^j, \qquad \boldsymbol{G}^i = G^{ij}\boldsymbol{G}_j. \tag{3.5a, b}$$

The coefficients G_{ij} and G^{ij} can be evaluated by the products:

$$G_{ij} = G_{ji} = \boldsymbol{G}_i \cdot \boldsymbol{G}_j, \qquad G^{ij} = G^{ji} = \boldsymbol{G}^i \cdot \boldsymbol{G}^j. \tag{3.6a, b}$$

The coefficients G_{ij} and G^{ij} are the covariant and contravariant components of the metric tensor of the *deformed* coordinate system, as g_{ij} and g^{ij} are the components of the metric tensor of the reference system, in accordance with equations (2.9) and (2.10). The contravariant components G^{ij} are determined by the covariant components, in the manner of (2.12):

$$G^{ij} = \frac{\text{cofactor of element } G_{ij} \text{ in matrix } [G_{ij}]}{|G_{ij}|}. \tag{3.7}$$

The components of the metric tensor, G_{ij} and G^{ij}, play the same roles in the deformed curvilinear system as their counterparts, g_{ij} and g^{ij}, in the reference system. In particular, the fundamental form of differential geometry determines the differential length dL in the deformed system:

$$dL^2 = d\boldsymbol{R} \cdot d\boldsymbol{R} = G_{ij}\, d\theta^i\, d\theta^j. \tag{3.8}$$

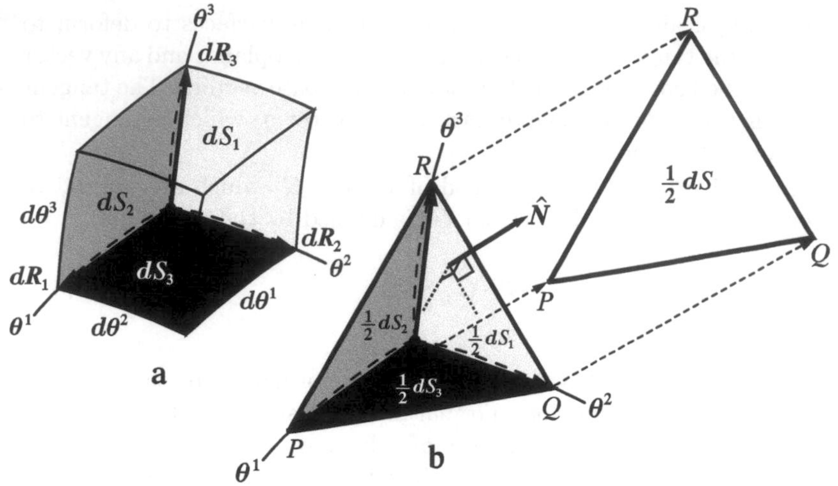

Figure 3.4 Deformed elemental parallelepiped (a) and elemental tetrahedron (b)

Just as $g_{\underline{ii}}$ determines the length along the undeformed θ^i line, $G_{\underline{ii}}$ provides the measure of length along the deformed θ^i line.

In accordance with (3.6a, b), the lengths of the vectors $\boldsymbol{G}_i$ and $\boldsymbol{G}^i$ are $\sqrt{G_{\underline{ii}}}$ and $\sqrt{G^{\underline{ii}}}$, respectively. The unit vectors in the directions of $\boldsymbol{G}_i$ and $\boldsymbol{G}^i$ are, respectively,

$$\hat{\boldsymbol{E}}_i = \frac{\boldsymbol{G}_i}{\sqrt{G_{\underline{ii}}}}, \qquad \hat{\boldsymbol{E}}^i = \frac{\boldsymbol{G}^i}{\sqrt{G^{\underline{ii}}}}. \tag{3.9a, b}$$

If $\boldsymbol{R} = X_i \hat{\boldsymbol{\imath}}_i$, where X_i are the Cartesian coordinates of the displaced particle, then the components G_{ij} are also given by an expression like (2.15):

$$G_{ij} = \frac{\partial X_k}{\partial \theta^i} \frac{\partial X_k}{\partial \theta^j}. \tag{3.10}$$

3.3 Dilation of Volume and Surface

An elemental volume of the deformed medium is bounded by the coordinate surfaces through the points $(\theta^1, \theta^2, \theta^3)$ and $(\theta^1 + d\theta^1, \theta^2 + d\theta^2, \theta^3 + d\theta^3)$ as depicted in Figure 3.4a. In the limit, the element approaches the parallelepiped defined by the edges $d\boldsymbol{R}_i = \boldsymbol{G}_i \, d\theta^{\underline{i}}$. The volume of the element is the triple scalar product of the right-hand triad:

$$dV = \boldsymbol{G}_1 \cdot (\boldsymbol{G}_2 \times \boldsymbol{G}_3) \, d\theta^1 \, d\theta^2 \, d\theta^3.$$

Following the notations of Section 2.3, we label the volume metric:

$$\sqrt{G} \equiv \boldsymbol{G}_1 \cdot (\boldsymbol{G}_2 \times \boldsymbol{G}_3). \tag{3.11a}$$

Thus, the volume of the differential element assumes the form:

$$dV = \sqrt{G} \, d\theta^1 \, d\theta^2 \, d\theta^3. \tag{3.11b}$$

The measure of the deformed volume $\sqrt{G}$ can be expressed in terms of the metric tensor G_{ij} as the measure of the undeformed volume $\sqrt{g}$ is expressed by equation (2.16a):

$$G = \left| G_{ij} \right|. \tag{3.11c}$$

The change of volume during deformation (called dilation) is given by the ratio

$$D \equiv \frac{dV}{dv} = \frac{\sqrt{G}}{\sqrt{g}}. \tag{3.12}$$

The tetrahedron of Figure 3.4b is bounded by the triangular elements of the deformed coordinate surfaces and the inclined plane with normal $\hat{\boldsymbol{N}} = N_i \boldsymbol{G}^i$. The area of a triangular element is $(1/2) \, dS_i$ and the area of the inclined surface is $(1/2) \, dS$ (see Section 2.5). In the limit

$$dS_i \, \hat{\boldsymbol{E}}^i = dS \, \hat{\boldsymbol{N}}, \tag{3.13}$$

where $\hat{\boldsymbol{E}}^i$ is the unit normal to the θ^i surface; see equation (3.9b). The area of the θ^3 coordinate surface is

$$dS_3 = \hat{\boldsymbol{E}}^3 \cdot (\boldsymbol{G}_1 \times \boldsymbol{G}_2) \, d\theta^1 \, d\theta^2.$$

In the manner of (2.40c),

$$dS_3 = \sqrt{G^{33}} \sqrt{G} \, d\theta^1 \, d\theta^2. \tag{3.14a}$$

Likewise,

$$dS_1 = \sqrt{G^{11}} \sqrt{G} \, d\theta^2 \, d\theta^3, \qquad dS_2 = \sqrt{G^{22}} \sqrt{G} \, d\theta^1 \, d\theta^3. \tag{3.14b, c}$$

From (3.13) and (3.14a–c), we obtain relationships similar to those appearing in equations (2.42a–c):

$$\sqrt{G}\, d\theta^2\, d\theta^3 = N_1\, dS,$$

$$\sqrt{G}\, d\theta^3\, d\theta^1 = N_2\, dS,$$

$$\sqrt{G}\, d\theta^1\, d\theta^2 = N_3\, dS.$$

Here $N_i \equiv \boldsymbol{G}_i \cdot \hat{\boldsymbol{N}}$ is the component of $\hat{\boldsymbol{N}} = N_i \boldsymbol{G}^i$. We observe that the elemental area of the θ^i surface experiences a change or surface dilation:

$$\frac{dS_{\underline{i}}}{ds_{\underline{i}}} = \sqrt{\frac{G^{\underline{ii}}}{g^{\underline{ii}}}} \sqrt{\frac{G}{g}}. \tag{3.15}$$

3.4 Vectors and Tensors Associated with the Deformed System

Again, but one system of coordinates θ^i are employed to identify the same particle in the initial ($t = 0$) and all subsequent (deformed) configurations. However, the geometry is altered; lines, surfaces and volumes are deformed. These changes are manifested in changes of the base vectors, metric tensor, and dilation of volume and surface. Specifically, the base vector $\boldsymbol{g}_i$ (or $\boldsymbol{g}^i$) is moved and deformed to the vector $\boldsymbol{G}_i$ (or $\boldsymbol{G}^i$); the metric g_{ij} (or g^{ij}) is changed to the metric G_{ij} (or G^{ij}); the volume metric $\sqrt{g}$ is dilated to the metric $\sqrt{G}$.

Since the initial (undeformed) system was arbitrary, every formulation for the initial system has a similar counterpart in the deformed system. Let us recast a few basic formulas. The vector products of $\boldsymbol{G}_i$ and $\boldsymbol{G}_j$, $\boldsymbol{G}^i$ and $\boldsymbol{G}^j$ have forms like (2.26a, b):

$$\boldsymbol{G}_i \times \boldsymbol{G}_j = E_{ijk}\boldsymbol{G}^k, \qquad \boldsymbol{G}^i \times \boldsymbol{G}^j = E^{ijk}\boldsymbol{G}_k, \tag{3.16a, b}$$

where, like the tensors e_{ijk} and e^{ijk} of (2.24a, b),

$$E_{ijk} \equiv \epsilon_{ijk}\sqrt{G}, \qquad E^{ijk} \equiv \frac{\epsilon^{ijk}}{\sqrt{G}}. \tag{3.17a, b}$$

Some care is required to avoid misunderstandings when components are associated with a different basis, i.e., deformed or undeformed. For example, one can express components of a vector $\boldsymbol{V}$ in either system:

$$\boldsymbol{V} = v^i \boldsymbol{g}_i = v_i \boldsymbol{g}^i,$$

or

$$\boldsymbol{V} = V^i \boldsymbol{G}_i = V_i \boldsymbol{G}^i.$$

Clearly $(v_i \neq V_i)$ and $(v^i \neq V^i)$. Moreover, v_i and v^i are associated via the undeformed metric, whereas V_i and V^i are associated via the deformed metric, viz.,

$$v^i = \boldsymbol{g}^i \cdot \boldsymbol{V} = g^{ij} v_j, \qquad V^i = \boldsymbol{G}^i \cdot \boldsymbol{V} = G^{ij} V_j. \tag{3.18a, b}$$

Unless otherwise noted, we associate components with the initial state, i.e, via the undeformed metric (g_{ij}, g^{ij}).

A tensor can also be associated with the initial or the deformed system. In certain instances it might be expedient to form components associated with the deformed system. For example, instead of the associated tensor of (2.61),

$$T^{pq} = g^{pi} g^{qj} T_{ij},$$

we can form the components associated with the deformed system:

$${}^*T^{pq} = G^{pi} G^{qj} T_{ij}.$$

Here the asterisk signals the deformed system. In such instances, one must exercise care to avoid misunderstandings, for clearly $({}^*T^{pq} \neq T^{pq})$. Since the initial system is usually prescribed, it is that metric (g_{ij} and g^{ij}) which is known, and consequently, association via the initial metric is usually the most practical.

Derivatives of vectors in the deformed system require similar precautions. Like the Christoffel symbols of (2.46a, b), we have the symbols of the deformed space:

$${}^*\Gamma_{ijk} \equiv \boldsymbol{G}_k \cdot \boldsymbol{G}_{i,j}, \qquad {}^*\Gamma^k_{ij} \equiv \boldsymbol{G}^k \cdot \boldsymbol{G}_{i,j} = -\boldsymbol{G}_i \cdot \boldsymbol{G}^k{}_{,j}. \tag{3.19a, b}$$

Like (2.47), the Christoffel symbols for the deformed state are given by

$${}^*\Gamma_{ijk} = \tfrac{1}{2}\left(G_{ik,j} + G_{jk,i} - G_{ij,k}\right). \tag{3.20}$$

Since our subject is suited to a common system of coordinates, in the initial and subsequent configurations, transformations pose no difficulties of notation, but follow the transformation laws given by (2.55), (2.57), and (2.58).

Covariant derivatives may also be expressed in the deformed system. Then, the Christoffel symbols in the formulations (2.63) to (2.65) must be replaced with the symbols of the deformed system which are formed from the metric of the latter (G_{ij} and G^{ij}).

If the integral transformations of Section 2.11 are performed with respect to the deformed system (through the deformed volume), then the volume metric $\sqrt{G}$ must replace the metric $\sqrt{g}$, and the component N_i of the unit normal vector must replace the component n_i. For example, equations (2.72) and (2.73) now take the forms:

$$\iiint_V A^i B_{,i}\, dV = \iint_S A^i B\, N_i\, dS - \iiint_V \frac{1}{\sqrt{G}}\left(A^i \sqrt{G}\right)_{,i} B\, dV$$

and

$$\iiint_V \boldsymbol{s}^i \cdot \delta \boldsymbol{V}_{,i}\, dV =$$

$$\iint_S (\boldsymbol{s}^i\, N_i) \cdot \delta \boldsymbol{V}\, dS - \iiint_V \frac{1}{\sqrt{G}}\left(\sqrt{G}\, \boldsymbol{s}^i\right)_{,i} \cdot \delta \boldsymbol{V}\, dV.$$

Again, tensors can be associated with the deformed system and all formulations can be cast in the deformed system. Then, the deformed vectors ($\boldsymbol{G}_i$ and $\boldsymbol{G}^i$) and the metric of the deformed configuration (G_{ij} and G^{ij}) provide the bases. Special *precautions* are needed if the systems are mixed.

3.5 Nature of Motion in Small Regions

Consider the neighboring particles P and Q of Figure 3.5. Let $\Delta \boldsymbol{r}$ denote the position of particle Q relative to P in the reference state and $\Delta \boldsymbol{R}$ the position of the same particle Q^* relative to P^* in a subsequent state. If the particles are sufficiently close, we need only the first approximations:

$$\Delta \boldsymbol{r} \doteq \boldsymbol{g}_i\,]_P\, \Delta \theta^i, \qquad \Delta \boldsymbol{R} \doteq \boldsymbol{G}_i\,]_{P^*}\, \Delta \theta^i, \tag{3.21a, b}$$

where $]_P$ denotes evaluation at P.

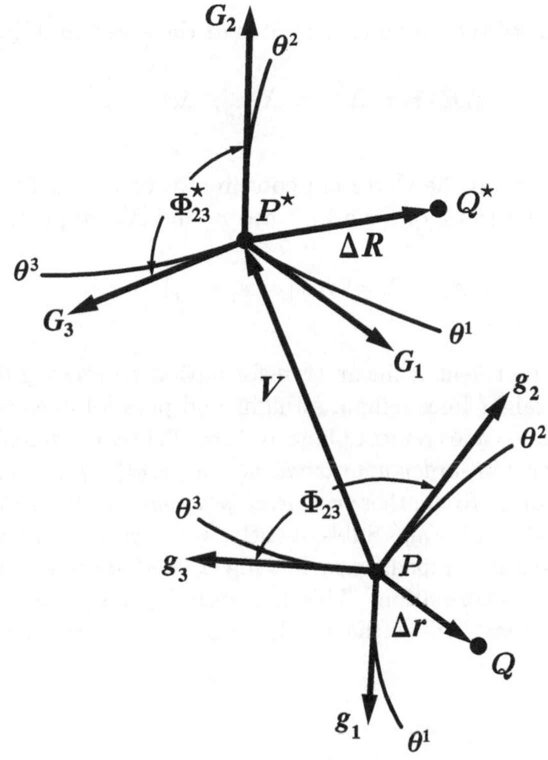

Figure 3.5 Motion in a small region

The new distance between the neighboring particles is therefore approximated by the metric tensors of the respective states. Specifically, in accordance with (2.13) and (3.8),

$$\Delta \ell^2 \doteq g_{ij}\,]_P\, \Delta\theta^i\, \Delta\theta^j, \qquad \Delta L^2 \doteq G_{ij}\,]_{P^*}\, \Delta\theta^i\, \Delta\theta^j. \tag{3.22}$$

The components of the metric tensor g_{ij} determine the initial length $\Delta\ell$ and the components G_{ij} determine the subsequent length ΔL for the same material line as it stretches from PQ to P^*Q^*. Although the metric tensors provide a description of the deformation, they provide *no* information about the displacement of the particles nor the rotation of the lines. Clearly, a given line may stretch without rotating or rotate without stretching. Throughout our analyses, careful distinctions between deformations and rigid motions are *paramount*.

If the approximation (3.21b) is expressed in terms of the *fixed* rectangular Cartesian system, then the coordinates of the deformed segment P^*Q^* are expressed in terms of the initial coordinates of the segment PQ:

$$\Delta \boldsymbol{R} \cdot \hat{\boldsymbol{\imath}}_i = \Delta X_i \doteq X_{i,j}\,]_P\, \Delta x_j, \tag{3.23}$$

where $\Delta x_i = \Delta \boldsymbol{r} \cdot \hat{\boldsymbol{\imath}}_i$. If the Cartesian coordinates of P and P^* are b_i and B_i, respectively, and those of Q and Q^* are x_i and X_i, respectively, then

$$X_i - B_i = X_{i,j}\,]_P\,(x_j - b_j). \tag{3.24}$$

These equations represent a linear transformation of rectangular coordinates in which straight lines remain straight and parallel lines remain parallel. It follows that planes remain plane and parallel planes remain parallel. Specifically, to the *first-order approximation*, a *parallelepiped* of a continuous medium deforms to another *parallelepiped*; *parallel lines* undergo the *same rotation* and *stretching*. Subsequently, we perceive an *infinitesimal* element, e.g., a parallelepiped, experiencing a rigid motion accompanied by a homogeneous deformation. This is a useful perception, but strictly valid *only in the limit* [2], as the neighboring edges and faces approach coincidence.

3.6 Strain

The deformation of a body is manifested by the extension of lines and the distortion of angles between lines. In general, the nature and severity of the deformation varies from point to point. A description of the deformation at a point requires a precise means to fully describe the changes of differential geometry. Any collection of geometrical quantities, which provides such description, constitutes a *measure of strain*. Many alternatives are possible. We explore two forms which are most useful in the mechanics of solids. Both are defined by geometrical properties of the convecting lines and surfaces. The first, the so-called Green or Cauchy-Green strain, is mathematically convenient. The second is the strain most commonly employed in engineering. The significance of these measures is apparent later (Chapters 5 and 6) when we examine the work expended by stress and when we express the internal potential of elastic materials.

Consider any neighboring particles at P and Q in the reference state and the same particles at P^* and Q^* in a deformed state, as depicted in Figure 3.5. The lengths of the vectors $\Delta \boldsymbol{r}$ and $\Delta \boldsymbol{R}$ joining the particles are denoted by $\Delta \ell$ and ΔL, respectively. We define an extensional strain of the line at P (and P^*) in the direction PQ:

$$\epsilon \equiv \lim_{\Delta\ell\to 0} \frac{1}{2}\left(\frac{\Delta L^2}{\Delta\ell^2} - 1\right) = \frac{1}{2}\left[\left(\frac{dL}{d\ell}\right)^2 - 1\right]. \tag{3.25a, b}$$

In accordance with (2.13) and (3.8),

$$\epsilon = \frac{1}{2}\left(\frac{d\boldsymbol{R}}{d\ell}\cdot\frac{d\boldsymbol{R}}{d\ell} - \frac{d\boldsymbol{r}}{d\ell}\cdot\frac{d\boldsymbol{r}}{d\ell}\right), \tag{3.26}$$

$$= \frac{1}{2}\left(G_{ij} - g_{ij}\right)\frac{d\theta^i}{d\ell}\frac{d\theta^j}{d\ell}. \tag{3.27}$$

The derivatives $(d\theta^i/d\ell)$ serve to establish the initial orientation of the line in question: If, for example, the unit vector $\hat{\mathbf{e}}$ defines the initial direction of the line, then from $d\boldsymbol{r} = \boldsymbol{g}_i\, d\theta^i = \hat{\mathbf{e}}\, d\ell$, it follows that

$$\frac{d\theta^i}{d\ell} = \boldsymbol{g}^i \cdot \hat{\mathbf{e}}. \tag{3.28}$$

Suppose that the particles P and Q lie on the θ^1 line. Then, in accordance with (3.28), (2.14a), and (2.6)

$$\frac{d\theta^2}{d\ell} = \frac{d\theta^3}{d\ell} = 0 \tag{3.29a, b}$$

and

$$\frac{d\theta^1}{d\ell} = \frac{1}{\sqrt{g_{11}}}. \tag{3.29c}$$

In this case, the extensional strain of (3.27) represents a strain of the θ^1 line:

$$\epsilon = \epsilon_{11} = \frac{1}{2}\left(\frac{G_{11}}{g_{11}} - 1\right).$$

In general, along a coordinate line θ^i, the extensional strain of (3.27) takes the form

$$\epsilon_{\underline{ii}} = \frac{1}{2}\left(\frac{G_{\underline{ii}}}{g_{\underline{ii}}} - 1\right). \tag{3.30}$$

Note that the extensional strain of (3.30) is nondimensional and has the simple geometrical interpretation of (3.25a, b). Consequently, the components $\epsilon_{\underline{ii}}$ are termed physical; they are determined by the components of the metric tensors of the undeformed and deformed states, $g_{\underline{ii}}$ and $G_{\underline{ii}}$, respectively.

According to (3.27), the extensional strain of *any* line is fully determined by the metric tensors of the deformed and undeformed states. The factors $(G_{ij}-g_{ij})/2$ of (3.27) are the covariant components of a tensor. They correspond to the components of the strain tensors identified with A. L. Cauchy [7] and G. Green [8] (see A. E. H. Love [1], p. 59 for further historical remarks). We label these strain components:

$$\gamma_{ij} \equiv \tfrac{1}{2}(G_{ij} - g_{ij}). \tag{3.31}$$

When the indices are equal, $(i = j)$, then according to (3.30) and the definition (3.31):

$$\epsilon_{\underline{ii}} = \frac{\gamma_{\underline{ii}}}{g_{\underline{ii}}}. \tag{3.32}$$

Equation (3.32) provides a simple relation between the tensorial component $\gamma_{\underline{ii}}$ of the strain tensor and the physical extensional strain $\epsilon_{\underline{ii}}$ of the θ^i coordinate line. It remains to attach geometrical meaning to the other three components $\gamma_{ij} = \gamma_{ji}$ $(i \neq j)$: By (2.14a) and (3.9a),

$$g_{ij} = \sqrt{g_{\underline{ii}}}\,\sqrt{g_{\underline{jj}}}\,(\hat{\mathbf{e}}_i \cdot \hat{\mathbf{e}}_j), \tag{3.33a}$$

$$G_{ij} = \sqrt{G_{\underline{ii}}}\,\sqrt{G_{\underline{jj}}}\,(\hat{\boldsymbol{E}}_i \cdot \hat{\boldsymbol{E}}_j). \tag{3.33b}$$

According to (3.30),

$$G_{\underline{ii}} = g_{\underline{ii}}\,(1 + 2\epsilon_{\underline{ii}}). \tag{3.34}$$

In view of the definition (3.31) and the equations (3.33a, b) and (3.34), we give a general definition of the physical[‡] strain components ϵ_{ij}:

$$\epsilon_{ij} = \frac{\gamma_{ij}}{\sqrt{g_{\underline{ii}}}\,\sqrt{g_{\underline{jj}}}}, \tag{3.35a}$$

$$= \tfrac{1}{2}\left[\sqrt{1+2\epsilon_{\underline{ii}}}\,\sqrt{1+2\epsilon_{\underline{jj}}}\,(\hat{\boldsymbol{E}}_i \cdot \hat{\boldsymbol{E}}_j) - (\hat{\mathbf{e}}_i \cdot \hat{\mathbf{e}}_j)\right]. \tag{3.35b}$$

[‡]*Note:* Some mathematicians have defined the "physical" component of a second-order covariant component (e.g., γ_{ij}) by the equation ${}^{*}_{P}\gamma_{ij} = \gamma_{ij}\,(g^{\underline{ii}} g^{\underline{jj}})^{1/2}$. In general, ${}^{*}_{P}\gamma_{ij}$ does not provide a simple physical interpretation, but does yield the appropriate physical dimensions.

The dot products $(\hat{e}_i \cdot \hat{e}_j)$ and $(\hat{\boldsymbol{E}}_i \cdot \hat{\boldsymbol{E}}_j)$ are the cosines of the angles between the coordinate lines θ^i and θ^j before and after the motion. If Φ_{ij} and Φ^*_{ij} denote the angles between the θ^i and θ^j lines of the initial and deformed states, respectively, then

$$\epsilon_{ij} = \tfrac{1}{2}\left(\sqrt{1+2\epsilon_{\underline{ii}}}\sqrt{1+2\epsilon_{\underline{jj}}}\cos\Phi^*_{ij} - \cos\Phi_{ij}\right). \tag{3.36}$$

Any quantity which measures the change of angle between lines is a *shear strain*. According to (3.36), ϵ_{ij} $(i \neq j)$ is a nondimensional quantity which serves to measure such change; it is the physical component of the shear strain.

Each tensorial component γ_{ij} of the Green strain tensor is related to a physical component ϵ_{ij} of strain by (3.35a, b). An extensional strain $(i = j)$ describes the stretching of the coordinate line and a shear strain $(i \neq j)$ determines the change of angle between two lines, as the change from Φ_{23} to Φ^*_{23} depicted in Figure 3.5. The collection of six components fully describes the state of strain at the point.

The foregoing definitions are most common, mathematically convenient and physically meaningful, but many alternatives are possible. Two deserve particular attention in the mechanics of solids. An extensional strain, which is commonly employed in engineering, is defined as follows:

$$\bar{h} \equiv \lim_{\Delta\ell\to 0}\left(\frac{\Delta L}{\Delta\ell} - 1\right) = \frac{dL}{d\ell} - 1. \tag{3.37a, b}$$

According to (3.25a, b) and (3.37a, b), the former is related to the latter by the equation:

$$\epsilon = \bar{h} + \tfrac{1}{2}\bar{h}^2. \tag{3.38}$$

In many practical problems of structures and machines, $\bar{h} \ll 1$. Then, the differences are often inconsequential. We return later to a definition of an engineering strain tensor (Section 3.12) and to some considerations of utility.

Another definition of strain has been advanced to describe severe deformations. An incremental extension of the line is compared to the *current* length $\Delta L'$, rather than the initial length $\Delta\ell$, and the differential of the natural strain ϵ' is defined as follows:

$$d\epsilon' = \lim_{\Delta L'\to 0}\frac{d(\Delta L')}{\Delta L'}. \tag{3.39}$$

If the differential is integrated from initial state $(t = \epsilon' = 0,\ \Delta L' = \Delta\ell)$ to any subsequent state (ϵ', $\Delta L' = \Delta L$), then

$$\epsilon' = \ln(1 + \bar{h}) = \tfrac{1}{2}\ln(1 + 2\epsilon). \tag{3.40a, b}$$

As a consequence of the relations (3.39) and (3.40a, b) the strain ϵ' is termed *logarithmic* or *natural* strain (see A. Nadai [9], vol. 1, p. 21).

3.7 Transformation of Strain Components

The strain components γ_{ij} of (3.31) are six covariant components of a symmetric tensor, just as the metric tensors G_{ij} and g_{ij}. It follows that the components in another system of coordinates $\bar{\theta}^i$ are obtained by the transformation of second-order tensors:

$$\bar{\gamma}_{ij} = \frac{\partial \theta^k}{\partial \bar{\theta}^i}\frac{\partial \theta^l}{\partial \bar{\theta}^j}\gamma_{kl}. \tag{3.41}$$

Although the covariant components γ_{ij} have the most immediate mathematical and physical meaning, according to (3.31), (3.32), and (3.35a, b), respectively, the mixed components γ^i_j and the contravariant components γ^{ij} also provide useful mathematical representations:

$$\gamma^i_j = g^{ik}\gamma_{kj}, \qquad \gamma^{ij} = g^{ik}g^{jl}\gamma_{kl}. \tag{3.42a, b}$$

The transformation and the components have simpler interpretations if the systems are Cartesian. Figure 3.6 shows two Cartesian systems x_i and $\bar{x}_i$. The direction cosines a_{ij} are given in the adjoining table; the unit vectors $\hat{\boldsymbol{\imath}}_i$ and $\hat{\boldsymbol{a}}_i$, in directions of lines x_i and $\bar{x}_i$, respectively, are therefore related simply by

$$\hat{\boldsymbol{a}}_i = a_{ij}\hat{\boldsymbol{\imath}}_j = \frac{\partial x_j}{\partial \bar{x}_i}\hat{\boldsymbol{\imath}}_j, \qquad \hat{\boldsymbol{\imath}}_i = \bar{a}_{ij}\hat{\boldsymbol{a}}_j = \frac{\partial \bar{x}_j}{\partial x_i}\hat{\boldsymbol{a}}_j.$$

Evidently,

$$a_{ij} = \hat{\boldsymbol{a}}_i \cdot \hat{\boldsymbol{\imath}}_j = \frac{\partial x_j}{\partial \bar{x}_i}, \qquad \bar{a}_{ij} = \hat{\boldsymbol{a}}_j \cdot \hat{\boldsymbol{\imath}}_i = \frac{\partial \bar{x}_j}{\partial x_i}. \tag{3.43a, b}$$

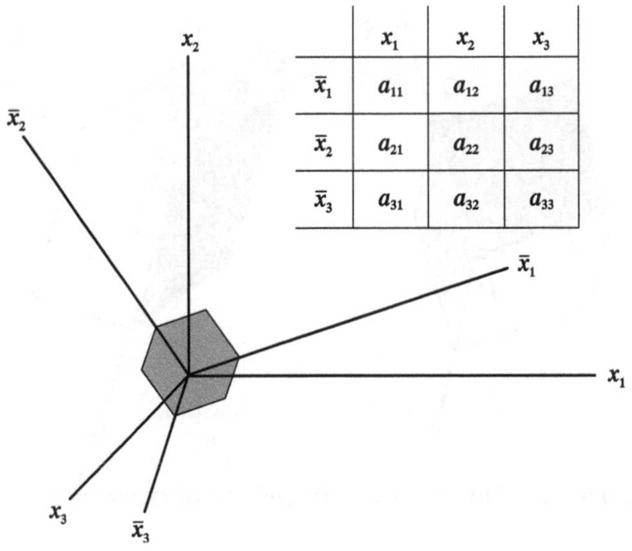

Figure 3.6 Direction cosines in a rectangular Cartesian system

According to (3.43a), the transformation (3.41) in the Cartesian system assumes the form

$$\bar{\gamma}_{ij} = a_{ik} a_{jl} \gamma_{kl}. \tag{3.44}$$

In the Cartesian system, the metric tensor has components $g_{ij} = \delta_{ij}$ and, according to (3.35a, b), the components of the Cartesian tensor are also the physical components of strain. Stated otherwise, the physical components may be regarded as the components of strain in a local Cartesian system.

3.8 Principal Strains

At every point of the undeformed body there are always three mutually perpendicular lines which remain perpendicular in the deformed body. The proof and the means of locating these lines are given in Section 3.10. The three directions are called *principal directions*; in general, they differ at each point. By definition, the shear strains associated with the principal directions are zero. This is illustrated by the small cube at P of Figure 3.7, which has edges along the principal directions; the deformed element at P^* is therefore rectangular. If the length of edge PQ is $\Delta\tilde{x}_1$, then the length of the deformed edge P^*Q^* is $\Delta\tilde{x}_1(1+2\tilde{\epsilon}_{11})^{1/2}$ to the first-order of approximation. When referred to principal directions, the array of components $\tilde{\epsilon}_{\underline{ii}}$

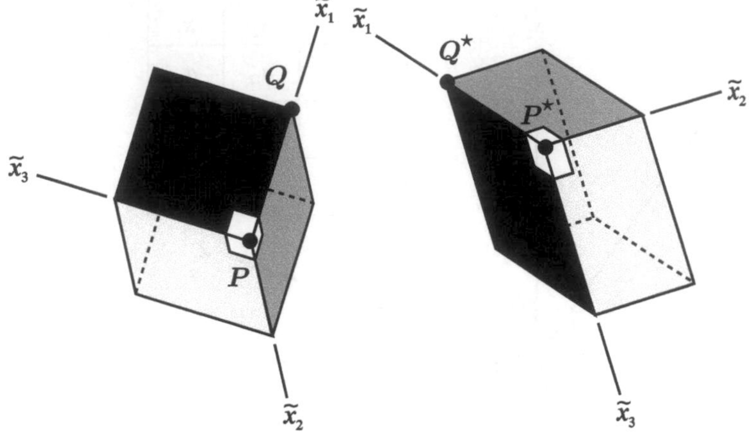

Figure 3.7 Deformation in principal directions

assumes a simpler form in a Cartesian system $\tilde{x}_i$, or $\tilde{\gamma}_{\underline{ii}}$, in a coincidental orthogonal system $\tilde{\theta}^i$:

$$\left.\begin{matrix} \tilde{\epsilon}_{11} & 0 & 0 \\ 0 & \tilde{\epsilon}_{22} & 0 \\ 0 & 0 & \tilde{\epsilon}_{33} \end{matrix}\right\}, \qquad \left.\begin{matrix} \tilde{\gamma}_{11} & 0 & 0 \\ 0 & \tilde{\gamma}_{22} & 0 \\ 0 & 0 & \tilde{\gamma}_{33} \end{matrix}\right\}. \tag{3.45a, b}$$

The extensional strains in the principal directions are termed the *principal strains.* They are the extremal values of the extensional strain at the point. In general, one principal strain is the *maximum* extensional strain at the point and one is the *minimum.*

Although, in general, there are three distinct principal strains and consequently, three distinct principal directions, certain special cases arise. The most special case arises if all three principal strains are equal:

$$\tilde{\epsilon}_{11} = \tilde{\epsilon}_{22} = \tilde{\epsilon}_{33}.$$

In this case, the extensional strain is the same in all directions and the shear strain vanishes for all directions. Every direction is a principal direction. An infinitesimal sphere of the medium would dilate (or contract) in a spherically symmetric manner as depicted in Figure 3.8. Such deformation is often referred to as *pure,* or *simple, dilatation.*

Another special case arises if two of the principal strains are equal: $\tilde{\epsilon}_{11} \neq \tilde{\epsilon}_{22} = \tilde{\epsilon}_{33}$. Then all lines in the $(\tilde{x}_2 - \tilde{x}_3)$ plane experience the same extensional strain and the shear strain vanishes for any directions in this plane. All directions in the $(\tilde{x}_2 - \tilde{x}_3)$ plane are principal directions. An infinitesimal cylinder with its axis along $\tilde{x}_1$ lengthens (or shortens) and dilates (or contracts) radially, but remains cylindrical (see Figure 3.9).

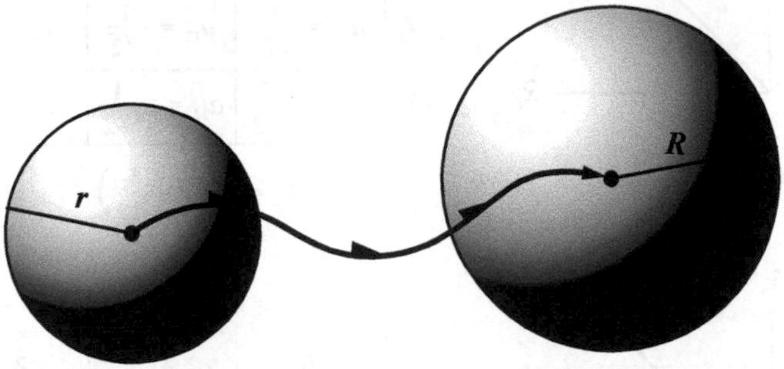

Figure 3.8 Pure dilation

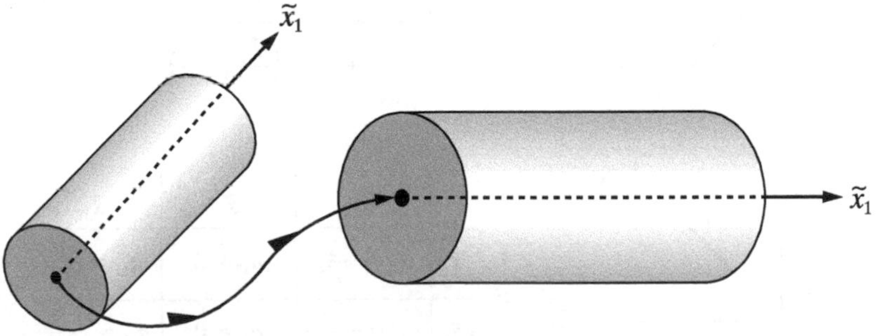

Figure 3.9 Two equal principal strains

3.9 Maximum Shear Strain

In Section 3.11, we show that the shear strain has a stationary value for the pairs of orthogonal lines which bisect the principal lines. These lines are the $\bar{x}_i$ lines of Figure 3.10; the principal lines are the $\tilde{x}_i$ lines. Each view in Figure 3.10 is accompanied by the appropriate table of direction cosines and the stationary value, $\bar{\epsilon}_{12}$, of the shear strain. The lower sign is associated with the negative $-\bar{x}_i$ line. One of the stationary values is the maximum shear strain and one the minimum. For example, if the principal strains $\tilde{\epsilon}_{\underline{ii}}$ take values such that $\tilde{\epsilon}_{11} > \tilde{\epsilon}_{22} > \tilde{\epsilon}_{33}$, then the maximum shear strain is $|\tilde{\epsilon}_{11} - \tilde{\epsilon}_{33}|/2$.

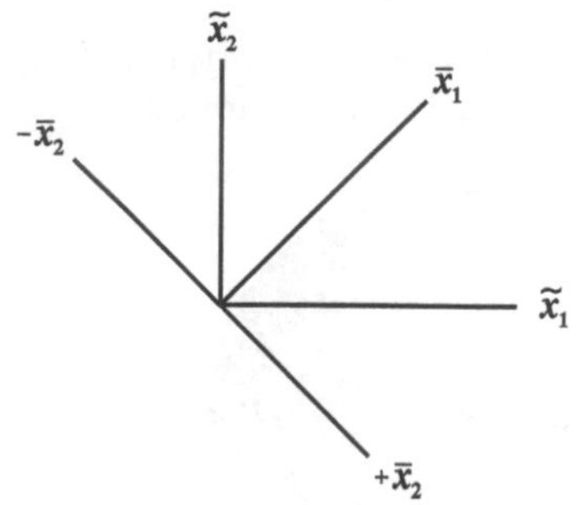

	$\tilde{x}_1$	$\tilde{x}_2$	$\tilde{x}_3$
$\bar{x}_1$	$a_{11} = \frac{1}{\sqrt{2}}$	$a_{12} = \frac{1}{\sqrt{2}}$	$a_{13} = 0$
$\pm\bar{x}_2$	$a_{21} = \pm\frac{1}{\sqrt{2}}$	$a_{22} = \mp\frac{1}{\sqrt{2}}$	$a_{23} = 0$

$$\bar{\epsilon}_{12} = \pm\left(\frac{\tilde{\epsilon}_{11} - \tilde{\epsilon}_{22}}{2}\right)$$

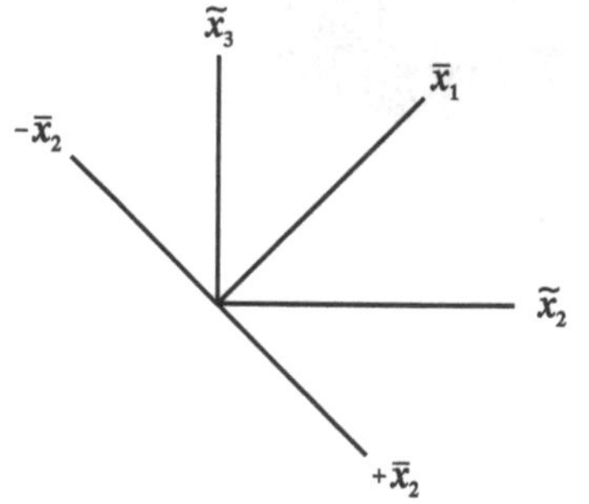

	$\tilde{x}_1$	$\tilde{x}_2$	$\tilde{x}_3$
$\bar{x}_1$	$a_{11} = 0$	$a_{12} = \frac{1}{\sqrt{2}}$	$a_{13} = \frac{1}{\sqrt{2}}$
$\pm\bar{x}_2$	$a_{21} = 0$	$a_{22} = \pm\frac{1}{\sqrt{2}}$	$a_{23} = \mp\frac{1}{\sqrt{2}}$

$$\bar{\epsilon}_{12} = \pm\left(\frac{\tilde{\epsilon}_{22} - \tilde{\epsilon}_{33}}{2}\right)$$

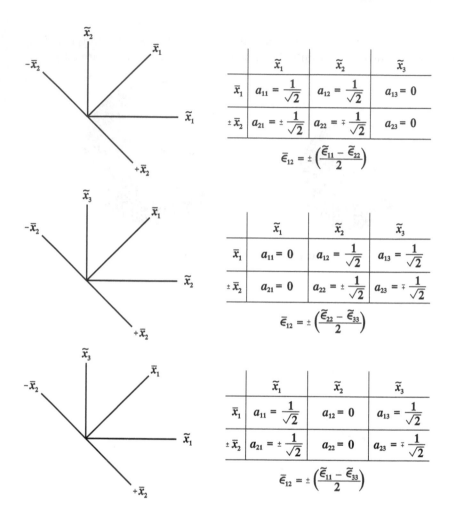

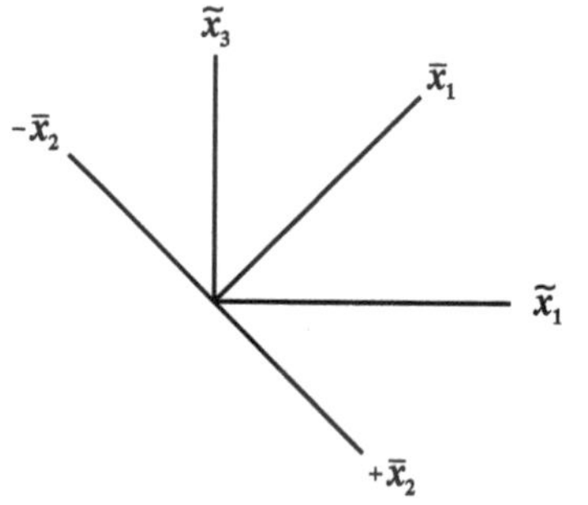

	$\tilde{x}_1$	$\tilde{x}_2$	$\tilde{x}_3$
$\bar{x}_1$	$a_{11} = \frac{1}{\sqrt{2}}$	$a_{12} = 0$	$a_{13} = \frac{1}{\sqrt{2}}$
$\pm\bar{x}_2$	$a_{21} = \pm\frac{1}{\sqrt{2}}$	$a_{22} = 0$	$a_{23} = \mp\frac{1}{\sqrt{2}}$

$$\bar{\epsilon}_{12} = \pm\left(\frac{\tilde{\epsilon}_{11} - \tilde{\epsilon}_{33}}{2}\right)$$

Figure 3.10 Directions and extremal values of maximum shear strain

To appreciate this kinematic relationship, examine the square element $ABCD$ of Figure 3.11a, which after deformation takes the form of the rectangle $A^*B^*C^*D^*$ of Figure 3.11b. Also, the inscribed square $MNOP$ is deformed to a skewed parallelogram $M^*N^*O^*P^*$. The shear strain associated with the larger angle $\varphi_{12(-)}$ (between $\bar{x}_1$ and $-\bar{x}_2$ lines) is the negative of the shear strain associated with the smaller angle φ_{12} (between the $\bar{x}_1$ and $+\bar{x}_2$ lines).

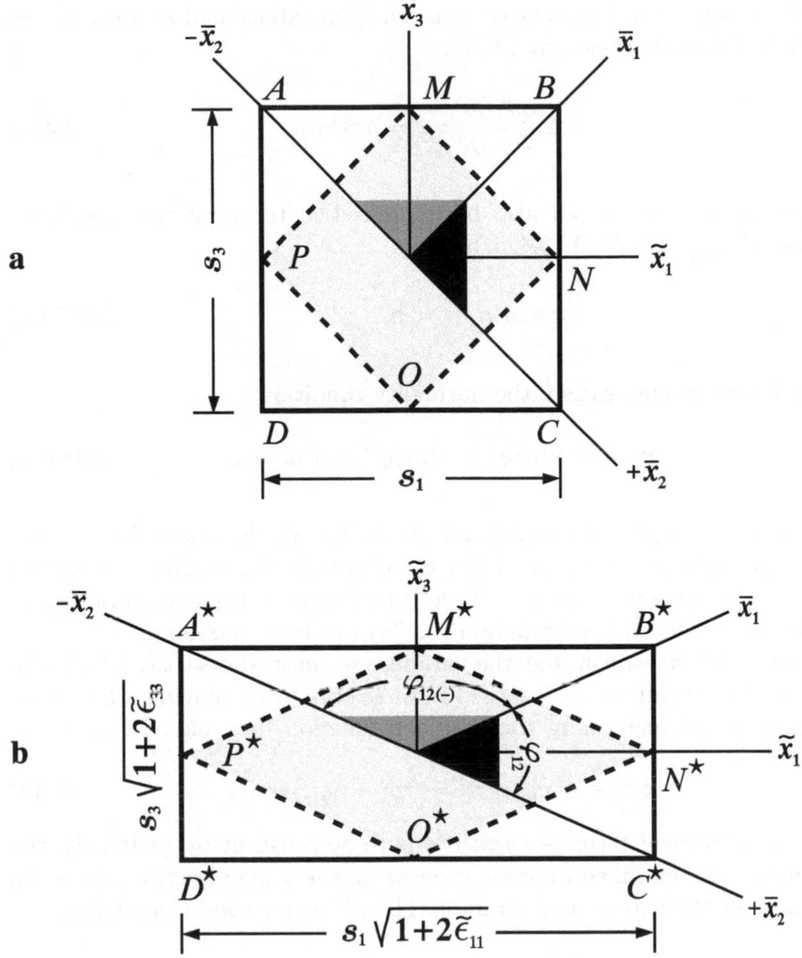

Figure 3.11 Maximum shear strain

3.10 Determination of Principal Strains and Principal Directions

Let $\hat{\boldsymbol{n}}$ denote a unit vector, and ℓ the initial arc length in the direction of $\hat{\boldsymbol{n}}$. The vector $\hat{\boldsymbol{n}}$ can be expressed in terms of an arbitrary system of coordinates θ^i; alternative forms follow:

$$\hat{\boldsymbol{n}} \equiv \frac{\partial \boldsymbol{r}}{\partial \ell} = \frac{\partial \theta^i}{\partial \ell}\boldsymbol{g}_i = n^i \boldsymbol{g}_i = n_i \boldsymbol{g}^i, \tag{3.46a–d}$$

wherein n_i and n^i are associated tensors. The extensional strain ϵ in the direction of $\hat{\boldsymbol{n}}$ is expressed by (3.27):

$$\epsilon \equiv \frac{\partial\theta^i}{\partial\ell}\frac{\partial\theta^j}{\partial\ell}\gamma_{ij} = n^i n^j \gamma_{ij}. \tag{3.47a}$$

The extensional strain can also be expressed in terms of the associated components (γ^i_j or γ^{ij}) of (3.42a, b):

$$\epsilon = n^i n_j \gamma^j_i = n_i n_j \gamma^{ij}. \tag{3.47b, c}$$

The unit vector must satisfy the normality condition:

$$\hat{\boldsymbol{n}} \cdot \hat{\boldsymbol{n}} = n^i n^j g_{ij} = n_i n_j g^{ij} = n^i n_i = 1. \tag{3.48a–d}$$

We now examine the extensional strain for all directions by varying the components (n^i or n_i) in (3.47a–c) subject to the auxiliary condition (3.48a–d). In particular, we ask: What direction gives the maximum extensional strain? The smooth function (3.47a) can be a maximum only if the derivatives $\partial\epsilon/\partial n^i$ vanish, but the variables n^i must also satisfy (3.48a–d). The subsidiary condition is imposed most readily by employing a Lagrange multiplier λ and examining the modified function (see, e.g., H. L. Langhaar [10]):

$$\phi \equiv \epsilon - \lambda g_{ij}\, n^i n^j = (\gamma_{ij} - \lambda g_{ij})\, n^i n^j. \tag{3.49}$$

Since the additional term $(-\lambda g_{ij} n^i n^j)$ is to be constant by (3.48a–d), the extremals of the modified function ϕ occur for the same directions n^i as the extremals for the extensional strain ϵ. The necessary conditions follow:

$$\frac{\partial\phi}{\partial n^i} = (\gamma_{ij} - \lambda g_{ij})\, n^j = 0. \tag{3.50a}$$

By the definition of associated components, the equations (3.50a) are equivalent to the following:

$$(\gamma^j_i - \lambda\delta^j_i)\, n_j = 0. \tag{3.50b}$$

Now, suppose that the components n_j define the direction of the extremal value, i.e., n_j satisfy (3.50b) and the normality condition (3.48a–d). Then, if we multiply (3.50b) by the corresponding component n^i, sum, and enforce (3.48a–d), we obtain

$$\epsilon = n^i n_j \gamma^j_i = \lambda\, n^j n_j = \lambda. \tag{3.51}$$

In words, the multiplier λ is that extremal value.

Equations (3.50b) are linear homogeneous equations in the variables (n_1, n_2, n_3). A nontrivial solution exists if, and only if, the determinant of the coefficients vanishes:

$$\left|\gamma_i^j - \lambda\delta_i^j\right| = 0. \tag{3.52a}$$

Expanding the determinant we obtain, according to (1.13) and (1.14a–c), the cubic equation:

$$\lambda^3 - I_1\lambda^2 + I_2\lambda - I_3 = 0, \tag{3.52b}$$

where

$$I_1 = \gamma_i^i, \tag{3.53a}$$

$$I_2 = \tfrac{1}{2}(\gamma_i^i\gamma_j^j - \gamma_j^i\gamma_i^j), \tag{3.53b}$$

$$I_3 = \tfrac{1}{6}(\gamma_i^i\gamma_j^j\gamma_k^k - 3\gamma_i^i\gamma_k^j\gamma_j^k + 2\gamma_j^i\gamma_k^j\gamma_i^k) = \gamma_1^i\gamma_2^j\gamma_3^k\epsilon_{ijk} = |\gamma_j^i|. \tag{3.53c}$$

The cubic equation (3.52b) has three real roots $\lambda_{(i)}$, the three principal strains. Here, the parentheses used to enclose the letter i serve to avoid confusing this label with an index. For each root, the simultaneous solution of (3.50b) and (3.48a–d) determines the corresponding principal direction $\hat{\boldsymbol{n}}_{(i)}$.

Suppose that $\lambda_{(1)}$ and $\lambda_{(2)}$ are distinct roots of (3.52b), and $n_{(1)}^i$ and $n_{(2)}^i$ are the corresponding solutions of (3.50b) and (3.48a–d). The roots, $\lambda_{(1)}$ and $\lambda_{(2)}$, are the principal strains (stationary values of extensional strain) in the directions of unit vectors $\hat{\boldsymbol{n}}_{(1)}$ $(= n_{(1)}^i\boldsymbol{g}_i)$ and $\hat{\boldsymbol{n}}_{(2)}$ $(= n_{(2)}^i\boldsymbol{g}_i)$. Again, the parentheses used to enclose the numbers (1) and (2) of $n_{(1)}^i$ and $n_{(2)}^i$ serve to avoid confusion with an index. Notice that $n_{(k)}^i$ are components of a tensor of first order. According to (3.50a), we can form the two equations:

$$(\gamma_{ij} - \lambda_{(1)}g_{ij})\, n_{(1)}^j n_{(2)}^i = 0, \qquad (\gamma_{ij} - \lambda_{(2)}g_{ij})\, n_{(2)}^j n_{(1)}^i = 0. \tag{3.54a, b}$$

Since $\gamma_{ij} = \gamma_{ji}$ and $g_{ij} = g_{ji}$, the difference of (3.54a) and (3.54b) is

$$(\lambda_{(2)} - \lambda_{(1)})\, n_{(2)}^i n_{(1)}^j g_{ij} = 0.$$

If the roots are distinct ($\lambda_{(2)} \neq \lambda_{(1)}$), then

$$n^j_{(1)} n^i_{(2)} g_{ij} = \hat{\boldsymbol{n}}_{(1)} \cdot \hat{\boldsymbol{n}}_{(2)} = 0.$$

The directions are *orthogonal*!

If two roots of (3.52b) are equal, say $\lambda_{(1)} = \lambda_{(2)} \neq \lambda_{(3)}$, then all directions normal to the distinct direction of $\hat{\boldsymbol{n}}_{(3)}$ have the same extensional strain. If all roots are equal, then all directions have the same extension strain; this is a *simple dilatation*. In any case, there exist three ($i = 1, 2, 3$) mutually orthogonal directions ($\hat{\boldsymbol{n}}_{(i)} = n^m_{(i)} \boldsymbol{g}_m$) and corresponding extensional strains $\lambda_{(i)}$ which satisfy (3.50a) and (3.52a). When these are distinct,

$$\hat{\boldsymbol{n}}_{(i)} \cdot \hat{\boldsymbol{n}}_{(j)} = n^k_{(i)} n^l_{(j)} g_{kl} = \delta_{ij}. \tag{3.55}$$

If $\tilde{x}_i$ denotes initial length along a principal line, then according to (3.46a–d)

$$\hat{\boldsymbol{n}}_{(i)} = n^m_{(i)}\, \boldsymbol{g}_m = \frac{\partial \theta^m}{\partial \tilde{x}_i} \boldsymbol{g}_m = n_{(i)m}\, \boldsymbol{g}^m = \frac{\partial \tilde{x}_i}{\partial \theta^m} \boldsymbol{g}^m.$$

It follows that

$$n^m_{(i)} = \boldsymbol{g}^m \cdot \hat{\boldsymbol{n}}_{(i)} = \frac{\partial \theta^m}{\partial \tilde{x}_i}, \qquad n_{(i)m} = \boldsymbol{g}_m \cdot \hat{\boldsymbol{n}}_{(i)} = \frac{\partial \tilde{x}_i}{\partial \theta^m}. \tag{3.56a–d}$$

The components of strain in principal directions follow from (3.41), (3.56b), (3.55), and (3.51):

$$\tilde{\gamma}_{ij} = \frac{\partial \theta^k}{\partial \tilde{x}_i} \frac{\partial \theta^l}{\partial \tilde{x}_j} \gamma_{kl} = n^k_{(i)} n^l_{(j)} \gamma_{kl} = \lambda_{(\underline{i})} \delta_{ij}. \tag{3.57}$$

Equation (3.57) implies that the shear strain components vanish in the principal directions. In other words, the material lines $\tilde{x}_i$, which are orthogonal in the initial state, remain orthogonal in the deformed state (see Figure 3.7). The extensional strains $\lambda_{(i)}$ are the stationary values; in general, one is the maximum and one is the minimum.

According to (3.41), (3.56d), and (3.57), components in an arbitrary system $\boldsymbol{\theta}^i$ are expressed in terms of the principal strains $\lambda_{(k)}$ and directions $n^i_{(k)}$ as follows:

$$\gamma_{ij} = \frac{\partial \tilde{x}_k}{\partial \theta^i} \frac{\partial \tilde{x}_l}{\partial \theta^j} \tilde{\gamma}_{kl} = \frac{\partial \tilde{x}_k}{\partial \theta^i} \frac{\partial \tilde{x}_k}{\partial \theta^j} \lambda_{(\underline{k})} = n_{(k)i} n_{(k)j} \lambda_{(\underline{k})}. \tag{3.58a, b}$$

The three principal directions are defined by the vectors $(\hat{\boldsymbol{n}}_{(1)}, \hat{\boldsymbol{n}}_{(2)}, \hat{\boldsymbol{n}}_{(3)})$, so-called eigenvectors of the tensor (γ_{ij}); the associated principal strains are the eigenvalues (see, e.g., the textbooks of B. Noble and J. W. Daniel [11] and A. Tucker [12]).

The components in a rectangular Cartesian system $\bar{x}_i$ are

$$\bar{\gamma}_{ij} \equiv \epsilon_{ij} = \frac{\partial \tilde{x}_k}{\partial \bar{x}_i} \frac{\partial \tilde{x}_k}{\partial \bar{x}_j} \lambda_{(\underline{k})} = n_{(k)i} n_{(k)j} \lambda_{(\underline{k})}, \tag{3.59a, b}$$

wherein $n_{(i)j}$ is the direction cosine between $\tilde{x}_i$ and $\bar{x}_j$ lines.

3.11 Determination of Extremal Shear Strain

If $\hat{\mathbf{a}}_1$ and $\hat{\mathbf{a}}_2$ are orthogonal unit vectors along lines of length $\bar{x}_1$ and $\bar{x}_2$, and $\hat{\boldsymbol{n}}_{(i)}$ are unit vectors in the direction of principal lines $\tilde{x}_i$, then in the previous notations:

$$\hat{\mathbf{a}}_\alpha \equiv \frac{\partial \boldsymbol{r}}{\partial \bar{x}_\alpha} = \frac{\partial \boldsymbol{r}}{\partial \tilde{x}_i} \frac{\partial \tilde{x}_i}{\partial \bar{x}_\alpha} = \frac{\partial \tilde{x}_i}{\partial \bar{x}_\alpha} \hat{\boldsymbol{n}}_{(i)} \equiv \bar{n}^i_{(\alpha)} \hat{\boldsymbol{n}}_{(i)} \qquad (\alpha = 1, 2)$$

and

$$\hat{\mathbf{a}}_\alpha \cdot \hat{\boldsymbol{n}}_{(i)} \equiv \bar{n}^i_{(\alpha)}. \tag{3.60}$$

According to (3.59a, b), the shear strain for the directions $(\bar{x}_1, \bar{x}_2)$ is given by

$$\bar{\gamma}_{12} \equiv \bar{\epsilon}_{12} = \frac{\partial \tilde{x}_i}{\partial \bar{x}_1} \frac{\partial \tilde{x}_i}{\partial \bar{x}_2} \lambda_{(\underline{i})} = \bar{n}^i_{(1)} \bar{n}^j_{(2)} \lambda_{(\underline{i})} \delta_{ij}. \tag{3.61a, b}$$

The vectors $\hat{\mathbf{a}}_\alpha$ satisfy the orthogonality and normality conditions:

$$\hat{\mathbf{a}}_1 \cdot \hat{\mathbf{a}}_2 = \bar{n}^i_{(1)} \bar{n}^j_{(2)} \delta_{ij} = 0, \tag{3.62a}$$

$$\hat{\mathbf{a}}_1 \cdot \hat{\mathbf{a}}_1 - 1 = \bar{n}^i_{(1)} \bar{n}^j_{(1)} \delta_{ij} - 1 = 0, \tag{3.62b}$$

$$\hat{\mathbf{a}}_2 \cdot \hat{\mathbf{a}}_2 - 1 = \bar{n}^i_{(2)} \bar{n}^j_{(2)} \delta_{ij} - 1 = 0. \tag{3.62c}$$

We seek the pair of directions $\bar{n}^i_{(1)}$ and $\bar{n}^i_{(2)}$ that render the maximum (or minimum) shear $\bar{\gamma}_{12}$. As before, we impose the stationary criteria upon a function ψ, which is the shear strain $\bar{\gamma}_{12}$ augmented by the auxiliary conditions (3.62a–c), using Lagrange multipliers μ, $\mu_{(1)}$, and $\mu_{(2)}$:

$$\psi = \bar{n}^i_{(1)}\bar{n}^j_{(2)}(\lambda_{(\underline{i})} - \mu)\,\delta_{ij}$$

$$-\mu_{(1)}(\bar{n}^i_{(1)}\bar{n}^j_{(1)}\delta_{ij} - 1) - \mu_{(2)}(\bar{n}^i_{(2)}\bar{n}^j_{(2)}\delta_{ij} - 1). \tag{3.63}$$

If the side conditions are enforced, then $\psi = \bar{\gamma}_{12}$. The necessary conditions for the extremal follow:

$$\frac{\partial\psi}{\partial\bar{n}^i_{(1)}} = (\lambda_{(\underline{i})} - \mu)\,\bar{n}^i_{(2)} - 2\mu_{(1)}\bar{n}^i_{(1)} = 0, \tag{3.64a}$$

$$\frac{\partial\psi}{\partial\bar{n}^i_{(2)}} = (\lambda_{(\underline{i})} - \mu)\,\bar{n}^i_{(1)} - 2\mu_{(2)}\bar{n}^i_{(2)} = 0. \tag{3.64b}$$

The multipliers μ, $\mu_{(1)}$, $\mu_{(2)}$ and the direction cosines $\bar{n}^i_{(1)}$ and $\bar{n}^i_{(2)}$ must be determined to satisfy (3.62a–c) and (3.64a, b).

If (3.64a) is multiplied by $\bar{n}^i_{(1)}$ and summed, and if (3.62a, b) are enforced, then

$$\mu_{(1)} = \tfrac{1}{2}\bar{n}^i_{(1)}\bar{n}^j_{(2)}\lambda_{(\underline{i})}\delta_{ij} = \tfrac{1}{2}\bar{\gamma}_{12}. \tag{3.65a}$$

Likewise, when we multiply (3.64b) by $\bar{n}^i_{(2)}$, sum, and enforce (3.62a, c), then

$$\mu_{(2)} = \tfrac{1}{2}\bar{n}^i_{(1)}\bar{n}^j_{(2)}\lambda_{(\underline{i})}\delta_{ij} = \tfrac{1}{2}\bar{\gamma}_{12}. \tag{3.65b}$$

Equations (3.65a, b) assert that the multipliers $\mu_{(1)}$ and $\mu_{(2)}$ are one half the shear strain in the directions of $\hat{\boldsymbol{a}}_1$ and $\hat{\boldsymbol{a}}_2$, respectively.

If (3.64a) is multiplied by $\bar{n}^i_{(2)}$, if the indicated summation is performed, and if (3.62a, c) are enforced, then the result follows:

$$\mu = \bar{n}^i_{(2)}\bar{n}^j_{(2)}\lambda_{(\underline{i})}\delta_{ij} = \bar{\epsilon}_{22}. \tag{3.66a}$$

Likewise, when we multiply (3.64b) by $\bar{n}^i_{(1)}$, sum, and enforce (3.62a, b), then

$$\mu = \bar{n}^i_{(1)}\bar{n}^j_{(1)}\lambda_{(\underline{i})}\delta_{ij} = \bar{\epsilon}_{11}. \tag{3.66b}$$

In words, equations (3.66a, b) state that the extensional strains in the directions of stationary shear strain (in the orthogonal directions of $\hat{\boldsymbol{a}}_1$ and $\hat{\boldsymbol{a}}_2$) are equal ($\bar{\epsilon}_{11} = \bar{\epsilon}_{22}$).

The expressions for the components of the vectors $\hat{\boldsymbol{a}}_1$ and $\hat{\boldsymbol{a}}_2$ [see (3.60)] satisfy (3.64a, b):

$$\hat{\boldsymbol{a}}_1 \cdot \hat{\boldsymbol{n}}_{(i)} = \bar{n}^i_{(1)} = \frac{\lambda_{(\underline{i})} - \mu}{2\mu_{(1)}} \bar{n}^i_{(2)},$$

$$\hat{\boldsymbol{a}}_2 \cdot \hat{\boldsymbol{n}}_{(i)} = \bar{n}^i_{(2)} = \frac{\lambda_{(\underline{i})} - \mu}{2\mu_{(2)}} \bar{n}^i_{(1)}.$$

In view of (3.65a, b), $\mu_{(1)} = \mu_{(2)} = \frac{1}{2}\bar{\gamma}_{12}$, so that

$$\bar{n}^i_{(1)} = \frac{\lambda_{(\underline{i})} - \mu}{\bar{\gamma}_{12}} \bar{n}^i_{(2)}, \qquad \bar{n}^i_{(2)} = \frac{\lambda_{(\underline{i})} - \mu}{\bar{\gamma}_{12}} \bar{n}^i_{(1)}. \tag{3.67a, b}$$

If the extensional strain $\bar{\epsilon}_{11} = \bar{\epsilon}_{22} = \mu$ does not equal a principal strain $\lambda_{(\underline{i})}$, then (3.67a, b) require that

$$\bar{n}^i_{(1)} = \pm \bar{n}^i_{(2)}.$$

In case $\bar{n}^i_{(1)} = \bar{n}^i_{(2)}$ for all i, then the orthogonality condition (3.62a) is violated. If

$$\bar{n}^1_{(1)} = \bar{n}^1_{(2)} = \bar{n}^2_{(1)} = -\bar{n}^2_{(2)} = k,$$

then (3.62a–c) require that

$$\bar{n}^3_{(1)} \bar{n}^3_{(2)} = 0,$$

$$2k^2 + (\bar{n}^3_{(1)})^2 = 1,$$

$$2k^2 + (\bar{n}^3_{(2)})^2 = 1.$$

Then, since $\bar{n}^3_{(1)} = \pm \bar{n}^3_{(2)}$,

$$\bar{n}^3_{(1)} = \bar{n}^3_{(2)} = 0, \qquad k = 1/\sqrt{2}.$$

This solution corresponds to directions which bisect the principal directions $\tilde{x}_1$ and $\tilde{x}_2$, as depicted in Figures 3.10 and 3.11. By substitution of the components $\bar{n}^i_{(\alpha)}$ into (3.61), we obtain the corresponding stationary value of the shear strain:

$$\bar{\gamma}_{12} \equiv \bar{\epsilon}_{12} = \frac{\tilde{\epsilon}_{11} - \tilde{\epsilon}_{22}}{2}. \tag{3.68}$$

If three distinct principal strains exist, then the three bisecting lines, as shown in Figure 3.10, are the directions of stationary shear. If, for example, the principal lines of *maximum* and *minimum* extensional strain are those labeled $\tilde{x}_1$ and $\tilde{x}_2$, then the bisecting lines $\bar{x}_1$ and $\bar{x}_2$ are the directions of *maximum* shear strain, given by (3.68). The lower sign of $\bar{\epsilon}_{12}$ in Figure 3.10 is associated with the negative $(-\bar{x}_2)$ line.

3.12 Engineering Strain Tensor

The extensional strain defined by (3.37a, b) is traditionally employed by engineers. From this definition we can construct a tensor: Let ℓ_A and L_A $(A = 1, 2, 3)$ denote the initial (undeformed) and subsequent (deformed) lengths of a line of principal strain. These principal lines constitute a local Cartesian system (ℓ_1, ℓ_2, ℓ_3 in the initial state) with the engineering components of strain:

$$\tilde{h}_{AB} \equiv \tilde{h}_{\underline{A}}\delta_{AB} = \left(\frac{dL_{\underline{A}}}{d\ell_{\underline{A}}} - 1\right)\delta_{AB}. \tag{3.69}$$

By the tensor transformations, we obtain the tensorial components h_{ij} of the engineering strain in an arbitrary system of coordinates θ^i:

$$h_{ij} = \frac{\partial \ell_A}{\partial \theta^i}\frac{\partial \ell_B}{\partial \theta^j}\tilde{h}_{AB}, \tag{3.70a}$$

$$= \frac{\partial \ell_A}{\partial \theta^i}\frac{\partial L_A}{\partial \theta^j} - \frac{\partial \ell_A}{\partial \theta^i}\frac{\partial \ell_A}{\partial \theta^j}, \tag{3.70b}$$

$$= \frac{\partial L_A}{\partial \theta^i}\frac{\partial \ell_A}{\partial \theta^j} - \frac{\partial \ell_A}{\partial \theta^i}\frac{\partial \ell_A}{\partial \theta^j}. \tag{3.70c}$$

Observe that the tensor is symmetric, $h_{ij} = h_{ji}$. Observe too that the final term in (3.70b, c) is the component of the metric g_{ij} according to (2.15).

The engineering strain tensor is especially relevant to the decomposition described in Section 3.14. It also plays a central role in the work/energy methods of Chapter 6.

3.13 Strain Invariants and Volumetric Strain

Three independent invariants can be formed from the components of the strain tensor. The simplest forms are

$$K_1 \equiv \gamma_i^i, \qquad K_2 \equiv \gamma_j^i \gamma_i^j, \qquad K_3 \equiv \gamma_j^i \gamma_k^j \gamma_i^k. \tag{3.71a–c}$$

Other invariants can be formed, but only three are independent. A notable example is found in the cubic equation (3.52b), wherein the coefficients I_1, I_2, and I_3 are invariants that are expressed in terms of K_1, K_2, and K_3 in accordance with (3.53a–c):

$$I_1 = K_1, \tag{3.72a}$$

$$I_2 = \tfrac{1}{2}(K_1^2 - K_2), \tag{3.72b}$$

$$I_3 = \tfrac{1}{6}(K_1^3 - 3K_1K_2 + 2K_3) = |\gamma_j^i|. \tag{3.72c}$$

The volume metrics $\sqrt{g}$ or $\sqrt{G}$ are also invariant. They can be expressed in terms of the invariants I_i or K_i. Specifically, we cite the following:

$$|g^{ik}G_{kj}| = |g^{ij}|\,|G_{ij}| = \frac{G}{g} = |2\gamma_j^i + \delta_j^i|.$$

In the manner of the Cauchy-Green strain, we define a volumetric strain:

$$'e \equiv \frac{1}{2}\left[\left(\frac{dV}{dv}\right)^2 - 1\right] = \frac{1}{2}\left(\frac{G}{g} - 1\right), \tag{3.73a}$$

$$= \gamma_i^i + \gamma_i^i\gamma_k^k - \gamma_j^i\gamma_i^j + \tfrac{2}{3}(\gamma_i^i\gamma_j^j\gamma_k^k + 2\gamma_j^i\gamma_k^j\gamma_i^k - 3\gamma_i^i\gamma_j^k\gamma_k^j)$$

$$= \gamma_i^i + \gamma_i^i\gamma_k^k - \gamma_j^i\gamma_i^j + 4\,|\gamma_j^i| = I_1 + 2I_2 + 4I_3. \tag{3.73b, c}$$

In the manner of engineering strain, we can define a volumetric strain:

$$'h \equiv \frac{dV}{dv} - 1 = \sqrt{\frac{G}{g}} - 1 = \bar{I}_1 + \bar{I}_2 + \bar{I}_3, \tag{3.74a–c}$$

where

$$\bar{I}_1 = h^i_i, \tag{3.75a}$$

$$\bar{I}_2 = \tfrac{1}{2}(h^i_i h^k_k - h^i_j h^j_i), \tag{3.75b}$$

$$\bar{I}_3 = \tfrac{1}{6}(h^i_i h^j_j h^k_k + 2h^i_j h^j_k h^k_i - 3h^i_i h^k_j h^j_k) = |h^i_j|. \tag{3.75c}$$

A volumetric strain, $'e$ or $'h$, is also termed a *dilatation*. If the strains are small ($h^i_j \ll 1$, $\gamma^i_j \ll 1$), then we have the approximations:

$$'h \doteq h^i_i, \qquad 'e \doteq \gamma^i_i, \qquad \gamma^i_i \doteq h^i_i, \qquad 'h \doteq 'e. \tag{3.76a–d}$$

To appreciate the role of the invariants, consider that the state of strain is described by the three principal strains and three angles which locate the principal directions. Since three invariants (I_1, I_2, I_3) determine the principal strains via the cubic equation (3.52b), the invariants and the angles also suffice to fully describe the state. In an isotropic material, physical properties are independent of directions and, consequently, the orientation of the principal lines is irrelevant. In short, the three strain invariants provide the essential information about the strained state of an isotropic material.

3.14 Decomposition of Motion into Rotation and Deformation

In many practical situations, specifically those involving thin structural components (beams, plates, and shells) elements undergo large rotations though strains remain very small. Such deformations are characteristic of the buckling and postbuckling of thin members. To achieve a better understanding of finite deformations, in general, and also the behavior of thin structural members, one must perceive the motion of a small element as comprised of a rigid motion (translation and rotation) and a deformation (strain). Here, we adopt that perception *and* develop the mathematical description of that decomposition (rigid motion and deformation).

3.14.1 Rotation Followed by Deformation

At each point of a continuous medium, the three principal lines of strain are very special, for we can install a local rectangular system along these lines and that system remains orthogonal in the deformed state. Let $\hat{\boldsymbol{n}}_A$ ($A = 1, 2, 3$) form a triad of unit vectors along the principal lines in the initial configuration, and let $\hat{\boldsymbol{N}}_A$ form the corresponding triad along the same lines in the deformed configuration. Let ℓ_A and L_A denote lengths along the same segments of the same principal line, i.e., distances between the same two neighboring particles. We define the *stretch* of a principal line:

$$\widetilde{C}_A \equiv \lim_{\Delta\ell_A \to 0} \frac{\Delta L_{\underline{A}}}{\Delta \ell_{\underline{A}}} = \frac{dL_{\underline{A}}}{d\ell_{\underline{A}}}.$$

As we defined the engineering strain tensor, so we define the *stretch tensor* in an arbitrary system of coordinates θ^i, namely, by the tensor transformation: In the local system of principal lines,

$$\widetilde{C}_{AB} \equiv \widetilde{C}_{\underline{A}}\, \delta_{AB}.$$

In the arbitrary system of coordinates θ^i,

$$C_{ij} = \frac{\partial \ell_A}{\partial \theta^i}\frac{\partial \ell_B}{\partial \theta^j}\widetilde{C}_{AB} = \frac{\partial L_A}{\partial \theta^i}\frac{\partial \ell_A}{\partial \theta^j} = \frac{\partial \ell_A}{\partial \theta^i}\frac{\partial L_A}{\partial \theta^j}. \tag{3.77a–c}$$

Note that the stretch tensor is also symmetric and since

$$g_{ij} = \frac{\partial \ell_A}{\partial \theta^i}\frac{\partial \ell_A}{\partial \theta^j},$$

the stretch tensor is related simply to the engineering strain of (3.70b, c) by

$$C_{ij} = h_{ij} + g_{ij}. \tag{3.78}$$

Using the transformation law (2.58), a mixed component of the stretch assumes the form:

$$C^i_j = \frac{\partial \theta^i}{\partial \ell_A}\frac{\partial \ell_B}{\partial \theta^j}\widetilde{C}^A_B = \frac{\partial \theta^i}{\partial \ell_A}\frac{\partial \ell_B}{\partial \theta^j}\widetilde{C}^A\delta_{AB} = \frac{\partial \theta^i}{\partial \ell_A}\frac{\partial L_A}{\partial \theta^j}, \tag{3.79a–c}$$

$$= g^{ik}C_{kj} = h^i_j + \delta^i_j. \tag{3.79d, e}$$

The triad of vectors $\hat{\boldsymbol{N}}_A$ can be viewed as a rigidly translated and rotated version of the triad $\hat{\boldsymbol{n}}_A$. Although we associate the two triads with different positions, the vectors differ only by the rigid rotation. Each vector of one triad can be expressed as a linear combination of the vectors of the other, i.e.,

$$\hat{\boldsymbol{N}}_A = r_A^{\bullet B}\hat{\boldsymbol{n}}_B. \tag{3.80}$$

Here, the position of the indices is arbitrary, but appropriate in hindsight [see (3.87a)]. The coefficients $r_A^{\bullet B}$ define the rigid rotation of the triad $\hat{\boldsymbol{n}}_A$ to $\hat{\boldsymbol{N}}_A$; these might be expressed in terms of three successive rotations or in terms of an angle of rotation about a prescribed axis (see, e.g., the monographs by J. L. Synge and B. A. Griffith [13] and L. A. Pars [14]). In any formulation, three quantities are necessary and sufficient to determine the coefficient $r_A^{\bullet B}$. By the definitions,

$$\hat{\boldsymbol{n}}_A = \frac{\partial \boldsymbol{r}}{\partial \ell_A}, \qquad \hat{\boldsymbol{N}}_A = \frac{\partial \boldsymbol{R}}{\partial L_A} = \frac{\partial \boldsymbol{R}}{\partial \ell_A}\frac{d\ell_{\underline{A}}}{dL_{\underline{A}}} \equiv \boldsymbol{G}_A \frac{d\ell_{\underline{A}}}{dL_{\underline{A}}}.$$

Now the vectors $\boldsymbol{r}_{,A} = \partial \boldsymbol{r}/\partial \ell_A \equiv \boldsymbol{g}_A$ and $\boldsymbol{R}_{,A} = \partial \boldsymbol{R}/\partial \ell_A \equiv \boldsymbol{G}_A$ are precisely the tangent base vectors of the local Cartesian systems. These are expressed in terms of the unit vectors:

$$\boldsymbol{g}_A \equiv \hat{\boldsymbol{n}}_A, \qquad \boldsymbol{G}_A = \left(\frac{dL_{\underline{A}}}{d\ell_{\underline{A}}}\right)\hat{\boldsymbol{N}}_A = \widetilde{C}_A^B \hat{\boldsymbol{N}}_B. \tag{3.81a, b}$$

In words, the base vector $\boldsymbol{G}_A$ is merely a stretched version of the rigidly rotated vector $\hat{\boldsymbol{N}}_A$. The base vectors for an arbitrary system of coordinates θ^i are also derived by the transformation rules for tensors. From equations (3.81a, b), we obtain

$$\boldsymbol{g}_i = \frac{\partial \ell_A}{\partial \theta^i}\hat{\boldsymbol{n}}_A, \qquad \hat{\boldsymbol{n}}_A = \boldsymbol{g}_A = \frac{\partial \theta^i}{\partial \ell_A}\boldsymbol{g}_i, \tag{3.82a–c}$$

$$\boldsymbol{G}_i \equiv \frac{\partial \boldsymbol{R}}{\partial \theta^i} = \frac{\partial \ell_A}{\partial \theta^i}\boldsymbol{G}_A = \frac{\partial \ell_A}{\partial \theta^i}\widetilde{C}_A^B \hat{\boldsymbol{N}}_B. \tag{3.83a–c}$$

Our transition from the arbitrary coordinates calls for the transformation:

$$\widetilde{C}_A^B = \frac{\partial \theta^m}{\partial \ell_A}\frac{\partial \ell_B}{\partial \theta^n} C_m^n. \tag{3.84}$$

Substitution of (3.84) into (3.83c) produces

$$\boldsymbol{G}_i = \frac{\partial \ell_B}{\partial \theta^n} C_i^n \hat{\boldsymbol{N}}_B.$$

By further substitution of the "rotation" (3.80) and then the transformation (3.82c), we obtain

$$\boldsymbol{G}_i = \frac{\partial \ell_B}{\partial \theta^n} C_i^n r_B^{\cdot A} \hat{\boldsymbol{n}}_A = C_i^n \left(\frac{\partial \ell_B}{\partial \theta^n} \frac{\partial \theta^j}{\partial \ell_A} r_B^{\cdot A} \right) \boldsymbol{g}_j. \qquad (3.85\text{a, b})$$

We *know* that the vectors $\boldsymbol{G}_i$ and $\boldsymbol{g}_i$ and the stretch C_i^n are tensors. We also know that the equation (3.85b) holds for an arbitrary state (strain and/or rotation). Then, the parenthetical term is also the component of a tensor, the rotation tensor in the system of arbitrary coordinates θ^i, as it is expressed by a tensor transformation from the local Cartesian system ℓ_A:

$$r_n^{\cdot j} = \frac{\partial \ell_B}{\partial \theta^n} \frac{\partial \theta^j}{\partial \ell_A} r_B^{\cdot A}. \qquad (3.85\text{c})$$

The reader can now see the merit in the positioning of the indices on "r," though the position on a *Cartesian* component is inconsequential. Equation (3.85b) is now expressed fully in terms of the tensors in the arbitrary coordinates:

$$\boldsymbol{G}_i = C_i^n r_n^{\cdot j} \boldsymbol{g}_j. \qquad (3.86\text{a})$$

Here, each component on the right side is associated with the metric g_{ij} or g^{ij} of the initial configuration. Accordingly, we have the alternative forms:

$$\boldsymbol{G}_i = C_{in} r^{nj} \boldsymbol{g}_j = C_{in} r^{n}_{\cdot j} \boldsymbol{g}^j. \qquad (3.86\text{b, c})$$

It is constructive to define intermediate triads:

$$\acute{\boldsymbol{g}}_n = r_n^{\cdot j} \boldsymbol{g}_j, \qquad \acute{\boldsymbol{g}}^n = r^{n}_{\cdot j} \boldsymbol{g}^j. \qquad (3.87\text{a, b})$$

The latter are rigidly rotated versions of the initial base vectors and are, otherwise similar. Specifically,

$$\acute{\boldsymbol{g}}_i \cdot \acute{\boldsymbol{g}}_j = g_{ij}, \qquad \acute{\boldsymbol{g}}_i \cdot \acute{\boldsymbol{g}}^j = \delta_i^j, \qquad \acute{\boldsymbol{g}}^i \cdot \acute{\boldsymbol{g}}^j = g^{ij}. \qquad (3.88\text{a–c})$$

According to (3.87a, b) and (3.88a–c), we have

$$\acute{\boldsymbol{g}}_n \cdot \acute{\boldsymbol{g}}^m = \delta_n^m = r_n^{\cdot j} r^{m}_{\cdot k} \boldsymbol{g}_j \cdot \boldsymbol{g}^k,$$

or the orthogonality condition:

$$\delta_n^m = r_n^{\bullet k} r^m_{\bullet k}. \tag{3.89}$$

Additionally, we obtain from (3.87a, b)

$$r_n^{\bullet j} = \boldsymbol{g}^j \cdot \acute{\boldsymbol{g}}_n, \qquad r^n_{\bullet j} = \boldsymbol{g}_j \cdot \acute{\boldsymbol{g}}^n, \tag{3.90a, b}$$

$$\boldsymbol{g}^j = r_n^{\bullet j} \acute{\boldsymbol{g}}^n, \qquad \boldsymbol{g}_j = r^n_{\bullet j} \acute{\boldsymbol{g}}_n, \tag{3.91a, b}$$

and, finally, from (3.91a, b)

$$\boldsymbol{g}_n \cdot \boldsymbol{g}^m = \delta_n^m = r^k_{\bullet n} r_k^{\bullet m}. \tag{3.92}$$

The equations (3.89) and (3.92) express the orthogonality of the rotation which *rigidly* rotates the vectors $\hat{\boldsymbol{n}}_\alpha$ to $\hat{\boldsymbol{N}}_\alpha$, the tangent base vectors $\boldsymbol{g}_i$ to $\acute{\boldsymbol{g}}_i$, and the normal base vectors $\boldsymbol{g}^i$ to $\acute{\boldsymbol{g}}^i$ by (3.80) and (3.87a, b). The subsequent deformation changes the triad $\acute{\boldsymbol{g}}_i$ or $\acute{\boldsymbol{g}}^i$ to the final deformed triad $\boldsymbol{G}_i$ via the stretch according to (3.86a–c):

$$\boldsymbol{G}_i = C_i^n \acute{\boldsymbol{g}}_n, \qquad \boldsymbol{G}_i = C_{in} \acute{\boldsymbol{g}}^n. \tag{3.93a, b}$$

From (3.93a, b), we also have

$$C_i^n = \boldsymbol{G}_i \cdot \acute{\boldsymbol{g}}^n, \qquad C_{in} = \boldsymbol{G}_i \cdot \acute{\boldsymbol{g}}_n = \boldsymbol{G}_n \cdot \acute{\boldsymbol{g}}_i = C_{ni}. \tag{3.94a, b}$$

The latter gives meaning to the components of the stretch tensor. For example, the component C_{12} is a component of the deformed vector $\boldsymbol{G}_1$ in the direction of the rotated base vector $\acute{\boldsymbol{g}}_2$, or, due to the symmetry of C_{ij}, the component of $\boldsymbol{G}_2$ in the direction of the vector $\acute{\boldsymbol{g}}_1$. It depends upon the shear strain ($h_{12} = h_{21}$); if the latter vanishes, then in accordance with (3.78), $C_{12} = g_{12}$.

The equations (3.31), (3.78), (3.79e), (3.93a), and (3.94b) provide the following:

$$\boldsymbol{G}_i = C_i^m \acute{\boldsymbol{g}}_m = (\delta_i^m + h_i^m) \acute{\boldsymbol{g}}_m, \tag{3.95}$$

$$h_{ij} = \tfrac{1}{2}(\boldsymbol{G}_i \cdot \acute{\boldsymbol{g}}_j + \boldsymbol{G}_j \cdot \acute{\boldsymbol{g}}_i) - g_{ij}, \tag{3.96}$$

$$\gamma_{ij} \equiv \tfrac{1}{2}(G_{ij} - g_{ij}) = h_{ij} + \tfrac{1}{2} h_i^m h_{mj}. \tag{3.97}$$

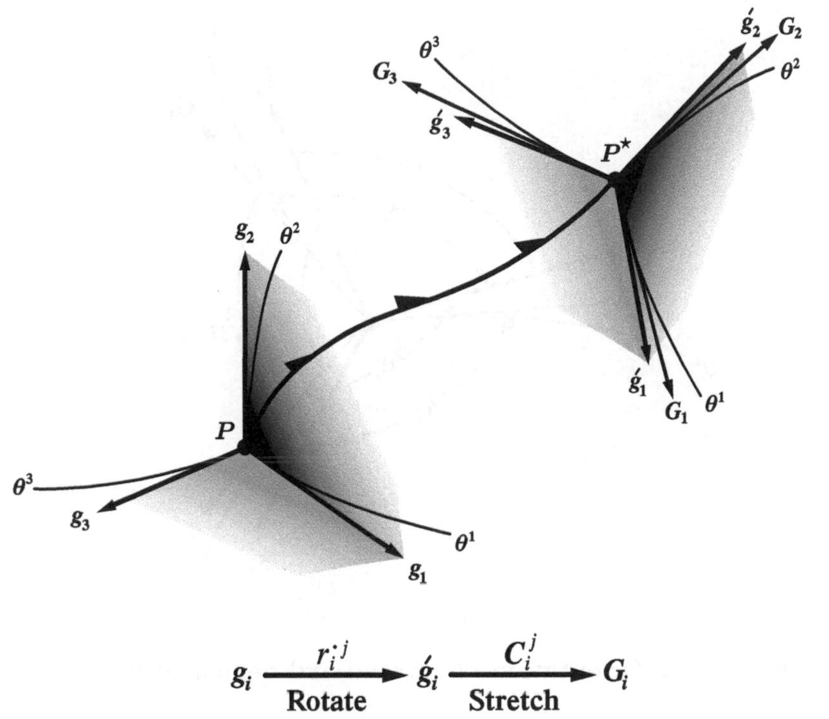

Figure 3.12 Decomposition of motion into rotation and strain

Note that h_{ij} and γ_{ij} differ in the principal values and not in the principal directions.

The decomposition expressed by (3.87a, b) and (3.93a, b) is illustrated by Figure 3.12. The initial triad $\boldsymbol{g}_i$ at a particle P is rigidly transported to P^* and rotated to the similar triad $\acute{\boldsymbol{g}}_i$ in accordance with (3.87a, b); the triad $\acute{\boldsymbol{g}}_i$ is then deformed to the final triad $\boldsymbol{G}_i$ in accordance with (3.93a, b).

Let us retrace the decomposition from another perspective: A rigid rotation (3.87a, b) carries the tangent and normal base vectors ($\boldsymbol{g}_i$ and $\boldsymbol{g}^i$) to similar triads ($\acute{\boldsymbol{g}}_i$ and $\acute{\boldsymbol{g}}^i$). A deformation carries the latter to the current deformed base vectors ($\boldsymbol{G}_i$ and $\boldsymbol{G}^i$). The last step is achieved via the stretch C_i^j such that [see (3.78) and (3.94b)]:

$$C_{ij} = \boldsymbol{G}_i \cdot \acute{\boldsymbol{g}}_j = \boldsymbol{G}_j \cdot \acute{\boldsymbol{g}}_i = h_{ij} + g_{ij}. \qquad (3.98\text{a–c})$$

It is useful to note that

$$h_{ij} = \boldsymbol{G}_i \cdot \acute{\boldsymbol{g}}_j - g_{ij} = \boldsymbol{G}_j \cdot \acute{\boldsymbol{g}}_i - g_{ij}. \qquad (3.99\text{a, b})$$

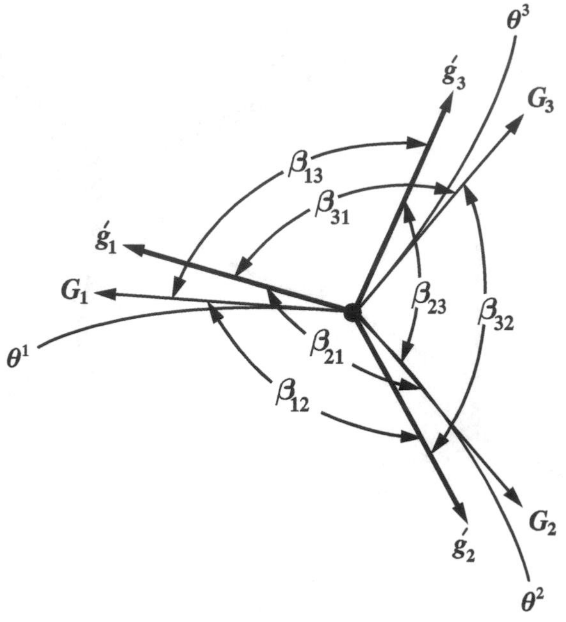

Figure 3.13 Orientation of rotated and deformed bases

The latter offers insight to the orientation of the rigidly rotated basis $\acute{\boldsymbol{g}}_i$: It lies in a position such that $h_{ij} = h_{ji}$; a shear strain is equally comprised of the two parts, i.e., $(\boldsymbol{G}_i \cdot \acute{\boldsymbol{g}}_j)$ and $(\boldsymbol{G}_j \cdot \acute{\boldsymbol{g}}_i)$ [see (3.96)]. If, for example, no stretch occurs (lengths of $\boldsymbol{G}_i$, $\acute{\boldsymbol{g}}_i$, and $\boldsymbol{g}_i$ are the same), then the angles β_{ij} of Figure 3.13 between the vectors $\boldsymbol{G}_i$ and $\acute{\boldsymbol{g}}_j$ are equal to that between the vectors $\boldsymbol{G}_j$ and $\acute{\boldsymbol{g}}_i$, i.e., $\beta_{ij} = \beta_{ji}$. Stated otherwise, the symmetry of the stretch C_{ij} and strain h_{ij} enforces an intermediate position of the rotated bases $\acute{\boldsymbol{g}}_i$ and $\acute{\boldsymbol{g}}^i$. Recall that if the coordinate lines θ^i lie in principal directions, the condition on the rotation requires that $\acute{\boldsymbol{g}}_i$ are coincident with $\boldsymbol{G}_i$, and, therefore, h_{ij} exhibits no shear components.

The intermediate (rotated) triad $\acute{\boldsymbol{g}}_j$ is characterized by the condition

$$\acute{\boldsymbol{g}}_i \cdot \boldsymbol{G}_j = \acute{\boldsymbol{g}}_j \cdot \boldsymbol{G}_i.$$

In any case, only the deformation by the symmetric tensor (stretch C_{ij} or strain h_{ij}) carries the rotated triad $\acute{\boldsymbol{g}}_j$ to the triad $\boldsymbol{G}_i$.

3.14.2 Deformation Followed by Rotation

The motion, which carries a vector $\boldsymbol{g}_i$ (or $\boldsymbol{g}^i$) to the vector $\boldsymbol{G}_i$ (or $\boldsymbol{G}^i$), can be decomposed in another way: First, we conceive of a stretching that

$$C_{ij} = \mathbf{G}_i \cdot \acute{\mathbf{g}}_j = \acute{\mathbf{G}}_i \cdot \mathbf{g}_j = \overline{C}_{ij}$$

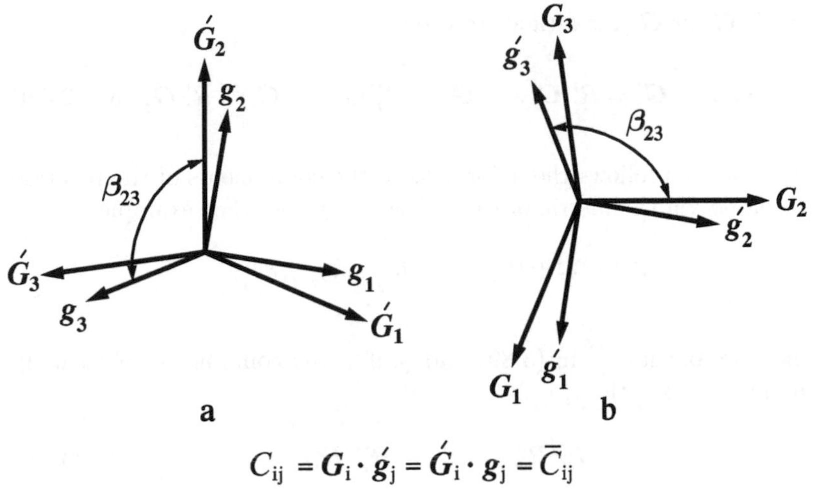

Figure 3.14 Deform → Rotate = Rotate → Deform

carries $\mathbf{g}_i$ (or $\mathbf{g}^i$) into an intermediate vector $\acute{\mathbf{G}}_i$ which is similar to the terminal vector $\mathbf{G}_i$, *but* has *not* experienced the rotation. Specifically,

$$\acute{\mathbf{G}}_m = \overline{C}^i_m \mathbf{g}_i = \overline{C}_{im} \mathbf{g}^i, \qquad (3.100\mathrm{a,b})$$

where

$$\acute{\mathbf{G}}_i \cdot \acute{\mathbf{G}}_j = G_{ij}, \qquad \acute{\mathbf{G}}^i \cdot \acute{\mathbf{G}}^j = G^{ij}, \qquad \acute{\mathbf{G}}^i \cdot \acute{\mathbf{G}}_j = \delta^i_j,$$

$$\overline{C}^i_m = \mathbf{g}^i \cdot \acute{\mathbf{G}}_m, \qquad \overline{C}_{im} = \mathbf{g}_i \cdot \acute{\mathbf{G}}_m. \qquad (3.101\mathrm{a,b})$$

Again, the prime symbol over the vectors $\mathbf{G}_m$ in (3.100a, b) signifies the first step of the decomposition.

The same deformation transforms $\mathbf{g}_i$ to $\acute{\mathbf{G}}_i$ and $\acute{\mathbf{g}}_i$ to $\mathbf{G}_i$. The two systems, ($\mathbf{g}_i$ and $\acute{\mathbf{G}}_i$) and ($\acute{\mathbf{g}}_i$ and $\mathbf{G}_i$), differ by a rigid rotation. The equivalence is evident in Figure 3.14; the two systems depicted at "a" and "b" differ by a rotation of 90° about the normal to the plane of the page. Hence,

$$\overline{C}^i_m = C^i_m = \frac{\partial \theta^i}{\partial \ell_A} \frac{\partial L_A}{\partial \theta^m}.$$

The motion, which carries the intermediate triad ($\acute{\mathbf{G}}_i$ or $\acute{\mathbf{G}}^i$) into the

final triad ($\boldsymbol{G}_i$ or $\boldsymbol{G}^i$), is a rigid rotation:

$$\boldsymbol{G}_i = R_i^{\boldsymbol{\cdot} j} \acute{\boldsymbol{G}}_j, \qquad \boldsymbol{G}^i = R^i_{\boldsymbol{\cdot} j} \acute{\boldsymbol{G}}^j, \qquad \acute{\boldsymbol{G}}^i = R_j^{\boldsymbol{\cdot} i} \boldsymbol{G}^j, \qquad \acute{\boldsymbol{G}}_i = R^j_{\boldsymbol{\cdot} i} \boldsymbol{G}_j. \text{(3.102a–d)}$$

Since the rotation follows the deformation, the components of the rotation are associated via the metric of the deformed system. For example,

$$R^{ij} = G^{mi} R_m^{\boldsymbol{\cdot}\, j}, \qquad R_{ij} = G_{mi} R^m_{\boldsymbol{\cdot}\, j}.$$

Like the components $r_i^{\boldsymbol{\cdot} j}$ in (3.89) and (3.92), the components $R_i^{\boldsymbol{\cdot} j}$ satisfy orthonormality conditions:

$$R_i^{\boldsymbol{\cdot} j} R^m_{\boldsymbol{\cdot}\, j} = \delta_i^m = R_j^{\boldsymbol{\cdot} m} R^j_{\boldsymbol{\cdot} i}. \qquad \text{(3.103a, b)}$$

The latter sequence, strain followed by rotation, can be useful in nonlinear problems of finite deformation that are analyzed by incremental methods, wherein each subsequent incremental motion is referred to a current (deformed) configuration. In many structural problems, the strains remain small though nonlinearities arise from large rotation; then the difference in the metric might be negligible (e.g., $\epsilon_{ij} \ll 1$, $G_{ij} \doteq g_{ij}$) and the difference in the components of rotation ($r_i^{\boldsymbol{\cdot} j}$ and $R_i^{\boldsymbol{\cdot} j}$) might also be negligible.

The decomposition expressed by (3.100a, b) and (3.102a, b) is illustrated by Figure 3.15. The initial triad $\boldsymbol{g}_i$ at particle P is deformed to the triad $\acute{\boldsymbol{G}}_i$ in accordance with (3.100a, b), then transported to P^* and rotated to the triad $\boldsymbol{G}_i$ in accordance with (3.102a, b).

From a purely mathematical point of view, the decomposition of motion into rotation and deformation [(3.87a, b) and (3.93a, b)], or deformation followed by rotation [(3.100a, b) and (3.102a, b)] is related to the multiplicative decomposition of a matrix (see, e.g., the textbook of B. Noble and J. W. Daniel [11] and Section 43 of the Appendix—*Tensor Fields* by J. L. Ericksen—to the treatise by C. Truesdell and R. A. Toupin [4]). This decomposition is known as the *polar representation* or *decomposition* of a matrix. Although this decomposition was already known in the second half of the nineteenth century, its physical significance and possible applications were recognized at a later time (see M. A. Biot [15], R. A. Toupin [16], and C. Truesdell and W. Noll [5]).

3.14.3 Increments of Rotation

To implement methods of treating finite deformations via successive linear increments, we need the relations between the increments of the rotations, i.e., between *tensors*, $\delta r_i^{\boldsymbol{\cdot} j}$ or $\delta R_i^{\boldsymbol{\cdot} j}$, and the incremental rotation

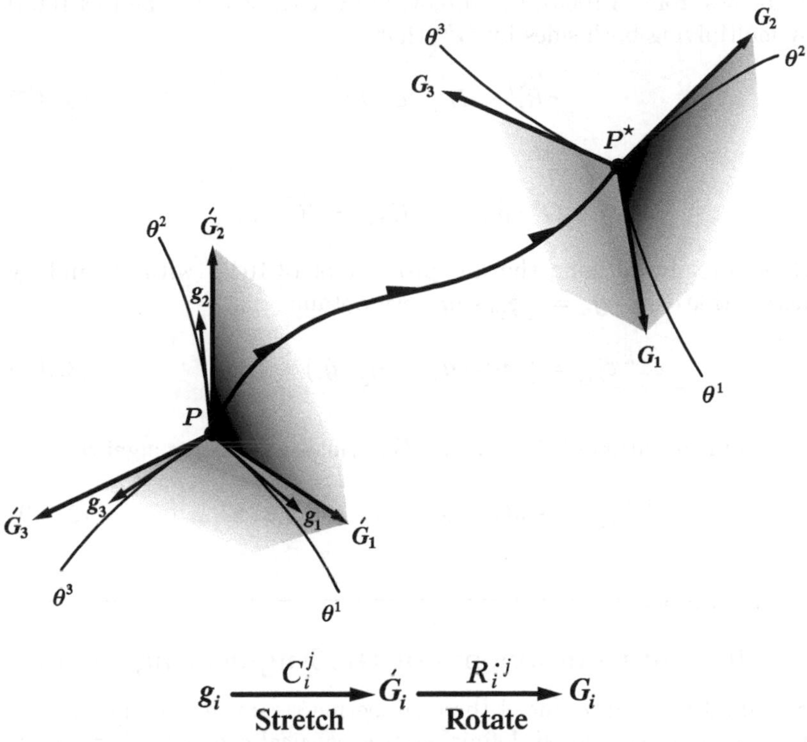

Figure 3.15 Alternative decomposition of motion into strain and rotation

vector, $\delta\boldsymbol{\omega} = \delta\bar{\omega}^i\, \acute{\boldsymbol{g}}_i = \delta\omega^i\, \boldsymbol{G}_i$. The increment of $\acute{\boldsymbol{g}}_i$, denoted by $\delta\acute{\boldsymbol{g}}_i$, can be expressed in the usual manner by the increments of the rotation vector or tensor:

$$\delta\acute{\boldsymbol{g}}_i = \delta\boldsymbol{\omega} \times \acute{\boldsymbol{g}}_i, \qquad \delta\acute{\boldsymbol{g}}_i = \delta r_i^{\cdot j}\, \boldsymbol{g}_j. \tag{3.104a, b}$$

Using (3.104a, b), (2.26a), and (3.87b) and multiplying both sides by $\boldsymbol{g}^j$, it follows that

$$\delta r_i^{\cdot j} = e_{kil}\, \delta\bar{\omega}^k\, r^{lj}, \tag{3.105}$$

where

$$\delta\bar{\omega}^k \equiv \acute{\boldsymbol{g}}^k \cdot \delta\boldsymbol{\omega}, \qquad e_{kil} = \sqrt{g}\, \epsilon_{kil}.$$

In a similar way, if the increment of the vector $\boldsymbol{G}_i$, $\delta\boldsymbol{G}_i$, assumes the forms:

$$\delta\boldsymbol{G}_i = \delta\boldsymbol{\omega} \times \boldsymbol{G}_i, \qquad \delta\boldsymbol{G}_i = \delta R_i^{\cdot j}\, \acute{\boldsymbol{G}}_j, \tag{3.106a, b}$$

then it follows from (3.106a), by employing (3.16a), (3.102b), and (3.106b) and by multiplying both sides by $\acute{\boldsymbol{G}}^m$ that

$$\delta R_i^{\bullet j} = E_{kil}\, \delta\omega^k\, R^{lj}, \tag{3.107}$$

where

$$\delta\omega^k \equiv \boldsymbol{G}^k \cdot \delta\boldsymbol{\omega}, \qquad E_{kil} = \sqrt{G}\, \epsilon_{kil}.$$

Furthermore, by forming the dot product of (3.104a) with $\acute{\boldsymbol{g}}_j$ and by considering that $\delta\acute{\boldsymbol{g}}_m \cdot \acute{\boldsymbol{g}}_n = -\acute{\boldsymbol{g}}_m \cdot \delta\acute{\boldsymbol{g}}_n$, we obtain:

$$\delta\bar{\omega}^k\, e_{kij} = \tfrac{1}{2}(\delta\acute{\boldsymbol{g}}_i \cdot \acute{\boldsymbol{g}}_j - \delta\acute{\boldsymbol{g}}_j \cdot \acute{\boldsymbol{g}}_i) \equiv \delta\bar{\omega}_{ji}. \tag{3.108}$$

Similarly, multiplication of (3.106a) by $\boldsymbol{G}_j$, yields after some algebra

$$\delta\omega^k\, E_{kij} = \tfrac{1}{2}(\delta\boldsymbol{G}_i \cdot \boldsymbol{G}_j - \delta\boldsymbol{G}_j \cdot \boldsymbol{G}_i) \equiv \delta\omega_{ji}. \tag{3.109}$$

3.15 Physical Components of the Engineering Strain

In Section 3.12 we introduced the engineering strain tensor. The tensor was based upon the accepted definition of a *physical extensional* strain. If ℓ_i and L_i denote initial and deformed lengths of a θ^i line, then

$$h_{\underline{ii}} = (dL_{\underline{i}}/d\ell_{\underline{i}}) - 1. \tag{3.110}$$

A consistent definition of all components can now be drawn from (3.96):

$${}_P h_{ij} = h_{ij} \Big/ \sqrt{g_{\underline{ii}}}\, \sqrt{g_{\underline{jj}}}. \tag{3.111}$$

A complete geometrical interpretation requires further examination:

From the definition of the extensional strain $h_{\underline{ii}}$ and the metric components, $G_{\underline{ii}}$ and $g_{\underline{ii}}$, we have

$$\boldsymbol{G}_i \equiv \sqrt{G_{\underline{ii}}}\, \hat{\boldsymbol{E}}_i = \sqrt{g_{\underline{ii}}}\, (1 + h_{\underline{ii}}) \hat{\boldsymbol{E}}_i. \tag{3.112a}$$

From (3.88a), we have

$$\acute{\boldsymbol{g}}_i = \sqrt{g_{\underline{ii}}}\, \hat{\boldsymbol{e}}'_i, \tag{3.112b}$$

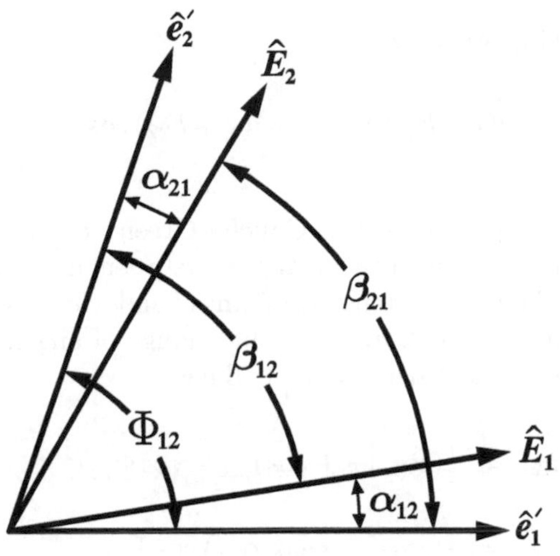

Figure 3.16 Physical components of the engineering strain tensor: change of angles

where $\hat{e}_i'$ is the unit vector in the direction of $\acute{g}_i$. From the definitions (3.96) and (3.111), we obtain

$$_Ph_{ij} = \frac{1}{2}\left[(1 + h_{\underline{ii}})(\hat{E}_i \cdot \hat{e}_j') + (1 + h_{\underline{jj}})(\hat{E}_j \cdot \hat{e}_i')\right] - (\hat{e}_i' \cdot \hat{e}_j'). \qquad (3.113)$$

In the case $i = j$, we have

$$_Ph_{\underline{ii}} = (1 + h_{\underline{ii}})(\hat{E}_{\underline{i}} \cdot \hat{e}_{\underline{i}}') - 1.$$

If the θ^i line lies in a principal direction, then $\hat{E}_i = \hat{e}_i'$, $\hat{E}_{\underline{i}} \cdot \hat{e}_{\underline{i}}' = 1$ (no shear strain) and $_Ph_{\underline{ii}} = h_{\underline{ii}}$. If the shear strain is small, $\hat{E}_{\underline{i}} \cdot \hat{e}_{\underline{i}}' \doteq 1$, then $_Ph_{\underline{ii}} \doteq h_{\underline{ii}}$.

In the case $i \neq j$, the strain in question is the shear strain. To appreciate the physical component of the engineering shear strain, consider a specific component ($i = 1$, $j = 2$). From Figure 3.16, we observe that

$$\hat{E}_1 \cdot \hat{e}_2' = \cos\beta_{12}, \qquad \hat{E}_2 \cdot \hat{e}_1' = \cos\beta_{21}.$$

The angles β_{12} and β_{21} are also depicted in Figure 3.13. Let $\Phi_{12} = \Phi_{21}$ denote the initial angle between the lines of coordinates θ^1 and θ^2, i.e.,

$$\hat{e}_1' \cdot \hat{e}_2' = \cos\Phi_{12}.$$

From equation (3.113), we have

$$_Ph_{12} = {}_Ph_{21} = \tfrac{1}{2}\left[(1+h_{11})\cos\beta_{12} + (1+h_{22})\cos\beta_{21}\right] - \cos\Phi_{12}.$$

Like the physical component of the Cauchy-Green strain [see equations (3.35a, b) and (3.36)], the change of angles between lines (shear) is not entirely decoupled from the extension of lines (and vice versa). In most instances the strains are small, $h_{\underline{ii}} \ll 1$; the change of angle, α_{12} and α_{21}, must be small of the same order, $\alpha_{ij} \ll 1$. Then

$$\begin{aligned}
_Ph_{12} = {}_Ph_{21} &\doteq \tfrac{1}{2}(\cos\beta_{12} + \cos\beta_{21} - \cos\Phi_{12}) \\
&= \tfrac{1}{2}(\cos\alpha_{12} + \cos\alpha_{21})\cos\Phi_{12} \\
&\quad + \tfrac{1}{2}(\sin\alpha_{12} + \sin\alpha_{21})\sin\Phi_{12} - \cos\Phi_{12} \\
&\doteq \tfrac{1}{2}(\alpha_{12} + \alpha_{21})\sin\Phi_{12}.
\end{aligned}$$

If the coordinates are initially orthogonal, then $\sin\Phi_{12} = 1$ and the approximation is one half of the change of angle, just as the physical approximation ϵ_{12} of the Cauchy-Green strain.

3.16 Strain-Displacement Relations

If $\boldsymbol{V}(\theta^1, \theta^2, \theta^3; t)$ denotes the displacement which carries a particle from the initial position $\boldsymbol{r}(\theta^1, \theta^2, \theta^3; 0)$ to the current position $\boldsymbol{R}(\theta^1, \theta^2, \theta^3; t)$, then

$$\boldsymbol{R} = \boldsymbol{r} + \boldsymbol{V}, \tag{3.114}$$

$$\boldsymbol{G}_i \equiv \boldsymbol{R}_{,i} = \boldsymbol{r}_{,i} + \boldsymbol{V}_{,i} \equiv \boldsymbol{g}_i + \boldsymbol{V}_{,i}, \tag{3.115a}$$

or using (3.91b)

$$\boldsymbol{G}_i = r^n_{\boldsymbol{\cdot}i}\, \acute{\boldsymbol{g}}_n + \boldsymbol{V}_{,i}. \tag{3.115b}$$

The components of the Cauchy-Green strain tensor are given by (3.31). By the substitution of (3.115a) into (3.31), we obtain

$$\gamma_{ij} = \tfrac{1}{2}(\boldsymbol{g}_i \cdot \boldsymbol{V}_{,j} + \boldsymbol{g}_j \cdot \boldsymbol{V}_{,i} + \boldsymbol{V}_{,i} \cdot \boldsymbol{V}_{,j}), \tag{3.116a}$$

or

$$\gamma_{ij} = \tfrac{1}{2}(\boldsymbol{G}_i \cdot \boldsymbol{V}_{,j} + \boldsymbol{G}_j \cdot \boldsymbol{V}_{,i} - \boldsymbol{V}_{,i} \cdot \boldsymbol{V}_{,j}). \tag{3.116b}$$

By combining (3.116a) and (3.116b), we also obtain

$$\gamma_{ij} = \tfrac{1}{4}[(\boldsymbol{g}_i + \boldsymbol{G}_i) \cdot \boldsymbol{V}_{,j} + (\boldsymbol{g}_j + \boldsymbol{G}_j) \cdot \boldsymbol{V}_{,i}]. \tag{3.116c}$$

The latter form is possibly useful in pursuing the computations for large deformations. The vector $\boldsymbol{V}$ is most usually expressed in terms of the initial basis:

$$\boldsymbol{V} = V^i \boldsymbol{g}_i. \tag{3.117}$$

Then (3.116a) assumes the form:

$$\gamma_{ij} = \tfrac{1}{2}(V_i|_j + V_j|_i + g^{mn} V_m|_i V_n|_j). \tag{3.118a}$$

Here, the vertical bar (|) signifies the covariant derivative with respect to the undeformed system (see Section 2.9). If the components of the displacement vector are referred to the deformed basis ($\boldsymbol{V} = V^i \boldsymbol{G}_i$), then (3.116b) yields:

$$\gamma_{ij} = \tfrac{1}{2}(V_i|_j^* + V_j|_i^* - G^{mn} V_m|_i^* V_n|_j^*). \tag{3.118b}$$

The asterisk next to the vertical line ($|^*$) in (3.118b) signifies covariant differentiation with respect to the deformed system. With the aid of (3.88a) and (3.115b), equation (3.96) yields the components of the engineering strain tensor:

$$h_{ij} = \tfrac{1}{2}[r_{ij} + r_{ji} + \acute{\boldsymbol{g}}_i \cdot (\boldsymbol{V}_{,j} - \acute{\boldsymbol{g}}_j) + \acute{\boldsymbol{g}}_j \cdot (\boldsymbol{V}_{,i} - \acute{\boldsymbol{g}}_i)], \tag{3.119a}$$

$$= \tfrac{1}{2}(\acute{\boldsymbol{g}}_i \cdot \boldsymbol{V}_{,j} + \acute{\boldsymbol{g}}_j \cdot \boldsymbol{V}_{,i}) + \tfrac{1}{2}(r_{ij} + r_{ji}) - g_{ij}. \tag{3.119b}$$

To provide some additional insight, we define the change $\Delta \boldsymbol{g}_i$ which carries vector $\boldsymbol{g}_i$ to $\acute{\boldsymbol{g}}_i$ as a consequence of rotation:

$$\Delta \boldsymbol{g}_i \equiv \acute{\boldsymbol{g}}_i - \boldsymbol{g}_i. \tag{3.120}$$

Then according to (3.91a, b)

$$\tfrac{1}{2}(r_{ij} + r_{ji}) = g_{ij} + \tfrac{1}{2}(\boldsymbol{g}_i \cdot \Delta\boldsymbol{g}_j + \boldsymbol{g}_j \cdot \Delta\boldsymbol{g}_i).$$

Equation (3.119b) assumes the form:

$$h_{ij} = \tfrac{1}{2}(\boldsymbol{g}_i \cdot \boldsymbol{V}_{,j} + \boldsymbol{g}_j \cdot \boldsymbol{V}_{,i}) + \tfrac{1}{2}(\boldsymbol{g}_i \cdot \Delta\boldsymbol{g}_j + \boldsymbol{g}_j \cdot \Delta\boldsymbol{g}_i)$$

$$+\tfrac{1}{2}(\Delta\boldsymbol{g}_i \cdot \boldsymbol{V}_{,j} + \Delta\boldsymbol{g}_j \cdot \boldsymbol{V}_{,i}). \tag{3.121}$$

The latter serves to decompose the strain h_{ij} into an initial term (linear in displacement), a second term (attributed to the rotation), and a final term (quadratic in the displacement).

3.17 Compatibility of Strain Components

The six equations (3.116a) express the *six* strain components γ_{ij} in terms of the derivatives of *three* components V_i of the displacement $\boldsymbol{V}$. If the strain components are prescribed, then six partial differential equations govern three functions V_i, *but* six equations are more than enough to determine the three functions. The equations are consistent only if the strain components are suitably interrelated; they must satisfy certain compatibility conditions.

The necessity for compatibility conditions might be seen from a simple geometrical viewpoint. Imagine two paths from particle P to particle Q in Figure 3.17. The path PAQ is formed of the column of square blocks from P to A and then the row from A to Q. Suppose that you knew the displacement $\boldsymbol{V}(P)$ and the rotation of the block initially at P and also the deformed shape of every other block along the path. You could then construct the new (deformed) path from P^* to Q^* and determine the displacement $\boldsymbol{V}(Q)$ of particle Q. Likewise, you might choose the path $PBCDQ$ shown dotted in Figure 3.17. Again, if you knew the deformed shapes of all blocks along the latter path, then you could construct the deformed path and again determine $\boldsymbol{V}(Q)$. If the deformations are compatible, so that the body remains intact, then you would arrive at the same point Q^* along both paths. Just as the shapes of the blocks cannot be assigned arbitrarily along the two paths, so the functions γ_{ij} cannot be arbitrarily prescribed.

Our approach to compatibility is not unlike the foregoing construction: We suppose that the displacement $\boldsymbol{V}]_P$ and the derivatives $\boldsymbol{V}_{,i}]_P$ are given

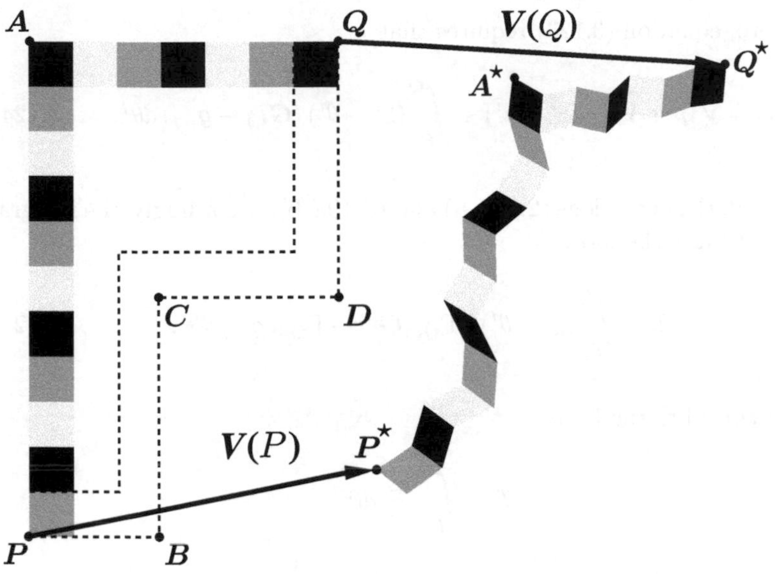

Figure 3.17 Compatibility of strain components

at a particle P with coordinates (a^1, a^2, a^3). Then the displacement at a particle Q with the coordinates (ξ^1, ξ^2, ξ^3) is given by the following integration:

$$\boldsymbol{V}]_Q = \boldsymbol{V}]_P + \int_P^Q d\boldsymbol{V}. \tag{3.122a}$$

Alternatively,

$$\boldsymbol{V}]_Q = \boldsymbol{V}]_P + \int_P^Q \boldsymbol{V}_{,i}\, d\theta^i = \boldsymbol{V}]_P + \int_P^Q \boldsymbol{V}_{,i}\, d(\theta^i - \xi^i). \tag{3.122b, c}$$

One integration by parts leads to the form:

$$\boldsymbol{V}]_Q = \boldsymbol{V}]_P + \boldsymbol{V}_{,i}]_P\,(\xi^i - a^i) + \int_P^Q (\xi^i - \theta^i)\, \boldsymbol{V}_{,ij}\, d\theta^j. \tag{3.123}$$

Now, by definition

$$\boldsymbol{V} = \boldsymbol{R} - \boldsymbol{r}, \qquad \boldsymbol{V}_{,i} = \boldsymbol{G}_i - \boldsymbol{g}_i.$$

Therefore, equation (3.123) requires that

$$\boldsymbol{V}]_Q = \boldsymbol{V}]_P + \boldsymbol{V}_{,i}]_P\,(\xi^i - a^i) + \int_P^Q (\xi^i - \theta^i)\,(\boldsymbol{G}_{i,j} - \boldsymbol{g}_{i,j})\,d\theta^j. \tag{3.124}$$

Now, recall the definitions (2.45a, b) and (3.19a, b); accordingly, the integral of (3.124) takes the form

$$\mathcal{I} \equiv \int_P^Q (\xi^i - \theta^i)({}^*\Gamma_{ijm}\boldsymbol{G}^m - \Gamma_{ijm}\,\boldsymbol{g}^m)\,d\theta^j. \tag{3.125}$$

This integral has the form

$$\mathcal{I} = \int_P^Q \mathcal{F}_j\,d\theta^j.$$

It is independent of the path (an exact integral $\mathcal{I}$ exists; $\mathcal{I}_{,j} = \mathcal{F}_j$) if, and only if,

$$\frac{\partial \mathcal{F}_j}{\partial \theta^n} = \frac{\partial \mathcal{F}_n}{\partial \theta^j}.$$

Applying the condition to the integral of (3.125), we obtain

$$\begin{aligned} &-({}^*\Gamma_{njm} - {}^*\Gamma_{jnm})\boldsymbol{G}^m + (\Gamma_{njm} - \Gamma_{jnm})\,\boldsymbol{g}^m \\ &+ (\xi^i - \theta^i)\big[({}^*\Gamma_{ijm}\boldsymbol{G}^m)_{,n} - ({}^*\Gamma_{inm}\boldsymbol{G}^m)_{,j} \\ &\qquad - (\Gamma_{ijm}\,\boldsymbol{g}^m)_{,n} + (\Gamma_{inm}\,\boldsymbol{g}^m)_{,j}\big] = 0. \end{aligned} \tag{3.126}$$

We recall the equations (2.47) and (3.20) and also the definition (3.31) of the strain component γ_{ij}:

$$\begin{aligned} \Gamma_{njm} &= \tfrac{1}{2}(g_{nm,j} + g_{jm,n} - g_{nj,m}), \\ {}^*\Gamma_{njm} &= \tfrac{1}{2}(G_{nm,j} + G_{jm,n} - G_{nj,m}) \\ &= (\gamma_{nm,j} + \gamma_{jm,n} - \gamma_{nj,m}) + \Gamma_{njm}. \end{aligned} \tag{3.127}$$

If the initial system of coordinates is regular (i.e., the derivatives $g_{ij,n}$ are continuous) and the derivatives of the strain components are continuous, then the first and second parenthetical terms of (3.126) vanish (i.e.,

$\Gamma_{njm} = \Gamma_{jnm}$ and ${}^*\Gamma_{njm} = {}^*\Gamma_{jnm}$). Then equation (3.126) is satisfied at each point, if and only if, the term in brackets vanishes, i.e.,

$$({}^*\Gamma_{ijm}\boldsymbol{G}^m)_{,n} - ({}^*\Gamma_{inm}\boldsymbol{G}^m)_{,j} - (\Gamma_{ijm}\,\boldsymbol{g}^m)_{,n} + (\Gamma_{inm}\,\boldsymbol{g}^m)_{,j} = 0. \quad (3.128a)$$

Employing again the earlier results of (2.46a, b) and (3.19a, b), we obtain

$$({}^*\Gamma_{ijm,n} - {}^*\Gamma_{inm,j} + {}^*\Gamma^p_{jm}\,{}^*\Gamma_{inp} - {}^*\Gamma^p_{mn}\,{}^*\Gamma_{ijp})\,\boldsymbol{G}^m$$

$$- (\Gamma_{ijm,n} - \Gamma_{inm,j} + \Gamma^p_{jm}\Gamma_{inp} - \Gamma^p_{mn}\Gamma_{ijp})\,\boldsymbol{g}^m = 0. \quad (3.128b)$$

The parenthetical terms are the components of the Riemann-Christoffel tensors of the deformed and undeformed system, respectively. In the notation of (2.68b), the equation (3.128b) can be written:

$${}^*R_{minj}\boldsymbol{G}^m - R_{minj}\boldsymbol{g}^m = 0.$$

As noted in Chapter 2, the components of the Riemann-Christoffel tensor vanish in a Euclidean space. Since the components of the initial system vanish, i.e., $R_{minj} = 0$, it remains to require that the components of the deformed system vanish, i.e.,

$${}^*R_{minj} = {}^*\Gamma_{ijm,n} - {}^*\Gamma_{inm,j} + ({}^*\Gamma_{inp}\,{}^*\Gamma_{jmq} - {}^*\Gamma_{ijp}\,{}^*\Gamma_{mnq})\,G^{pq} = 0. \quad (3.129)$$

In accordance with (3.127), each Christoffel symbol of the deformed system can be expressed in terms of the Christoffel symbol of the undeformed configuration and the strain components.

If the initial system is Cartesian, then the Christoffel symbols of the initial system vanish in equation (3.127); then the equation (3.129) assumes the form:

$$\begin{aligned}{}^*R_{minj} &= \gamma_{jm,in} - \gamma_{ij,mn} - \gamma_{nm,ij} + \gamma_{in,mj}\\ &\qquad + (\gamma_{ip,n} + \gamma_{np,i} - \gamma_{in,p})(\gamma_{mq,j} + \gamma_{jq,m} - \gamma_{mj,q})\,G^{pq}\\ &\qquad - (\gamma_{ip,j} + \gamma_{jp,i} - \gamma_{ij,p})(\gamma_{mq,n} + \gamma_{nq,m} - \gamma_{mn,q})\,G^{pq}\\ &= 0. \end{aligned} \quad (3.130)$$

In an arbitrary system of coordinates, the compatibility conditions can be derived from (3.129) by employing the relations (2.64) and (3.127), the definition (3.31), and by considering that

$$\tfrac{1}{2}(g_{jm,in} - g_{ij,mn} - g_{mn,ij} + g_{in,mj}) = \Gamma^p_{mn}\Gamma_{ijp} - \Gamma^p_{mj}\Gamma_{inp}.$$

In that case, they assume the form:

$$\begin{aligned} {}^*R_{minj} &= \gamma_{jm}|_{in} - \gamma_{ij}|_{mn} - \gamma_{nm}|_{ij} + \gamma_{in}|_{mj} \\ &\quad + [({}^*\Gamma^p_{in} - \Gamma^p_{in})({}^*\Gamma^q_{mj} - \Gamma^q_{mj})\, G_{pq}] \\ &\quad - [({}^*\Gamma^p_{ij} - \Gamma^p_{ij})({}^*\Gamma^q_{mn} - \Gamma^q_{mn})\, G_{pq}] = 0. \end{aligned} \tag{3.131}$$

Note that the bracketed terms of (3.131) are nonlinear in the components of strain. Note too that the components G^{pq} can be expressed in terms of the covariant components G_{pq} via (3.7) and the latter in terms of the strains via (3.31).

In view of the symmetry of the metric and strain tensors, some of the equations are identical while others vanish identically. Indeed, there are but six distinct components of the Riemann-Christoffel tensor and, therefore, six distinct equations represented by (3.131); specifically

$$\left.\begin{array}{lll} {}^*R_{2112} = 0, & {}^*R_{3113} = 0, & {}^*R_{2332} = 0, \\ {}^*R_{2113} = 0, & {}^*R_{2331} = 0, & {}^*R_{1223} = 0. \end{array}\right\} \tag{3.132a–f}$$

The six distinct equations of (3.132a–f) are conditions for the compatibility of the strain components. In view of the symmetry, the six distinct components can be represented by the components of a symmetrical second-order tensor:

$$T^{kl} \equiv \epsilon^{kmi}\epsilon^{lnj}\,{}^*R_{minj}. \tag{3.133}$$

In a simpler form, the compatibility conditions require that

$$T^{ij} = 0. \tag{3.134}$$

If the initial system is Cartesian ($\Gamma_{ijm} = 0$) and the strains vanish (${}^*\Gamma_{ijm} = 0$), then the integral $\mathcal{I}$ of (3.124) or (3.125) vanishes. Then the displacement of the particle Q is composed of two terms: The displacement of the particle at P and the relative rotation caused by a rigid-body rotation; these are expressed by the first and second terms of (3.124).

The six equations (3.132a–f) or (3.134) insure the compatibility of strains in a simply connected region. If the region is multiply connected, i.e., if the body contains holes, then one additional condition is required for each hole: To insure a single-valued displacement, the integral (3.122a) about a closed path enclosing each hole must vanish.

An interesting treatment of compatibility is presented in the monograph by B. Fraeijs de Veubeke [17]. In particular, he presents two theorems that establish the compatibility of *small* strains, if only three components of (3.134) are satisfied in the interior of the body and three vanish on the boundary.

3.18 Rates and Increments of Strain and Rotation

Whenever the properties of a medium are altered by deformation, it is essential that we follow the same particles and examine the geometrical and physical changes of the same material lines, surfaces, and volume. Accordingly, we have adopted a single system of coordinates, affixed a single label to each particle, which is unaltered with the passage of time. Consequently, during any interval of time Δt, a particle is transported from one position P to another P^* without changes in the coordinates θ^i. The incremental changes of strain are without ambiguity; they are the extensions or shears of the same links embedded in the material. We conceive of a *very* small increment of the strain $\Delta\gamma_{ij}$ or obtain the rate $\dot{\gamma}_{ij}$ if we divide by Δt and pass to the limit ($\Delta t \to 0$). As long as the increment is sufficiently small, the consequences are the same; therefore, the *very* small increment *or* the rate are given by

$$\dot{\gamma}_{ij} = \tfrac{1}{2}(\boldsymbol{G}_i \cdot \boldsymbol{G}_j - \boldsymbol{g}_i \cdot \boldsymbol{g}_j)^{\boldsymbol{\cdot}} = \tfrac{1}{2}(\boldsymbol{G}_i \cdot \boldsymbol{G}_j)^{\boldsymbol{\cdot}} = \tfrac{1}{2}(\boldsymbol{G}_i \cdot \dot{\boldsymbol{G}}_j + \boldsymbol{G}_j \cdot \dot{\boldsymbol{G}}_i). \tag{3.135}$$

From the equations of transformation (3.41), it follows that the rate $\dot{\bar{\gamma}}_{ij}$ in another system of coordinates $\bar{\theta}_i$ is given by

$$\dot{\bar{\gamma}}_{ij} = \frac{\partial \theta^k}{\partial \bar{\theta}^i} \frac{\partial \theta^l}{\partial \bar{\theta}^j} \dot{\gamma}_{kl}. \tag{3.136}$$

Although the strain increment $\dot{\gamma}_{ij}$ has mathematical and physical properties, which are similar to the strain, it is distinct. In particular, the principal directions of the rate $\dot{\gamma}_{ij}$ and strain γ_{ij} are generally different.

It is most likely that the strain rate $\dot{\gamma}_{ij}$ would be employed in analyses of inelastic solids. In the event of finite deformations, such increments might be associated with the deformed state; then components might also be mathematically associated via the deformed metric (G^{ij} or G_{ij}).

In accordance with (3.115a), the rate of (3.135) can be expressed in terms of the velocity $\dot{\boldsymbol{V}}$

$$\dot{\gamma}_{ij} = \tfrac{1}{2}(\boldsymbol{G}_i \cdot \dot{\boldsymbol{V}}_{,j} + \boldsymbol{G}_j \cdot \dot{\boldsymbol{V}}_{,i}). \tag{3.137}$$

By contrast, one can form a skewsymmetric tensor:

$$\dot{\omega}_{ij} \equiv \tfrac{1}{2}(\boldsymbol{G}_i \cdot \dot{\boldsymbol{G}}_j - \boldsymbol{G}_j \cdot \dot{\boldsymbol{G}}_i), \tag{3.138a}$$

$$= \tfrac{1}{2}(\boldsymbol{G}_i \cdot \dot{\boldsymbol{V}}_{,j} - \boldsymbol{G}_j \cdot \dot{\boldsymbol{V}}_{,i}). \tag{3.138b}$$

It follows from (3.135) and (3.138a) and from (3.137) and (3.138b), respectively, that

$$\dot{\boldsymbol{G}}_j = \dot{\boldsymbol{V}}_{,j} = (\dot{\gamma}_{ij} + \dot{\omega}_{ij})\, \boldsymbol{G}^i. \tag{3.139}$$

Equation (3.139) serves to decompose the rate $\dot{\boldsymbol{G}}_j$ into two parts: a deformational part ($\dot{\gamma}_{ij}\boldsymbol{G}^i$) and a rotational part ($\dot{\omega}_{ij}\boldsymbol{G}^i$). The former fully describes the rate of deformation; the latter represents the rigid-body "spin" and, since it is a small increment, it can also be expressed by a "spin" vector $\dot{\boldsymbol{\omega}}$ in the usual manner:

$$\dot{\omega}_{ij}\boldsymbol{G}^i = \dot{\boldsymbol{\omega}} \times \boldsymbol{G}_j = \dot{\omega}^i E_{ijk} \boldsymbol{G}^k. \tag{3.140}$$

It follows that

$$\dot{\omega}_{ij} = E_{kji}\dot{\omega}^k, \qquad \dot{\omega}^k = \tfrac{1}{2}E^{ijk}\dot{\omega}_{ji}. \tag{3.141a, b}$$

Recall that the rotation is determined by the tensor R_{ij} of (3.102a, b), wherein the vector $\acute{\boldsymbol{G}}_i$ shown in Figure 3.15 is perceived as the consequence of the prior deformation ($\boldsymbol{g}_i \to \acute{\boldsymbol{G}}_i$). In accordance with (3.138a) and (3.102a, d),

$$\dot{\omega}_{ij} = \tfrac{1}{2}(R_{ik}\dot{R}_j^{\cdot k} - R_{jk}\dot{R}_i^{\cdot k}). \tag{3.142}$$

Recall too that the rotation in question is the rotation that carries the initial principal lines to their current positions; indeed

$$(\hat{\boldsymbol{N}}_\alpha)^{\boldsymbol{\cdot}} \equiv \dot{\boldsymbol{\omega}} \times \hat{\boldsymbol{N}}_\alpha. \tag{3.143}$$

The vector $\dot{\boldsymbol{G}}_j$ is decomposed in (3.139), wherein the rate of deformation is expressed by the strain rate $\dot{\gamma}_{ij}$. Alternatively, the vector $\dot{\boldsymbol{G}}_j$ is expressed in terms of the strain rate $\dot{h}_{ij}$ by differentiating (3.93a):

$$\dot{\boldsymbol{G}}_j = (C_j^n \acute{\boldsymbol{g}}_n)^{\boldsymbol{\cdot}} = \dot{C}_j^n \acute{\boldsymbol{g}}_n + C_j^n (\acute{\boldsymbol{g}}_n)^{\boldsymbol{\cdot}}. \tag{3.144}$$

Recall that according to (3.79e), the increment of the engineering strain component, $\dot{h}_{ij}$, is

$$\dot{h}^i_j = \dot{C}^i_j. \tag{3.145}$$

Recall also that the motion of $\acute{\boldsymbol{g}}_n$ is only rigid rotation. Thus, the increment of the vector $\acute{\boldsymbol{g}}_n$ is produced solely by rotation:

$$(\acute{\boldsymbol{g}}_n)^{\boldsymbol{\cdot}} = \dot{\boldsymbol{\omega}} \times \acute{\boldsymbol{g}}_n = \dot{\bar{\omega}}^i (\acute{\boldsymbol{g}}_i \times \acute{\boldsymbol{g}}_n) = \dot{\bar{\omega}}^i e_{ink}\, \acute{\boldsymbol{g}}^k. \tag{3.146}$$

In place of (3.139) we have

$$\dot{\boldsymbol{G}}_j = \dot{h}^n_j \acute{\boldsymbol{g}}_n + C^k_j\, \dot{\bar{\omega}}^i\, e_{ikn} \acute{\boldsymbol{g}}^n. \tag{3.147}$$

We note that the rotational increment has alternative bases in the equations (3.140) and (3.147), viz.,

$$\dot{\boldsymbol{\omega}} = \dot{\omega}^i \boldsymbol{G}_i = \dot{\bar{\omega}}^i \acute{\boldsymbol{g}}_i.$$

The rate of dilatation $\dot{D}$ follows from (3.12), (3.16a), (3.17a), (1.12c), and (3.139) :

$$\dot{D} \equiv \left(\sqrt{\frac{G}{g}}\right)^{\boldsymbol{\cdot}} = \frac{1}{6\sqrt{g}} \left\{\epsilon^{ijk} \left[\boldsymbol{G}_k \cdot (\boldsymbol{G}_i \times \boldsymbol{G}_j)\right]^{\boldsymbol{\cdot}}\right\}$$

$$= \sqrt{\frac{G}{g}}\, G^{ms} \left(\dot{\gamma}_{ms} + \dot{\omega}_{ms}\right)$$

$$= D\, G^{ij} \dot{\gamma}_{ij}. \tag{3.148}$$

3.19 Eulerian Strain Rate

An ideal *fluid*, unlike a solid, has no memory of a prior state. Then strain has no meaning, but behavior does depend upon strain rate which describes changes from a current state to an immediate subsequent state. Expressed otherwise, one does not follow particles during any finite motion (so-called Lagrangean viewpoint) but fixes on a position and perceives the passing of particles (so-called Eulerian viewpoint).

The Eulerian extensional *rate* of a line of current length dX_1, in the rectilinear direction X_1, can be defined by analogy with (3.26):

$$\dot{E}_{11} \equiv \frac{1}{2} \lim_{\Delta X_1 \to 0} \frac{(\Delta \boldsymbol{R}_1 \cdot \Delta \boldsymbol{R}_1)^{\cdot}}{\Delta X_1 \Delta X_1}.$$

The Eulerian shear *rate* between orthogonal lines of current lengths dX_1 and dX_2 can be defined likewise:

$$\dot{E}_{12} \equiv \frac{1}{2} \lim_{\substack{\Delta X_1 \to 0 \\ \Delta X_2 \to 0}} \frac{(\Delta \boldsymbol{R}_1 \cdot \Delta \boldsymbol{R}_2)^{\cdot}}{\Delta X_1 \Delta X_2}.$$

Here $\Delta \boldsymbol{R}_i$ is tangent to the material line in direction X_i, the Cartesian coordinates of the *current* position of the observation. Unlike the initial coordinates θ^i, the coordinates X_i of any given particle are functions of time. In alternative forms, the Eulerian rates are given by

$$\dot{E}_{ij} = \frac{1}{2} \left(\frac{\partial \boldsymbol{R}}{\partial \theta^k} \cdot \frac{\partial \boldsymbol{R}}{\partial \theta^l} \right)^{\cdot} \frac{\partial \theta^k}{\partial X_i} \frac{\partial \theta^l}{\partial X_j} \tag{3.149a}$$

$$= \frac{1}{2} (\boldsymbol{G}_k \cdot \boldsymbol{G}_l)^{\cdot} \frac{\partial \theta^k}{\partial X_i} \frac{\partial \theta^l}{\partial X_j} = \frac{1}{2} \dot{G}_{kl} \frac{\partial \theta^k}{\partial X_i} \frac{\partial \theta^l}{\partial X_j}$$

$$= \frac{\partial \theta^k}{\partial X_i} \frac{\partial \theta^l}{\partial X_j} \dot{\gamma}_{kl}. \tag{3.149b}$$

Mathematically, the relation appears quite innocent, a tensor transformation of a symmetric tensor $\dot{\gamma}_{ij}$ in one system θ^i to another symmetric tensor $\dot{E}_{ij}$ in a second system X_i. However, to have physical meaning, the rates on the left and right sides must pertain to the same material. In any case, these rates apply to the material at the current position prescribed by coordinates X_i. If any finite time (and deformation) has elapsed, then the equation is useful only if the initial coordinates θ^i are known in terms of the current position $\theta^i(X_i)$; that unlikely prospect renders the equation a mathematical curiosity. If the current state is also the initial state (no time lapse), or if no appreciable motion has occurred (in a practical sense), then the relationship $\theta^i(X_i)$ is available and (3.149b) provides the transformation from the tensorial components $\dot{\gamma}_{ij}$ in the arbitrary coordinates θ^i to the physical components $\dot{E}_{ij}$ in the Cartesian coordinates X_i.

3.20 Strain Deviator

The inelastic deformation of a solid is usually accompanied by changes in its physical attributes; each successive increment of inelastic strain, however slight, alters the properties depending on the nature of that strain. Accordingly, such behavior can only be described in terms of incremental strains or rates. At any given state, the current properties are dependent on the entire history of deformation, the complete path of the strain as traced by the successive increments.

It is known that some materials, most notably certain metals, are little affected by changes of volume (dilation). Indeed, some materials experience insignificant changes of volume; other materials exhibit only elastic (not inelastic) dilation. For these reasons, certain behavior is best described by a *strain deviator*. This may be viewed as a measure of strain that does not exhibit the dilation. Customarily, the deviatoric strain is defined by components

$$\eta_{ij} \equiv \gamma_{ij} - \tfrac{1}{3}\gamma_k^k g_{ij}. \tag{3.150}$$

Recall that according to (3.74a–c) and (3.76a–d), the term γ_k^k is an approximation of the volumetric strain *when* the strains are small, viz.,

$$'h \equiv \frac{dV}{dv} - 1 = \sqrt{\frac{G}{g}} - 1 \doteq \gamma_i^i.$$

In accordance with (3.71a–c), the invariants of the deviatoric strain η_{ij} follow:

$$\acute{K}_1 \equiv \eta_i^i = 0, \tag{3.151a}$$

$$\acute{K}_2 \equiv \eta_j^i \eta_i^j = K_2 - \tfrac{1}{3}K_1^2, \tag{3.151b}$$

$$\acute{K}_3 \equiv \eta_j^i \eta_k^j \eta_i^k = K_3 - K_1 K_2 + \tfrac{2}{9}K_1^3. \tag{3.151c}$$

In accordance with (3.71a–c) to (3.73a–c), the dilation associated with such deviatoric strain is small $O(\gamma^3)$. In practical circumstances, the strains may be small enough to regard such dilation as inconsequential.

Since the inelastic behavior and relevant properties are dependent on prior deformations, it is logical to utilize a deviatoric increment *associated* with the *current state*, viz.,

$$\dot{\bar{\eta}}_{ij} \equiv \dot{\gamma}_{ij} - \tfrac{1}{3}\dot{\gamma}_{kl} G^{kl} G_{ij}. \tag{3.152}$$

Then the associated dilation rate of (3.148) is $\dot{D} = 0$.

3.21 Approximation of Small Strain

If the principal strains are small compared to unity ($\tilde{\epsilon}_{\underline{ii}} \ll 1$), then it follows from (3.58a, b) that all components of strain are small compared to unity. The Cauchy-Green strain $\gamma_{\underline{ii}}$ defined by (3.31) and the engineering strain $h_{\underline{ii}}$ of (3.96) are equivalent for most *practical* purposes [see equation (3.97)]. The physical component of the shear strain ϵ_{ij} (3.36) is approximately

$$\epsilon_{ij} \doteq \tfrac{1}{2}(\cos\Phi^*_{ij} - \cos\Phi_{ij}).$$

If the θ^i and θ^j lines are initially orthogonal, then the shear strain is essentially half the reduction of the initial right angle:

$$\epsilon_{ij} \doteq \frac{1}{2}\left(\frac{\pi}{2} - \Phi^*_{ij}\right).$$

This is one half of the customary shear strain in engineering usage. From (3.97), it follows too that the components of the engineering strain tensor h_{ij} of equation (3.96) are practically the same as the components of the Cauchy-Green tensor γ_{ij} given by (3.31). It is very important to note that these approximations are independent of the magnitudes of the motion and especially the rotation. Many practical problems of flexible bodies, thin beams, plates, and shells (e.g., buckling) involve large rotations and displacements, although strains remain very small.

From (3.87a), (3.95), and (3.97), it follows that

$$\boldsymbol{G}_i = (\delta_i^m + h_i^m)\,\acute{\boldsymbol{g}}_m \doteq (\delta_i^m + \gamma_i^m)\,\acute{\boldsymbol{g}}_m = (\delta_i^m + \gamma_i^m)\,r_m^{\cdot\,j}\boldsymbol{g}_j. \qquad (3.153\text{a–c})$$

Since the strain components are small compared to unity, the inverse of (3.153b) is approximated by

$$\acute{\boldsymbol{g}}_i \doteq (\delta_i^m - \gamma_i^m)\,\boldsymbol{G}_m. \qquad (3.154)$$

By forming the dot product of equation (3.154) with $\boldsymbol{g}^k$ and by using (3.87a), we obtain

$$r_n^{\cdot\,k} \doteq (\delta_n^m - \gamma_n^m)\,\boldsymbol{G}_m \cdot \boldsymbol{g}^k. \qquad (3.155)$$

Furthermore, multiplication of both sides of equation (3.154) by $r^{i}_{\bullet k}$ and in view of (3.91b), it follows that

$$\boldsymbol{g}_k = r^{m}_{\bullet k}\, \acute{\boldsymbol{g}}_m \doteq r^{m}_{\bullet k}\, (\delta^n_m - \gamma^n_m)\, \boldsymbol{G}_n. \tag{3.156}$$

In equation (3.153c), $r_m^{\bullet j}$ defines the rotation that transforms the triad $\boldsymbol{g}_j$ to $\acute{\boldsymbol{g}}_j$ and γ_i^m defines the deformation that transforms the triad $\acute{\boldsymbol{g}}_i$ to $\boldsymbol{G}_i$. Equation (3.156) gives the inverse transformation that successively deforms $\boldsymbol{G}_i$ to $\acute{\boldsymbol{g}}_i$ and then rotates $\acute{\boldsymbol{g}}_i$ to $\boldsymbol{g}_i$.

3.22 Approximations of Small Strain and Moderate Rotation

We take the view that a moderate rotation admits an approximation as a vector like the spin $\dot{\boldsymbol{\omega}}$ of (3.140). Of course, a rotation of any magnitude can be "represented" as a vector; direction signifies the axis of turning, the length denotes magnitude, and an arrowhead connotes sense. Such representation does not admit commutative addition as a vector. As the rotation becomes smaller, the accuracy of the *vectorial* treatment improves. Our practical interests reside in certain problems of thin flexible bodies (beams, plates, and shells) which often entail very small strains [at most $O(10^{-2})$], but moderate rotations [possibly $O(10^{-1})$ or about 6°].

Recall the decomposition of (3.86a–c) and (3.87a, b), wherein the motion is perceived as a rotation ($\boldsymbol{g}_i \to \acute{\boldsymbol{g}}_i$) and subsequent deformation ($\acute{\boldsymbol{g}}_i \to \boldsymbol{G}_i$). Specifically, if we accept the aforementioned approximations, then (3.87a) and (3.95) yield:

$$\acute{\boldsymbol{g}}_i = r_i^{\bullet n}\boldsymbol{g}_n \doteq \boldsymbol{g}_i + \boldsymbol{\omega} \times \boldsymbol{g}_i, \tag{3.157}$$

$$\boldsymbol{G}_i = (\delta_i^n + h_i^n)\, \acute{\boldsymbol{g}}_n \doteq (\delta_i^n + \gamma_i^n)\, \acute{\boldsymbol{g}}_n. \tag{3.158}$$

The rotation vector $\boldsymbol{\omega}$ of (3.157) can be expressed by alternative components:

$$\boldsymbol{\omega} = \Omega^i\, \boldsymbol{g}_i = \bar{\omega}^i\, \acute{\boldsymbol{g}}_i = \omega^i\, \boldsymbol{G}_i.$$

We define

$$e_{ij} \equiv \tfrac{1}{2}(\boldsymbol{g}_i \cdot \boldsymbol{G}_j + \boldsymbol{g}_j \cdot \boldsymbol{G}_i) - g_{ij}, \tag{3.159}$$

$$\Omega_{ij} \equiv \tfrac{1}{2}(\boldsymbol{g}_i \cdot \boldsymbol{G}_j - \boldsymbol{g}_j \cdot \boldsymbol{G}_i). \tag{3.160}$$

We note that (3.159) is actually the linear approximation of the strain γ_{ij} defined by (3.31). If the approximations (3.158) and (3.157) are substituted into (3.159) and (3.160), respectively, then

$$e_{ij} = \gamma_{ij} + \tfrac{1}{2}(\gamma_j^n \, \Omega^k e_{kni} + \gamma_i^n \, \Omega^k e_{knj}), \tag{3.161}$$

$$\Omega_{ij} = \Omega^k e_{kji} + \tfrac{1}{2}(\gamma_j^n \, \Omega^k e_{kni} - \gamma_i^n \, \Omega^k e_{knj}). \tag{3.162}$$

Since the strain is small, it is evident that e_{ij} is small of the same order of magnitude and that

$$\Omega_{ij} \doteq \Omega^k e_{kji}. \tag{3.163}$$

Therefore, the approximations of (3.157) and (3.158) have the further simplifications

$$\acute{\boldsymbol{g}}_i \doteq \boldsymbol{g}_i + \Omega_{ki} \, \boldsymbol{g}^k, \tag{3.164}$$

$$\boldsymbol{G}_i \doteq \boldsymbol{g}_i + (e_{ik} + \Omega_{ki}) \, \boldsymbol{g}^k. \tag{3.165}$$

The latter (3.165) differs from (3.158) in terms $O(\gamma\Omega)$. Then

$$\gamma_{ij} \equiv \tfrac{1}{2}(\boldsymbol{G}_i \cdot \boldsymbol{G}_j - g_{ij})$$

$$\doteq e_{ij} + \tfrac{1}{2}(e_{ik} + \Omega_{ki})(e_j^k + \Omega^k_{\cdot j}), \tag{3.166a}$$

$$\gamma_{ij} \doteq e_{ij} + \tfrac{1}{2}\Omega_{ki}\Omega^k_{\cdot j}. \tag{3.166b}$$

The final approximation is consistent with the omission of those terms $O(\gamma\Omega)$ in (3.165). The approximation (3.166b) is most frequently employed in the analyses involving small strains of thin bodies which may experience "moderate" rotations.

Since our approximation of small strain implies $\gamma_{ij} \doteq h_{ij}$ [see equation (3.97)], we can examine the circumstance of moderate rotation via the equation (3.121), viz.,

$$h_{ij} = \tfrac{1}{2}(\boldsymbol{g}_i \cdot \boldsymbol{V}_{,j} + \boldsymbol{g}_j \cdot \boldsymbol{V}_{,i}) + \tfrac{1}{2}(\boldsymbol{g}_i \cdot \Delta\boldsymbol{g}_j + \boldsymbol{g}_j \cdot \Delta\boldsymbol{g}_i)$$

$$+ \tfrac{1}{2}(\Delta\boldsymbol{g}_i \cdot \boldsymbol{V}_{,j} + \Delta\boldsymbol{g}_j \cdot \boldsymbol{V}_{,i}). \tag{3.167}$$

In accordance with (3.95), (3.115a, b), and (3.120),

$$\boldsymbol{G}_i = \boldsymbol{V}_{,i} + \boldsymbol{g}_i$$

$$= (\delta_i^j + h_i^j)\acute{\boldsymbol{g}}_j, \tag{3.168}$$

$$\Delta \boldsymbol{g}_i \equiv \acute{\boldsymbol{g}}_i - \boldsymbol{g}_i. \tag{3.169}$$

It follows that

$$\boldsymbol{V}_{,i} = (\delta_i^j + h_i^j)\Delta \boldsymbol{g}_j + h_i^j \boldsymbol{g}_j. \tag{3.170}$$

Again, the motion of $\boldsymbol{g}_i$ to $\acute{\boldsymbol{g}}_i$ is rigid rotation. To a first-order approximation:

$$\Delta \boldsymbol{g}_i = \boldsymbol{\omega} \times \boldsymbol{g}_i = \Omega^k e_{kij} \boldsymbol{g}^j, \tag{3.171a}$$

$$= \Omega_{ji} \boldsymbol{g}^j. \tag{3.171b}$$

The substitution of (3.170) and (3.171a, b) into the final term of (3.167) yields all terms $O(\Omega^2)$, $O(h\Omega)$, and higher:

$$\tfrac{1}{2}(\Delta \boldsymbol{g}_i \cdot \boldsymbol{V}_{,j} + \Delta \boldsymbol{g}_j \cdot \boldsymbol{V}_{,i}) \doteq \Omega_{ki} \Omega_j^k. \tag{3.172}$$

The second term of (3.167) is linear in $\Delta \boldsymbol{g}_i$; hence, the first-order approximation of (3.171a, b) does not yield terms of order Ω^2. One must examine the finite rotation expressed by (3.87a, b), viz., $\acute{\boldsymbol{g}}_n = r_n^{\cdot j} \boldsymbol{g}_j$, and retain terms of order Ω^2. Successive rotations, Ω_1, Ω_2, Ω_3, in any sequence (see L. A. Pars [14], p. 104), serves to establish the result:

$$\tfrac{1}{2}(\boldsymbol{g}_i \cdot \Delta \boldsymbol{g}_j + \boldsymbol{g}_j \cdot \Delta \boldsymbol{g}_i) \doteq -\tfrac{1}{2}\Omega_{ki}\Omega_j^k. \tag{3.173}$$

The substitution of the approximations (3.172) and (3.173) into (3.167) gives the approximation of strain h_{ij} in the form (3.166b).

Finally, using (3.87a) and (3.164), the components of the rotation tensor $r_i^{\cdot j}$ assume the form

$$r_i^{\cdot j} \doteq \delta_i^j + \Omega_{\cdot i}^j. \tag{3.174}$$

3.23 Approximations of Small Strain and Small Rotation

If the rotations are small of the same order of magnitude as the strains, e.g., $\Omega \approx O(10^{-2})$, then (3.166b) has the approximation:

$$\gamma_{ij} \doteq e_{ij} \equiv \tfrac{1}{2}(\boldsymbol{g}_i \cdot \boldsymbol{G}_j + \boldsymbol{g}_j \cdot \boldsymbol{G}_i) - g_{ij}. \tag{3.175a}$$

From (3.115a), we have the form:

$$\gamma_{ij} \doteq \tfrac{1}{2}(\boldsymbol{g}_i \cdot \boldsymbol{V}_{,j} + \boldsymbol{g}_j \cdot \boldsymbol{V}_{,i}). \tag{3.175b}$$

In place of (3.131), we have the linear approximation of the compatibility equations:

$$^*\overline{R}_{minj} \doteq \gamma_{jm}|_{in} - \gamma_{ij}|_{mn} - \gamma_{nm}|_{ij} + \gamma_{in}|_{mj} = 0. \tag{3.176}$$

Like $^*R_{minj}$, $^*\overline{R}_{minj}$ represents one of six distinct components. The engineer understands that our knowledge of material behavior seldom permits us to predict strains in structural and mechanical components with an accuracy better than the foregoing approximations admit. Problems of severe inelastic deformations, e.g., forging or extruding, or the elastic deformations of rubber-like materials require more precise analyses.

Chapter 4

Stress

4.1 Stress Vector

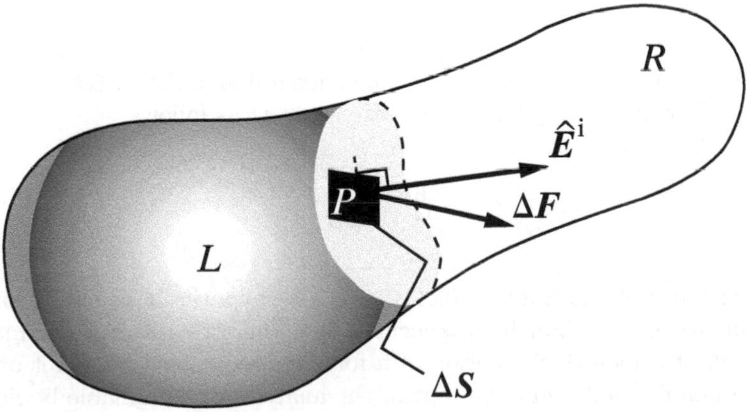

Figure 4.1 Internal force/stress

Previously we examined the kinematics of a continuous medium and introduced the concept of strain to describe the deformation at a point. Here, we explore the distribution of forces interacting between elemental portions of the medium and introduce the concept of stress to describe the intensity and character of internal forces at a point. We must note emphatically that our notions of stress in a solid body cannot be addressed without acknowledging the deformations. Excepting only the most extraordinary circumstances, we observe stresses in the deformed state. Then, we must deal with the convected coordinates, a curvilinear system; the metric is generally unknown a priori, but constitutes a part of the solution.

The body of Figure 4.1 is divided into parts L and R by a surface S. The part R exerts a force $\boldsymbol{F}$ upon the surface S of part L (and L exerts a

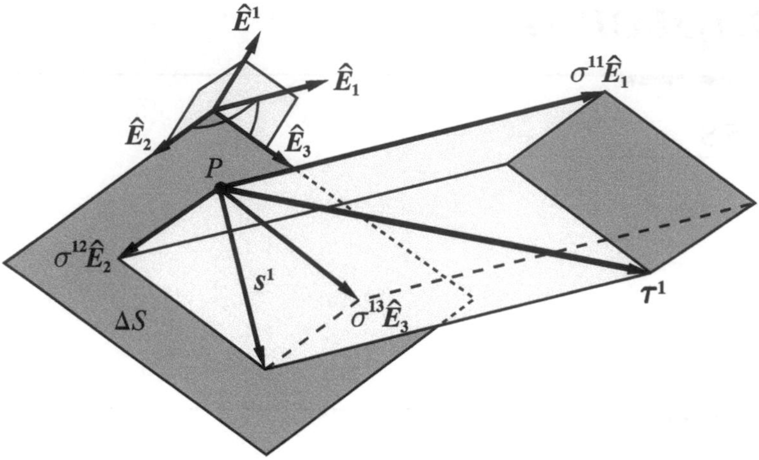

Figure 4.2 Stress vector and physical components

force $-\boldsymbol{F}$ on R). The force acting on the elemental area ΔS is $\Delta\boldsymbol{F}$; it acts at a point P of ΔS. We define the stress vector $\boldsymbol{\tau}^i$ as follows:

$$\boldsymbol{\tau}^i = \lim_{\Delta S \to 0} \frac{\Delta \boldsymbol{F}}{\Delta S}. \tag{4.1}$$

The existence of the limit in (4.1) is a basic hypothesis in the theory of continuous media, though, in special cases, the stress at a point may become infinite. Indeed, the notion of a force concentrated at a point of a body is a familiar and useful concept in the analyses of deformable bodies. However, it is a mathematical tool, not a physical reality, and special care should be exercised when interpreting results involving such singularities.

In general, there is a different stress vector for every surface through P; the vector is meaningless unless we also stipulate the orientation of the surface ΔS. The unit normal $\hat{\boldsymbol{E}}^i$ to the surface ΔS serves to define the orientation of ΔS and a superscript (^{i}) on $\boldsymbol{\tau}^i$ identifies the stress vector with the surface. Note that part L exerts the stress $-\boldsymbol{\tau}^i$ on the surface S of R at P, in accordance with Newton's third law.

The stress vector $\boldsymbol{\tau}^1$ of Figure 4.2 can be decomposed into a component σ^{11} in the direction of a unit vector $\hat{\boldsymbol{E}}_1$ (not in ΔS) and a component s^1 tangent to ΔS. This is illustrated in Figure 4.2. The component s^1 can be decomposed into components σ^{12} and σ^{13} in the directions of $\hat{\boldsymbol{E}}_2$ and $\hat{\boldsymbol{E}}_3$, both tangent to the surface at P. Then,

$$\boldsymbol{\tau}^1 = \sigma^{1i}\hat{\boldsymbol{E}}_i. \tag{4.2}$$

The components σ^{1i} are called the *physical components* of the stress vector. *If* $\hat{\boldsymbol{E}}_1 = \hat{\boldsymbol{E}}^1$, the unit normal to S at P, then σ^{11} is called the *normal* stress component and s^1 the *shear* stress, while σ^{12} and σ^{13} are components of the shear stress. The most convenient directions for the triad $\hat{\boldsymbol{E}}_i$ are those of the tangent vectors $\boldsymbol{G}_i$ [see equation (3.9a)].

4.2 Couple Stress

The system of forces distributed on the area ΔS of Figure 4.2 can be approximated by a force $\Delta \boldsymbol{F}$ at a given point P and a couple $\Delta \boldsymbol{M}$. The latter system is said to be equipollent to the actual distributed system; the couple $\Delta \boldsymbol{M}$ has the moment of the distributed forces about the point. If the forces are continuously distributed, then we anticipate that $\Delta \boldsymbol{M}$ vanishes in the limit $\Delta S \to 0$. This is a basic assumption of most theories of continuous media. However, it is conceivable that the microstructure of the real material has properties such that the couple $\Delta \boldsymbol{M}$ need not vanish ([18] to [19]). If a theory of a continuous medium is to describe such materials, then it must admit a *couple stress*. The couple stress vector $\boldsymbol{c}^i$ is defined as the stress $\boldsymbol{\tau}^i$ of (4.1); that is

$$\boldsymbol{c}^i = \lim_{\Delta S \to 0} \frac{\Delta \boldsymbol{M}}{\Delta S}. \tag{4.3}$$

The physical components of the couple stress are defined as the components of stress by (4.2); that is, $\boldsymbol{c}^i = c^{ik} \hat{\boldsymbol{E}}_k$.

The distribution of force acting upon the surface can be better approximated by specifying moments of higher order, in addition to the first-order moments c^{ik}. It is also conceivable that the higher-order moments are needed in the limit if the continuous media is to represent certain uncommon materials. Theory which includes higher-order moments has been termed *multipolar continuum mechanics* ([18] to [19]).

Couple stresses and multipolar theories are not included in our subsequent developments. Nonetheless, the reader should be aware of such possibilities.

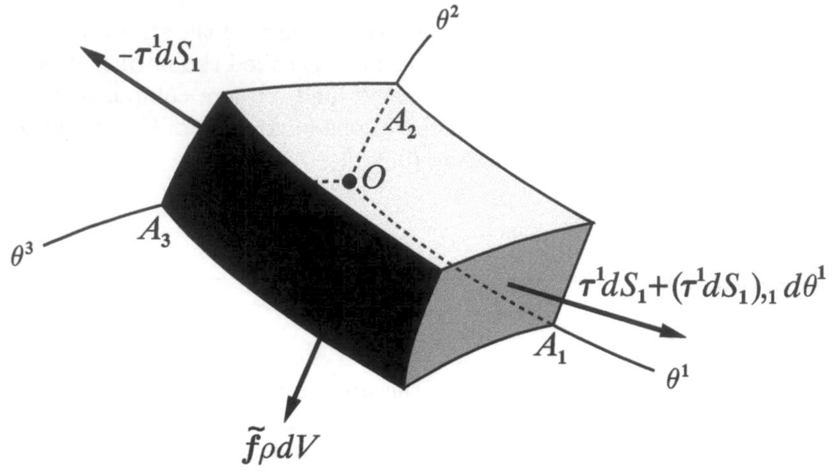

Figure 4.3 Actions upon an infinitesimal element

4.3 Actions Upon an Infinitesimal Element

Since the behavior of a solid depends upon the entire history of deformation, the solid is said to have memory. Because it has memory, it is essential that we follow the same particles, the lines, and the surfaces composed of the same particles as the body deforms. In other words, we adopt the "material" viewpoint. We identify the particles, lines, and surfaces with the coordinates of a reference state, usually the undeformed or unloaded state. The forms of these material lines and surfaces in the deformed body are not known a priori. Consequently, we study the stresses within the deformed body on the assumption that the coordinate lines and surfaces are deformed to arbitrary, but continuous and smooth, curves and surfaces.

An octahedral element of the body is depicted in Figure 4.3. The faces lie on the coordinate surfaces θ^i and neighboring surfaces $(\theta^i + d\theta^i)$. In the limit $(d\theta^i \to 0)$ the element approaches a parallelepiped (see Sections 2.5 and 3.5). Henceforth, we refer to the infinitesimal "parallelepiped."

The forces on the hidden faces of the parallelepiped are

$$- \boldsymbol{\tau}^i \, dS_{\underline{i}}, \tag{4.4}$$

and the forces on the visible faces must be augmented by a differential as follows:

$$\boldsymbol{\tau}^i \, dS_{\underline{i}} + (\boldsymbol{\tau}^{\underline{i}} \, dS_{\underline{i}})_{,i} \, d\theta^{\underline{i}} + O[(d\theta^i)^4]. \tag{4.5}$$

Because we are only concerned with relations in the limit, $\overline{\boldsymbol{OA}}_i \to 0$, and because the forces are continuous, we may consider the forces to act at the centers of the faces. To simplify the picture, only the forces on the θ^1 surfaces are shown in Figure 4.3.

In addition to tractions exerted on the faces of the element, a force $(\tilde{\boldsymbol{f}} \rho\, dV)$ is exerted by external agencies. Because we intend to pass to the limit, $\overline{\boldsymbol{OA}}_i \to 0$, this force is considered to be at the center of the parallelepiped. Here, ρ denotes the mass density of the deformed medium and $\tilde{\boldsymbol{f}}$ the force per unit mass. The force may be a gravitational force, electromagnetic force, or an inertial force, i.e., $\tilde{\boldsymbol{f}} = -\ddot{\boldsymbol{V}}$. Often the force $\tilde{\boldsymbol{f}}$ is called a *body force* to distinguish it from forces exerted via direct contact. If the theory for the medium includes couple stresses, then the element has couples $\boldsymbol{c}^i\, dS_{\underline{i}}$ acting upon the faces and differing from face to face, much as the force $\boldsymbol{\tau}^i\, dS_{\underline{i}}$ of Figure 4.3. Moreover, a medium which resists couple stresses can also resist a body couple $\rho\widetilde{\boldsymbol{C}}$ exerted by external agencies as the body force $\rho\tilde{\boldsymbol{f}}$ is applied to the element of Figure 4.3. Since such actions are uncommon, we do not include them here.

With the addition of the body force $(\tilde{\boldsymbol{f}} \rho\, dV)$ to the forces of (4.4) and (4.5), we obtain the sum (the net force on the body):

$$(\boldsymbol{\tau}^{\underline{i}}\, dS_{\underline{i}})_{,i}\, d\theta^i + \tilde{\boldsymbol{f}} \rho\, dV + O[(d\theta^i)^4]. \tag{4.5}$$

4.4 Equations of Motion

Recall the equations (3.14a–c) and (3.11b) for the deformed area dS_i and volume dV:

$$dS_i = \sqrt{G}\,\sqrt{G^{\underline{ii}}}\, d\theta^j\, d\theta^k \qquad (i \neq j \neq k), \tag{4.6}$$

$$dV = \sqrt{G}\, d\theta^1\, d\theta^2\, d\theta^3. \tag{4.7}$$

Also note that the vector $\overline{\boldsymbol{OA}}_i$ has the first-order approximation $(A_i \to O)$:

$$\overline{\boldsymbol{OA}}_i \doteq \boldsymbol{R}_{,\underline{i}}\, d\theta^i = \boldsymbol{G}_{\underline{i}}\, d\theta^i. \tag{4.8}$$

An equation of motion for the element requires that the sum (4.5) of all forces equals the mass $\times$ acceleration $(\rho\ddot{\boldsymbol{R}}\, dV)$:

$$(\boldsymbol{\tau}^{\underline{i}}\, dS_{\underline{i}})_{,i}\, d\theta^i + \tilde{\boldsymbol{f}} \rho\, dV + O[(d\theta^i)^4] = \rho\ddot{\boldsymbol{R}}\, dV. \tag{4.9}$$

When the expressions (4.6) and (4.7) are substituted into (4.9) and the result is divided by $dV = \sqrt{G}\, d\theta^1\, d\theta^2\, d\theta^3$, we obtain (in the limit $\theta^i \to 0$):

$$\frac{1}{\sqrt{G}} \left(\sqrt{G}\, \sqrt{G^{\underline{ii}}}\, \boldsymbol{\tau}^i \right)_{,i} + \rho \tilde{\boldsymbol{f}} = \rho \ddot{\boldsymbol{R}}. \tag{4.10}$$

Another equation of motion requires that the moment equals the moment-of-inertia $\times$ the angular acceleration. That moment-of-inertia has an order-of-magnitude $O[(d\theta^i)^5]$. The moment of stresses has a first approximation:

$$\overline{\boldsymbol{OA}}_i \times (\boldsymbol{\tau}^i\, dS_{\underline{i}}) \doteq \boldsymbol{G}_i \times (\boldsymbol{\tau}^i\, dS_{\underline{i}})\, d\theta^{\underline{i}}.$$

Note that

$$(d\theta^i\, dS_{\underline{i}}) = \sqrt{G}\, \sqrt{G^{\underline{ii}}}\, d\theta^1\, d\theta^2\, d\theta^3.$$

Therefore, irrespective of the motion, we obtain (in the limit $\theta^i \to 0$):

$$\boldsymbol{G}_i \times \left(\sqrt{G}\, \sqrt{G^{\underline{ii}}}\, \boldsymbol{\tau}^i \right) = \mathbf{o}. \tag{4.11}$$

In a primitive form, the two equations (4.10) and (4.11) are the equations of motion which govern the stresses $\boldsymbol{\tau}^i$; the components comprise six scalar equations.

As alternatives, we can employ a stress $\boldsymbol{\sigma}^i$ per unit of undeformed area and mass density ρ_0 per unit of undeformed volume. Then, instead of the equations (4.10) and (4.11), we obtain

$$\frac{1}{\sqrt{g}} \left(\sqrt{g}\, \sqrt{g^{\underline{ii}}}\, \boldsymbol{\sigma}^i \right)_{,i} + \rho_0 \tilde{\boldsymbol{f}} = \rho_0 \ddot{\boldsymbol{R}}, \tag{4.12}$$

$$\boldsymbol{G}_i \times \left(\sqrt{g}\, \sqrt{g^{\underline{ii}}}\, \boldsymbol{\sigma}^i \right) = \mathbf{o}. \tag{4.13}$$

The components of vectors $\boldsymbol{\tau}^i$ or $\boldsymbol{\sigma}^i$ can be given in terms of the base vectors $\boldsymbol{G}_i$ or $\acute{\boldsymbol{g}}_i$; both exclude the rigid motion. To express components in initial directions (i.e., $\boldsymbol{g}_i$) is not a useful option because the actions upon the material (and responses) must be related to the current (i.e., convected) orientations of the lines and surfaces. Accordingly, we have the alternatives:

$$\boldsymbol{\tau}^i = \sigma^{ij}\, \hat{\boldsymbol{E}}_j, \tag{4.14a}$$

$$\boldsymbol{\sigma}^i = \overset{*}{\sigma}{}^{ij}\, \hat{\boldsymbol{E}}_j = \bar{\sigma}^{ij}\, \hat{\boldsymbol{e}}'_j. \tag{4.14b, c}$$

As before

$$\hat{\boldsymbol{E}}_j = \frac{\boldsymbol{G}_j}{\sqrt{G_{\underline{jj}}}}, \qquad \hat{\boldsymbol{e}}'_j = \frac{\acute{\boldsymbol{g}}_j}{\sqrt{g_{\underline{jj}}}}.$$

It follows that

$$\sigma^{ij} = \sqrt{G_{\underline{jj}}}\, \boldsymbol{G}^j \cdot \boldsymbol{\tau}^i, \tag{4.15a}$$

$$\overset{*}{\sigma}^{ij} = \sqrt{G_{\underline{jj}}}\, \boldsymbol{G}^j \cdot \boldsymbol{\sigma}^i, \tag{4.15b}$$

$$\bar{\sigma}^{ij} = \sqrt{g_{\underline{jj}}}\, \acute{\boldsymbol{g}}^j \cdot \boldsymbol{\sigma}^i. \tag{4.15c}$$

We note that these components (σ^{ij}, $\overset{*}{\sigma}^{ij}$, $\bar{\sigma}^{ij}$) are *not*, in general, tensorial but physical components.

4.5 Tensorial and Invariant Forms of Stress and Internal Work

To identify the tensorial components of the alternative stresses and the *associated* components of strain, we turn to an examination of an all-important *invariant*, namely, the work expended by those stresses per unit of mass. We consider the increment (or rate) of work done upon the elemental parallelepiped ($dV = \sqrt{G}\, d\theta^1\, d\theta^2\, d\theta^3$) of Figure 4.3 under the displacement $\dot{\boldsymbol{R}}$: We anticipate the limit ($d\theta^i \to 0$) and consider the increment $\dot{\boldsymbol{R}}$ at the left (θ^1) and $\dot{\boldsymbol{R}} + \dot{\boldsymbol{R}}_{,1}\, d\theta^1$ at the right ($\theta^1 + d\theta^1$), etc.; we recall also the expression (4.6) for the area (e.g., $dS_1 = \sqrt{G}\sqrt{G^{11}}\, d\theta^2\, d\theta^3$), etc. In addition to the work $\dot{w}_s$ of the stresses upon the faces, we have the work $\dot{w}_f$ of a body force $\tilde{\boldsymbol{f}}$. Per unit volume:

$$\dot{w}_s + \dot{w}_f = \frac{1}{\sqrt{G}}\left(\sqrt{G}\,\sqrt{G^{\underline{ii}}}\,\boldsymbol{\tau}^i\right) \cdot \dot{\boldsymbol{R}}_{,i}$$

$$+ \frac{1}{\sqrt{G}}\left(\sqrt{G}\,\sqrt{G^{\underline{ii}}}\,\boldsymbol{\tau}^i\right)_{,i} \cdot \dot{\boldsymbol{R}} + \rho\tilde{\boldsymbol{f}} \cdot \dot{\boldsymbol{R}}. \tag{4.16}$$

Conservation of work and energy requires that the work done equals the change in kinetic energy $\dot{k} = \rho\ddot{\boldsymbol{R}} \cdot \dot{\boldsymbol{R}}$ (where ρ denotes mass density) and energy otherwise stored (e.g., as elastic potential) or dissipated (e.g., as heat). Let us signify the latter, collectively, as $\dot{u}$. Then

$$\dot{w}_s + \dot{w}_f = \dot{u} + \dot{k}. \tag{4.17a}$$

Upon rearranging terms, we obtain

$$\frac{1}{\sqrt{G}}\left[\left(\sqrt{G}\,\sqrt{G^{\underline{ii}}}\,\boldsymbol{\tau}^i\right)_{,i} + \rho\tilde{\boldsymbol{f}} - \rho\ddot{\boldsymbol{R}}\right]\cdot\dot{\boldsymbol{R}} + \left(\sqrt{G^{\underline{ii}}}\,\boldsymbol{\tau}^i\right)\cdot\dot{\boldsymbol{R}}_{,i} = \dot{u}. \tag{4.17b}$$

The first term on the left corresponds solely to work done upon the rigid translation $\dot{\boldsymbol{R}}$; stated otherwise, energy can be neither stored (elastically) or dissipated (inelastically) by virtue of such nondeformational motion. It is no more or less than the *principle of virtual work* to assert the vanishing of the bracketed term; this is the equation of motion expressed in the form (4.10). In terms of stress $\boldsymbol{\sigma}^i$, force $\tilde{\boldsymbol{f}}$, and mass density ρ_0, per units of *initial* area and *initial* volume, the equation takes the form (4.12).

Let us now return to the remaining term of (4.17b), i.e., $\dot{u}$, or it's counterpart $\dot{u}_0$ (per unit initial volume):

$$\dot{u} = \sqrt{G^{\underline{ii}}}\,\boldsymbol{\tau}^i\cdot\dot{\boldsymbol{R}}_{,i} \equiv \sqrt{G^{\underline{ii}}}\,\boldsymbol{\tau}^i\cdot\dot{\boldsymbol{G}}_i, \tag{4.18a, b}$$

$$\dot{u}_0 = \sqrt{g^{\underline{ii}}}\,\boldsymbol{\sigma}^i\cdot\dot{\boldsymbol{R}}_{,i} \equiv \sqrt{g^{\underline{ii}}}\,\boldsymbol{\sigma}^i\cdot\dot{\boldsymbol{G}}_i. \tag{4.18c, d}$$

Since the vector $\dot{\boldsymbol{G}}_i$ transforms as the component of a covariant tensor, the other factor in the invariant must transform as the component of a contravariant tensor. Therefore, by employing the definitions

$$\boldsymbol{t}^i \equiv \sqrt{G^{\underline{ii}}}\,\boldsymbol{\tau}^i \equiv \tau^{ij}\boldsymbol{G}_j, \tag{4.19a, b}$$

$$\boldsymbol{s}^i \equiv \sqrt{g^{\underline{ii}}}\boldsymbol{\sigma}^i \equiv s^{ij}\boldsymbol{G}_j \equiv t^{ij}\acute{\boldsymbol{g}}_j, \tag{4.19c–e}$$

and in view of (4.14a–c) and (4.15a–c), we obtain the tensorial components of stress

$$\tau^{ij} \equiv \boldsymbol{G}^j\cdot\boldsymbol{\tau}^i\sqrt{G^{\underline{ii}}} = \sqrt{G^{\underline{ii}}/G_{\underline{jj}}}\,\sigma^{ij}, \tag{4.20a, b}$$

$$s^{ij} \equiv \boldsymbol{G}^j\cdot\boldsymbol{\sigma}^i\sqrt{g^{\underline{ii}}} = \sqrt{g^{\underline{ii}}/G_{\underline{jj}}}\,\overset{*}{\sigma}{}^{ij}, \tag{4.20c, d}$$

$$t^{ij} \equiv \acute{\boldsymbol{g}}^j\cdot\boldsymbol{\sigma}^i\sqrt{g^{\underline{ii}}} = \sqrt{g^{\underline{ii}}/g_{\underline{jj}}}\,\bar{\sigma}^{ij}. \tag{4.20e, f}$$

Note: $\boldsymbol{\tau}^i$ and σ^{ij} are force per unit deformed area; $\boldsymbol{\sigma}^i$, $\overset{*}{\sigma}{}^{ij}$ and $\bar{\sigma}^{ij}$ are force per unit *un*deformed area; $\boldsymbol{G}_i$ is tangent to the convected coordinate lines while $\acute{\boldsymbol{g}}_i$ is rigidly rotated as the principal lines of strain. In subsequent formulations of work and energy, we can appreciate advantages of the latter components. Most noteworthy are the forms of the invariants; from (4.18a–d) and (4.19a–e) we have

$$\dot{u} = \tau^{ij}\boldsymbol{G}_j \cdot \dot{\boldsymbol{G}}_i, \tag{4.21a}$$

$$\dot{u}_0 = s^{ij}\boldsymbol{G}_j \cdot \dot{\boldsymbol{G}}_i = t^{ij}\acute{\boldsymbol{g}}_j \cdot \dot{\boldsymbol{G}}_i. \tag{4.21b, c}$$

We recall the expressions of the strain components, the Cauchy-Green strain γ_{ij} and the engineering strain h_{ij}, equations (3.31) and (3.96); we note also that both are symmetric tensors. For our immediate purpose, we recall the alternative equations (3.139) and (3.147) for the increment,

$$\dot{\boldsymbol{G}}_i = (\dot{\gamma}_{ij} + \dot{\omega}_{ji})\,\boldsymbol{G}^j = (\dot{h}_{ij} + C_i^k\,\dot{\bar{\omega}}^l e_{lkj})\,\acute{\boldsymbol{g}}^j. \tag{4.22a, b}$$

The corresponding forms of (4.21a–c) follow:

$$\dot{u} = \tau^{ij}\dot{\gamma}_{ij} + \tau^{ij}\dot{\omega}_{ji}, \tag{4.23a}$$

$$\dot{u}_0 = s^{ij}\dot{\gamma}_{ij} + s^{ij}\dot{\omega}_{ji}, \tag{4.23b}$$

$$= t^{ij}\dot{h}_{ij} + t^{ij}C_i^k\,\dot{\bar{\omega}}^l e_{lkj}. \tag{4.23c}$$

Again, the concept of virtual work precludes work upon the rigid-body rotation $\dot{\boldsymbol{\omega}}$. If follows that

$$\tau^{ij} = \tau^{ji}, \qquad s^{ij} = s^{ji}, \qquad t^{ij}C_i^k = t^{ik}C_i^j. \tag{4.24a–c}$$

These are the conditions (4.11) and (4.13) expressed in terms of the tensorial components of the stresses. By employing (3.79e), the last equation (4.24c) assumes the form

$$t^{ij}(\delta_i^k + h_i^k) = t^{ik}(\delta_i^j + h_i^j). \tag{4.24d}$$

The literature abounds with various names for the stress tensors, τ^{ij}, s^{ij}, and t^{ij}. It appears that the history can be traced to A. L. Cauchy [21], [22] (1823 and 1829, respectively) with subsequent works by G. Piola [23]

(1833), G. Kirchhoff [24] (1852), and E. Trefftz [25] (1933). The components s^{ij} of the stress tensor have been variously termed the "second" Piola-Kirchhoff components and the Kirchhoff-Trefftz components. The work of E. Trefftz [25] gives a graphic description. The components t^{ij} appear in the work of G. Jaumann [26] (1918). It is *most important* that we understand their relation to the *physical* components, their *mathematical* properties (tensorial and symmetrical properties) and their *associations* with the appropriate strains as embodied in the work rates (4.23a–c).

We can infer from the invariant forms (4.23a–c) that these components transform as contravariant components, i.e., if θ^i and $\bar{\theta}^i$ are alternative regular coordinates, then

$$\bar{\tau}^{ij}(\bar{\theta}^n) = \frac{\partial\bar{\theta}^i}{\partial\theta^k}\frac{\partial\bar{\theta}^j}{\partial\theta^l}\,\tau^{kl}(\theta^n), \tag{4.25a}$$

$$\bar{s}^{ij} = \frac{\partial\bar{\theta}^i}{\partial\theta^k}\frac{\partial\bar{\theta}^j}{\partial\theta^l}\,s^{kl}, \qquad \bar{t}^{ij} = \frac{\partial\bar{\theta}^i}{\partial\theta^k}\frac{\partial\bar{\theta}^j}{\partial\theta^l}\,t^{kl}. \tag{4.25b, c}$$

Characteristics of the alternative stresses are evident in (4.21a–c). Each of the following vectors transforms as a contravariant component:

$$\boldsymbol{t}^i \equiv \tau^{ij}\boldsymbol{G}_j, \qquad \boldsymbol{s}^i \equiv s^{ij}\boldsymbol{G}_j \equiv t^{ij}\acute{\boldsymbol{g}}_j, \tag{4.26a–c}$$

Epilogue

We recognize a school that advocates the "one theory fits all" approach to the mechanics of continua (any fluid or solid). Invariably, followers encounter difficulties in the treatment of constitutive laws. As previously noted, these media possess profoundly different properties; a perfect fluid has no memory whereas a perfect solid has total recall. Consequently, a description of the solid which does not adhere to a Lagrangean view poses a great mathematical challenge: It leads to subsidiary conditions to insure so-called "material objectivity"; such conditions serve to insure the physical consistency of the associated stress-strain relations. Our approach avoids such mathematical agony; our association of the consistent components of stress and strain exemplify the advantage.

Suppose that we attempt to express the active stress $\boldsymbol{s}^i$ in terms of the initial system $\boldsymbol{g}_i$; instead of $\boldsymbol{s}^i = t^{ij}\acute{\boldsymbol{g}}_j$, let us employ $\boldsymbol{s}^i = {'t}^{ij}\boldsymbol{g}_j$. Then the incremental work must assume the form:

$${'t}^{ij}\boldsymbol{g}_j \cdot \dot{\boldsymbol{R}}_{,i} = {'t}^{ij}\boldsymbol{g}_j \cdot \dot{\boldsymbol{G}}_i.$$

Note that $\dot{\boldsymbol{R}}_{,i} = \dot{\boldsymbol{G}}_{,i} = \dot{\boldsymbol{V}}_{,i}$, the rate of the displacement gradient $\boldsymbol{V}_{,i}$.

Indeed, some of the aforementioned advocates have attempted to develop constitutive "laws" in terms of the displacement gradients. But that is *not* a strain; the gradient $\boldsymbol{V}_{,i}$ includes rigid rotation which *must* be purged from any constitutive "law." Hence the agonies of "material objectivity"! Alternatively, one might perform the rotation [see equation (3.91b)]:

$$\boldsymbol{g}_j = r^n_{\boldsymbol{\cdot} j} \acute{\boldsymbol{g}}_n .$$

Then the incremental work assumes the form

$${}'t^{ij} \boldsymbol{g}_j \cdot \dot{\boldsymbol{G}}_i = {}'t^{ij} r^n_{\boldsymbol{\cdot} j}\, \acute{\boldsymbol{g}}_n \cdot \dot{\boldsymbol{G}}_i = ({}'t^{ij} r^n_{\boldsymbol{\cdot} j}) \dot{h}_{in} .$$

Now, the latter expresses that rate in terms of a bona fide strain ("materially objective") but demonstrates that ${}'t^{ij}$ is *not* a bona fide stress. To obtain a stress t^{ij} which is "materially objective," one must introduce the rigid rotation, viz.,

$$t^{ij} = ({}'t^{ij} r^n_{\boldsymbol{\cdot} j}).$$

4.6 Transformation of Stress—Physical Basis

Alternatively, one can appeal to the physical meanings and the conditions of motion to deduce the linear transformations (4.25a–c). We suppose that one requires the stress vector $\boldsymbol{\tau}^i$ (or $\boldsymbol{\sigma}^i$), or its components, on an arbitrary surface at a point O^*. Such surface is depicted in Figure 4.4 where it passes *near* O^* and slices through the coordinate surfaces along lines $A_2^* A_3^*$, $A_3^* A_1^*$, and $A_1^* A_2^*$ to form the tetrahedron. Let $d\bar{S}_1$ denote the area of the surface $(A_1^* A_2^* A_3^*)$ and dS_i the area of the face on the coordinate surface θ^i.

As before, we denote the unit normals to the coordinate surfaces by $\hat{\boldsymbol{E}}^i$; similarly, we denote the normal to surface $d\bar{S}_1$ by $\hat{\boldsymbol{A}}^1$. We anticipate passing to the limit; our surface approaching the point O^* and the vectors in question approaching values at that site. The facial areas are then related as follows:

$$\hat{\boldsymbol{A}}^1\, d\bar{S}_1 - \hat{\boldsymbol{E}}^i\, dS_i = \mathbf{o}. \tag{4.27}$$

We suppose that surface $d\bar{S}_1$ is a coordinate surface, $\bar{\theta}^1$, of an alternative system $\bar{\theta}^i$ so that

$$\hat{\boldsymbol{A}}^i = \overline{\boldsymbol{G}}^i \Big/ \sqrt{\overline{G}^{\underline{ii}}}. \tag{4.28}$$

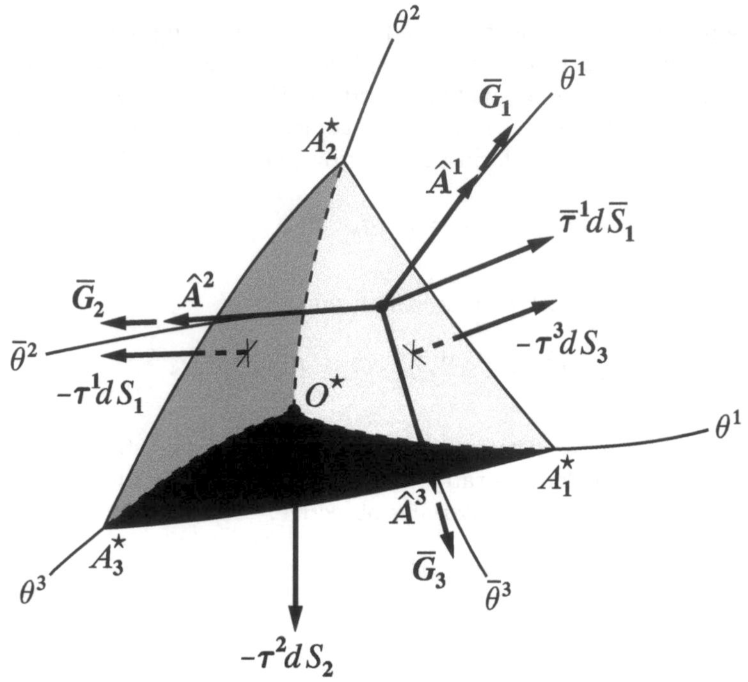

Figure 4.4 Transformation of stress

Then, by equation (4.27),

$$\frac{dS_i}{d\bar{S}_1} = \sqrt{\frac{G^{\underline{ii}}}{\overline{G}^{11}}}\, \overline{\boldsymbol{G}}^1 \cdot \boldsymbol{G}_i. \tag{4.29}$$

The force upon the $\bar{\theta}^1$ face is $\bar{\boldsymbol{\tau}}^1\, d\bar{S}_1$ and that on the θ^i face $\boldsymbol{\tau}^i\, dS_{\underline{i}}$. If this element is in motion, the sum of all external forces is $dm\, \ddot{\boldsymbol{R}}$, where dm denotes the mass; $dm = \rho\, dV$. The equation of motion approaches

$$\bar{\boldsymbol{\tau}}^1\, d\bar{S}_1 - \boldsymbol{\tau}^i\, dS_i = \rho\, dV \ddot{\boldsymbol{R}}.$$

Since $dV/d\bar{S}_1 \to 0$ *in the limit,*

$$\bar{\boldsymbol{\tau}}^1 = \boldsymbol{\tau}^i \frac{dS_i}{d\bar{S}_1}. \tag{4.30a}$$

By means of (4.20a), the latter is expressed in terms of the components $\bar{\tau}^{1j}$

and τ^{ij}:

$$\bar{\tau}^{1j} = \overline{\boldsymbol{G}}^j \cdot \boldsymbol{G}_m \sqrt{\frac{\overline{G}^{11}}{G^{\underline{ii}}}} \frac{dS_i}{d\bar{S}_1} \tau^{im}. \tag{4.30b}$$

Then, according to (4.29),

$$\bar{\tau}^{1j} = (\overline{\boldsymbol{G}}^j \cdot \boldsymbol{G}_m)(\overline{\boldsymbol{G}}^1 \cdot \boldsymbol{G}_i)\tau^{im}. \tag{4.30c}$$

By the chain-rule for partial differentiation

$$\boldsymbol{G}_m = \overline{\boldsymbol{G}}_k \frac{\partial \bar{\theta}^k}{\partial \theta^m}.$$

The final version of the *equation of motion* follows:

$$\bar{\tau}^{1j} = \frac{\partial \bar{\theta}^1}{\partial \theta^i} \frac{\partial \bar{\theta}^j}{\partial \theta^m} \tau^{im}. \tag{4.30d}$$

However, $\bar{\theta}^1$ could as well be any coordinate $\bar{\theta}^i$; therefore

$$\bar{\tau}^{ij} = \frac{\partial \bar{\theta}^i}{\partial \theta^l} \frac{\partial \bar{\theta}^j}{\partial \theta^m} \tau^{lm}. \tag{4.30e}$$

Had we referred to the stress $\boldsymbol{\sigma}^i$ (per unit of original area), then we would have obtained either of the forms

$$\bar{s}^{ij} = \frac{\partial \bar{\theta}^i}{\partial \theta^l} \frac{\partial \bar{\theta}^j}{\partial \theta^m} s^{lm}, \qquad \bar{t}^{ij} = \frac{\partial \bar{\theta}^i}{\partial \theta^l} \frac{\partial \bar{\theta}^j}{\partial \theta^m} t^{lm}. \tag{4.30f, g}$$

Again we note that the origin is the equation of *motion*; hence the transformation applies in dynamic or static circumstances.

Now, suppose that the surface $(A_1^* A_2^* A_3^*)$ has the unit normal

$$\hat{\boldsymbol{N}} \equiv \hat{\boldsymbol{A}}^1 \equiv n^i \boldsymbol{G}_i \equiv n_i \boldsymbol{G}^i. \tag{4.31a–c}$$

Also, suppose that the stress applied at a point of the surface is

$$\boldsymbol{T} \equiv \bar{\boldsymbol{\tau}}^1. \tag{4.32}$$

According to (4.30a), (4.14a), (4.28), (4.29), (4.31c), and (4.20b), it follows that

$$\boldsymbol{T} = \sigma^{ij} \sqrt{\frac{G^{\underline{ii}}}{G_{\underline{jj}}}}\, n_i \boldsymbol{G}_j = \tau^{ij} n_i \boldsymbol{G}_j. \qquad (4.33\text{a, b})$$

4.7 Properties of a Stressed State

Several practical and mathematical questions arise about any stressed state, namely: What is the maximum (or minimum) value of the normal stress and what is the orientation of the surface of that stress? What is the maximum value of the shear stress and the orientation of those surfaces? Additionally, we may require invariants which characterize the state of stress. These various properties are extremely useful in the treatment of materials: Fracture of brittle materials is often associated with the maximum normal stress. Failure of certain materials is related to the maximum shear and to the orientation of those surfaces (e.g., along the grain of wood or the oriented filaments of a composite). Invariants play an important role in the treatment of isotropic media.

Mathematically, the state of stress is defined by a second-order tensor, as the state of strain. Note, however, that the analogy is mathematical, *not* physical. Our use of covariant components of strain (γ_{ij}) and contravariant components of stress (s^{ij}) can be attributed to their close relationships with their physical counterparts [see equations (3.35a, b) and (4.20c, d)] and also to the invariant $s^{ij}\dot{\gamma}_{ij}$ (rate of work). The components s^{ij} are often the more useful; however, our general remarks apply as well to the alternative components, τ^{ij} and t^{ij}. Mathematically, the determinations of the maximum and minimum, normal and shear stresses are accomplished as their kinematical counterparts (extensional and shear strains). Hence, we need only restate the results (see Sections 3.8 to 3.11):

1. At each point of a continuous medium there exist three mutually perpendicular directions in which the shear stresses vanish. These are the principal directions of stress. Note that the principal directions of strain are not necessarily the same. The normal stresses in these directions are the principal stresses ($\tilde{s}^1, \tilde{s}^2, \tilde{s}^3$).

2. One of the principal stresses is the maximum normal stress at that point and one is the minimum.

3. The maximum shear stress occurs on the surfaces that bisect the directions of maximum and minimum normal stress (see Figure 4.5).

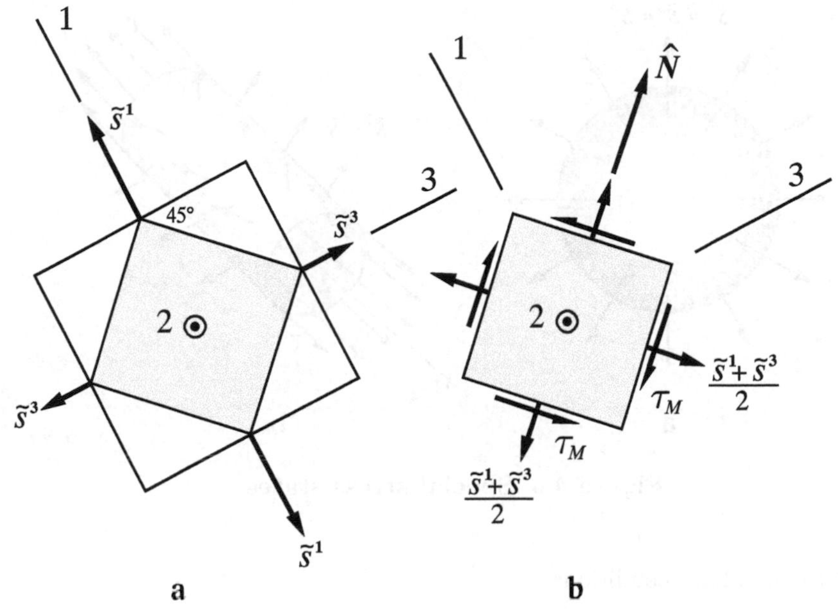

Figure 4.5 Principal stresses/maximum shear stress

The magnitude of that stress is

$$\tau_M = (\tilde{s}^1 - \tilde{s}^3)/2, \tag{4.34}$$

wherein $\tilde{s}^1$ and $\tilde{s}^3$ signify the maximum and minimum principal stresses.

The principal values, directions, and maximum shear are determined as the counterparts of a strained state. First, one obtains the principal values, the real roots of a cubic equation [like (3.52b)]:

$$s^3 - J_1 s^2 + J_2 s - J_3 = 0, \tag{4.35}$$

where the coefficients are three invariants:

$$J_1 = s^i_i, \qquad J_2 = \tfrac{1}{2}(s^i_i s^j_j - s^i_j s^j_i), \qquad J_3 = |s^i_j|. \tag{4.36a–c}$$

The directions are defined by a unit vector $\hat{\boldsymbol{A}} = A^i \boldsymbol{G}_i$, which is determined by the linear equations

$$(s^j_i - s\,\delta^j_i)A_j = 0, \tag{4.37}$$

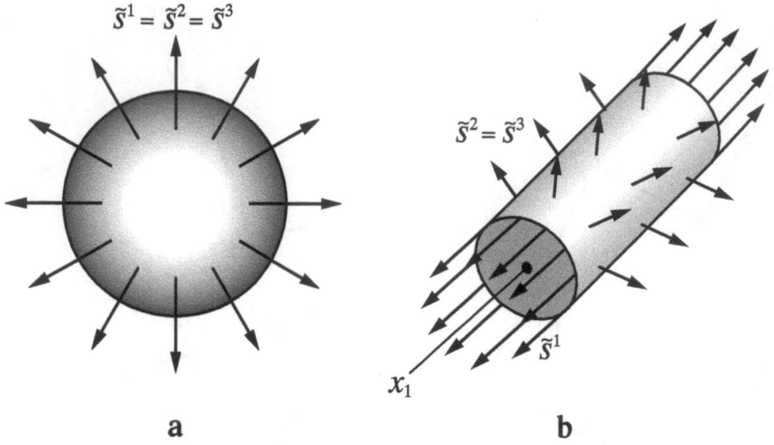

Figure 4.6 Special stress states

with the auxiliary condition:

$$A^i A_i = 1. \tag{4.38}$$

As always, we want a clear physical/geometrical description as provided by Figure 4.5. Here the directions 1 and 3 are the directions of maximum and minimum normal stress (principal values); $\hat{\boldsymbol{N}}$ is normal to a surface of maximum shear stress. It is a matter of interest that the normal stress s_M on the surface of maximum shear is

$$s_M = (\tilde{s}^1 + \tilde{s}^3)/2. \tag{4.39}$$

In special circumstances:

1. $\tilde{s}^1 = \tilde{s}^2 = \tilde{s}^3$; *No* shear occurs on any surface. This is a so-called "hydrostatic" state (Figure 4.6a).
2. $\tilde{s}^1 \neq \tilde{s}^2 = \tilde{s}^3$; *Only* normal stresses act upon a cylindrical surface with axis x_1 as shown in Figure 4.6b.
3. $\tilde{s}^1 = -\tilde{s}^2$, $\tilde{s}^3 = 0$; *No* normal stress occurs on the surfaces of maximum shear. This is a state of simple or "pure" shear (Figure 4.7a, b).

As previously noted, components of stress are inherently related to surfaces and directions in the *deformed* medium. It follows that the principal surfaces are orthogonal in the deformed (stressed) state. Unlike, the directions

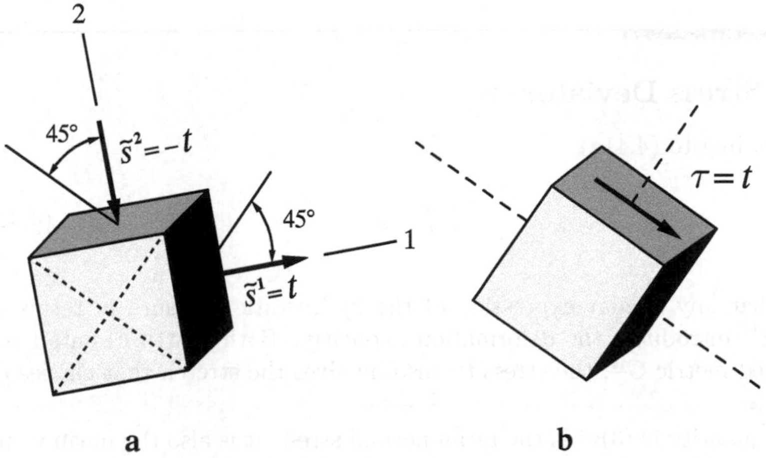

Figure 4.7 Simple or "pure" shear state

of principal strain, the directions (surfaces) of principal stress need *not* be orthogonal in the initial state.

4.8 Hydrostatic Stress

If the stress on every surface at a point is the same, then it is a normal stress

$$\boldsymbol{\tau}^i = p\,\hat{\boldsymbol{E}}^i, \qquad \boldsymbol{\sigma}^i = p_0\hat{\boldsymbol{E}}^i, \tag{4.40a, b}$$

or, by (4.20a, c, e),

$$\tau^{ij} = p\,G^{ij}, \tag{4.41a}$$

$$s^{ij} = p_0\sqrt{\frac{g^{\underline{ii}}}{G^{\underline{ii}}}}\,G^{ij}, \tag{4.41b}$$

$$t^{ij} = p_0\sqrt{\frac{g^{\underline{ii}}}{G^{\underline{ii}}}}\,\acute{\boldsymbol{g}}^j \cdot \boldsymbol{G}^i. \tag{4.41c}$$

This state of stress is known as *hydrostatic*, as it is the state that exists in a liquid (hydro) at rest (static). It is the only state wherein the same stress exists on every surface. This is depicted in Figure 4.6a.

4.9 Stress Deviator

According to (4.41a)

$$\tfrac{1}{3}\tau^i_i = p. \tag{4.42}$$

Note that any similar expression of the hydrostatic pressure in terms of s^{ij} or t^{ij} introduces the deformation explicitly: Both (4.41b, c) entail the deformed metric $G^{\underline{ii}}$; the stress t^{ij} also involves the stretch that carries $\acute{\boldsymbol{g}}^i$ to $\boldsymbol{G}^i$.

The quantity $(1/3)\tau^i_i$ is the mean normal stress; it is also the mean value of the physical components $(1/3)\sigma^i_i$ on the deformed surfaces. A state of stress wherein $\tau^i_i = 0$ is called *deviatoric.*

For any state of stress, we can *define*, as the *deviatoric* part, or the *stress deviator*:

$$'\tau^{ij} = \tau^{ij} - \tfrac{1}{3}\tau^k_k G^{ij}, \qquad '\tau^i_j = \tau^i_j - \tfrac{1}{3}\tau^k_k \delta^i_j. \tag{4.43a, b}$$

The deviator plays a role in describing the behavior of a material that is insensitive to a hydrostatic state, i.e., pressure.

4.10 Alternative Forms of the Equations of Motion

With the results (4.19a–e), the equations (4.10) or (4.12) can be expressed in terms of the vectors ($\boldsymbol{t}^i$, $\boldsymbol{s}^i$) or the tensorial components (τ^{ij}, s^{ij}, t^{ij}):

$$\frac{1}{\sqrt{G}}\left(\sqrt{G}\,\boldsymbol{t}^i\right)_{,i} + \rho\tilde{\boldsymbol{f}} = \rho\ddot{\boldsymbol{R}}, \tag{4.44a}$$

$$\frac{1}{\sqrt{g}}\left(\sqrt{g}\,\boldsymbol{s}^i\right)_{,i} + \rho_0\tilde{\boldsymbol{f}} = \rho_0\ddot{\boldsymbol{R}}. \tag{4.44b}$$

or

$$\frac{1}{\sqrt{G}}\left(\sqrt{G}\,\tau^{ij}\boldsymbol{G}_j\right)_{,i} + \rho\tilde{\boldsymbol{f}} = \rho\ddot{\boldsymbol{R}}, \tag{4.45a}$$

$$\frac{1}{\sqrt{g}}\left(\sqrt{g}\,s^{ij}\boldsymbol{G}_j\right)_{,i} + \rho_0\tilde{\boldsymbol{f}} = \rho_0\ddot{\boldsymbol{R}}, \tag{4.45b}$$

$$\frac{1}{\sqrt{g}}\left(\sqrt{g}\,t^{ij}\acute{\boldsymbol{g}}_j\right)_{,i} + \rho_0\tilde{\boldsymbol{f}} = \rho_0\ddot{\boldsymbol{R}}. \tag{4.45c}$$

4.11 Significance of Small Strain

If strain components are small $\epsilon_{ij} \ll 1$, then $g_{ij} \doteq G_{ij}$, etc. Then, for most practical purposes:

$$\boldsymbol{\tau}^i \doteq \boldsymbol{\sigma}^i, \qquad \sigma^{ij} \doteq \overset{*}{\sigma}{}^{ij}, \qquad \tau^{ij} \doteq s^{ij}.$$

These approximations are quite evident from the definitions [see also equations (4.15a, b) and (4.20a–d)]. Recall too that the differences between vectors $\acute{\boldsymbol{g}}_i$ and $\boldsymbol{G}_i$ are merely the engineering strains; both undergo the rotation of the principal lines—however large! Therefore,

$$\sigma^{ij} \doteq \overset{*}{\sigma}{}^{ij} \doteq \bar{\sigma}^{ij}, \qquad \tau^{ij} \doteq s^{ij} \doteq t^{ij}. \tag{4.46a, b}$$

This also means that

$$p \doteq p_0 \doteq \tfrac{1}{3}\tau^i_i \doteq \tfrac{1}{3}s^i_i \doteq \tfrac{1}{3}t^i_i.$$

Then one can speak practically of the *deviatoric* components:

$$'s^{ij} \equiv s^{ij} - \tfrac{1}{3}s^k_k g^{ij}, \qquad 't^{ij} \equiv t^{ij} - \tfrac{1}{3}t^k_k g^{ij}. \tag{4.47a, b}$$

Again, the rotations may still be large. As an example, we cite the very large rotations of the blade in a band saw, which experiences rotations of 180° although strains remain small everywhere .

4.12 Approximation of Moderate Rotations

The notion of a moderate rotation $\boldsymbol{\omega}$ and its role in structural problems (e.g., buckling of thin bodies) are introduced in Section 3.22. As the consequence of very small strains:

$$\acute{\boldsymbol{g}}_i \doteq \boldsymbol{G}_i, \qquad \boldsymbol{t}^i \doteq \boldsymbol{s}^i, \qquad \tau^{ij} \doteq s^{ij} \doteq t^{ij}.$$

Also, according to (3.164), moderate rotations admit the approximation:

$$\acute{\boldsymbol{g}}_i \doteq \boldsymbol{G}_i \doteq \boldsymbol{g}_i + \Omega_{ki}\boldsymbol{g}^k.$$

Then, from any one of the equations (4.45a–c), we obtain the approximate equation of motion:

$$\frac{1}{\sqrt{g}}\left(\sqrt{g}\, s^{ij}\boldsymbol{g}_j + \sqrt{g}\, s^{ij}\Omega_{kj}\boldsymbol{g}^k\right)_{,i} + \rho_0\tilde{\boldsymbol{f}} = \rho_0\ddot{\boldsymbol{R}}. \tag{4.48}$$

Inherent in this approximation is the rotation (components Ω_{ij}) and the derivative $\Omega_{kj,i}$. The structural engineer may recognize the latter as the change-of-curvature which plays a major role in the equations of a buckled configuration (see the example presented in Section 4.14).

4.13 Approximations of Small Strains and Small Rotations; Linear Theory

When strains *and* rotations are small compared to unity ($\epsilon \ll 1, \Omega \ll 1$), then (4.48) assumes the form:

$$\frac{1}{\sqrt{g}}\left(\sqrt{g}\, s^{ij}\boldsymbol{g}_j\right)_{,i} + \rho_0\tilde{\boldsymbol{f}} = \rho_0\ddot{\boldsymbol{R}}. \tag{4.49}$$

This is linear in the dependent variables (s^{ij} and $\boldsymbol{R}$). It is the form most frequently employed in mechanical and structural problems involving small elastic deformations and in some circumstances of *small* inelastic deformations.

4.14 Example: Buckling of a Beam

Moderate rotations play an important role in the buckling of thin structural elements. This can be illustrated by the following elementary example

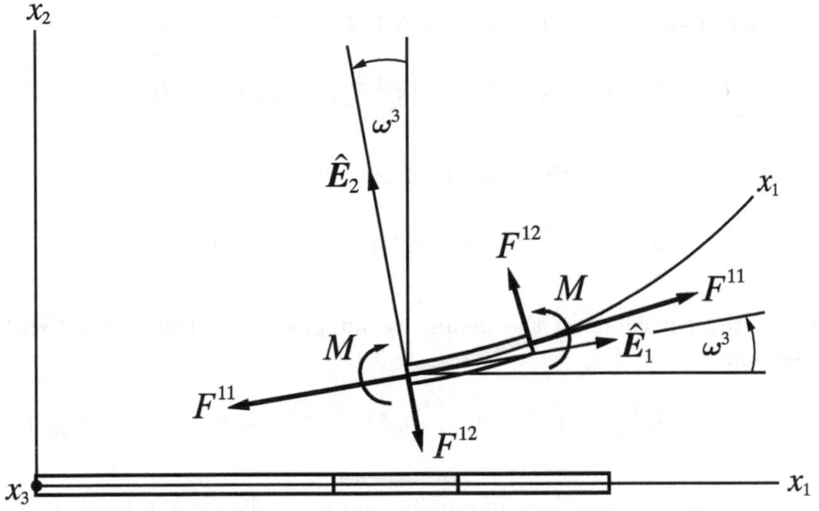

Figure 4.8 Illustrative example: Buckling of a beam

of a beam.

A slender beam with axis $\theta^1 = x_1$ is symmetrical about the $x_1 - x_2$ plane of rectangular coordinates (x_1, x_2, x_3) as illustrated in Figure 4.8. Our beam bends only in the plane of symmetry according to the Bernoulli hypothesis, i.e., axis and cross-section experience the same moderate rotation. Tangent vectors $\boldsymbol{g}_1 = \hat{\boldsymbol{\imath}}_1$ and $\boldsymbol{g}_2 = \hat{\boldsymbol{\imath}}_2$ are carried to

$$\hat{\boldsymbol{E}}_1(x_1) \doteq \hat{\boldsymbol{\imath}}_1 + \boldsymbol{\omega}(x_1) \times \hat{\boldsymbol{\imath}}_1, \qquad \hat{\boldsymbol{E}}_2(x_1) \doteq \hat{\boldsymbol{\imath}}_2 + \boldsymbol{\omega}(x_1) \times \hat{\boldsymbol{\imath}}_2.$$

If κ denotes the curvature of the deformed beam, then

$$\hat{\boldsymbol{E}}_{1,1} \doteq \kappa \hat{\boldsymbol{E}}_2 \doteq \omega^3{}_{,1} \hat{\boldsymbol{\imath}}_2, \qquad \hat{\boldsymbol{E}}_{2,1} \doteq -\kappa \hat{\boldsymbol{E}}_1 \doteq -\omega^3{}_{,1} \hat{\boldsymbol{\imath}}_1.$$

By the usual assumptions, the only nonzero components of stress are s^{11} and s^{12}, and these are functions only of coordinates x_1 and x_2 within a cross-section; moreover,

$$s^{11} \doteq \frac{F^{11}}{A} - \frac{M x_2}{I_2}, \qquad F^{12} \equiv \iint_A s^{12}\, dA, \qquad I_2 = \iint_A (x_2)^2\, dA.$$

Additionally, we place the axis at the centroid and assume that the lateral surfaces are stress-free.

Recasting (4.45b) and setting $g \doteq G \doteq 1$, $\boldsymbol{G}_i \doteq \hat{\boldsymbol{E}}_i$, we have

$$\boldsymbol{s}^i{}_{,i} = (s^{11}\hat{\boldsymbol{E}}_1 + s^{12}\hat{\boldsymbol{E}}_2)_{,1} + (s^{21}\hat{\boldsymbol{E}}_1)_{,2} + \rho_0\tilde{\boldsymbol{f}} = \rho_0\ddot{\boldsymbol{R}}.$$

If no body forces act and the beam is at rest, then

$$(s^{11}{}_{,1} - \kappa s^{12})\hat{\boldsymbol{E}}_1 + (\kappa s^{11} + s^{12}{}_{,1})\hat{\boldsymbol{E}}_2 = 0.$$

To obtain the equations of the beam, we integrate over the *symmetrical* cross-section:

$$F^{11}{}_{,1} - \kappa F^{12} = 0, \qquad \kappa F^{11} + F^{12}{}_{,1} = 0. \tag{4.50a, b}$$

These are consistent equations of equilibrium for a slice of the beam with *no* lateral loading and *small* strains.

In addition to the equilibrium of force, we require that the moment vanish on a slice ($dx_i \to 0$):

$$\left(M_{,1} + F^{12} + \iint_A x_2\,\kappa s^{12}\,dA\right)\hat{\boldsymbol{E}}_3 = 0, \qquad M \equiv -\iint_A x_2 s^{11}\,dA.$$

Since the beam is thin ($x_2\,\kappa \ll 1$), we neglect the final term and accept the approximation:

$$M_{,1} + F^{12} \doteq 0. \tag{4.51}$$

We recognize that the equations (4.50a, b) are the conditions for equilibrium of force components in the absence of lateral loading, tangential (x_1) and lateral (x_2); whereas the equation (4.51) is the condition for equilibrium of moment. We see the role of the moderate rotation and consequent curvature $\kappa \doteq \omega^3{}_{,1}$.

Typically, one presumes a thrust $F^{11} = -P$ (constant) upon a straight column; i.e.,

$$\kappa = \omega^3 = F^{12} = M = 0.$$

A slight perturbation causes changes

$$F^{11} = -P + \epsilon\overline{P}(x_1), \qquad \omega^3 = \epsilon\bar{\omega}(x_1), \qquad M = \epsilon\overline{M}(x_1), \qquad F^{12} = \epsilon\overline{F}.$$

Substituting these variables into the equilibrium equations, we obtain

$$\epsilon\overline{P}_{,1} - \epsilon^2\bar{\omega}_{,1}\overline{F} = 0,$$

$$-\epsilon\,\bar{\omega}_{,1}P + \epsilon^2\bar{\omega}_{,1}\overline{P} + \epsilon\overline{F}_{,1} = 0,$$

$$\epsilon\overline{M}_{,1} + \epsilon\overline{F} = 0.$$

Terms $O(\epsilon^2)$ are neglected to obtain

$$\overline{P}_{,1} \doteq 0, \qquad \overline{F}_{,1} - \bar{\omega}_{,1}P = 0, \qquad \overline{M}_{,1} + \overline{F} = 0.$$

Eliminating the transverse shear from the last two equations, we obtain the familiar equation:

$$\overline{M}_{,11} + P\bar{\omega}_{,1} = 0. \tag{4.52}$$

We can also recall (3.160) and (3.163) to obtain

$$\bar{\omega} = \omega^3 = \omega_{21} = \tfrac{1}{2}(\hat{\boldsymbol{\imath}}_2 \cdot \boldsymbol{V}_{,1} - \hat{\boldsymbol{\imath}}_1 \cdot \boldsymbol{V}_{,2}),$$

$$\bar{e}_{21} = \tfrac{1}{2}(\hat{\boldsymbol{\imath}}_2 \cdot \boldsymbol{V}_{,1} + \hat{\boldsymbol{\imath}}_1 \cdot \boldsymbol{V}_{,2}) \doteq \bar{\gamma}_{21} = 0.$$

Here $\boldsymbol{V} = V^1\hat{\boldsymbol{\imath}}_1 + V^2\hat{\boldsymbol{\imath}}_2$ so that

$$\hat{\boldsymbol{\imath}}_2 \cdot \boldsymbol{V}_{,1} = V^2_{\ ,1} = -V^1_{\ ,2} = -\hat{\boldsymbol{\imath}}_1 \cdot \boldsymbol{V}_{,2}.$$

Therefore,

$$\bar{\omega} = V^2_{\ ,1}.$$

By the Bernoulli theory

$$\overline{M} = EI\,\bar{\kappa} = EI\bar{\omega}_{,1} = EI\,V^2_{\ ,11}.$$

Equation (4.52) takes the final form:

$$V^2_{\ ,1111} + \frac{P}{EI}V^2_{\ ,11} = 0.$$

The buckling loads are the characteristic values (P/EI) for the solutions $V^2(x_1)$ subject to four end conditions on the displacement and derivatives (V^2, $\bar{\omega} = V^2_{\ ,1}$, $\overline{M} = EI\,V^2_{\ ,11}$, and $\overline{F} = -EI\,V^2_{\ ,111}$).

Chapter 5

Behavior of Materials

5.1 Introduction

All of the preceding text, the kinematics in Chapter 3 and the descriptions of stresses in Chapter 4, are applicable to any media that are cohesive and continuous. Strictly speaking, no real materials qualify, but certainly exhibit some irregularities, if only microscopic. All are composed of molecules in various assemblies and/or crystalline structures (e.g., plastics and metals); some exhibit macroscopic discontinuities (e.g., filament reinforced materials). Still the preceding descriptions are often applicable, provided that the relevant properties are essentially continuous or, stated otherwise, the phenomena in question are not contingent upon those irregularities (e.g., inter- or intra-crystalline mechanisms of deformation). Indeed, most analyses of machine parts and structural components are adequately accomplished via the concept of a continuous medium. Precautions are warranted at locations of discontinuity or at sites of microstructural activity.

The states of local deformation and internal force are characterized by the variables of strain and stress, respectively; additionally, the former are fully determined by the displacement field. Still a solution to any practical problem (e.g., the consequences of an externally imposed displacement or force upon any body) remains indeterminate without a description of the properties of the media, i.e., a relationship between strain and stress; for elastic media an expression of internal energy in terms of strain also suffices. Such relations, termed *constitutive equations*, characterize the deformational resistance of the material.

Like a mathematical description of any physical phenomenon, the constitutive equations are approximations derived from experimental observations and established principles. Such approximations are always a compromise between accuracy and utility, since precise formulations are worthless if they are unworkable. Accordingly, we do not attempt a general treatment of continuous media, but a systematic development of those established theories which have proven most useful in the practical analyses of solid bodies. Our development rests upon classical concepts of thermodynamics and mechanics of materials.

5.2 General Considerations

Physical phenomena may be classified as mechanical, thermal, electromagnetic, chemical, etc., depending on the quantities that characterize a change of state. Fortunately, most deformations can be adequately described by mechanical and thermal quantities. Accordingly, we restrict our attention to the thermomechanical behavior of materials. The absolute *temperature* and the *displacement* are the primitive independent variables which characterize the thermomechanical state of a body. Other relevant quantities, for example, stress components, heat flow, and internal energy are supposedly determined by the history of the motion and the temperature.

The constitutive equations are to describe the behavior of a material, not the behavior of a body. Consequently, the equations should involve variables that characterize the *local* state, for example, an equation that relates stress, strain, and temperature at a *point*.

From experimental observation and practice, we know that the local thermomechanical behavior of a material is unaffected by rigid-body motions. In our study of solids, we preclude such effects by adopting the *material viewpoint*, that is, we follow the convecting lines and surfaces; in effect, we move with the material and perceive only the deformation. Moreover, we restrict our attention to simple materials, wherein the *strain* and *stress components* and the *strain* and *stress rates* are the only mechanical variables entering the constitutive equations. At the same time, our viewpoint enables us to account for any *directional properties* associated with material lines and surfaces. For example, the crystalline structure of a metal or the laminations in a composite require symmetries with respect to certain convecting surfaces.

The constitutive equations rest upon the established principles of physics: The essential mechanical principles are set forth in the preceding chapters. It remains to review certain principles of thermodynamics.

5.3 Thermodynamic Principles

A deformed body may possess energy in various forms which may be characterized by electrical, chemical, thermal, and mechanical quantities. Here, we presume that the motion involves only thermal and mechanical changes; that is, we neglect other effects, such as electrical and chemical. Moreover, since our immediate interest is the behavior of the material, our attention is directed to the elemental mass rather than the entire body, to energy density (per unit mass) rather than the total energy of the body.

Independent variables that determine the energy density are a matter of experimental observation. We know that the total-energy density $\mathcal{E}$ includes kinetic energy $\mathcal{K}$ (per unit mass) which depends on the velocity as follows:‡

$$\mathcal{K} = \tfrac{1}{2}\dot{\boldsymbol{V}} \cdot \dot{\boldsymbol{V}}. \tag{5.1}$$

In addition, the energy $\mathcal{E}$ includes internal energy $\mathcal{I}$ which depends on the thermal and deformational state of the medium:

$$\mathcal{E} = \mathcal{K} + \mathcal{I}. \tag{5.2}$$

The absolute temperature T and the strain components ϵ_{ij} are among the variables that determine the internal energy $\mathcal{I}$. The strain might be expressed by the physical components ϵ_{ij}, or, by appropriate tensorial components (e.g., γ_{ij} or h_{ij}).

All the *independent variables*, which influence thermal and mechanical changes, are needed to define a *thermodynamic state*. Any changes in these variables constitute a *thermodynamic process*, and any process beginning and ending with the same state constitutes a *thermodynamic cycle*.

The *first law of thermodynamics* asserts that any change in the energy content of a mechanical system is the sum of the work done by applied forces and the heat supplied. If Ω and Q denote the work done and heat supplied (per unit mass), then the first law takes the form

$$\dot{\mathcal{I}} + \dot{\mathcal{K}} = \dot{\Omega} + \dot{Q}. \tag{5.3}$$

‡A dot signifies a material time derivative like the strain rate of (3.135), that is, a rate associated with a specific element of material.

The *second law of thermodynamics* is expressed in the inequality‡

$$\oint \frac{\delta Q}{T} = \oint \frac{\dot{Q}}{T}\,dt \leq 0, \tag{5.4}$$

wherein the loop on the integral sign signifies that integration extends through a complete cycle and the δQ signifies an increment, not necessarily an exact differential. The equality holds in (5.4) if, and only if, the process is reversible. This means that during any reversible process, the integral depends only on the initial and final states; *then*, there exists a function S, called the *entropy* density:

$$S \equiv \int \frac{\delta Q}{T}. \tag{5.5a}$$

In any reversible cycle $\Delta S = 0$, dS is an exact differential

$$dS = \frac{\delta Q}{T}, \qquad \dot{S} = \frac{\dot{Q}}{T}. \tag{5.5b, c}$$

5.4 Excessive Entropy

At times, it proves helpful to express $\dot{Q}/T$ as a sum of two parts: the derivative of the entropy function S associated with reversible processes and the quantity

$$_D\dot{S} \equiv \dot{S} - \frac{\dot{Q}}{T}. \tag{5.6}$$

The quantity $_DS$ might be called *excessive entropy*.† Since S is a function of the state variables, the second law (5.4) takes the form

$$\oint {_D\dot{S}}\,dt \geq 0. \tag{5.7}$$

‡Inequality (5.4) is applicable to a homogeneous state and applies to a simple substance. The validity of (5.4) is open to question if the local behavior is influenced by thermal gradients.

†The quantity $T\,{_D\dot{S}}$ has the dimension of energy rate and is often called the *internal dissipation*.

The inequality applies to every irreversible process and the integrand vanishes in a reversible process. It follows that

$$ {}_{D}\dot{S} \geq 0. \tag{5.8} $$

In words, the excess entropy can only increase. The inequality (5.8) is attributed to R. J. E. Clausius.

5.5 Heat Flow

Let $\dot{\boldsymbol{q}}$ denote the heat flux (per unit deformed area). Then the rate of heat flowing across an elemental coordinate surface dS_i is

$$ \dot{\boldsymbol{q}} \cdot \hat{\boldsymbol{E}}^i \, dS_{\underline{i}} = \dot{\boldsymbol{q}} \cdot \boldsymbol{G}^i \sqrt{G} \, d\theta^j \, d\theta^k \qquad (i \neq j \neq k \neq i). $$

An element formed by the coordinate surfaces θ^i and $\theta^i + d\theta^i$ has heat *supplied* by surroundings at the rate:

$$ -\frac{\partial}{\partial \theta^i} (\dot{\boldsymbol{q}} \cdot \boldsymbol{G}^i \sqrt{G}) \, d\theta^1 \, d\theta^2 \, d\theta^3. $$

In addition to the heat flowing into the element, heat may be generated *within* the element. If $\dot{h}$ denotes the rate of heat generated (per unit mass), then the net heat rate (per unit mass) is

$$ \dot{Q} = \dot{h} - \frac{1}{\rho\sqrt{G}} (\dot{q}^i \sqrt{G})_{,i}, \tag{5.9} $$

wherein

$$ \dot{q}^i \equiv \dot{\boldsymbol{q}} \cdot \boldsymbol{G}^i. \tag{5.10} $$

5.6 Entropy, Entropy Flux, and Entropy Production

The property/variable known as "entropy" seems to perplex many (the authors included) upon initial encounter. This may be attributed to its intangible character as compared to heat and other forms of energy. These latter quantities possess physical attributes or, at least, manifestations;

e.g., we perceive heat flowing from the coals to the kettle to the soup, and we observe kinetic energy converting to heat via friction, etc. Our natural inclination to ascribe such physical characteristics has led to various interpretations and terms ("excess" entropy, "flux," and "production"). Perhaps, it is simpler to accept the mathematical/theoretical view that there exists an integrating factor ($1/T$), differential ($dS \equiv dQ/T$), and a consequent integral (S) for the *reversible* process. Then, our further efforts at physical identity are merely attempts to retain and explain entropy in *irreversible* processes. In what follows, we offer some additional descriptions that appear in the literature.

By analogy with heat flux, one can define an *entropy flux*

$$\dot{\mathbf{s}} \equiv \frac{\dot{\mathbf{q}}}{T} \tag{5.11}$$

and a rate of entropy generation

$$\dot{e} \equiv \frac{\dot{h}}{T}. \tag{5.12}$$

Then, in the manner of (5.9), one can define a pseudoentropy rate

$${}_P\dot{S} \equiv \dot{e} - \frac{1}{\rho\sqrt{G}}\left(\dot{s}^i\sqrt{G}\right)_{,i}. \tag{5.13a}$$

It follows from (5.9), (5.11), and (5.12) that

$${}_P\dot{S} = \frac{\dot{Q}}{T} + \frac{\dot{q}^i}{\rho T^2}T_{,i}. \tag{5.13b}$$

The entropy production is

$$\dot{\eta} \equiv \dot{S} - {}_P\dot{S}. \tag{5.14a}$$

It follows from (5.6), (5.13b), and (5.14a) that

$$\dot{\eta} = {}_D\dot{S} - \frac{\dot{q}^i}{\rho T^2}T_{,i}. \tag{5.14b}$$

The heat flux $\dot{q}^i$ must have the opposite sign of the temperature gradient $T_{,i}$, since heat must flow from a hotter to a colder region. Therefore, the final term of (5.14b) is always positive and, in view of (5.8),

$$\dot{\eta} \geq 0. \tag{5.15}$$

In words, the entropy production is never negative. The inequality (5.15) is often termed the local Clausius-Duhem inequality and is used for nonhomogeneous states of a continuous medium.

5.7 Work of Internal Forces (Stresses)

We recall several basic relations from Section 4.5; specifically, equation (4.17a), which asserts the conservation of work and energy for an elemental *volume*; equation (4.17a) is but another form of (5.3):

$$\frac{\dot{u}}{\rho} = \frac{1}{\rho}(\dot{w}_s + \dot{w}_f - \dot{k}) = \dot{\Omega} - \dot{\mathcal{K}} = \dot{\mathcal{I}} - \dot{\mathcal{Q}}. \tag{5.16}$$

In words, the mechanical work $\dot{\Omega}$, less the increase in kinetic energy $\dot{\mathcal{K}}$ is manifested in the increase of internal energy $\dot{\mathcal{I}}$, less the heat supplied $\dot{\mathcal{Q}}$. We recall that equation (4.17b) in view of the equations of motion (4.10) [or (4.44a, b) and (4.45a–c)] leads to equation (4.18a) or (4.18d), viz.,

$$\dot{u} = \sqrt{G^{\underline{ii}}}\boldsymbol{\tau}^i \cdot \dot{\boldsymbol{G}}_i = \boldsymbol{t}^i \cdot \dot{\boldsymbol{G}}_i, \qquad \dot{u}_0 = \sqrt{g^{\underline{ii}}}\boldsymbol{\sigma}^i \cdot \dot{\boldsymbol{G}}_i = \boldsymbol{s}^i \cdot \dot{\boldsymbol{G}}_i.$$

Here, the subscript "0" signifies the energy per unit *initial* volume. In terms of the alternative components of stress $(\tau^{ij}, s^{ij}, t^{ij})$, these equations take the forms:

$$\dot{u} = \tau^{ij}\dot{\gamma}_{ij}, \tag{5.17a}$$

$$\dot{u}_0 = s^{ij}\dot{\gamma}_{ij} = t^{ij}\dot{h}_{ij}. \tag{5.17b, c}$$

These are equations (4.23a–c), wherein the final term (work upon the *rigid* rotation $\dot{\boldsymbol{\omega}}$) is deleted. In accordance with (5.16), we have the alternative forms

$$\frac{1}{\rho}\tau^{ij}\dot{\gamma}_{ij} = \frac{1}{\rho_0}s^{ij}\dot{\gamma}_{ij} = \frac{1}{\rho_0}t^{ij}\dot{h}_{ij} = \dot{\Omega} - \dot{\mathcal{K}}, \tag{5.17d}$$

where ρ_0 and ρ denote the initial and current mass densities, respectively.

5.8 Alternative Forms of the First and Second Laws

Equations (5.17a–d) yield the following forms of the first law [(5.3), (5.16)]:

$$\frac{\dot{u}}{\rho} = \frac{\dot{u}_0}{\rho_0} = \dot{\Omega} - \dot{\mathcal{K}} = \dot{\mathcal{I}} - \dot{Q},$$

$$\frac{1}{\rho}\tau^{ij}\dot{\gamma}_{ij} = \frac{1}{\rho_0}s^{ij}\dot{\gamma}_{ij} = \frac{1}{\rho_0}t^{ij}\dot{h}_{ij} = \dot{\mathcal{I}} - \dot{Q}. \tag{5.18a–c}$$

Again, these equations assert that the rate of mechanical work and thermal energy (heat) supplied, $(\dot{\Omega} + \dot{Q})$, equals the change in internal and kinetic energy $(\dot{\mathcal{I}} + \dot{\mathcal{K}})$. Here the rates can also be viewed as increments in a moment δt of time; e.g.,

$$\frac{1}{\rho_0}s^{ij}\,\delta\gamma_{ij} = \delta\mathcal{I} - \delta Q. \tag{5.18d}$$

According to (5.6), (5.8), and (5.18a–c), the second law takes the form:

$${}_D\dot{W} \equiv T\,{}_D\dot{S} = T\dot{S} - \dot{\mathcal{I}} + \frac{1}{\rho_0}s^{ij}\dot{\gamma}_{ij} \geq 0. \tag{5.19}$$

The final term represents the work expended by the stresses; it has the alternative forms:

$${}_\rho\dot{w}_s \equiv \frac{1}{\rho_0}s^{ij}\dot{\gamma}_{ij} = \frac{1}{\rho_0}t^{ij}\dot{h}_{ij} = \frac{1}{\rho}\tau^{ij}\dot{\gamma}_{ij}. \tag{5.20a–c}$$

The practical choice depends on the circumstances of any given application; e.g., whether it is more convenient to couch a formulation in the variables of the initial or current state. The term $T\dot{S}$ represents the heat supplied in a reversible process (e.g., elastic deformation). Together, ${}_\rho\dot{w}_s$ and $T\dot{S}$, represent the energy available in a reversible process; $\dot{\mathcal{I}}$ represents the increase in internal energy. The difference ${}_D\dot{W} \equiv T\,{}_D\dot{S}$ is the energy dissipated in the irreversible process. Finally, we note that ${}_D\dot{W} = 0$ in the reversible process (e.g., elastic deformation).

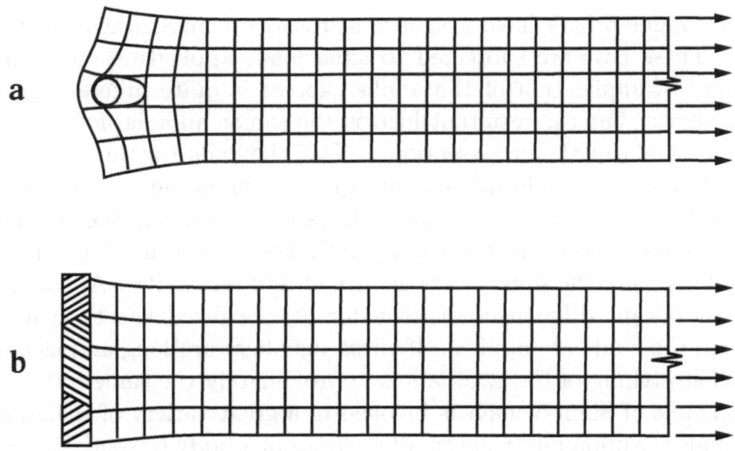

Figure 5.1 Saint-Venant's Principle

5.9 Saint-Venant's Principle

Any mathematical description of the behavior of a material requires verification which, in turn, necessitates physical experimentation. Our theories are largely mechanical and are expressed in terms of strains and stresses. These are continuous variables which require evaluation, strictly speaking, at a *point*. At best, one can only evaluate an average strain in some finite "gage" length. Likewise, one can only measure the force upon some finite area, i.e., an average stress. Still the uses of such *averages* depend on some insights to the distributions, preferably a uniformity. To that end, we rely on a *"principle"* enunciated by Barré de Saint-Venant in 1855 [27].

In essence, Saint-Venant's principle states that *two different distributions of force acting on the same small portion of a body have essentially the same effects on parts of the body which are sufficiently far from the region of application, provided that these forces have the same resultant*. Stated otherwise, any self-equilibrated system of forces upon a small region has diminishing effects at a distance.

According to Saint-Venant's principle, one could augment the loading on a small region by any equipollent system with little effect at a remote region. This means that the precise distribution is not practically relevant to effects, or measurements thereof, at a distant site. Of course, the actual load distribution can significantly alter the *local* effects, but responses, and measurements, at a distance are little affected by the precise manner of loading. As an illustration, consider the situations shown in Figures 5.1a, b.

Similar rectangular bars have a square gridwork of lines drawn on their exteriors. These bars are subjected to axial loads applied near the ends. The manner of application of the applied forces is quite different in the two cases shown, but the resultant load is the same: an axial force acting along the centroid of the cross-sections. Near the ends (in the vicinity of the applied loads), the deformations are quite different; severe distortions of the small squares are evident in both cases. However, the observed deformations are essentially the same throughout the central portion of these two bars; here the squares are deformed into rectangles. Apparently, the only significant differences in behavior are localized effects near the ends, where the loads are applied. If these members are long and slender, the overall stretching of the two bars is approximately the same.

The principle of Saint-Venant is invoked in a great variety of situations, wherein loads act upon relatively small portions of a body (essentially "concentrated" loads). These include many practical circumstances, especially those involving thin structural elements, such as struts, beams, springs, plates, and shells. The principle is implicit in most correlations between theories and experiments on mechanical behavior. An immediate case is the simple tensile test, as described in the next section; the distribution of tractions, as exerted by the jaws of the machine upon the bar, defy mathematical description.

The reader is forewarned that localized deformations and stresses can be paramount in some instances. Fracture and fatigue failures are usually manifestations of local disturbances. Clearly, local distributions of stress are crucial when analyzing such phenomena.

The interested reader will find additional justifications of Saint-Venant's principle in Section 5.23.

5.10 Observations of Simple Tests

Before proceeding to the formulation of constitutive equations, let us recall the behavior of some typical materials subjected to a simple state of stress; for example, a relation between normal stress and extensional strain is obtained from the test of a prismatic bar in axial tension or compression. A similar relation between shear stress and strain is obtained from torsion tests of thin tubes. In either case, the quantities observed and measured are not stress and strain, but applied loads and displacements. Indeed, it is quite impossible to measure the stress and strain components within a deformed body. At best we may measure average values and, from our observations, postulate relations between stress and strain components. The

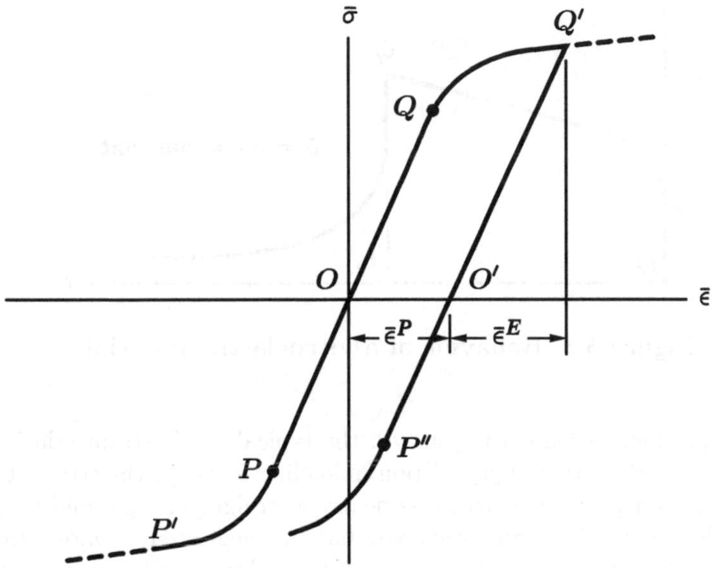

Figure 5.2 Stress-strain diagram from a static-load test

validity of such postulates hinges on the successful application in predicting the response of real bodies.

A *stress-strain diagram* from a static-load test of a typical ductile metal is shown in Figure 5.2. The load is slowly applied and the response is essentially independent of time. An average stress $\bar{\sigma}$ is plotted versus the corresponding strain $\bar{\epsilon}$. The relation between the stress $\bar{\sigma}$ and strain $\bar{\epsilon}$ is essentially linear in the limited range PQ; the slope is a constant E. The stress at P or Q is termed the *proportional limit* in compression or tension, respectively. Within this range unloading (Q to O) retraces the loading curve (O to Q); there is no permanent deformation. Slight discrepancies may be detected, but they are effects which may be neglected in most analyses. Within the range PQ, the behavior is *linear elastic*; that is, $\bar{\sigma} = E\bar{\epsilon}$. Although most materials that exhibit a linear stress-strain relation are elastic, some elastic materials show decidedly nonlinear relations. In general, elasticity does *not* imply linearity.

When the metal is subjected to a stress slightly beyond the proportional limit, permanent deformations are observed. This stress is termed the *elastic limit*. The distinction between the proportional limit and elastic limit is slight and somewhat indefinite because the transition is so gradual. Usually it suffices to assume that the same condition characterizes the transition from linear elastic to inelastic behavior. The condition that characterizes the transition is the *initial-yield condition*.

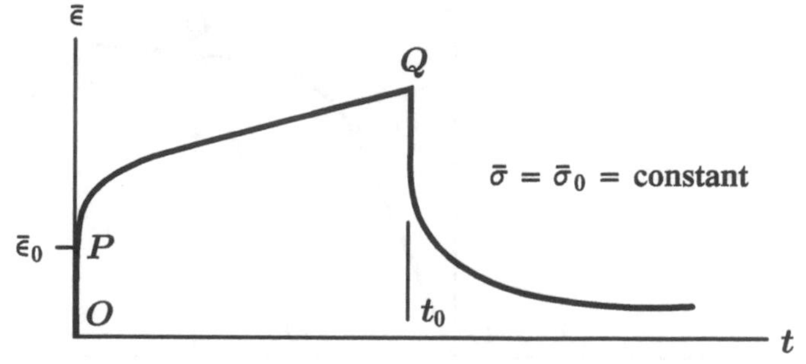

Figure 5.3 Behavior of a viscoelastic material

If the specimen is loaded beyond Q, the typical stress-strain relation is represented by the curve OQQ'. Upon unloading from Q', the stress-strain relation traces a path $Q'O'$ which is nearly a straight line parallel to QO. Again, slight deviations are observed, but are neglected in most analyses. Then the total strain incurred during loading to Q' consists of an *elastic* (recoverable) part $\bar{\epsilon}^E$ and a *plastic* (irrecoverable) part $\bar{\epsilon}^P$, that is, $\bar{\epsilon} = \bar{\epsilon}^E + \bar{\epsilon}^P$.

When the virgin specimen is subjected to the reverse loading, the stress-strain relation is represented by the curve OPP'. This may not be similar to the curve OQQ', although frequently the relation for reverse loading is quite similar.

If the material is permanently deformed by loading along OQQ', the subsequent yield condition may be altered. For example, reverse loading along $Q'O'P''$ results in yielding at a stress of different magnitude than the initial-yield stress of the virgin material.

Often the response of a material depends upon the duration of loading. For example, if a constant stress acts for a period of time, the strain may increase with time. In Figure 5.3 a stress $\bar{\sigma}_0$ is abruptly applied at $t = 0$ and is accompanied by an instantaneous elastic strain $\bar{\epsilon}_0$ followed by continuing strain like the flow of a viscous fluid. Such progressive deformation is termed *creep*. If the load is abruptly removed at $t = t_0$, an elastic part of the strain is recovered quickly, more may be recovered in time, and a part of the strain may be irrecoverable.

Materials that possess some attributes of elasticity and some characteristics of viscous fluids are called *viscoelastic*. The constitutive equations for viscoelastic materials must depend on time and time derivatives.

An elastic material always recovers its initial configuration; we say it remembers one reference state. An ideal fluid is totally indifferent to previous deformation; it has no memory. Most real materials have a limited memory. Some are influenced by recent events, but tend to ignore ancient history. The attributes of memory must be included in the constitutive equations to the extent that they may influence the particular response. For example, the tendency to creep gradually under sustained loads may be unimportant when studying the response to rapidly oscillating loads. Other effects, such as strain-hardening during gradual loading, are even less dependent on the rate of loading. Those inelastic deformations which are insensitive to rates are often termed "plastic deformations." Time is often neglected in the constitutive equations for elasto-plastic deformation even though the history has a vital role.

5.11 Elasticity

An elastic body can be deformed and restored to an initial state without performing work upon the body or supplying heat from its surroundings. In other words, the deformation of an elastic body is a reversible process. It follows too that internal forces of interaction (stresses) are conservative, that is, depend only on the existing state.

The thermodynamic state of an elastic material may be defined by the strain components and the temperature.‡ Since elastic deformations are reversible, we presume the existence of the entropy density:

$$S = S(\gamma_{ij}, T). \tag{5.21}$$

If the derivatives of S are continuous and form a nonvanishing determinant, then (5.21) implies that

$$T = T(\gamma_{ij}, S). \tag{5.22}$$

In words, the absolute temperature and entropy can be regarded as alternative independent variables. Then, we may regard the internal-energy density as a function of the strain components and the entropy:

$$\mathcal{I} = \mathcal{U}(\gamma_{ij}, S). \tag{5.23}$$

According to the equality (5.19)

$$d\mathcal{U} = \frac{1}{\rho_0} s^{ij}\, d\gamma_{ij} + T\, dS, \tag{5.24a}$$

‡Since the gradients are usually inconsequential, we neglect them at the outset to simplify our development.

but according to (5.23)

$$d\mathcal{U} = \frac{\partial \mathcal{U}}{\partial \gamma_{ij}} d\gamma_{ij} + \frac{\partial \mathcal{U}}{\partial S} dS. \tag{5.24b}$$

Since the variables γ_{ij} and S are independent in (5.24a, b), it follows that

$$s^{ij} = \rho_0 \frac{\partial \mathcal{U}}{\partial \gamma_{ij}}, \qquad T = \frac{\partial \mathcal{U}}{\partial S}. \tag{5.25a, b}$$

If the derivatives of s^{ij} and T are continuous and form a nonvanishing determinant, then equations (5.25a, b) imply that

$$\gamma_{ij} = \gamma_{ij}(s^{ij}, T), \qquad S = S(s^{ij}, T). \tag{5.26a, b}$$

In words, s^{ij} and γ_{ij} are alternative independent variables as S and T. Then, we can define a complementary-energy density, the *thermodynamic potential of Gibbs* or *Gibbs potential*,‡

$$\mathcal{G}(s^{ij}, T) = \mathcal{U} - TS - \frac{1}{\rho_0} s^{ij} \gamma_{ij}. \tag{5.27}$$

The differential is

$$d\mathcal{G} = d\mathcal{U} - T\,dS - S\,dT - \frac{1}{\rho_0} s^{ij}\,d\gamma_{ij} - \frac{1}{\rho_0}\gamma_{ij}\,ds^{ij}. \tag{5.28a}$$

In view of (5.24a)

$$d\mathcal{G} = -\frac{1}{\rho_0}\gamma_{ij}\,ds^{ij} - S\,dT. \tag{5.28b}$$

‡The function $\mathcal{G}$ is the negative of the usual complementary energy of mechanical theory, but defined here in accordance with thermodynamic theory; see I. I. Gol'denblat ([28], p. 201). The transformation from the variables $\mathcal{U}$, γ_{ij}, and S of (5.23) and (5.25a, b) to the variables $\mathcal{G}$, s^{ij}, and T of (5.27) and (5.29a, b) is known as a Legendre transformation. The transformation and the existence of the function $\mathcal{G}$ requires that the Hessian determinant of the function $\mathcal{U}$ does not vanish; see C. Lanczos [29], p. 161 of the first or fourth edition.

In the manner of (5.25a, b), since the variables T and s^{ij} are independent in (5.28b), we have

$$\gamma_{ij} = -\rho_0 \frac{\partial \mathcal{G}}{\partial s^{ij}}, \qquad S = -\frac{\partial \mathcal{G}}{\partial T}. \tag{5.29a, b}$$

The *free-energy*[‡] *density* is

$$\mathcal{F}(\gamma_{ij}, T) \equiv \mathcal{U} - TS. \tag{5.30}$$

The differential follows from (5.30) and (5.24a)

$$d\mathcal{F} = \frac{1}{\rho_0} s^{ij}\, d\gamma_{ij} - S\, dT. \tag{5.31}$$

In the manner of (5.25a, b) and (5.29a, b), we obtain

$$s^{ij} = \rho_0 \frac{\partial \mathcal{F}}{\partial \gamma_{ij}}, \qquad S = -\frac{\partial \mathcal{F}}{\partial T}. \tag{5.32a, b}$$

The *enthalpy* (*heat function*) *density* is

$$\mathcal{H}(s^{ij}, S) = \mathcal{U} - \frac{1}{\rho_0} s^{ij} \gamma_{ij}. \tag{5.33}$$

Again, the differential follows from (5.33) and (5.24a):

$$d\mathcal{H} = -\frac{1}{\rho_0} \gamma_{ij}\, ds^{ij} + T\, dS.$$

The counterparts of (5.25a, b), (5.29a, b), and (5.32a, b) follow:

$$\gamma_{ij} = -\rho_0 \frac{\partial \mathcal{H}}{\partial s^{ij}}, \qquad T = \frac{\partial \mathcal{H}}{\partial S}. \tag{5.34a, b}$$

For convenience, the potential functions $\mathcal{U}$, $\mathcal{F}$, $\mathcal{G}$, and $\mathcal{H}$ are given in Table 5.1. The functions $\mathcal{U}$ and $\mathcal{F}$ are so-called elastic strain-energy functions while $\mathcal{G}$ and $\mathcal{H}$ are complementary-energy functions, as the terms are used in the discipline of elasticity.

[‡]In essence, (5.30) and (5.33) accomplish partial Legendre transformations. The former transforms from the variable S to T but retains the variable γ_{ij} while the latter transforms from the variable γ_{ij} to s^{ij} but retains the variable S.

Table 5.1

Internal energy: $\mathcal{U}(\gamma_{ij}, S)$

$$s^{ij} = \rho_0 \frac{\partial \mathcal{U}}{\partial \gamma_{ij}}, \qquad T = \frac{\partial \mathcal{U}}{\partial S} \tag{5.25a, b}$$

Free energy: $\mathcal{F}(\gamma_{ij}, T) = \mathcal{U} - TS$

$$s^{ij} = \rho_0 \frac{\partial \mathcal{F}}{\partial \gamma_{ij}}, \qquad S = -\frac{\partial \mathcal{F}}{\partial T} \tag{5.32a, b}$$

Gibbs potential: $\mathcal{G}(s^{ij}, T) = \mathcal{U} - TS - (s^{ij}\gamma_{ij})/\rho_0$

$$\gamma_{ij} = -\rho_0 \frac{\partial \mathcal{G}}{\partial s^{ij}}, \qquad S = -\frac{\partial \mathcal{G}}{\partial T} \tag{5.29a, b}$$

Enthalpy function: $\mathcal{H}(s^{ij}, S) = \mathcal{U} - (s^{ij}\gamma_{ij})/\rho_0$

$$\gamma_{ij} = -\rho_0 \frac{\partial \mathcal{H}}{\partial s^{ij}}, \qquad T = \frac{\partial \mathcal{H}}{\partial S} \tag{5.34a, b}$$

If a deformation occurs very rapidly, there is little time for the material to reach thermal equilibrium with its surroundings. Then the process is essentially *adiabatic*, that is, $dQ = dS = 0$, and

$$d\mathcal{U} = \frac{1}{\rho_0} s^{ij}\, d\gamma_{ij}, \qquad d\mathcal{H} = -\frac{1}{\rho_0}\gamma_{ij}\, ds^{ij}, \tag{5.35a, b}$$

$$s^{ij} = \rho_0 \frac{\partial \mathcal{U}}{\partial \gamma_{ij}}, \qquad \gamma_{ij} = -\rho_0 \frac{\partial \mathcal{H}}{\partial s^{ij}}. \tag{5.36a, b}$$

If a deformation is very slow, the body is maintained at the temperature of its surroundings. Then the process is *isothermal* (that is, $dT = 0$), and

$$d\mathcal{F} = \frac{1}{\rho_0} s^{ij}\, d\gamma_{ij}, \qquad d\mathcal{G} = -\frac{1}{\rho_0}\gamma_{ij}\, ds^{ij}, \tag{5.37a, b}$$

$$s^{ij} = \rho_0 \frac{\partial \mathcal{F}}{\partial \gamma_{ij}}, \qquad \gamma_{ij} = -\rho_0 \frac{\partial \mathcal{G}}{\partial s^{ij}}. \qquad (5.38\text{a, b})$$

In either case, adiabatic or isothermal, there is a potential function $\mathcal{U}$ or $\mathcal{F}$ which gives the stress components according to (5.36a) or (5.38a) and a complementary function $\mathcal{H}$ or $\mathcal{G}$ which gives the strain components according to (5.36b) or (5.38b). In either case, the potential function and its complementary function are similarly related, namely,

$$\mathcal{H} = \mathcal{U} - \frac{1}{\rho_0} s^{ij}\gamma_{ij}, \qquad \mathcal{G} = \mathcal{F} - \frac{1}{\rho_0} s^{ij}\gamma_{ij}. \qquad (5.39\text{a, b})$$

The foregoing formulations are expressed in terms of the tensorial components γ_{ij} and s^{ij}. These are often the most natural; in part, because the stress components are based upon areas of the reference (undeformed) state. Since all invariants are densities per unit mass, which is presumably conserved, the corresponding forms are readily expressed in terms of the alternative components (h_{ij} and t^{ij}; γ_{ij} *and* τ^{ij}).

As an example, we cite Mooney elasticity [30], wherein the free energy is taken as a quadratic form:

$$\rho_0 \mathcal{F} = 2C_1 \gamma_i^i + 2C_2(2\gamma_i^i + \gamma_i^i \gamma_j^j - \gamma_j^i \gamma_i^j),$$

where $\gamma_i^i \equiv g^{ij}\gamma_{ij}$. This provides the stress-strain relations:

$$s^{ij} = 2(C_1 + 2C_2)g^{ij} + 2C_2(2g^{ij}\gamma_k^k - g^{kj}\gamma_k^i - g^{ki}\gamma_k^j).$$

It must be noted that the undeformed state is not *necessarily* unstressed. This is a consequence of the incompressibility of the Mooney (rubber-like) material; e.g., $\gamma_{ij} \equiv 0$ may be accompanied by an arbitrary hydrostatic stress $s^{ij} = pg^{ij}$.

5.12 Inelasticity

The occurrence of inelastic strains γ_{ij}^N ($N = 1, \ldots, M$) characterizes an irreversible process. These strains may be a consequence of time-dependent (viscous) or time-independent (plastic) behavior. Then, work is dissipated by nonconservative stresses s_N^{ij} upon the inelastic strains; the rate of dissipation follows:

$$_D\dot{W} = \frac{1}{\rho_0} s_N^{ij} \dot{\gamma}_{ij}^N \quad \text{(sum also on majuscule indices)}. \tag{5.40}$$

In accordance with the second law of thermodynamics [see Sections 5.3 and 5.4, equation (5.6), and inequality (5.7)],

$$_D\dot{W} \equiv T\,_D\dot{S} = T\dot{S} - \dot{Q} \geq 0. \tag{5.41}$$

The first law of thermodynamics [see Sections 5.3 and 5.8, and equations (5.18a–d) and (5.23)] asserts that the work performed plus heat supplied are manifested in internal energy:

$$\frac{1}{\rho_0} s^{ij} \dot{\gamma}_{ij} + \dot{Q} = \dot{\mathcal{U}}. \tag{5.42}$$

In accordance with (5.42), the inequality (5.41) takes the form:

$$\frac{1}{\rho_0} s^{ij} \dot{\gamma}_{ij} + T\dot{S} - \dot{\mathcal{U}} = {}_D\dot{W} \geq 0. \tag{5.43}$$

Also, from (5.40) we obtain

$$\frac{1}{\rho_0} s^{ij} \dot{\gamma}_{ij} - \frac{1}{\rho_0} s_N^{ij} \dot{\gamma}_{ij}^N + T\dot{S} - \dot{\mathcal{U}} = 0. \tag{5.44}$$

The free energy $\mathcal{F}$ must depend also on the additional strains:

$$\mathcal{F} = \mathcal{F}(\gamma_{ij}, \gamma_{ij}^N, T). \tag{5.45}$$

In accordance with (5.30),

$$\dot{\mathcal{U}} = \dot{\mathcal{F}} + S\dot{T} + \dot{S}T. \tag{5.46}$$

By substituting (5.46) into (5.44), we obtain

$$\frac{1}{\rho_0} s^{ij} \dot{\gamma}_{ij} - \frac{1}{\rho_0} s_N^{ij} \dot{\gamma}_{ij}^N - \dot{\mathcal{F}} - S\dot{T} = 0. \tag{5.47a}$$

Alternatively,

$$\left(\frac{1}{\rho_0}s^{ij} - \frac{\partial \mathcal{F}}{\partial \gamma_{ij}}\right)\dot{\gamma}_{ij} + \left(-\frac{1}{\rho_0}s_N^{ij} - \frac{\partial \mathcal{F}}{\partial \gamma_{ij}^N}\right)\dot{\gamma}_{ij}^N - \left(S + \frac{\partial \mathcal{F}}{\partial T}\right)\dot{T} = 0. \quad (5.47b)$$

If we assume that our model is adequate, that the material is described by these *independent* variables $(\gamma_{ij}, \gamma_{ij}^N, T)$,‡ then it follows from (5.47b) that

$$s^{ij} = \rho_0 \frac{\partial \mathcal{F}}{\partial \gamma_{ij}}, \qquad s_N^{ij} = -\rho_0 \frac{\partial \mathcal{F}}{\partial \gamma_{ij}^N}, \qquad S = -\frac{\partial \mathcal{F}}{\partial T}. \quad (5.48a–c)$$

The inelastic strains γ_{ij}^N play the same role as the internal variables of L. Onsager [31], [32], J. Meixner [33], and M. A. Biot [34]. They provide means to accommodate physical alterations of microstructure, e.g., dislocations and slip in crystalline materials [35], [36].

Finally, we note that complementary functionals, such as the Gibbs potential [see Section 5.11, equation (5.27)] can also be employed in the present context. The reader is referred to the publication [37] which provides models to simulate the various types of inelastic deformations.

5.13 Linearly Elastic (Hookean) Material

Since one can always conceive of a Cartesian/rectangular system and always transform to another, we can simplify our presentation by casting our formulations in the former (Cartesian/rectangular). Then,

$$\gamma_{ij} = \epsilon_{ij} = \gamma_j^i = \gamma^{ij}.$$

If the strains are small enough, then the behavior of an elastic material can be approximated by a linear relation between stress components, strain components, and the temperature change

$$s^{ij} \doteq E^{ijkl}\epsilon_{kl} - \alpha^{ij}(T - T_0). \quad (5.49)$$

A material which obeys the linear relation (5.49) is termed *Hookean* after R. Hooke† who proposed the proportionality of stress and strain.

‡More precisely, the state, hence $\mathcal{F}$, is a functional of inelastic strain history, the strain path, and not merely the prevailing state of inelastic strain.

†R. Hooke. *De Potentia Restitutiva (Of Spring)*. London, 1678.

Let us assume that the free energy of (5.30) can be expanded in a power series about a reference state ($\epsilon_{ij} = 0$, $T = T_0$):

$$\mathcal{F} = \mathcal{F}\Big]_0 + \frac{\partial \mathcal{F}}{\partial \epsilon_{ij}}\Big]_0 \epsilon_{ij} + \frac{1}{2}\frac{\partial^2 \mathcal{F}}{\partial \epsilon_{ij}\,\partial \epsilon_{kl}}\Big]_0 \epsilon_{ij}\epsilon_{kl} + \cdots + \frac{\partial^2 \mathcal{F}}{\partial \epsilon_{ij}\,\partial T}\Big]_0 \epsilon_{ij}(T - T_0) + \cdots .$$

Now, let us measure the free energy from the reference state (i.e., $\mathcal{F}\,]_0 = 0$), since the additive constant is irrelevant. In accordance with (5.32a) and (5.49), the stress vanishes in the reference state and $(\partial \mathcal{F}/\partial \epsilon_{ij})]_0 = 0$. It follows that

$$\mathcal{F} = \frac{1}{2}\frac{\partial^2 \mathcal{F}}{\partial \epsilon_{ij}\,\partial \epsilon_{kl}}\Big]_0 \epsilon_{ij}\epsilon_{kl} + \cdots + \frac{\partial^2 \mathcal{F}}{\partial \epsilon_{ij}\,\partial T}\Big]_0 \epsilon_{ij}(T - T_0) + \cdots . \tag{5.50}$$

According to (5.32a) only the terms shown explicitly in (5.50) contribute linear terms to the stress and, therefore, any additional terms play no role in the stress-strain relation of the Hookean material.

It follows from (5.32a, b), (5.49), and (5.50) that

$$E^{ijkl} = \rho_0 \frac{\partial^2 \mathcal{F}}{\partial \epsilon_{ij}\,\partial \epsilon_{kl}}\Big]_0, \qquad \alpha^{ij} = -\rho_0 \frac{\partial^2 \mathcal{F}}{\partial \epsilon_{ij}\,\partial T}\Big]_0. \tag{5.51a, b}$$

Then (5.50) has the form

$$\rho_0\,\mathcal{F} = \tfrac{1}{2}E^{ijkl}\epsilon_{ij}\epsilon_{kl} - \alpha^{ij}\epsilon_{ij}(T - T_0), \tag{5.52}$$

and from (5.51a, b), it follows that

$$E^{ijkl} = E^{klij} = E^{jikl} = E^{ijlk}, \qquad \alpha^{ij} = \alpha^{ji}. \tag{5.53a, b}$$

The inverse of (5.49), viz., the strain-stress relations, are given by the Gibbs energy:

$$\rho_0 \mathcal{G} = -\tfrac{1}{2}D_{ijkl}s^{ij}s^{kl} - \beta_{ij}s^{ij}(T - T_0). \tag{5.54}$$

From (5.29a), we obtain

$$\epsilon_{ij} = D_{ijkl}s^{kl} + \beta_{ij}(T - T_0), \tag{5.55a}$$

wherein

$$D_{ijkl}E^{ijmn} = \delta_k^m \delta_l^n, \qquad \beta_{ij} = D_{ijmn}\alpha^{mn}. \tag{5.55b, c}$$

From the preceding relations, it follows that

$$\alpha^{ij} = E^{ijmn}\beta_{mn}. \tag{5.55d}$$

In view of the symmetry properties (5.53a) of the coefficient E^{ijkl}, only 21 coefficients are needed to characterize the general anisotropic Hookean material; they form the array:

$$\begin{array}{cccccc} E^{1111} & E^{1122} & E^{1133} & E^{1123} & E^{1113} & E^{1112} \\ & E^{2222} & E^{2233} & E^{2223} & E^{2213} & E^{2212} \\ & & E^{3333} & E^{3323} & E^{3313} & E^{3312} \\ & & & E^{2323} & E^{2313} & E^{2312} \\ & & & & E^{1313} & E^{1312} \\ & & & & & E^{1212}. \end{array} \tag{5.56}$$

In the following sections, some particular cases of additional symmetries are discussed. Further details are given in A. E. H. Love [1], Chapter VI. The work of A. E. Green and J. E. Adkins [38] provides additional information about the strain-energy functions of nonlinear elasticity and about the symmetries associated with particular crystalline structures.

Finally, we note that Hookean behavior seldom occurs unless the strains are also small, i.e., $\epsilon_{ij} \ll 1$; then $s^{ij} \doteq \overset{*}{\sigma}{}^{ij}$, the physical component.

5.14 Monotropic Hookean Material

An element of material is elastically symmetric with respect to a coordinate surface, e.g., the surface x^3, if the form of the energy function $\mathcal{F}$ is unchanged when x^3 is replaced by $-x^3$. Physically, this means that the elastic properties appear the same whether the element is viewed from the position (x^1, x^2, a) or $(x^1, x^2, -a)$. Since the symmetry exists with respect to one direction, we term it *monotropic*.

To determine the consequences of such symmetry, consider the transformation of *rectangular* coordinates:

$$\bar{x}^1 = x^1, \qquad \bar{x}^2 = x^2, \qquad \bar{x}^3 = -x^3.$$

The strain components are transformed according to the array (5.57):

$$\begin{array}{lll} \bar{\epsilon}_{11} = \epsilon_{11}, & \bar{\epsilon}_{12} = \epsilon_{12}, & \bar{\epsilon}_{13} = -\epsilon_{13}, \\ & \bar{\epsilon}_{22} = \epsilon_{22}, & \bar{\epsilon}_{23} = -\epsilon_{23}, \\ & & \bar{\epsilon}_{33} = \epsilon_{33}. \end{array} \tag{5.57}$$

Since $\bar{\epsilon}_{13} = -\epsilon_{13}$ and $\bar{\epsilon}_{23} = -\epsilon_{23}$, any term of (5.52) which contains either component alone must change sign; therefore, the coefficients of such terms must vanish to maintain the symmetry, that is,

$$E^{1123} = E^{1113} = E^{2223} = E^{2213} = E^{2312} = E^{1312} = E^{3323} = E^{3313} = 0,$$

and

$$\alpha^{13} = \alpha^{23} = 0.$$

In short a coefficient must vanish if the index 3 appears one or three times. The remaining 13 elastic coefficients follow:

$$\begin{array}{cccccc} E^{1111} & E^{1122} & E^{1133} & \cdots & \cdots & E^{1112} \\ & E^{2222} & E^{2233} & \cdots & \cdots & E^{2212} \\ & & E^{3333} & \cdots & \cdots & E^{3312} \\ & & & E^{2323} & E^{2313} & \cdots \\ & & & & E^{1313} & \cdots \\ & & & & & E^{1212}. \end{array}$$

Symmetry with respect to one direction is a useful approximation for shell-like bodies, wherein properties are symmetric with respect to the normal to a surface.

5.15 Orthotropic Hookean Material

If the material is elastically symmetric with respect to two orthogonal directions, e.g., to the x^2 and x^3 coordinates, then additional coefficients vanish, namely

$$E^{1112} = E^{3312} = E^{2313} = E^{2212} = 0, \qquad \alpha^{12} = 0.$$

The remaining nine elastic coefficients form the array (5.58):

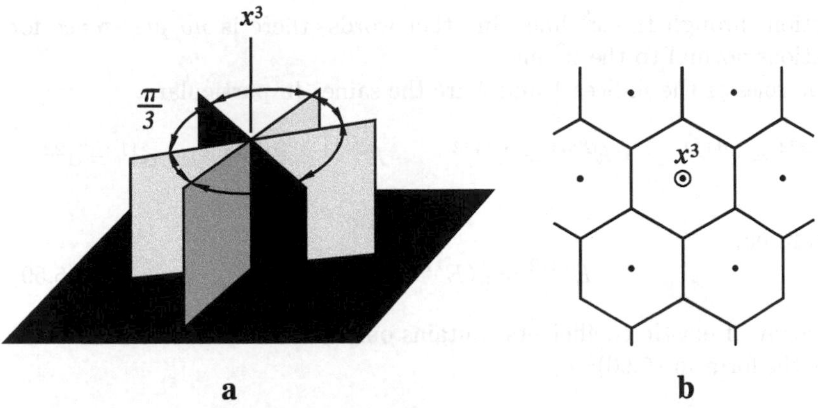

Figure 5.4 Hexagonal symmetry

$$
\begin{array}{cccccc}
E^{1111} & E^{1122} & E^{1133} & \cdots & \cdots & \cdots \\
 & E^{2222} & E^{2233} & \cdots & \cdots & \cdots \\
 & & E^{3333} & \cdots & \cdots & \cdots \\
 & & & E^{2323} & \cdots & \cdots \\
 & & & & E^{1313} & \cdots \\
 & & & & & E^{1212}.
\end{array}
\tag{5.58}
$$

The orthotropic continuum is often a useful approximation for bodies which are stiffened equally by filaments intersecting at right angles, for example, plastics which are reinforced by glass, metal, or Kevlar fibers [39].

5.16 Transversely Isotropic (Hexagonally Symmetric) Hookean Material

A material is hexagonally symmetric with respect to a line, say, the x^3 axis, if it is symmetric with respect to the x^3 axis and all directions passing through the axis and forming angles $\pi/3$. The planes of symmetry are depicted in Figure 5.4a.

The coefficients of the hexagonally symmetric material are obtained by further restrictions upon the orthotropic material; that is, the energy potential must be invariant under a rotation $\pi/3$ about an x^3 line. The reduction of coefficients is given by A. E. H. Love ([1], Section 105, p. 151). Here, we note the essential features, but omit the details. It is a curious, but important fact, that the result provides symmetry with respect to *every* direction through the x^3 line. In other words, there is *no preference* for directions normal to the x^3 line.

The roles of the indices 1 and 2 are the same. In particular,

$$E^{2222} = E^{1111}, \qquad E^{2233} = E^{1133}, \qquad E^{2323} = E^{1313}, \qquad \alpha^{11} = \alpha^{22}.$$

In addition,

$$E^{1212} = \tfrac{1}{2}(E^{1111} - E^{1122}). \tag{5.59}$$

The array of elastic coefficients contains only five independent values and takes the form in (5.60):

$$\begin{matrix} E^{1111} & E^{1122} & E^{1133} & \cdots & \cdots & \cdots \\ & E^{1111} & E^{1133} & \cdots & \cdots & \cdots \\ & & E^{3333} & \cdots & \cdots & \cdots \\ & & & E^{1313} & \cdots & \cdots \\ & & & & E^{1313} & \cdots \\ & & & & & E^{1212}. \end{matrix} \tag{5.60}$$

One can also express all five elastic coefficients by the form:

$$E^{ijkl} = \lambda\delta_{ij}\delta_{kl} + G(\delta_{ik}\delta_{jl} + \delta_{il}\delta_{jk}) + \alpha(\delta_{ij}a_k a_l + \delta_{kl}a_i a_j)$$

$$+ \beta(\delta_{il}a_k a_j + \delta_{jk}a_i a_l + \delta_{ik}a_j a_l + \delta_{jl}a_i a_k) + \gamma(a_i a_j a_k a_l),$$

where the arbitrary direction of transverse isotropy is defined by the vector $\hat{\boldsymbol{a}} = a_i \hat{\boldsymbol{\imath}}_i$ and the quantities (α, β, γ, λ, G) are five independent elastic properties.

Hexagonal symmetry provides a continuum approximation for composites formed like the honeycomb shown in Figure 5.4b. Such honeycombs are employed as an intermediate layer in sandwich plates to provide strength and reduce weight.

5.17 Isotropic Hookean Material

If there are no preferred directions, then the roles of the indices 1, 2, and 3 can be fully interchanged. It follows that

$$E^{1111} = E^{2222} = E^{3333}, \qquad E^{1122} = E^{1133} = E^{2233},$$

$$E^{1212} = E^{1313} = E^{2323}, \qquad \alpha^{11} = \alpha^{22} = \alpha^{33}.$$

In view of (5.59),

$$E^{1212} = E^{1313} = E^{2323} = \tfrac{1}{2}(E^{1111} - E^{1122}).$$

Only two elastic coefficients and one thermal coefficient remain. All coefficients can be expressed in terms of three as follows:

$$E^{ijkl} = G(\delta^{ik}\delta^{jl} + \delta^{il}\delta^{jk}) + \lambda\delta^{ij}\delta^{kl}, \tag{5.61}$$

$$\alpha^{ij} = \alpha(3\lambda + 2G)\delta_{ij}. \tag{5.62}$$

Equations (5.61) and (5.62) serve to express the two elastic coefficients and one thermal coefficient in terms of the established coefficients, G, λ, and α: The coefficient G is known as the *shear modulus*, λ is the *Lamé coefficient*, and α the *coefficient of thermal expansion.*

In arbitrary coordinates, θ^i, equations (5.61) and (5.62) assume the form

$$E^{ijkl} = G(g^{ik}g^{jl} + g^{il}g^{jk}) + \lambda g^{ij}g^{kl},$$

$$\alpha^{ij} = \alpha(3\lambda + 2G)g^{ij}.$$

The free energy (5.52) takes the form

$$\rho_0\mathcal{F} = G\epsilon_{ij}\epsilon_{ij} + \frac{\lambda}{2}\epsilon_{ii}\epsilon_{jj} - \alpha(3\lambda + 2G)(T - T_0)\epsilon_{kk}. \tag{5.63}$$

In view of the isotropy, the free energy, as well as the internal energy, can be expressed in terms of the principal strains $(\tilde{\epsilon}_1, \tilde{\epsilon}_2, \tilde{\epsilon}_3)$, or in terms of the strain invariants (I_1, I_2, I_3 or K_1, K_2, K_3; see Section 3.13):

$$\rho_0\mathcal{F} = \left(G + \frac{\lambda}{2}\right) I_1^2 - 2GI_2 - \alpha(3\lambda + 2G)(T - T_0)I_1.$$

Likewise, the Gibbs potential and the enthalpy can be expressed in terms of the stress invariants. The stress (5.49) takes the form:

$$s^{ij} = 2G\epsilon_{ij} + \lambda\epsilon_{kk}\delta_{ij} - \alpha(3\lambda + 2G)(T - T_0)\delta_{ij}. \tag{5.64}$$

The mean normal stress follows from (5.64):

$$\tfrac{1}{3}s^{ii} = \tfrac{1}{3}(3\lambda + 2G)[\epsilon_{kk} - 3\alpha(T - T_0)]. \tag{5.65}$$

Solving (5.65) for ϵ_{ii} and substituting the result into (5.64), we obtain

$$2G\epsilon_{ij} = s^{ij} - \frac{\lambda}{3\lambda + 2G}s^{kk}\delta_{ij} + 2G\alpha(T - T_0)\delta_{ij}. \tag{5.66}$$

Notice that the only effect of a temperature change on a stress-free isotropic body is a dilatation; that is, equal extensional strains $\epsilon_{ij} = \alpha(T - T_0)\delta_{ij}$ in every direction.

5.18 Heat Conduction

Fourier's law provides the simplest description of heat flow. In a thermally isotropic medium the heat flux is assumed proportional to the temperature gradient; that is,

$$\dot{\boldsymbol{q}} = -\Lambda \frac{\partial T}{\partial \theta^i} \boldsymbol{G}^i, \tag{5.67}$$

where Λ is a constant depending upon conductivity of the medium. It follows from (5.9) and (5.10) that the heat supplied by *conduction* is

$$\dot{Q} = \frac{1}{\rho\sqrt{G}}\left(\Lambda G^{ij} T_{,i} \sqrt{G}\right)_{,j}, \tag{5.68a}$$

or, since we presume the conservation of mass,

$$\dot{Q} = \frac{1}{\rho_0\sqrt{g}}\left(\Lambda G^{ij} T_{,i} \sqrt{G}\right)_{,j}. \tag{5.68b}$$

5.19 Heat Conduction in the Hookean Material

In practice Hooke's law is applicable primarily to small strains, wherein $\sqrt{G} \doteq \sqrt{g}$ and $G^{ij} \doteq g^{ij}$. As before (Sections 5.13 to 5.17), the equations can be cast in a rectangular Cartesian system x^i. Then,

$$\gamma_{ij} = \epsilon_{ij}, \qquad g^{ij} = g_{ij} = \delta^{ij}, \qquad \sqrt{g} = 1,$$

and equation (5.68b) assumes the simpler form:‡

$$\dot{Q} = \frac{1}{\rho_0}(\Lambda T_{,i})_{,i}. \tag{5.69}$$

The entropy rate $\dot{S}$ is obtained from (5.32b) as follows:

$$\dot{S} \doteq -\frac{\partial^2 \mathcal{F}}{\partial \epsilon_{ij} \partial T}\dot{\epsilon}_{ij} - \frac{\partial^2 \mathcal{F}}{\partial T^2}\dot{T}. \tag{5.70}$$

Substituting (5.70) into (5.5c) and using the notation (5.51b), we have

$$\dot{Q} = T\frac{\alpha^{ij}}{\rho_0}\dot{\epsilon}_{ij} - T\frac{\partial^2 \mathcal{F}}{\partial T^2}\dot{T}. \tag{5.71}$$

If the strain rate vanishes, then the heat rate results from the temperature change, that is, from the second term on the right side of (5.71); hence, the so-called heat capacity C_v at constant volume is defined as follows:

$$C_v \equiv -T\frac{\partial^2 \mathcal{F}}{\partial T^2}. \tag{5.72}$$

Then, (5.71) assumes the form:

$$\dot{Q} = T\frac{\alpha^{ij}}{\rho_0}\dot{\epsilon}_{ij} + C_v\dot{T}. \tag{5.73}$$

In practice, the strain rates are often small enough that the first term on the right side of (5.73) can be neglected; then, the approximation of (5.73) and (5.69) provide the approximation

$$(\Lambda T_{,i})_{,i} \doteq \rho_0\, C_v \dot{T}. \tag{5.74}$$

‡The approximation (5.69) removes terms which couple the heat conduction and deformation.

Equation (5.74) implies that the heat conduction is independent of the deformation; it may be invalid in certain dynamic problems. A better approximation results if we set $T = T_0 + \Delta T$ where T_0 denotes the temperature of the reference state; then, if ΔT is sufficiently small, (5.73) is replaced by the approximation

$$\dot{Q} = \frac{T_0}{\rho_0}\alpha^{ij}\dot{\epsilon}_{ij} + C_v\dot{T}. \tag{5.75}$$

Instead of (5.74) we have the approximation

$$(\Lambda T_{,i})_{,i} \doteq T_0\,\alpha^{ij}\dot{\epsilon}_{ij} + \rho_0\,C_v\dot{T}. \tag{5.76}$$

Equations (5.75) and (5.76) provide a linear theory of *coupled thermoelasticity*; the thermal and mechanical effects are coupled by the first term on the right side of (5.76).

5.20 Coefficients of Isotropic Elasticity

A linear, elastic, isotropic, and homogeneous material is characterized by two elastic constants, the Lamé coefficient λ and the shear modulus G in (5.61). This conclusion was reached by G. Green (1830) and precipitated a celebrated controversy since it conflicted with the earlier (one constant) theory of C. L. M. H. Navier [40]. The latter theory had been accepted by eminent elasticians (for example, A. L. Cauchy [41], S. D. Poisson [42], B. F. E. Clayperon [43], and G. Lamé [44]).‡

To give the elastic coefficients physical meaning we consider a few simple situations of isothermal deformations; in each, the strains are small.

5.20.1 Simple Tension

Consider the state of simple tension:

$$s^{11} = \bar{\sigma}, \qquad s^{12} = s^{13} = s^{22} = s^{23} = s^{33} = 0.$$

It follows from (5.66) that

$$\epsilon_{11} = \frac{(\lambda + G)}{G(3\lambda + 2G)}\bar{\sigma},$$

‡A brief historical account is presented by A. E. H. Love [1]. Details are contained in the works of I. Todhunter and K. Pearson [45] and S. P. Timoshenko [46].

$$\epsilon_{22} = \epsilon_{33} = -\frac{\lambda}{2G(3\lambda + 2G)}\bar{\sigma},$$

$$\epsilon_{12} = \epsilon_{13} = \epsilon_{23} = 0.$$

In this situation, the more convenient coefficients are the *Young's modulus* E and *Poisson's ratio* ν:

$$\bar{\sigma} = E\epsilon_{11}, \qquad \epsilon_{22} = \epsilon_{33} = -\nu\epsilon_{11}. \tag{5.77}$$

It follows that

$$E = \frac{(3\lambda + 2G)}{(\lambda + G)}G, \tag{5.78}$$

$$\nu = \frac{\lambda}{2(\lambda + G)}. \tag{5.79}$$

The state of simple tension is one that satisfies equilibrium conditions. The consequent strains correspond to simple extension in direction x^1 and equal contraction in every transverse direction (e.g., x^2, x^3). Indeed, we suppose that such state exists in a long isotropic specimen subjected to a tension test as described in Section 5.10. One can measure the average extension ϵ_{11} in a gage length and even the lateral contraction ($\epsilon_{22}, \epsilon_{33}$). Practically, that is the means to determine the modulus E.

5.20.2 Simple Shear

Consider the state of simple shear:

$$s^{12} = \bar{\tau}, \qquad s^{11} = s^{22} = s^{33} = s^{13} = s^{23} = 0. \tag{5.80}$$

According to (5.66)

$$2\epsilon_{12} \equiv \Phi'_{12} = \bar{\tau}/G, \qquad \epsilon_{11} = \epsilon_{22} = \epsilon_{33} = \epsilon_{13} = \epsilon_{23} = 0. \tag{5.81}$$

This serves to identify the coefficient G as the *shear modulus* of elasticity. Note that Φ'_{12} is the change of angle between lines (x^1 and x^2) and is often employed as the shear strain.

The state of simple shear satisfies equilibrium; the consequent strain is compatible. The state is not so easily realized, but closely simulated when

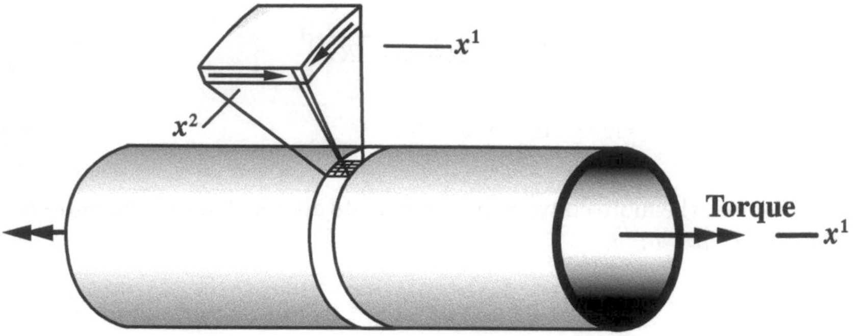

Figure 5.5 Shear strain via torsion

a thin-walled isotropic cylinder is subjected to simple torsion. Then, such state occurs in the axial x^1 and circumferential x^2 directions as depicted in Figure 5.5.

5.20.3 Uniform Hydrostatic Pressure

If a body is subjected to a uniform pressure p,

$$s^{ij} = -p\delta_{ij}.$$

Then, according to (5.66) the dilatation (see Section 3.13) is proportional to the pressure:

$$\epsilon_{ii} = -\frac{3}{3\lambda + 2G}p. \tag{5.82}$$

Alternatively,

$$p = -K\epsilon_{ii}, \tag{5.83}$$

where

$$K = \lambda + \tfrac{2}{3}G. \tag{5.84}$$

The coefficient K is called the *bulk modulus.*

According to (5.78), (5.79), and (5.84),

$$\lambda = \frac{\nu E}{(1+\nu)(1-2\nu)} = \frac{2\nu G}{1-2\nu}, \tag{5.85}$$

$$K = \frac{E}{3(1-2\nu)}, \qquad G = \frac{E}{2(1+\nu)}. \qquad (5.86), (5.87)$$

Note that $K \to \infty$ as $\nu \to \frac{1}{2}$; in words, the material approaches incompressibility.

The free energy (5.63) for a simple isothermal dilatation ($\epsilon_{ij} = \epsilon\delta_{ij}$) has the form

$$\rho_0\mathcal{F} = \tfrac{9}{2}K\epsilon^2. \tag{5.88}$$

For an isothermal distortion ($\epsilon_{ij} = \eta_{ij}$, $\epsilon_{ii} = 0$)

$$\rho_0\mathcal{F} = G\eta_{ij}\eta_{ij}. \tag{5.89}$$

For an isothermal state of simple tension [($\epsilon_{11} = \epsilon$, $\epsilon_{22} = -\nu\epsilon$, $\epsilon_{33} = -\nu\epsilon$) and ($\epsilon_{12} = \epsilon_{23} = \epsilon_{13} = 0$)]

$$\rho_0\mathcal{F} = \frac{E}{2}\epsilon^2. \tag{5.90}$$

The free-energy function $\mathcal{F}$ must be a minimum in the unstrained state; that is, positive work is required to cause any deformation. Therefore, it follows from (5.88) to (5.90) that

$$K > 0, \qquad G > 0, \qquad E > 0. \tag{5.91}$$

From (5.86), (5.87), and (5.91) it follows that

$$-1 < \nu < 1/2. \tag{5.92}$$

5.21 Alternative Forms of the Energy Potentials

Engineers tend to use the coefficients ν and E (or G) rather than λ. In terms of ν and E, the free-energy function of (5.63) takes the form:

$$\rho_0\mathcal{F} = \frac{E}{2(1+\nu)}\left(\epsilon_{ij}\epsilon_{ij} + \frac{\nu}{1-2\nu}\epsilon_{ii}\epsilon_{jj}\right) - \frac{E}{1-2\nu}\alpha(T-T_0)\epsilon_{ii}. \tag{5.93}$$

The stress components are expressed in terms of the strain components and temperature as follows:

$$s^{ij} = \frac{E}{1+\nu}\left(\epsilon_{ij} + \frac{\nu}{1-2\nu}\epsilon_{kk}\delta_{ij}\right) - \frac{E}{1-2\nu}\alpha(T-T_0)\delta_{ij}. \tag{5.94}$$

The Gibbs potential has the form

$$\rho_0 \mathcal{G} = -\frac{1+\nu}{2E}\left(s^{ij}s^{ij} - \frac{\nu}{1+\nu}s^{ii}s^{jj}\right) - \alpha(T-T_0)s^{ii}. \tag{5.95}$$

The strain components are expressed in terms of the stress components and temperature as follows:

$$\epsilon_{ij} = \frac{1+\nu}{E}\left(s^{ij} - \frac{\nu}{1+\nu}s^{kk}\delta_{ij}\right) + \alpha(T-T_0)\delta_{ij}. \tag{5.96}$$

In subsequent studies of incipient plastic flow, it is convenient to express the energy functions in terms of strain and stress deviators η_{ij} and $'s^{ij}$. From (3.150), (4.43a), (5.93), and (5.95), we obtain

$$\rho_0 \mathcal{F} = G\eta_{ij}\eta_{ij} + \frac{K}{2}\epsilon_{ii}\epsilon_{jj} - 3K\alpha(T-T_0)\epsilon_{ii}, \tag{5.97}$$

$$\rho_0 \mathcal{G} = -\frac{1}{4G}{}'s^{ij}\,{}'s^{ij} - \frac{1}{18K}s^{ii}s^{jj} - \alpha(T-T_0)s^{ii}. \tag{5.98}$$

Notice too that

$${}'s^{ij} = 2G\eta_{ij}. \tag{5.99}$$

5.22 Hookean Behavior in Plane-Stress and Plane-Strain

In many practical circumstances, one stress (say $\boldsymbol{s}^3 = s^{3i}\hat{\boldsymbol{\imath}}_i$) is negligible. This is often a valid assumption in thin bodies (beams, plates, or shells), wherein x^3 denotes the coordinate normal to the surfaces. Then, no work is expended upon the associated strains ($s^{3i}\epsilon_{3i} \doteq 0$). In this case, the stress-strain relation [see equation (5.49)] assumes the form:

$$s^{\alpha\beta} = C^{\alpha\beta\gamma\eta}\epsilon_{\gamma\eta} - \alpha^{\alpha\beta}(T-T_0). \tag{5.100}$$

In general, the Hookean behavior of plane stress is governed by four elastic coefficients $C^{\alpha\beta\gamma\eta}$ and three thermal coefficients $\alpha^{\alpha\beta}$. The internal-energy density takes the form:

$$\mathcal{U} = \tfrac{1}{2}C^{\alpha\beta\gamma\eta}\epsilon_{\alpha\beta}\epsilon_{\gamma\eta} - \alpha^{\alpha\beta}\epsilon_{\alpha\beta}(T-T_0). \tag{5.101}$$

If the material is isotropic,

$$C^{\alpha\beta\gamma\eta} = \frac{E}{2(1+\nu)}\left(\delta^{\alpha\gamma}\delta^{\beta\eta} + \delta^{\alpha\eta}\delta^{\beta\gamma} + \frac{2\nu}{1-\nu}\delta^{\alpha\beta}\delta^{\gamma\eta}\right), \qquad (5.102)$$

$$\alpha^{\alpha\beta} = \frac{E}{1-\nu}\alpha\delta^{\alpha\beta}. \qquad (5.103)$$

In certain other practical situations, the constraints inhibit strain components in one direction, say x^3, i.e.,

$$\epsilon_{i3} = 0, \qquad \epsilon_{\alpha\beta} = \epsilon_{\alpha\beta}(x^1, x^2).$$

That is the circumstance of *plane-strain*. For a Hookean material, equations (5.49) take the form:

$$s^{ij} = E^{ij\alpha\beta}\epsilon_{\alpha\beta} - \alpha^{ij}(T - T_0),$$

which for an isotropic material yield:

$$s^{\alpha\beta} = 2G\left(\epsilon_{\alpha\beta} + \frac{\nu}{1-2\nu}\delta^{\alpha\beta}\,\gamma_{\lambda\lambda}\right) - \frac{E}{1-2\nu}\alpha(T - T_0)\delta_{\alpha\beta}, \qquad (5.104a)$$

$$s^{33} = \frac{E}{1-2\nu}\left[\frac{\nu}{1+\nu}\epsilon_{\lambda\lambda} - \alpha(T - T_0)\right], \qquad (5.104b)$$

$$s^{\alpha 3} = 0. \qquad (5.104c)$$

For arbitrary coordinates θ^i, the constitutive equations for plane-stress expressed in terms of the elastic coefficients E and ν assume the form:

$$s^{\alpha\beta} = \frac{E}{1+\nu}\left[\gamma^{\alpha\beta} + \frac{\nu}{1-\nu}g^{\alpha\beta}\,\gamma^{\gamma}_{\gamma} - \frac{1+\nu}{1-\nu}\alpha g^{\alpha\beta}(T - T_0)\right], \qquad (5.105a)$$

$$\gamma_{\alpha\beta} = \frac{1+\nu}{E}\left(s_{\alpha\beta} - \frac{\nu}{1+\nu}g_{\alpha\beta}\,s^{\gamma}_{\gamma}\right) + \alpha g_{\alpha\beta}(T - T_0). \qquad (5.105b)$$

The plane-strain equations have the following form:

$$s^{\alpha\beta} = \frac{E}{1+\nu}\left[\gamma^{\alpha\beta} + \frac{\nu}{1-2\nu}g^{\alpha\beta}\,\gamma^{\gamma}_{\gamma} - \frac{1+\nu}{1-2\nu}\alpha g^{\alpha\beta}(T - T_0)\right], \qquad (5.106a)$$

$$\gamma_{\alpha\beta} = \frac{1+\nu}{E}\left[s_{\alpha\beta} - \nu g_{\alpha\beta}\, s^{\gamma}_{\gamma} + \alpha E g_{\alpha\beta}(T - T_0)\right]. \tag{5.106b}$$

The plane stress, or strain, equations, e.g., (5.105a, b) or (5.106a, b), do not preclude nonhomogeneity; in either case, the coefficients might be functions of position (coordinates θ^i).

5.23 Justification of Saint-Venant's Principle

As previously noted (see Section 5.9), the principle of Saint-Venant is crucial to engineering practice, especially in structural design, wherein an accurate description of loadings is seldom possible. Although generally accepted and tacitly invoked, the principle is also supported by theoretical arguments and demonstrated by specific solutions.

The rational arguments of J. N. Goodier (1937) [47] are quite convincing yet devoid of mathematical frills. The essential arguments follow: It is presumed that the body is Hookean and subjected to equipollent (self-equilibrated) tractions *of order* p upon a small region; the maximum linear dimension of that region is denoted by ϵ. Such system does work only by virtue of the *relative* displacements between the sites of application (e.g., between two opposing forces of equal magnitude). If the material has modulus *of order* E, then the strain is *of order* p/E and a relative displacement is *of order* $p\epsilon/E$. The force upon any small surface δs is *of order* $p\,\delta s$ and the work done upon the relative displacement is *of order* $p^2\epsilon\,\delta s/E$. Since $\delta s \leq \epsilon^2$, the total work is *of order* $p^2\epsilon^3/E$. If a stress is *of order* p, then the consequent strain energy density is *of order* p^2/E and the total energy in a region of volume $\tilde{v}$ is *of order* $p^2\tilde{v}/E$. However, the total work done by the equipollent tractions is manifested in the strain energy; it is *of order* $p^2\epsilon^3/E$. One is thereby led to the conclusion that most of the internal energy, and hence the associated deformations and stresses are confined to the region of loading.

Another mathematical argument can be traced to the works of O. Zanaboni (1937) [48], [49] and the subsequent works of P. Locatelli (1940 and 1941, respectively) [50], [51]. These are precise arguments, but also founded upon energy assessments; as such, they demonstrate the decay of certain averages of the stresses.

The evaluation of stresses produced by specific loadings on Hookean bodies provide more specific information regarding their decay with distances from the site of such loads. The interested reader can consult the original works of R. von Mises [52] and E. Sternberg [53].

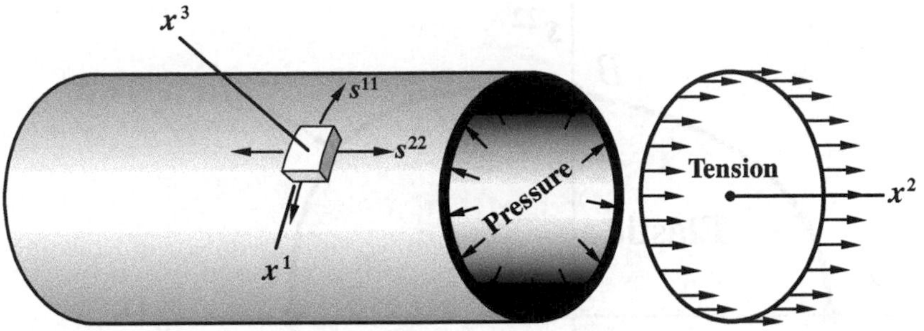

Figure 5.6 Simulation of combined stresses s^{11} and s^{22}

The principle of Saint-Venant plays an essential role in structural mechanics. Most practical theories of structural components (e.g., struts, beams, plates, and shells) do not account for the actual distributions of local loading, nor the specifics of interactions at supports and connections. Again, the practitioner must be wary of localized effects and their role in the specific structure.

Although the various mathematical arguments presume elastic behavior, our experiences and observations indicate that Saint-Venant's principle applies to inelastic behavior as well. Indeed, the yielding which might occur at a site of local loading, at a support or connection, serves to relieve concentrated stresses, i.e., reduce the local equipollent components.

The reader will find a more comprehensive account of Saint-Venant's principle in the book by Y. C. Fung [54].

5.24 Yield Condition

The term *yielding* denotes the occurrence of a permanent deformation; it is also called *plastic flow*. A mathematical statement which defines all states of incipient plastic flow is called a *yield condition*. It marks the transition from elastic to plastic deformation.

In the case of simple tension the yield condition is expressed by[‡]

$$s^{11} = Y,$$

[‡]Engineers usually compute the stress $\overset{*}{\sigma}{}^{11}$ [see (4.14b) and (4.15b)] based on the original area. However, the deformations which precede yielding are usually so small that the differences between components s^{ij}, τ^{ij}, $\overset{*}{\sigma}{}^{ij}$, and σ^{ij} are insignificant.

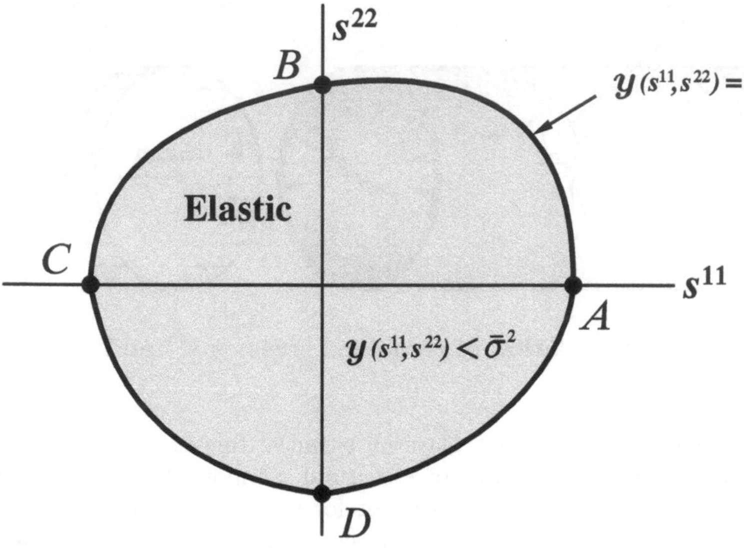

Figure 5.7 Yield condition for combined stresses s^{11} and s^{22}

where Y denotes the tensile yield stress. When $s^{11} < Y$, the deformation is elastic; when $s^{11} = Y$, plastic flow is imminent.

To extend this idea we proceed to a plane state of stress. For example, suppose that a thin circular tube is subjected to the combined action of axial tension and internal pressure as illustrated by Figure 5.6. If the wall is thin the radial normal stress is negligible and the wall is subjected to normal stresses s^{11} and s^{22} as shown. In a series of experiments, it is possible to obtain many combinations of s^{11} and s^{22} which initiate yielding; they plot a curve AB on the $(s^{11} - s^{22})$ plane of Figure 5.7.

Reversing the loads would produce negative values and a plot of the closed curve $ABCDA$. The curve is called a *yield curve.* The equation of that curve is a yield condition; it has the form

$$\mathcal{Y}(s^{11}, s^{22}) = \bar{\sigma}^2,$$

in which $\bar{\sigma}$ is a real constant. The various two-dimensional states of stress plot a point in the $(s^{11} - s^{22})$ plane. If the point lies within the curve, that is, if

$$\mathcal{Y}(s^{11}, s^{22}) < \bar{\sigma}^2,$$

then the material behaves elastically. Notice that the stresses at points A, B, C, D are the yield stresses in simple tension or compression for the circumferential and axial directions; notice too that these magnitudes may be quite different.

Plastic flow is initiated by the combined action of all stress components. In general, the yield condition must be a function of the six stress components;‡ that is,

$$\mathcal{Y}(s^{11}, s^{22}, s^{33}, s^{12}, s^{13}, s^{23}) = \bar{\sigma}^2. \tag{5.107}$$

Some theory of yielding is needed to establish the yield condition because it is not possible to perform an experiment for every conceivable state of stress. The theory sets forth the quantities which affect yielding and the role each plays. Some experiments are also needed to indicate the form of the function and to determine the magnitude of constants ascribed to specific materials.

Extensive experiments by P. W. Bridgman ([55] to [57]) have indicated that the superposition of a hydrostatic pressure has little influence on the plastic flow of most metals. Consequently, most theories disregard the mean normal (hydrostatic) part of the stress components. Then the yield condition involves only the deviator or, stated otherwise, it does not depend on the first invariant of the stress tensor.

5.25 Yield Condition for Isotropic Materials

As an alternative to (5.107) the yield condition may be expressed as a function of the principal stresses together with three variables, for example, angles, which specify the principal directions. If the material is isotropic, the orientation of the principal directions need not enter the picture. Then, the yield condition can take the form

$$\mathcal{Y}(\tilde{s}^1, \tilde{s}^2, \tilde{s}^3) = \bar{\sigma}^2, \tag{5.108}$$

where $\tilde{s}^1, \tilde{s}^2, \tilde{s}^3$ denote the principal stresses. Since the principal stresses can be expressed in terms of the stress invariants, the latter may supplant the principal stresses in the yield condition.

For an isotropic material, the state of stress is described adequately by three variables, the principal stresses $(\tilde{s}^1, \tilde{s}^2, \tilde{s}^3)$. Then a state of stress can be depicted as a point in a space† with rectangular coordinates $(\tilde{s}^1, \tilde{s}^2, \tilde{s}^3)$. This is illustrated by Figure 5.8. Also, the state of stress is represented by

‡We consider only isothermal states.

†This representation was introduced by B. P. Haigh [58] and H. M. Westergaard [59].

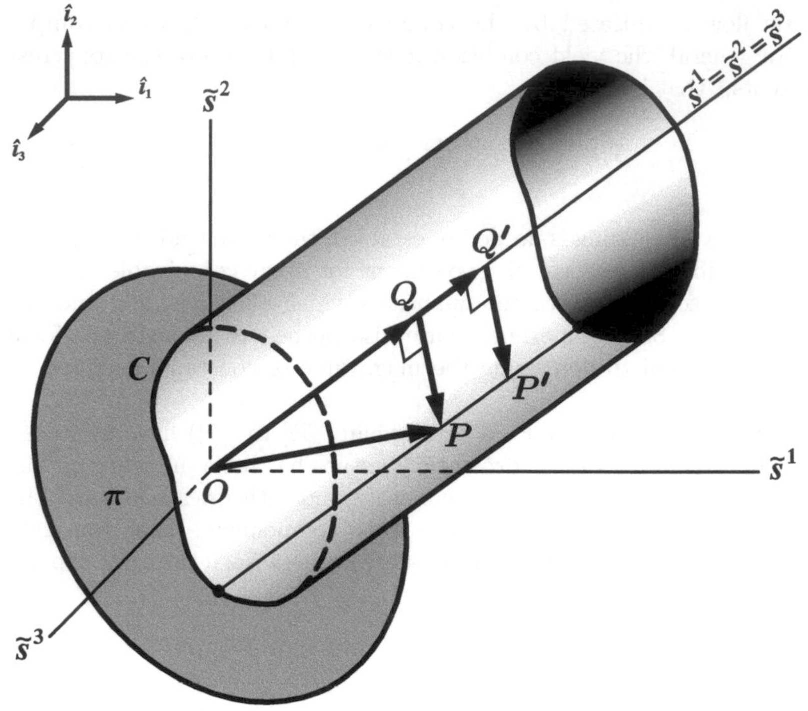

Figure 5.8 Cylindrical yield surface in the space of principal stresses $(\tilde{s}^1, \tilde{s}^2, \tilde{s}^3)$

the vector $\overline{\boldsymbol{OP}}$ which can be decomposed into a component $\overline{\boldsymbol{OQ}}$ along the line $(\tilde{s}^1 = \tilde{s}^2 = \tilde{s}^3)$ and a component $\overline{\boldsymbol{QP}}$ perpendicular to $\overline{\boldsymbol{OQ}}$.

If $p = \frac{1}{3}(\tilde{s}^1 + \tilde{s}^2 + \tilde{s}^3)$ denotes the mean normal stress, then

$$\overline{\boldsymbol{OQ}} = p(\hat{\boldsymbol{\imath}}_1 + \hat{\boldsymbol{\imath}}_2 + \hat{\boldsymbol{\imath}}_3), \tag{5.109}$$

$$\overline{\boldsymbol{QP}} = (\tilde{s}^1 - p)\hat{\boldsymbol{\imath}}_1 + (\tilde{s}^2 - p)\hat{\boldsymbol{\imath}}_2 + (\tilde{s}^3 - p)\hat{\boldsymbol{\imath}}_3, \tag{5.110a}$$

$$= {}'\tilde{s}^i\, \hat{\boldsymbol{\imath}}_i, \tag{5.110b}$$

where ${}'\tilde{s}^i$ are the principal values of the stress deviator tensor ${}'s^{ij}$. The vector $\overline{\boldsymbol{OQ}}$ represents a hydrostatic state of stress while $\overline{\boldsymbol{QP}}$ represents the deviator.

The yield condition (5.108) is represented by a surface in the stress space of Figure 5.8. If the stress components satisfy (5.108), then the vector $\overline{\boldsymbol{OP}}$

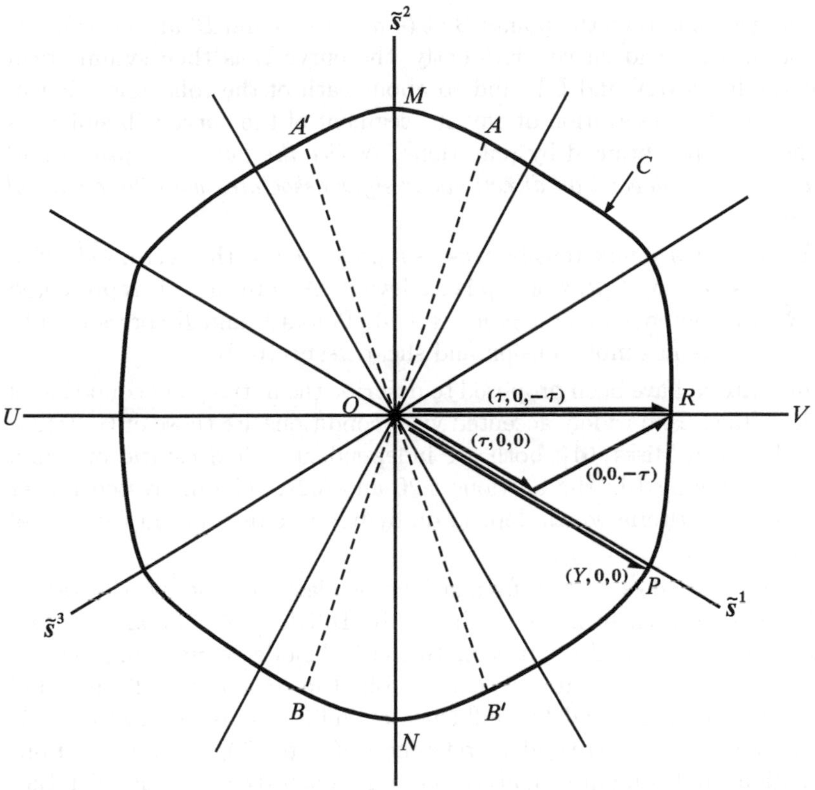

Figure 5.9 Cross-sectional curve C of the yield surface

terminates on the yield surface. If yielding is unaffected by a hydrostatic state of stress, then the components $(\tilde{s}^1 + A, \tilde{s}^2 + A, \tilde{s}^3 + A)$ satisfy (5.108) too, and the vector $\overline{\boldsymbol{OP}} + A(\hat{\boldsymbol{\imath}}_1 + \hat{\boldsymbol{\imath}}_2 + \hat{\boldsymbol{\imath}}_3)$ also terminates on the yield surface. Geometrically, if P is a point on the yield surface, then the point P', obtained by adding the vector $\overline{\boldsymbol{QQ'}} = A(\hat{\boldsymbol{\imath}}_1 + \hat{\boldsymbol{\imath}}_2 + \hat{\boldsymbol{\imath}}_3)$—a hydrostatic stress of magnitude A—is also on the surface. Evidently, the surface is a cylinder with generators parallel to the line $\tilde{s}^1 = \tilde{s}^2 = \tilde{s}^3$. All that is needed to define the surface is the shape of a cross-section.

The plane π in Figure 5.8 is perpendicular to the line $(\tilde{s}^1 = \tilde{s}^2 = \tilde{s}^3)$. The yield cylinder intersects the π plane along a curve C which defines the yield condition. One such curve is shown in Figure 5.9; the axes $\tilde{s}^i$ appear foreshortened in this view of the π plane.

Some symmetry properties of curve C are worth noting. Since we are concerned with isotropic behavior, interchanging the roles of $\tilde{s}^1$, $\tilde{s}^2$, and $\tilde{s}^3$ has no effect. For example, if point A at a stress state (a, b, c) is on C, then so is A' at (c, b, a) in Figure 5.9. If the sense of the stresses does not influence yielding, then the points B at $(-a, -b, -c)$ and B' at $(-c, -b, -a)$ are also on the yield curve. Evidently, the curve C is then symmetrical about the lines MN and UV and so about each of the solid lines shown. It follows that a description of any 30° segment of the curve will suffice as all others can be obtained by reflections. *Notice the two assumptions that yielding does not depend on directions or signs associated with the principal stresses.*

In Figure 5.9, a simple tensile stress is represented by the vector $\overline{\boldsymbol{OP}}$ with components $(Y+p,\ p,\ p)$, where p is irrelevant. A pure shear is represented by $\overline{\boldsymbol{OR}}$ with components $(\tau+p,\ p,\ -\tau+p)$. Points P and R correspond to the yield points in simple tension and shear, respectively.

Many criteria have been proposed to describe the initial yield condition of ductile metals. Two widely accepted yield conditions are those of H. Tresca [60] and R. von Mises [61]; both are independent of hydrostatic pressure. These are examined in the Sections 5.26 and 5.27. The interested reader will find other criteria which depend on hydrostatic pressure in the text of J. Lubliner [62].

For further studies on the foundations of plasticity theory, the reader is referred to the early monographs by R. Hill [63], A. Nadai [64], and L. M. Kachanov [65]. Alternative mathematical models concerning the plastic behavior of various materials can be found in references [62], [66], and [67]. Dynamic effects and the influence of finite strains on elasto-plastic material behavior are treated in references [68] to [70]. Finally, various formulations and computational aspects relevant to the solution of initial- and boundary-value problems are presented in works [71] and [72].

5.26 Tresca Yield Condition

In 1864, H. Tresca [60] proposed that yielding occurs when the maximum shear stress attains a certain magnitude. This can be expressed as follows:

$$\tilde{s}^1 - \tilde{s}^3 = \pm 2\tau \begin{Bmatrix} + & \text{if } \tilde{s}^1 \geq \tilde{s}^2 \geq \tilde{s}^3 \\ - & \text{if } \tilde{s}^1 \leq \tilde{s}^2 \leq \tilde{s}^3 \end{Bmatrix}, \tag{5.111a}$$

$$\tilde{s}^2 - \tilde{s}^3 = \pm 2\tau \begin{Bmatrix} + & \text{if } \tilde{s}^2 \geq \tilde{s}^1 \geq \tilde{s}^3 \\ - & \text{if } \tilde{s}^2 \leq \tilde{s}^1 \leq \tilde{s}^3 \end{Bmatrix}, \tag{5.111b}$$

$$\tilde{s}^1 - \tilde{s}^2 = \pm 2\tau \begin{Bmatrix} + & \text{if } \tilde{s}^1 \geq \tilde{s}^3 \geq \tilde{s}^2 \\ - & \text{if } \tilde{s}^1 \leq \tilde{s}^3 \leq \tilde{s}^2 \end{Bmatrix}. \tag{5.111c}$$

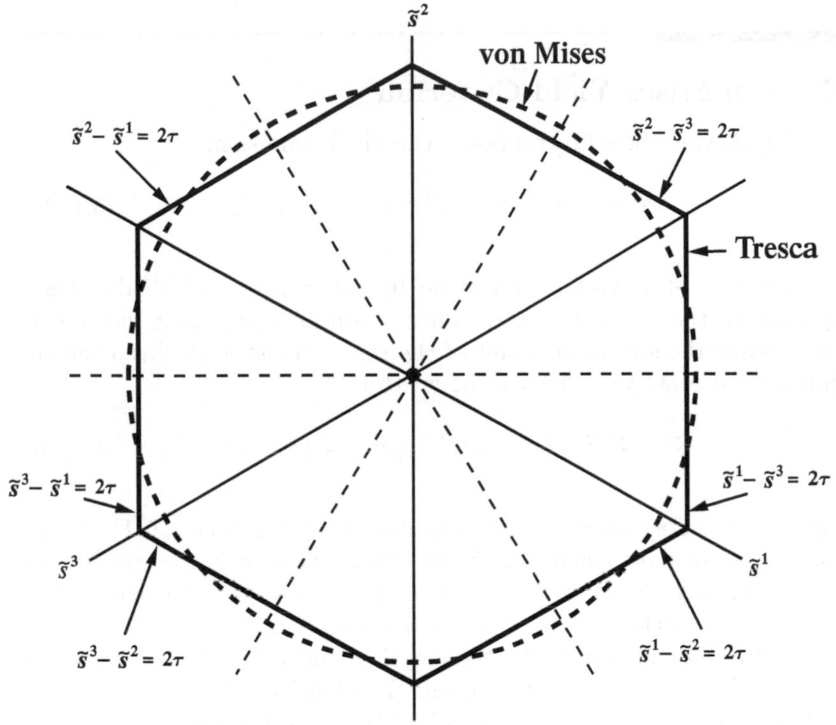

Figure 5.10 Tresca and von Mises yield curves

In equations (5.111a–c), τ is a positive constant to be determined experimentally. These are the equations of six planes which form a regular hexagonal cylinder parallel to the line $\tilde{s}^1 = \tilde{s}^2 = \tilde{s}^3$. The yield curve C (the cross-section of the cylinder) in the π plane is shown in Figure 5.10.

The Tresca yield condition is also given by

$$\left[(\tilde{s}^1 - \tilde{s}^3)^2 - (2\tau)^2\right]\left[(\tilde{s}^2 - \tilde{s}^3)^2 - (2\tau)^2\right]\left[(\tilde{s}^1 - \tilde{s}^2)^2 - (2\tau)^2\right] = 0. \tag{5.111d}$$

When the relative magnitudes of the principal stresses are known a priori, the Tresca condition takes a simple form: it is expressed by the equation of one plane; otherwise it assumes a mathematically complicated form.

5.27 von Mises Yield Criterion

In 1913, R. von Mises [61] proposed the yield condition:

$$'s^{ij}\,'s^{ij} = ('\tilde{s}^1)^2 + ('\tilde{s}^2)^2 + ('\tilde{s}^3)^2 = \bar{\sigma}^2, \tag{5.112a}$$

where $\bar{\sigma}$ is a constant to be determined by experiment and $'s^{ij}$ denotes a component of the stress deviator. This condition states that yielding is initiated when the second invariant of the stress deviator attains a certain magnitude. Alternatively, this can be written

$$(\tilde{s}^1 - \tilde{s}^2)^2 + (\tilde{s}^2 - \tilde{s}^3)^2 + (\tilde{s}^1 - \tilde{s}^3)^2 = 3\bar{\sigma}^2. \tag{5.112b}$$

Equation (5.112a) expresses the condition that the vector $\overline{\boldsymbol{QP}}$ of Figure 5.8 has a constant magnitude $\bar{\sigma}$. In other words, it is the equation of a circular cylinder with axis along the line ($\tilde{s}^1 = \tilde{s}^2 = \tilde{s}^3$). The curve C in the π plane is a circle of radius $\bar{\sigma}$ as shown dotted in Figure 5.10.

It is natural to inquire as to the physical meaning of the von Mises condition. Two interpretations prevail: H. Hencky [73], in 1924, noted that the elastic energy of distortion is $'s^{ij}\,'s^{ij}/4G$ [see equation (5.98)] and interpreted the von Mises condition as stating that yielding is initiated when the elastic energy of distortion reaches a critical value. Later, A. Nadai [9] (1937) observed that the net shear stress on the octahedral plane (its normal trisects the principal directions) has magnitude $('s^{ij}\,'s^{ij}/3)^{1/2}$. Accordingly, he interpreted the von Mises condition to mean that yielding is initiated when the octahedral shear stress reaches a critical value. The invariant $('s^{ij}\,'s^{ij})^{1/2}$ is proportional to the so-called *generalized* or *effective stress*.

Independently, F. Schleicher [74], [75] (1925 and 1926, respectively) and R. von Mises [76] (see the brief discussion of the paper by F. Schleicher [75]) generalized the von Mises yield condition by proposing that the constant $\bar{\sigma}$ be a function of the mean normal stress. This introduces the influence of hydrostatic stress. Such yield condition is represented graphically by a surface of revolution about the line ($\tilde{s}^1 = \tilde{s}^2 = \tilde{s}^3$), e.g., a cone.

The constants τ and $\bar{\sigma}$ of (5.111a–d) and (5.112a, b) may be determined by one experiment, for example, a simple tension test. If Y denotes the yield stress in simple tension, from (5.111a–d) and (5.112a, b)

$$\tau = \frac{Y}{2}, \qquad \bar{\sigma} = \sqrt{\frac{2}{3}}\,Y.$$

Then, the von Mises circle circumscribes the Tresca hexagon of Figure 5.10

and the yield stresses predicted for simple shear will differ by about 15%. Similarly, if τ and $\bar{\sigma}$ are chosen to give agreement in simple shear, discrepancies arise for other states of stress. A compromise is obtained by choosing the constants so that

$$\bar{\sigma} = \sqrt{2}\, m\tau,$$

where

$$1 < m < 2/\sqrt{3}.$$

The greatest difference between the yield stress components predicted by the two conditions is always less than 8% if m is suitably chosen.

5.28 Plastic Behavior

Once yielding is initiated, the plastic flow may or may not persist. It is necessary to examine the requirements for subsequent plastic deformation and the stress-strain relations which govern them. The relations are of two types, those for *ideally plastic* and those for *strain-hardening* materials. The former is characterized by a yield condition which is unaltered by plastic deformation; as the term implies, this is an idealized material, though a few (notably, structural steel) do exhibit nearly ideal plasticity. Most materials are altered by inelastic deformations and, specifically, the yield condition is modified; these are usually termed strain-hardening, as their resistance to yielding is increased.

Without loss of generality, we can suppose that the dependent variables are cast in a Cartesian system x^i, wherein $\gamma_{ij} = \epsilon_{ij}$. Moreover, we couch our formulations in conventional terms with commensurate limitations, viz., strains remain small such that prior and subsequent elastic strains obey Hooke's law. Although we eventually require relations in terms of all six components of stress and strain, we begin by examining the behavior in simple situations.

In the case of *simple tension*, or *shear*, the essential quantities are a single stress component s and the corresponding strain component ϵ. The behavior is then described by the familiar stress-strain diagram. The diagram in Figure 5.11 is typical of a strain-hardening metal. It is linear elastic if the stress does not violate the yield condition, $-\overline{Y}_2 < s < \overline{Y}_1$, and has a positive slope everywhere. Continuous loading from O to B causes a plastic deformation. Unloading traces the curve BO' parallel to AO. The material is again stress-free but has been altered by the plastic strain $\epsilon^P = OO'$. Notice that the strain when at B consists of a plastic part ϵ^P and an elastic part $\epsilon^E = O'P = Y_1/E$ recovered during unloading. Reloading from O'

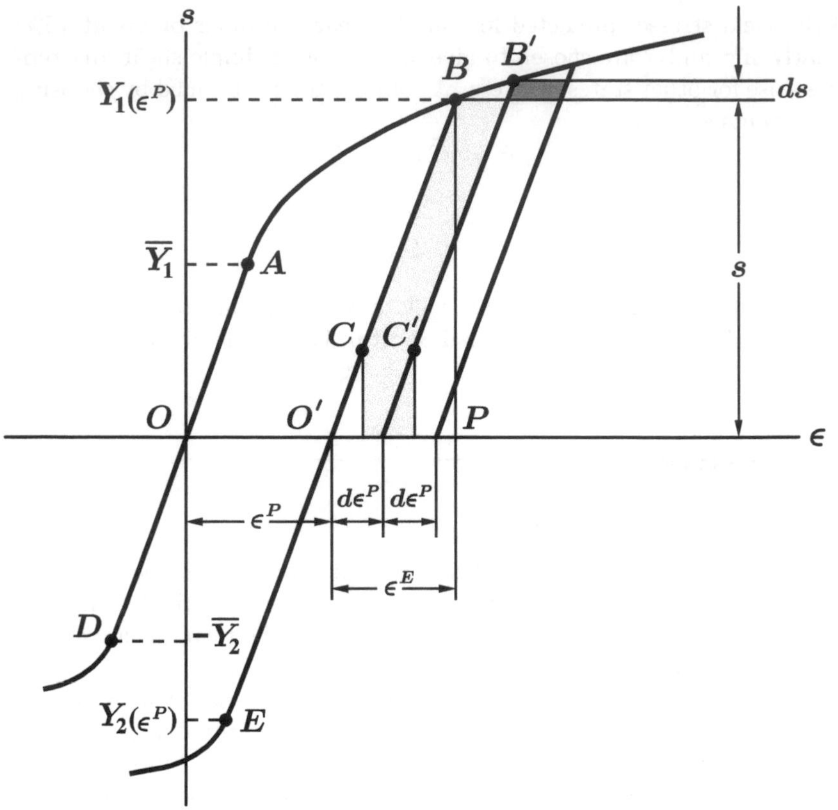

Figure 5.11 Simple stress-strain behavior of a typical metal

produces elastic strain (follows EB) if the stress does not violate the new yield condition Y_1. It is characteristic of a strain-hardening material that *each increment* of plastic flow results in a new condition for subsequent plastic flow. The prevailing yield condition depends on the entire history of plastic deformation; mathematically, the yield stress is a functional, for example,

$$Y = \int_0^{\epsilon^P} f(\epsilon^P)\, d\epsilon^P,$$

where Y signifies the current yield stress in tension or compression.

From our observations of the simple tension test, we can assert that plastic flow in tension will occur only if

$$s = Y, \tag{5.113a}$$

and

$$ds > 0. \tag{5.113b}$$

Then

$$ds = k\, d\epsilon^P, \tag{5.114a}$$

where k is a positive scalar. In any case, a change in stress is accompanied by an elastic strain

$$ds = E\, d\epsilon^E. \tag{5.114b}$$

The small strain increment is the sum of the elastic and plastic parts:

$$d\epsilon = d\epsilon^E + d\epsilon^P = \left(\frac{1}{E} + \frac{1}{k}\right) ds. \tag{5.115}$$

Bear in mind that it is essential to know the entire history of plastic strain as it determines the prevailing yield condition.

A general stress-strain relation involves six stress components and the corresponding six strain components. It cannot be represented by a single diagram, nor can the strain-hardening material be characterized by the positive slope of a diagram. Some additional concepts are needed, and to this end we examine the energy expended during plastic flow. Suppose that the material is in a deformed state represented by point C of Figure 5.11. It is then subjected to a cycle of loading and unloading in which the stress is just sufficient to enforce an incremental plastic strain $d\epsilon^P$. The energy expended is

$$s\, d\epsilon^P > 0.$$

It is represented by the "shaded" area of Figure 5.11. During a second cycle which causes a second increment of plastic strain, the energy expended is

$$s\, d\epsilon^P + ds\, d\epsilon^P > s\, d\epsilon^P.$$

Therefore,

$$ds\, d\epsilon^P > 0, \qquad d\epsilon^P \neq 0. \tag{5.116a, b}$$

More and more energy is expended to enforce each successive plastic strain increment. The additional work is $(ds\, d\epsilon^P)$. It is represented graphically by the area of the "solid" parallelogram in Figure 5.11.

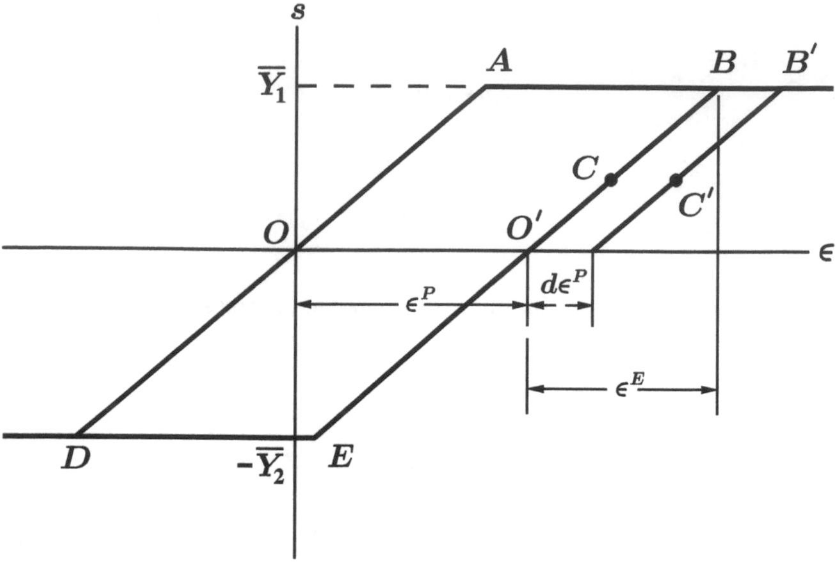

Figure 5.12 Stress-strain diagram of a linear elastic, ideally plastic material

A linear elastic, ideally plastic material has the simple stress-strain diagram of Figure 5.12. It differs from the diagram for a strain-hardening material in that the yield condition is not altered by plastic deformation. In other respects it is similar; the behavior is linear and elastic if the stress does not violate the yield condition, $-\overline{Y}_2 < s < \overline{Y}_1$. Plastic flow in tension will occur only if

$$s = Y, \tag{5.117a}$$

and

$$ds = 0. \tag{5.117b}$$

Notice that the plastic strain has an indefinite magnitude. As before, any change of stress is accompanied by an elastic strain; $d\epsilon^E = ds/E$. In contrast with the strain-hardening material, no additional work is needed to cause successive increments of plastic strain; (5.116a, b) is supplanted by

$$ds\, d\epsilon^P = 0, \qquad d\epsilon^P \neq 0. \tag{5.118a, b}$$

For any simple state of stress, the stress-strain relation is particularly

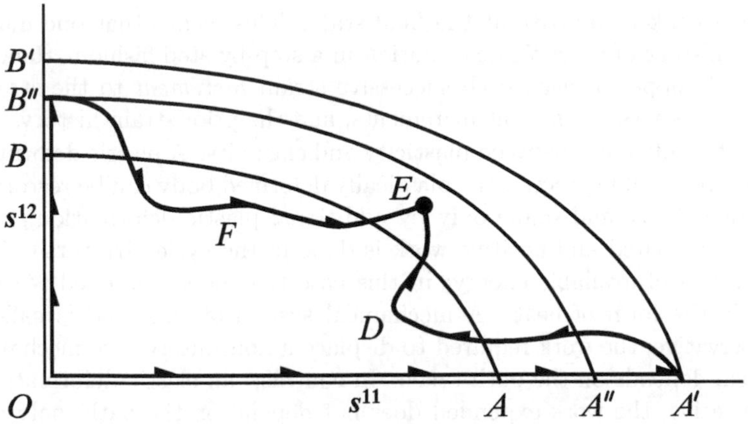

Figure 5.13 Path dependence in strain-hardening plasticity

simple *if* there is no unloading (stress varies monotonically). Indeed, then there is no need to distinguish between elastic and plastic behavior. However, when unloading occurs there is no longer a unique stress-strain relation. There are many possible strains for any given stress; the correct value will depend upon the previous plastic deformation.

To illustrate some features of plastic strain-stress relations, let us consider a thin tube (as depicted in Figure 5.5) subjected to torsion (s^{12}) *and also* tension (s^{11}). Let x^1, x^2, x^3 denote the axial, circumferential, and radial coordinates at a point of the wall. The initial yield condition is represented by curve AB in Figure 5.13. Now suppose that the tube is subjected to tension s^{11}, along path OA' in Figure 5.13, which causes yielding and establishes the new yield curve $A'B'$. If the material is isotropic and plastically incompressible, then the plastic strains are

$$\epsilon_{11}^P = \epsilon, \qquad \epsilon_{22}^P = \epsilon_{33}^P = -\epsilon/2, \qquad \epsilon_{12}^P = 0.$$

Now, if the tension s^{11} is reduced and the torsion s^{12} is applied, the tube may be loaded according to path $A'DE$ in Figure 5.13. Suppose that instead of loading according to path $OA'DE$, the tube were loaded according to path $OB''FE$ where the plastic deformation establishes the yield curve $B''A''$. The plastic strain produced by this loading is essentially simple shear:

$$\epsilon_{11}^P = \epsilon_{22}^P = \epsilon_{33}^P = 0, \qquad \epsilon_{12}^P = \gamma.$$

Notice that paths $A'DE$ and $B''FE$ are accompanied by elastic deformations only. Thus a state of stress E may be accompanied by entirely different plastic strains and a different yield condition may prevail depending on the path taken to arrive at the final state. This means that one must trace the history of a plastic deformation in a step-by-step fashion. Hence, we can only hope to relate each successive strain *increment* to the stress components, stress-component increments, and the prior strain history.

Notice the difference between plasticity and elasticity. A plastic deformation is an irreversible process; the plastically deformed body can be restored to its original size and shape only by additional plastic deformation, but energy is dissipated and positive work is done in the cycle. Irreversibility implies a loss of available energy; in this case the energy supplied is dissipated in the form of heat. A mechanical system of this kind is called nonconservative; the work required to displace a nonconservative mechanical system depends on the path taken. In contrast, an elastic deformation is conservative; the work expended does not depend on the path, only on the initial and final states.

Two types of stress-strain relations should be distinguished. A *total stress-strain relation* denotes a relation between the stress and *total strain components*, while an *incremental stress-strain relation* denotes a relation between stress (and stress increments) and *strain increments*. The former does not take account of the path and, for this reason, it is not wholly acceptable as a description of plastic behavior.

5.29 Incremental Stress-Strain Relations

In general, yielding can occur only if the state of stress satisfies the prevailing yield condition. The condition is expressed mathematically by an equation involving six stress components:

$$\mathcal{Y}(s^{ij}) = \bar{\sigma}^2, \tag{5.119}$$

where $\bar{\sigma}$ is a parameter which characterizes the yield strength of the material.

Most materials are altered by permanent strain and the yield condition is continually changing during plastic deformation. If the material strain hardens, then the function $\mathcal{Y}$ and the parameter $\bar{\sigma}$ change during plastic deformation; the change is termed *cold-working*. Only the ideally plastic material is unaffected by plastic strain and only then the function $\mathcal{Y}$ and the parameter $\bar{\sigma}$ are fixed.

The conditions for subsequent yielding and the dependence of the yield criterion on prior deformations are defined by so-called *hardening rules*. Two forms are distinct: *isotropic hardening* and the *kinematic hardening*.

The former is the simplest hardening model and implies that the yield surface expands with no change of shape and no shift of the origin. However, the usefulness of this model in predicting real material behavior is limited. The latter, kinematic hardening, corresponds to translation of the yield surface with no change of size or shape. Combination of the aforementioned hardening laws is also possible (*mixed hardening* rules). Alternative models and generalizations may be found, e.g., in [62] and [65] .

In general, the condition (5.119) depends on the plastic-strain *history*. However, we cannot expect to express the yield condition as a function of the stress and plastic-strain components. If, for example, a specimen were plastically extended and subsequently compressed to its original size and shape, no permanent strain would remain; yet, in all likelihood the yield condition would be altered. Mathematically speaking, the yield condition must be expressed as a *functional* of the plastic strain; that is, an integral extending throughout the plastic strain interval $(0, \epsilon_{ij}^P)$.

In most cases, incremental changes in the state of stress are accompanied by unique increments of strain and an incremental change of the yield condition $\mathcal{Y}$ is implied, namely,

$$d\mathcal{Y} \equiv \frac{\partial \mathcal{Y}}{\partial s^{ij}}\, ds^{ij}. \tag{5.120a}$$

Again, the functional $\mathcal{Y}$ may depend on the loading path in stress space, and the increment $d\mathcal{Y}$ need not be an exact differential.

The yield condition may be regarded as a surface in stress space, it changes when the material strain hardens. When the stress point lies within the yield surface, $\mathcal{Y}(s^{ij}) < \bar{\sigma}^2$, the behavior is elastic. Yielding occurs only when $\mathcal{Y}(s^{ij}) = \bar{\sigma}^2$. Moreover, the strain-hardening material yields only if the stress point moves outside of the prevailing yield surface and thereby changes the yield condition:

$$d\mathcal{Y} \equiv \frac{\partial \mathcal{Y}}{\partial s^{ij}}\, ds^{ij} > 0. \tag{5.120b}$$

If the material is ideally plastic, the yield condition does not change:

$$d\mathcal{Y} = 0. \tag{5.120c}$$

Notice that $\mathcal{Y}$ here plays a role similar to s in equations (5.113a, b) and (5.117a, b) and $\bar{\sigma}$ has a role similar to Y. The work expended to enforce the plastic strain increment is

$$s^{ij}\, d\epsilon_{ij}^P > 0, \qquad d\epsilon_{ij}^P \neq 0. \tag{5.121a, b}$$

A strain-hardening material can suffer an incremental plastic strain if, and only if, the stress at the yield condition is augmented by a suitable stress increment. Following D. C. Drucker [77], we imagine that an external agency applies and removes the necessary stress increment ds^{ij} and does positive work during the resulting plastic flow; that is,

$$ds^{ij}\, d\epsilon_{ij}^{P} > 0, \qquad d\epsilon_{ij}^{P} \neq 0. \tag{5.122}$$

Again, we can interpret this to mean that additional work must be expended on the strain-hardening material during successive increments of plastic flow. If the material is ideally plastic,

$$ds^{ij}\, d\epsilon_{ij}^{P} = 0, \qquad d\epsilon_{ij}^{P} \neq 0. \tag{5.123}$$

Collecting the conditions for plastic flow ($d\epsilon_{ij}^{P} \neq 0$), we have

$$\mathcal{Y}(s^{ij}) = \bar{\sigma}^2, \tag{5.124a}$$

$$d\mathcal{Y} \equiv \frac{\partial \mathcal{Y}}{\partial s^{ij}}\, ds^{ij} \begin{cases} > 0, & (5.124\text{b}) \\ = 0, & (5.124\text{c}) \end{cases}$$

$$ds^{ij}\, d\epsilon_{ij}^{P} \begin{cases} > 0, & d\epsilon_{ij}^{P} \neq 0, \quad (5.125\text{a}) \\ = 0, & d\epsilon_{ij}^{P} \neq 0, \quad (5.125\text{b}) \end{cases}$$

where the inequality or equality holds accordingly as the material is strain-hardening or ideally plastic. The yield condition (5.124a) is a generalization of (5.113a); the inequalities (5.124b) and (5.125a) are generalizations of (5.113b) and (5.116a); the equalities (5.124c) and (5.125b) are generalizations of (5.117b) and (5.118a).

If the function $\mathcal{Y}$ defines a smooth surface, the conditions (5.124b, c) and (5.125a, b) are fulfilled if, and only if,‡

$$d\epsilon_{ij}^{P} = d\lambda \frac{\partial \mathcal{Y}}{\partial s^{ij}}, \tag{5.126}$$

where $d\lambda$ is a positive scalar. Equation (5.126), also known as *normality condition*, was originally proposed by R. von Mises [78] (1928) and later extended by W. Prager [79] (1949). The case in which $\partial\mathcal{Y}/\partial s^{ij}$ is not

‡Since ds^{ij} is *any* increment satisfying (5.124b, c), one can always find a ds^{ij} which violates (5.125a, b) unless $d\epsilon_{ij}^{P}$ is given by (5.126).

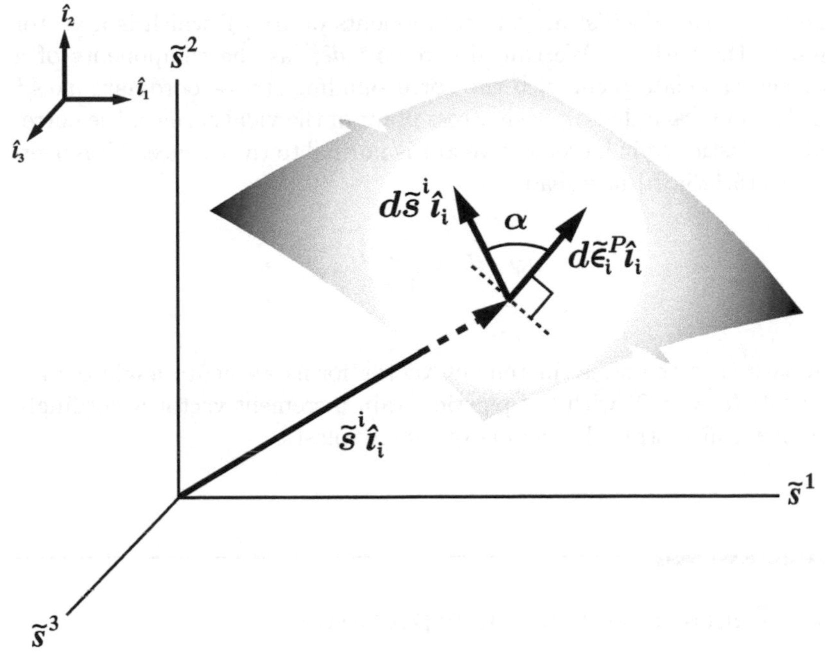

Figure 5.14 Normality of the plastic-strain increment

defined must receive special attention; this occurs at the corner of a yield surface. This circumstance has been addressed by various authors (see, e.g., W. T. Koiter [80], J. L. Sanders [81], and J. J. Moreau [82]).

According to (5.126), the stress-strain relation is linked to the yield condition since it depends on $\mathcal{Y}$. Moreover, if the yield condition is independent of the invariant s_i^i, then $\partial\mathcal{Y}/\partial s^{ij} = \partial\mathcal{Y}/\partial\,'s^{ij}$, and (5.126) involves only the stress-deviator components.

5.30 Geometrical Interpretation of the Flow Condition

For simplicity, let us apply the flow-criteria to the isotropic condition (5.108). As before, $\tilde{s}^i$ denotes a principal stress; now, $\tilde{\epsilon}_i$ denotes the corresponding component of strain. Then, (5.126) assumes the form:

$$d\tilde{\epsilon}_i^P = d\lambda \frac{\partial \mathcal{Y}}{\partial \tilde{s}^i}. \tag{5.127}$$

Now, view $\mathcal{Y}(\tilde{s}^i) = \bar{\sigma}^2$ as a surface in stress space shown in Figure 5.14.

The derivatives $\partial\mathcal{Y}/\partial\tilde{s}^i$ are the components of grad $\mathcal{Y}$ which is a vector normal to the surface. We can also regard $d\tilde{\epsilon}_i^P$ as the components of a vector and associate them with the corresponding stress-coordinate axes.‡ Then (5.127) means that, at each stress point of the yield surface, the corresponding plastic strain increment vector is normal to the surface. Moreover, equations (5.125a, b) now read

$$d\tilde{s}^i\, d\tilde{\epsilon}_i^P = \begin{cases} > 0, \\ = 0, \end{cases}$$

which says that the stress-increment vector forms an acute angle α or a right angle ($\alpha = \pi/2$) with the plastic-strain increment vector accordingly as the material is strain-hardening or ideally plastic.

5.31 Thermodynamic Interpretation

The foregoing derivation of the plastic flow relation (5.126) follows its historical development. However, the basic hypothesis (5.121a) is, in fact, an immediate consequence of the second law of thermodynamics:

According to (5.19), the energy dissipated during an incremental strain $d\epsilon_{ij}$ must satisfy the inequality

$$s^{ij}\, d\epsilon_{ij} - \rho_0(dI - T\, dS) \geq 0. \tag{5.128a}$$

The internal energy $\mathcal{I} = \mathcal{U}(\epsilon_{ij}, S)$ can be eliminated in favor of the free energy according to (5.30). Then the inequality assumes the form

$$\rho_0\, \delta_D W \equiv s^{ij}\, d\epsilon_{ij} - \rho_0\, d\mathcal{F} - \rho_0 S\, dT \geq 0. \tag{5.128b}$$

Now, in case of a reversible deformation ($d\epsilon_{ij} = d\epsilon_{ij}^E$), the *equality* holds in (5.128b) and the differential $d\mathcal{F}$ is determined in the form of (5.31):

$$d\mathcal{F} = \frac{1}{\rho_0} s^{ij}\, d\epsilon_{ij}^E - S\, dT. \tag{5.129}$$

‡This representation serves only for purposes of graphic illustration. One must not infer, for example, that principal directions of stress and strain increments are coincident.

Substituting (5.129) into (5.128b), we obtain the inequality in the form (5.121a):

$$\rho_0 \, \delta_D W = s^{ij}(d\epsilon_{ij} - d\epsilon^E_{ij}) = s^{ij} \, d\epsilon^P_{ij} \geq 0. \tag{5.130}$$

In effect, the inequality is a restatement of the second law. It asserts that the energy dissipated (per unit of undeformed volume) is always positive during plastic flow.

5.32 Tangent Modulus of Elasto-plastic Deformations

If the material strain hardens, then the yield condition is a functional of the plastic strains. Stated otherwise, an incremental plastic strain produces a first-variation of the yield condition. In accordance with (5.124a–c) to (5.126), we introduce a parameter G^P:

$$d\mathcal{Y} \equiv G^P \, d\lambda = \frac{\partial \mathcal{Y}}{ds^{ij}} \, ds^{ij} \geq 0. \tag{5.131}$$

The parameter G^P characterizes strain-hardening; it is a functional of the plastic strain history:‡

$$G^P > 0 \quad \Longleftrightarrow \quad \text{strain-hardening,}$$

$$G^P = 0 \quad \Longleftrightarrow \quad \text{ideal plasticity.}$$

A strain increment includes the elastic increment (5.55a) and the inelastic increment (5.126); for the isothermal conditions

$$d\epsilon_{ij} = D_{ijkl} \, ds^{kl} + \frac{\partial \mathcal{Y}}{\partial s^{ij}} \, d\lambda. \tag{5.132}$$

According to (5.132), (5.49), and (5.55a),

$$ds^{mn} = E^{ijmn} \, d\epsilon_{ij} - E^{ijmn} \frac{\partial \mathcal{Y}}{\partial s^{ij}} \, d\lambda. \tag{5.133}$$

‡The parameter G^P is related to the "modulus" E^P of equation (5.144) (Section 5.33).

To eliminate $d\lambda$, we multiply (5.132) by $[E^{ijmn}(\partial\mathcal{Y}/\partial s^{mn})]$, sum, and employ (5.55b) and (5.131):

$$E^{ijmn}\frac{\partial\mathcal{Y}}{\partial s^{mn}}\,d\epsilon_{ij} = \frac{\partial\mathcal{Y}}{\partial s^{mn}}\,ds^{mn} + E^{ijmn}\frac{\partial\mathcal{Y}}{\partial s^{ij}}\frac{\partial\mathcal{Y}}{\partial s^{mn}}\,d\lambda$$

$$= \left(G^P + E^{ijmn}\frac{\partial\mathcal{Y}}{\partial s^{ij}}\frac{\partial\mathcal{Y}}{\partial s^{mn}}\right)d\lambda,$$

or

$$d\lambda = \left(E^{ijmn}\frac{\partial\mathcal{Y}}{\partial s^{mn}}\,d\epsilon_{ij}\right)\Big/\left(G^P + E^{ijmn}\frac{\partial\mathcal{Y}}{\partial s^{ij}}\frac{\partial\mathcal{Y}}{\partial s^{mn}}\right). \tag{5.134}$$

Introducing (5.134) into (5.133), we obtain

$$ds^{ij} = E_T^{ijmn}\,d\epsilon_{mn}, \tag{5.135}$$

where E_T^{ijmn} is the so-called "tangent modulus":

$$E_T^{ijmn} = E^{ijmn} - \frac{E^{ijpq}E^{klmn}\dfrac{\partial\mathcal{Y}}{\partial s^{pq}}\dfrac{\partial\mathcal{Y}}{\partial s^{kl}}}{G^P + E^{ijmn}\dfrac{\partial\mathcal{Y}}{\partial s^{ij}}\dfrac{\partial\mathcal{Y}}{\partial s^{mn}}}. \tag{5.136}$$

5.33 The Equations of Saint-Venant, Lévy, Prandtl, and Reuss

In 1870, B. de Saint-Venant [83] proposed that the principal directions of the plastic-strain increment coincide with the principal directions of stress. In 1871, M. Lévy [84], [85] proposed relations between the components of the plastic-strain increment and the components of the stress deviator; these implied the coincidence of their principal directions. Independently, R. von Mises [61] in 1913 proposed the same equations, namely,

$$d\epsilon_{ij}^P = d\lambda\,{}'s^{ij}. \tag{5.137}$$

L. Prandtl [86] (1924) extended these by taking account of the elastic strain as well. In 1930, A. Reuss [87] generalized the equations of L. Prandtl.

Notice that (5.137) implies no permanent change of volume. By comparing (5.126) with (5.137), we have

$$\frac{\partial \mathcal{Y}}{\partial s^{ij}} = {}'s^{ij}.$$

We see that the stress-strain relation (5.137) is to be associated with the von Mises yield condition, namely,

$$\mathcal{Y} = \tfrac{1}{2}\,{}'s^{ij}\,{}'s^{ij} = \tfrac{1}{2}\bar{\sigma}^2. \qquad (5.138\text{a, b})$$

We define a strain-increment invariant $d\bar{\epsilon}^P$ analogous to $\bar{\sigma}$ as follows:

$$(d\bar{\epsilon}^P)^2 \equiv d\epsilon^P_{ij}\,d\epsilon^P_{ij}. \qquad (5.139)$$

Substituting (5.137) into (5.139) and using (5.138), we obtain

$$d\bar{\epsilon}^P = d\lambda\,\bar{\sigma}, \qquad d\lambda = \frac{d\bar{\epsilon}^P}{\bar{\sigma}}. \qquad (5.140\text{a, b})$$

Inserting this expression for $d\lambda$ into (5.137), we have

$$d\epsilon^P_{ij} = \frac{1}{\bar{\sigma}} d\bar{\epsilon}^P\,{}'s^{ij}. \qquad (5.141)$$

Since $d\epsilon^P_{ii} = 0$, the work expended (5.130) to cause the inelastic strain $d\epsilon^P_{ij}$ is

$$\rho_0\,d\,{}_DW = s^{ij}\,d\epsilon^P_{ij} = \left({}'s^{ij} + \frac{s^{kk}}{3}\delta_{ij}\right) d\epsilon^P_{ij}$$

$$= {}'s^{ij}\,d\epsilon^P_{ij}. \qquad (5.142)$$

According to (5.137), (5.138b), (5.140a), and (5.142)

$$\rho_0\,d\,{}_DW = d\lambda\,{}'s^{ij}\,{}'s^{ij} = d\lambda\bar{\sigma}^2 = \bar{\sigma}\,d\bar{\epsilon}^P. \qquad (5.143)$$

Let us assume that $\bar{\sigma}$ is a function of the net work expended, that is,‡

$$\bar{\sigma} = H\left(\int \rho_0\,d\,{}_DW\right) = H\left(\int \bar{\sigma}\,d\bar{\epsilon}^P\right).$$

‡This assumption and its physical basis are discussed by R. Hill [63].

Since the work is path-dependent, we *cannot* expect a solution in the form $\bar{\sigma} = \bar{\sigma}(\bar{\epsilon}^P)$. However, loading implies the existence of a positive modulus:

$$E^P = \frac{d\bar{\sigma}}{d\bar{\epsilon}^P}. \tag{5.144}$$

Then, using (5.144) we can rewrite (5.141) as

$$d\epsilon^P_{ij} = \frac{d\bar{\sigma}}{E^P\bar{\sigma}}\,{}'s^{ij}. \tag{5.145}$$

The scalar E^P plays a role analogous to the elastic modulus E. It may be determined from a simple stress-strain test. For example, if s and ϵ^P denote a simple tension stress and the corresponding plastic extensional strain,

$$\bar{\sigma} = \sqrt{\frac{2}{3}}\, s$$

and, if the material is plastically incompressible,

$$d\bar{\epsilon}^P = \sqrt{\frac{3}{2}}\, d\epsilon^P.$$

Therefore,‡

$$E^P = \frac{2}{3}\frac{ds}{d\epsilon^P}.$$

If the material is ideally plastic, $\bar{\sigma}$ is constant, E^P is zero, and the scalar $d\lambda$ (or $d\bar{\epsilon}^P$) is indeterminate. In reality, the extent of the plastic flow is determined by the constraints imposed by boundaries or by adjacent material of the body.

When the elastic deformations are neglected, the behavior is termed *rigid-plastic.* Then equations (5.137) or (5.145) are the complete stress-strain relations. They are associated with the names of M. Lévy and R. von Mises. When elastic deformations are taken into account, they are described by

$$d\epsilon^E_{ij} = \frac{d\,{}'s^{ij}}{2G} + \frac{ds^{kk}}{9K}\delta_{ij}. \tag{5.146}$$

Then, the stress-strain relations are usually associated with the names of L. Prandtl and A. Reuss.

‡ E^P is related to the hardening parameter G^P by $G^P = \bar{\sigma}^2 E^P$.

5.34 Hencky Stress-Strain Relations

In 1924, H. Hencky [73] proposed a total stress-strain relationship of the form

$$\epsilon_{ij}^P = \phi\, 's^{ij}, \tag{5.147}$$

where ϕ is a scalar, positive or zero: In the case of strain-hardening, $\phi > 0$ if, and only if, $d\mathcal{Y} > 0$; $\phi = 0$ if $d\mathcal{Y} \leq 0$. In the case of ideal plasticity, $\phi \geq 0$ if $d\mathcal{Y} = 0$, $\phi = 0$ if $d\mathcal{Y} < 0$. We can show that this is an integrated version of (5.145).

Suppose we apply the incremental stress-strain relation (5.145) to a so-called *radial* loading path. Such a path follows a straight line emanating from the origin in Figure 5.14. Then, the stress components are maintained in fixed rations:

$$'s^{ij} = C\, 's_0^{ij},$$

where $'s_0^{ij}$ are constants and C is a scalar. According to (5.138a, b), (5.139), and (5.145),

$$d\epsilon_{ij}^P = \frac{'s_0^{ij}}{E^P}\, dC,$$

$$d\bar{\epsilon}^P = \frac{\bar{\sigma}_0}{E^P}\, dC,$$

where $\bar{\sigma}_0^2 \equiv 's_0^{ij}\, 's_0^{ij}$. Eliminating dC/E^P, we have

$$d\epsilon_{ij}^P = \frac{'s_0^{ij}}{\bar{\sigma}_0}\, d\bar{\epsilon}^P,$$

and integrating, we obtain

$$\epsilon_{ij}^P = \frac{\bar{\epsilon}^P}{\bar{\sigma}_0}\, 's_0^{ij} = \frac{\bar{\epsilon}^P}{\bar{\sigma}}\, 's^{ij}.$$

This is equation (5.147), wherein

$$\phi = \bar{\epsilon}^P / \bar{\sigma}.$$

The Hencky relations take no account of differing paths of loading.

5.35 Plasticity without a Yield Condition; Endochronic Theory

The yield criteria of classical plasticity signal an abrupt transition from the conditions of elasticity to inelasticity. Consequently, portions of the body are governed by the constitutive equations of elasticity and adjoining portions by the equations of inelasticity. Moreover, the interface moves as yielding progresses. In practice, different computational procedures are required for the elastic and inelastic parts; additionally, a procedure is needed to follow the movement of the interface as loading progresses. Of course, such abrupt yielding is not characteristic of all materials. Hence, various authors have advanced theories that admit a gradual evolution of inelastic strain from the onset of loading. Here, we present rudimentary features of such theory and cite works which offer numerous refinements and alternatives. As before, we confine our attention to isotropic behavior.

Theories of inelasticity typically define a variable that increases monotonically with changes of state. A. C. Pipkin and R. S. Rivlin [88] introduced a second invariant of the strain increment

$$d\gamma^2 = d\epsilon_{ij}\, d\epsilon^{ij}. \tag{5.148}$$

The variable γ can be viewed as an arc-length in strain space; it has been called "time" since it measures the events which alter the material. A generalization was given by K. C. Valanis [89] who introduced an "intrinsic time measure" ζ:

$$d\zeta^2 = a^2 p^{ijkl}\, d\epsilon_{ij}\, d\epsilon_{kl}. \tag{5.149}$$

In addition, K. C. Valanis introduced an "intrinsic time scale" $z = z(\zeta)$ such that

$$\frac{dz}{d\zeta} > 0 \qquad (0 < \zeta < \infty). \tag{5.150}$$

The latter offers a means to fit the stress-strain relation of a given material. If $z = \zeta$ and $a^2 p^{ijkl} = g^{ik} g^{jl}$, then $\zeta = \gamma$, the arc-length (5.148).

As a simple example, one might adopt certain assumptions: (i) additivity of elastic and inelastic strain increments, (ii) small strain, (iii) no permanent (inelastic) volumetric strain, and (iv) Hookean elastic response. Then,

$$d\,{}'s^{ij} = 2G\, d\eta^E_{ij} = 2G(d\eta_{ij} - d\epsilon^P_{ij}), \tag{5.151}$$

$$ds^i_i = 3K\, d\epsilon^i_i. \tag{5.152}$$

Recall that $'s^{ij}$ denote the components of the stress deviator tensor. In a form, similar to the Lévy relation (5.141), let

$$d\epsilon^P_{ij} = \frac{dz}{\lambda}\,'s^{ij} \qquad \left(\frac{dz}{\lambda} > 0\right). \tag{5.153}$$

The elimination of the inelastic increment in (5.151) and (5.153) gives the differential equation:

$$\frac{d\,'s^{ij}}{dz} + \frac{2G}{\lambda}\,'s^{ij} = 2G\frac{d\eta_{ij}}{dz}. \tag{5.154}$$

Formally, a "solution" is expressed by the "hereditary" integral:‡

$$'s^{ij} = 2G\int_{z_0}^{z} e^{-(2G/\lambda)(z-\tau)}\,d\eta_{ij}(\tau). \tag{5.155}$$

The form (5.155) is encompassed by Valanis' "endochronic" theory. He proposed a particularly simple relation between the "intrinsic time" z and the "strain measure" ζ, specifically,

$$\frac{d\zeta}{dz} = (1+\beta\zeta) \qquad (\beta > 0), \tag{5.156}$$

wherein

$$d\zeta^2 = d\epsilon^{ij}\,d\epsilon_{ij}. \tag{5.157}$$

Although the relation (5.153) has the appearance of the Lévy equation (5.141), the present theory exhibits marked differences. Most notably, some inelastic deformation occurs at the onset of loading, though the initial response (at $\zeta = z = 0$) is elastic, viz.,

$$d\,'s^{ij} = 2G\,d\eta_{ij}, \qquad ds^i_i = 3K\,d\epsilon^i_i.$$

Some attributes of the theory are evident in a simple stress-strain relation: Consider, for example, monotonic loading in simple shear, i.e., $(\eta_{12} \equiv \eta = \zeta)$ and $('s^{12} = s^{12} \equiv s)$; from the initial state $(z = \zeta = s = 0)$, we have for the solution of (5.154):

$$s = \frac{2G(1+\beta\zeta)}{\alpha+\beta}\left[1-(1+\beta\zeta)^{-[1+(\alpha/\beta)]}\right], \tag{5.158}$$

‡Note that z is not strictly independent, but depends on the strain η_{ij} via (5.149) and (5.153) [see also (5.156) and (5.157)].

where $\alpha \equiv 2G/\lambda$. At large strain η, the function (5.158) approaches the line ($\eta_{12} \equiv \eta = \zeta$)

$$s = \frac{2G}{\alpha + \beta}(1 + \beta\eta). \tag{5.159}$$

As noted by K. C. Valanis [89], the parameter $\beta > 0$ characterizes hardening. During unloading ($d\eta = -d\zeta$) from the stressed state $s(\zeta_1) = s_1$, the stress follows:

$$s = \frac{2G(1+\beta\zeta)}{\alpha+\beta}\left[-(1+\beta\zeta)^{-[1+(\alpha/\beta)]} - 1 + 2\left(\frac{1+\beta\zeta_1}{1+\beta\zeta}\right)^{[1+(\alpha/\beta)]}\right]. \tag{5.160}$$

It is interesting to note the moduli at the sites $s = 0$ and $s = s_1$, during loading (5.158) and unloading (5.160): Initially,

$$\left.\frac{ds}{d\eta}\right]_{\zeta=0} = 2G. \tag{5.161}$$

At $s = s_1$, during loading,

$$\left.\frac{ds}{d\eta}\right]_{\zeta_1} = 2G - \frac{\alpha s_1}{1+\beta\zeta_1}, \tag{5.162}$$

and during unloading,

$$\left.\frac{ds}{d\eta}\right]_{\zeta_1} = -\left.\frac{ds}{d\zeta}\right]_{\zeta_1} = 2G + \frac{\alpha s_1}{1+\beta\zeta_1}. \tag{5.163}$$

The latter is graphically evident at the point P of Figure 5.15, where the solid line follows the loading and unloading paths according to (5.158) and (5.160), respectively; unloading at P exhibits the greater modulus of (5.163). Since we anticipate stepwise linear computational procedures, we can identify unloading via the criterion:

$$'s^{ij}\dot{\epsilon}^P_{ij} < 0.$$

During such unloading, we can replace the equations of the endochronic theory, e.g., equation (5.160), by the equations of elasticity, linear or nonlinear. That modification produces the dotted trace of Figure 5.15.

The foregoing example describes a simple model which exhibits basic features of the so-called endochronic theory. Additional aspects and elaborations are given by K. C. Valanis ([90] to [92]), by Z. P. Bažant and

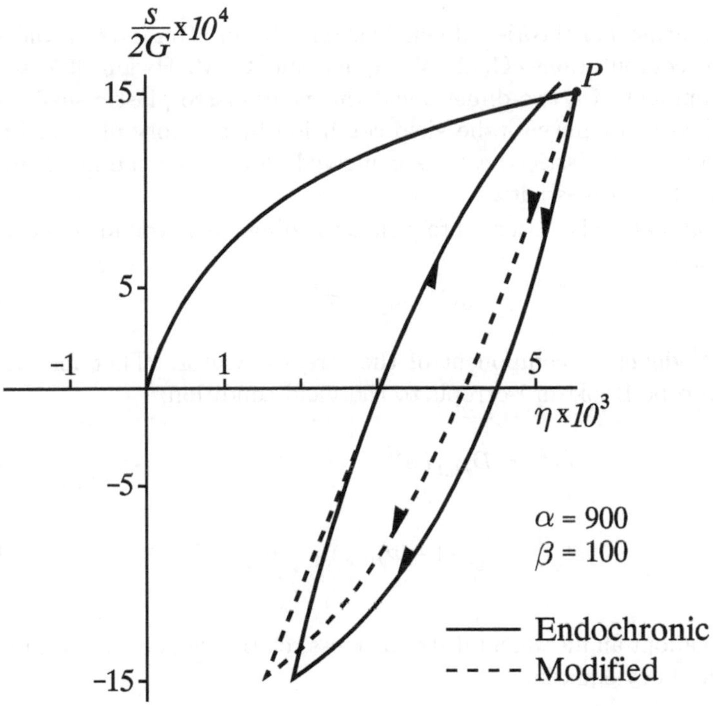

Figure 5.15 Endochronic theory: simple shear example

P. Bhat [93], and by Z. P. Bažant [94]. These also include time-dependent effects; actual time replaces "intrinsic time" to accommodate viscous behavior. A visco-elasto-plastic model is described by G. A. Wempner and J. Aberson [37]. The latter also provides thermodynamic interpretations, correlations with classical plasticity, and graphic presentations of the stress-strain relations.

5.36 An Endochronic Form of Ideal Plasticity

The bending of elasto-plastic shells is yet another important problem, wherein the classical theory predicts the initiation of yielding at a surface and the subsequent progression of inelastic layers. Practically, this can be accommodated, but only with procedures that monitor and modify the computations at numerous stations through the thickness. In effect, the shell must be treated via procedures of three dimensions as opposed to the two-dimensional theories of elastic shells. In an attempt to avoid such additional computations, G. A. Wempner and C.-M. Hwang [95] sought the development of a two-dimensional theory of elasto-plastic shells. One derived theory circumvents the yield condition by a theory of endochronic type. Here, the inelastic theory is contrived to emulate the more abrupt transition of ideal plasticity.

Our ideal material is assumed to yield according to the von Mises criterion at stress ${}_0s^{ij}$:

$$ {}_0{}'s^{ij}\, {}_0{}'s_{ij} \equiv \bar{\sigma}_0^2, \tag{5.164}$$

where ${}_0{}'s^{ij}$ denote a component of the stress deviator. That material is supposed to be Hookean isotropic to the yield condition:

$$ d\epsilon^E_{ij} = D_{ijkl}\, ds^{kl}, \tag{5.165}$$

$$ D_{ijkl} = \frac{1}{E}\big[(1+\nu)\delta_{ik}\delta_{jl} - \nu\delta_{ij}\delta_{kl}\big]. \tag{5.166}$$

Again, we adopt an incremental strain-stress relation which has the appearance of the Lévy equation:

$$ d\epsilon_{ij} = d\epsilon^E_{ij} + d\epsilon^P_{ij} \tag{5.167}$$

$$ = D_{ijkl}\, ds^{kl} + \frac{1}{E}\,{}'s^{ij}\, d\lambda. \tag{5.168}$$

Now, the relation (5.168) departs from the classical theory: ϵ^P_{ij} and $d\lambda$ are not initiated abruptly at the yield condition [i.e., (5.164)], but grow, albeit gradually, with an "intrinsic strain" ζ. To emulate the ideal plasticity, we employ the form:

$$ d\lambda = \frac{E}{\bar{\sigma}_0}\sigma^n\, d\zeta, \tag{5.169}$$

where

$$ \sigma^2 \equiv \frac{1}{\bar{\sigma}_0^2}\,{}'s^{ij}\,{}'s_{ij}, \qquad d\zeta^2 = d\epsilon^{ij}\, d\epsilon_{ij}. \tag{5.170a, b}$$

For simplicity, let us introduce the following nondimensional variables:

$$ e_{ij} \equiv \frac{E}{\bar{s}_0}\epsilon_{ij}, \qquad \sigma^{ij} \equiv \frac{s^{ij}}{\bar{s}_0}, \qquad {}'\sigma^{ij} \equiv \frac{{}'s^{ij}}{\bar{s}_0}, \tag{5.171a–c}$$

$$\bar{s}_0^2 \equiv {}_0s^{ij}{}_0s_{ij}, \qquad dz^2 \equiv \frac{E^2}{\bar{s}_0^2}\, d\zeta^2 = de_{ij}\, de^{ij}. \qquad (5.171\text{d–f})$$

If we multiply (5.168) by $(E/\bar{s}_0)$ and employ the nondimensional variables (5.171a–f), we obtain

$$de_{ij} = [(1+\nu)\delta_{ik}\delta_{jl} - \nu\delta_{ij}\delta_{kl}]\, d\sigma^{kl} + {}'\sigma_{ij}\sigma^n\, dz. \qquad (5.172)$$

During loading $d\sigma = {}'\sigma^{ij}\, d'\sigma_{ij} > 0$, elastic and inelastic strains evolve according to equation (5.172). During unloading ${}'\sigma^{ij}\, d'\sigma_{ij} < 0$, or equivalently ${}'\sigma^{ij}\, de^P_{ij} < 0$, the behavior is elastic. Then, the final term, $de^P_{ij} = {}'\sigma_{ij}\sigma^n\, dz$, is deleted from (5.172). Any subsequent unloading progresses according to (5.165); this exhibits *no* hardening.

To appreciate this model, we examine the simple states:
(i) In simple tension, we set $\epsilon_{22} = \epsilon_{33} = -(\epsilon_{11}/2)$, since inelastic strain predominates in the measure dz. Furthermore,

$${}'\sigma^{11} = \sqrt{\frac{2}{3}}\,\sigma^{11}, \qquad \sigma = \sigma^{11}, \qquad dz = \sqrt{\frac{3}{2}}\, de_{11}, \qquad (5.173\text{a–c})$$

$$de_{11} = d\sigma^{11} + \left(\sigma^{11}\right)^{n+1} de_{11}. \qquad (5.173\text{d})$$

(ii) In simple shear s^{12}:

$${}'\sigma^{12} = \sigma^{12}, \qquad \sigma = \sqrt{2}\,\sigma^{12}, \qquad dz = \sqrt{2}\, de_{12}, \qquad (5.174\text{a–c})$$

$$de_{12} = (1+\nu)\, d\sigma^{12} + \left(\sqrt{2}\,\sigma^{12}\right)^{n+1} de_{12}. \qquad (5.174\text{d})$$

Here, we set $\sqrt{2}\,\sigma^{12} \equiv \tau^{12}$; then

$$de_{12} = \frac{(1+\nu)}{\sqrt{2}}\, d\tau^{12} + (\tau^{12})^{n+1}\, de_{12}. \qquad (5.174\text{e})$$

Equations (5.173d) and (5.174e) assume the similar forms:

$$de = \frac{k}{1-\sigma^{n+1}}\, d\sigma, \qquad (5.175)$$

where $e = e_{11}$ or e_{12}, $k = 1$ or $(1+\nu)/\sqrt{2}$, and $\sigma = \sigma^{11}$ or τ^{12}. The simple

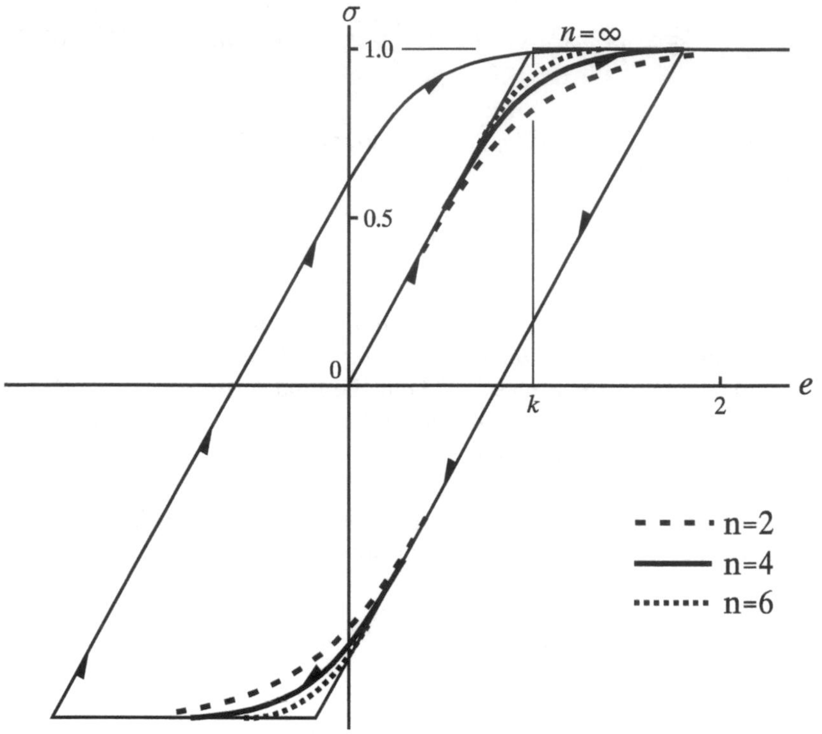

Figure 5.16 Endochronic form of ideal plasticity; response to a simple loading cycle

stress-strain relation of our model has the following properties:

$$\left.\frac{d\sigma}{de}\right]_{\sigma=0} = \frac{1}{k}, \qquad \lim_{\sigma\to 1}\frac{d\sigma}{de} = 0, \qquad \lim_{\substack{n\to\infty \\ \sigma<1}}\frac{d\sigma}{de} = \frac{1}{k}.$$

Figure 5.16 depicts loadings for various powers n and the limiting case $n = \infty$ (ideal plasticity). A cycle of loading-unloading-reloading traces the solid line ($n = 4$). Note that the material is unaltered by prior loading.

An application of this theory is included in the subsequent treatment of elasto-plastic shells (see Subsection 10.8.5).

5.37 Viscous Behavior

Some structural materials exhibit attributes of viscous fluids, particularly at elevated temperatures. For this reason, we consider briefly the essential characteristics of a viscous medium.

One striking difference between a solid and a fluid stems from their respective ability and inability to remember their pasts. An elastic solid tends to always return to its initial configuration; a viscous fluid tends to retain its volume but is otherwise indifferent to previous configurations. Indeed, the particles of a fluid can move in a manner that destroys the continuity of lines and surfaces.

Although the state of a viscous fluid does not depend on the relative positions of particles, the motion does depend on the relative velocities. In other words, the strain rates, rather than the strain components, play the important role in the description of viscous fluids.

We confine our attention to an ideal fluid, often called a *Stokesian fluid*. The ideal fluid is motionless only if the stress is hydrostatic; that is, if

$$\tau^{ij} = -pG^{ij}.$$

Here, we use the stress components τ^{ij} related to the force $\boldsymbol{t}^i$ by (4.26a). The use of $\boldsymbol{t}^i$ (force/current area) seems more meaningful since in the fluid we have no "undeformed state" (initial state or area have no meaning). In any motion the stress components depend on the strain rates.

To describe a viscous fluid, let us assume that the internal energy is a function of the dilatation and temperature.[‡] Then, the counterpart of (5.23) is

$$\mathcal{I} = \mathcal{I}(D, S). \tag{5.176}$$

In accordance with (3.148),

$$\dot{\mathcal{I}} = \frac{\partial \mathcal{I}}{\partial S}\dot{S} + \frac{\partial \mathcal{I}}{\partial D} D\, G^{ij} \dot{\gamma}_{ij}. \tag{5.177}$$

We recall the first law of thermodynamics in the form,

$$\dot{\mathcal{I}} = \dot{\Omega} + \dot{Q} - \dot{\mathcal{K}}, \tag{5.178}$$

[‡]In classical thermodynamics, it is customary to regard the internal energy as a function of pressure p and temperature T. Here, we replace the thermodynamic variable p by the dilatation D, and the temperature T by the entropy S, to parallel the development of elasticity.

wherein the derivatives are *material derivatives*; that is, the increments are associated with the same elemental mass.

In the manner of (5.17d), the power transmitted by external forces to the elemental mass is

$$\dot{\Omega} = \frac{1}{\rho}\tau^{ij}\dot{\gamma}_{ij} + \dot{\mathcal{K}}. \tag{5.179}$$

Eliminating $\dot{\Omega}$ and $\dot{K}$ from (5.178) and (5.179), we obtain

$$\dot{Q} = \dot{\mathcal{I}} - \frac{1}{\rho}\tau^{ij}\dot{\gamma}_{ij}. \tag{5.180}$$

Substituting $\dot{\mathcal{I}}$ of (5.177) into (5.180), we have

$$\dot{Q} + \frac{1}{\rho}\tau^{ij}\dot{\gamma}_{ij} = \frac{\partial \mathcal{I}}{\partial S}\dot{S} + \frac{\partial \mathcal{I}}{\partial D} D\, G^{ij}\dot{\gamma}_{ij}. \tag{5.181}$$

If the process is reversible, then according to the second law,

$$\dot{Q} = T\dot{S}. \tag{5.182}$$

The stress component of the reversible process is labeled τ_R^{ij}. With the latter notation and with (5.182), equation (5.181) takes the form:

$$T\dot{S} + \frac{1}{\rho}\tau_R^{ij}\dot{\gamma}_{ij} = \frac{\partial \mathcal{I}}{\partial S}\dot{S} + \frac{\partial \mathcal{I}}{\partial D} DG^{ij}\dot{\gamma}_{ij}. \tag{5.183}$$

Recall that the entropy and strain components are supposedly independent variables. Therefore, from (5.183) we have

$$\tau_R^{ij} = \rho D \frac{\partial \mathcal{I}}{\partial D} G^{ij} = \rho_0 \frac{\partial \mathcal{I}}{\partial D} G^{ij}, \tag{5.184a}$$

$$T = \frac{\partial \mathcal{I}}{\partial S}. \tag{5.184b}$$

Equation (5.184a) describes a hydrostatic state of stress which derives from the energy potential $\mathcal{I}$. It is called the *thermodynamic pressure.*

For an irreversible process, equation (5.182) must be replaced by the inequality

$$T\dot{S} > \dot{Q}.$$

In the manner of (5.6), we can define

$$ {}_D\dot{Q} \equiv T\dot{S} - \dot{Q}. \tag{5.185} $$

In keeping with (5.8)

$$ {}_D\dot{Q} > 0. \tag{5.186} $$

The quantity ${}_D\dot{Q}$ is the rate of heat generated by the irreversible process; for example, heat is produced if the motion is opposed by internal friction.

In accordance with (5.182) and (5.185), we define

$$ {}_R\dot{Q} \equiv \dot{Q} + {}_D\dot{Q} = T\dot{S}. \tag{5.187} $$

The quantity ${}_R\dot{Q}$ represents the rate of heat supplied during a reversible process which effects the given change of state. In accordance with (5.180)

$$ {}_R\dot{Q} = \dot{\mathcal{I}} - \frac{1}{\rho}\tau_R^{ij}\dot{\gamma}_{ij}. \tag{5.188} $$

Eliminating ${}_R\dot{Q}$ from (5.187) and (5.188), we obtain

$$ {}_D\dot{Q} = (\dot{\mathcal{I}} - \dot{Q}) - \frac{1}{\rho}\tau_R^{ij}\dot{\gamma}_{ij}. $$

But, the parenthetical term on the right is given by (5.180) and, therefore,

$$ {}_D\dot{Q} = \frac{1}{\rho}(\tau^{ij} - \tau_R^{ij})\dot{\gamma}_{ij}. \tag{5.189} $$

Recall that τ_R^{ij} is the conservative stress component, the pressure given by (5.184a). The difference

$$ \tau_D^{ij} \equiv \tau^{ij} - \tau_R^{ij} \tag{5.190} $$

is the nonconservative stress component, otherwise termed the *dissipative stress component*.

In accordance with the second law (5.186),

$$ {}_D\dot{Q} = \frac{1}{\rho}\tau_D^{ij}\dot{\gamma}_{ij} > 0. \tag{5.191} $$

The motion of a viscous fluid is irreversible and the inequality sign applies for any nonvanishing strain rate (with the possible exception of a simple dilatation). The left side of (5.191) represents the rate of energy dissipation.

To describe a simple viscous fluid, we assume that the rate of energy dissipation ${}_D\dot{Q}$ is a function of the strain rates and the absolute temperature; that is,

$$ {}_D\dot{Q} = \frac{1}{\rho}\tau_D^{ij}\dot{\gamma}_{ij} = {}_DW(\dot{\gamma}_{ij}, T). \qquad (5.192\text{a, b})$$

Equation (5.192b) is one way of saying that the dissipative components of stress depend upon the strain rates and temperature:

$$ \tau_D^{ij} = \tau_D^{ij}(\dot{\gamma}_{ij}, T). \qquad (5.193)$$

5.38 Newtonian Fluid

A *Newtonian fluid* is characterized by a linear relation between the stress components and the strain-rate components. Since the dissipative stress components vanish with the strain-rate components, equation (5.193) takes the form‡

$$ \tau_D^{ij} = \mu^{ijkl}\dot{\gamma}_{kl}. \qquad (5.194)$$

If $\dot{E}_{ij}$ are the Eulerian strain rates associated with the rectangular x^i coordinates of the current position of observation, then, according to (3.149b)

$$ \dot{E}_{ij} = \frac{\partial\theta^k}{\partial x^i}\frac{\partial\theta^l}{\partial x^j}\dot{\gamma}_{kl}, $$

where θ^i is a convected (curvilinear) coordinate and $\dot{\gamma}_{ij}$ is the associated strain rate.

If Σ^{ij} denote the rectangular components of stress associated with the x^i lines and τ^{ij} the components associated with the convected θ^i lines, then, as a special case of (4.25a) [viz., $\Sigma^{ij}(x^n)$ replaces $\bar{\tau}^{ij}(\bar{\theta}^n)$],

$$ \Sigma^{ij} = \frac{\partial x^i}{\partial\theta^k}\frac{\partial x^j}{\partial\theta^l}\tau^{kl}, \qquad \tau^{ij} = \frac{\partial\theta^i}{\partial x^k}\frac{\partial\theta^j}{\partial x^l}\Sigma^{kl}. $$

‡Note that the coefficients of (5.194) to (5.196) are temperature-dependent.

A Newtonian fluid is characterized by a linear relation between the physical (rectangular) components of stress and strain; namely,

$$\Sigma^{ij} = \bar{\mu}^{ijkl} \dot{E}_{kl},$$

wherein the coefficients $\bar{\mu}^{ijkl}$ are constants for a homogeneous isotropic medium.

By the foregoing tensorial transformations

$$\tau^{ij} = \mu^{ijkl} \dot{\gamma}_{kl},$$

where

$$\mu^{ijkl} = \frac{\partial \theta^i}{\partial x^m} \frac{\partial \theta^j}{\partial x^n} \frac{\partial \theta^k}{\partial x^p} \frac{\partial \theta^l}{\partial x^q} \bar{\mu}^{mnpq}.$$

Since these are rectangular Cartesian tensors, three invariants assume the forms:

$$\overline{\mathcal{K}}_1 = \dot{E}_{ii}, \qquad \overline{\mathcal{K}}_2 = \dot{E}_{ij} \dot{E}_{ij}, \qquad \overline{\mathcal{K}}_3 = \dot{E}_{ij} \dot{E}_{ik} \dot{E}_{jk}.$$

In the arbitrary curvilinear system

$$\mathcal{K}_1(\theta^1, \theta^2, \theta^3) = \frac{\partial \theta^i}{\partial x^l} \frac{\partial x^k}{\partial \theta^i} \dot{E}_l^k = \dot{\gamma}_i^i = G^{mn} \dot{\gamma}_{mn},$$

$$\mathcal{K}_2 = \dot{\gamma}_j^i \dot{\gamma}_i^j, \qquad \mathcal{K}_3 = \dot{\gamma}_j^i \dot{\gamma}_i^k \dot{\gamma}_k^j.$$

If the fluid is isotropic, we can define a dissipation function ${}_D W$, analogous to the strain energy of an elastic medium. As the linear stress-strain relation derives from a quadratic strain energy, so the linear (Newtonian) fluid is associated with a quadratic dissipation function ${}_D W$:

$${}_D W = 2 \frac{G^{pi} G^{qj}}{\rho} (\mu \dot{\gamma}_{pq} \dot{\gamma}_{ij} + \frac{\overset{*}{\lambda}}{2} \dot{\gamma}_{pi} \dot{\gamma}_{qj}). \tag{5.195}$$

In accord with (5.192) and (5.195), the dissipative stress components are related to the strain rates, as follows:

$$\tau_D^{ij} = G^{pi} G^{qj} \, (2\mu \dot{\gamma}_{pq} + \overset{*}{\lambda} G^{kl} G_{pq} \dot{\gamma}_{kl}). \tag{5.196}$$

In a homogeneous fluid, μ is a constant of proportionality between the shear strain rate and the corresponding shear stress; μ is called the *coefficient of viscosity.*

According to (3.148) and (5.196),

$$\frac{1}{3}(\tau_D)^i_i = \frac{1}{3}G_{ij}\tau^{ij}_D = \left(\frac{2}{3}\mu + \overset{*}{\lambda}\right)\dot{\gamma}^i_i$$

$$= \left(\frac{2}{3}\mu + \overset{*}{\lambda}\right)\frac{\dot{D}}{D}. \tag{5.197}$$

The quantity $(\frac{2}{3}\mu + \overset{*}{\lambda})$ is termed the *coefficient of bulk viscosity*; it is analogous to the elastic (bulk) modulus K of (5.84).

5.39 Linear Viscoelasticity

As the name implies, viscoelasticity describes a medium that exhibits attributes of viscous fluids and elastic solids. Most real materials are, in fact, aggregates of discrete elementary parts. The behavior of the composite manifests the properties of the various constituents. In this instance, we conceive of a material with a microstructure of viscous and elastic elements, and then extend the concepts to a continuous medium by assuming a continuous distribution of the resultant properties.

The following sections provide the reader with a concise introduction to linear viscoelasticity. For a more comprehensive study, the interested reader is referred to other work. For example, the monograph by R. M. Cristensen [96] emphasizes the mathematical aspects of the subject. The book by W. Flügge [97] is an introduction to the theory of linear viscoelasticity, whereas Y. M. Haddad [98] emphasizes applications, such as the use of viscoelasticity in the analysis of composites. Finally, a theoretical development including both transient and dynamic aspects of linear viscoelasticity is presented in the monograph by R. S. Lakes [99].

Eventually, we seek constitutive relations involving a general state of stress and strain. However, to appreciate the salient features, let us first examine the simple state of stress and the corresponding deformation.

A linear viscoelastic behavior was proposed by J. C. Maxwell [100]. The Maxwell material experiences viscous (permanent) strain ϵ^P and elastic strain ϵ^E under the action of the simple stress τ. The net strain ϵ is the sum

$$\epsilon = \epsilon^P + \epsilon^E. \tag{5.198}$$

In the manner of a viscous fluid, the permanent strain-*rate* $\dot{\epsilon}^P$ is assumed proportional to the stress τ; that is,

$$\dot{\epsilon}^P = \frac{\tau}{\mu}, \tag{5.199}$$

where μ is a coefficient of viscosity.‡ The elastic strain ϵ^E is proportional to the stress τ; that is,

$$\epsilon^E = \frac{\tau}{E}, \tag{5.200}$$

where E is a modulus of elasticity. In accordance with (5.198) to (5.200), the net strain rate is

$$\begin{aligned} \dot{\epsilon} &\equiv \dot{\epsilon}^P + \dot{\epsilon}^E \\ &= \frac{\tau}{\mu} + \frac{\dot{\tau}}{E}. \end{aligned} \tag{5.201}$$

Note: Here we assume that the same stress τ enforces the viscous and elastic strains.

The Maxwell material is analogous to the mechanical model depicted in Figure 5.17. The assembly consists of a viscous and an elastic element; both transmit the same force τ. The extensions of the viscous and elastic element are ϵ^P and ϵ^E, respectively, and the extension of the assembly is the sum.

Three situations are particularly interesting:

1. The transient strain resulting from the abrupt application of a constant stress, $\tau = \tau_0 H(t)$.
2. The transient stress resulting from a suddenly enforced strain, $\epsilon = \epsilon_0 H(t)$.
3. The steady-state strain caused by a sinusoidally oscillating stress, $\tau = \tau_0 \sin \omega t$.

Here τ_0 and ϵ_0 are constants and $H(t)$ denotes the unit step at $t = 0$. In cases (1) and (2), there is to be no initial ($t = 0^-$) stress or strain.

‡Neither μ in (5.199) nor E in (5.200) necessarily symbolize the same coefficients as appear in the three-dimensional theories of viscosity or elasticity.

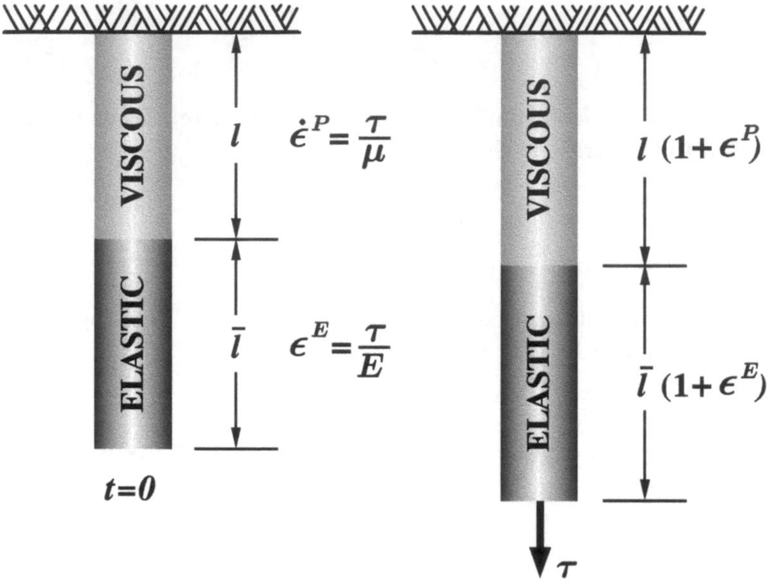

Figure 5.17 Model for the Maxwell material

In case 1, the initial conditions are

$$\tau = \tau_0 H(t), \qquad \epsilon(0^-) = 0. \tag{5.202}$$

Then, the solution of (5.201) is

$$\epsilon = \left(\frac{1}{E} + \frac{t}{\mu}\right)\tau_0 H(t). \tag{5.203}$$

The strain increases abruptly to the value τ_0/E, then increases linearly with time as depicted in Figure 5.18a. This linear increase of strain is the simplest form of creep.

The response to the step stress is determined by the parenthetical term of (5.203); the function in parentheses is called the *creep compliance.*

In case 2, the Maxwell body is subjected to a sudden strain

$$\epsilon = \epsilon_0 H(t), \qquad \tau(0^-) = 0. \tag{5.204}$$

Then, the solution of (5.201) is

$$\tau = \left(E e^{-(E/\mu)t}\right) \epsilon_0 H(t). \tag{5.205}$$

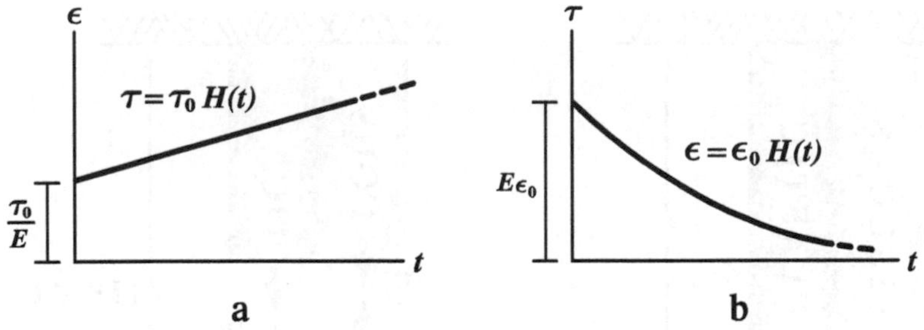

Figure 5.18 Responses of the Maxwell material

The stress τ jumps abruptly to the value $E\epsilon_0$, then gradually decays and eventually vanishes as depicted in Figure 5.18b.

The response to the step strain is determined by the parenthetical term of (5.205); the function in parentheses is called the *relaxation modulus*.

In case 3, a sinusoidal stress is the imaginary part of

$$\tau = \tau_0 e^{i\omega t}, \tag{5.206}$$

where $i = \sqrt{-1}$. Following the current convention,‡ we consider the steady-state solution of (5.201) with the complex stress of (5.206). That solution is

$$\epsilon = \left(\frac{1}{E} - \frac{i}{\mu\omega}\right)\tau_0 e^{i\omega t}. \tag{5.207}$$

The response to a sinusoidal loading is the imaginary part of (5.207), a sinusoidal strain which lags the sinusoidal stress. The time lag and amplitude are determined by the parenthetical factor, called the *complex compliance.*

A different type of viscoelasticity was proposed by H. Jeffreys [101]. The net stress τ is the sum of a dissipative part τ_D and a conservative part τ_R; that is,

$$\tau = \tau_D + \tau_R. \tag{5.208}$$

The former τ_D is associated with viscous flow and is proportional to the strain rate:

$$\tau_D = \mu\dot{\epsilon}, \tag{5.209}$$

‡This is common practice in the study of steady-state oscillations of linear systems. It is understood that the real and imaginary parts of the solution are the responses to the cosine and sine functions, respectively.

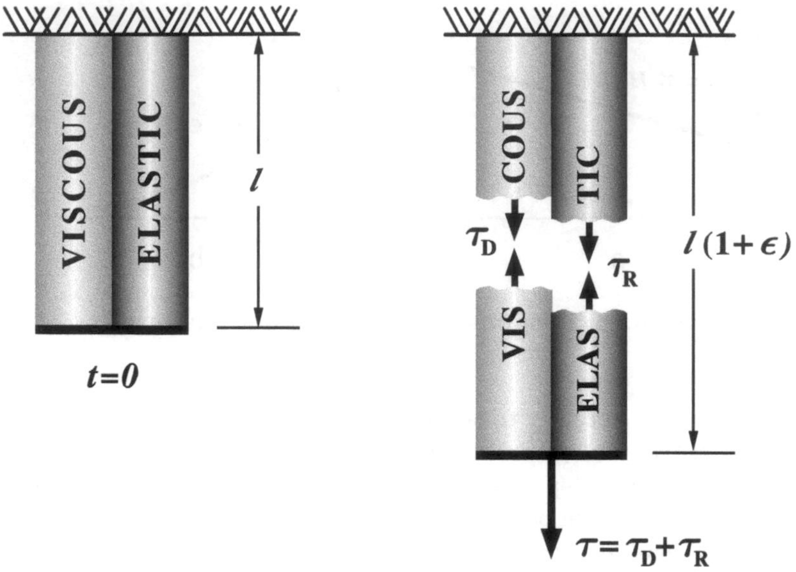

Figure 5.19 Model for the Kelvin material

where μ is a coefficient of viscosity. The latter τ_R is derived from an elastic deformation and is proportional to the strain:

$$\tau_R = E\epsilon, \tag{5.210}$$

where E is a modulus of elasticity. In accordance with (5.208) to (5.210)

$$\tau \equiv \tau_D + \tau_R = \mu\dot{\epsilon} + E\epsilon. \tag{5.211}$$

A material obeying the differential equation (5.211) is commonly called a Kelvin solid. The Kelvin material is analogous to the assembly of Figure 5.19. The viscous and elastic elements are assembled such that both suffer the same extension ϵ. The net force on the assembly is the sum of two forces τ_D and τ_R, transmitted by the viscous and elastic elements, respectively.

The creep compliance, relaxation modulus, and complex compliance of the Kelvin material are

$$\left(1 - e^{-(E/\mu)t}\right)/E, \tag{5.212}$$

$$E + \mu\delta(t), \tag{5.213}$$

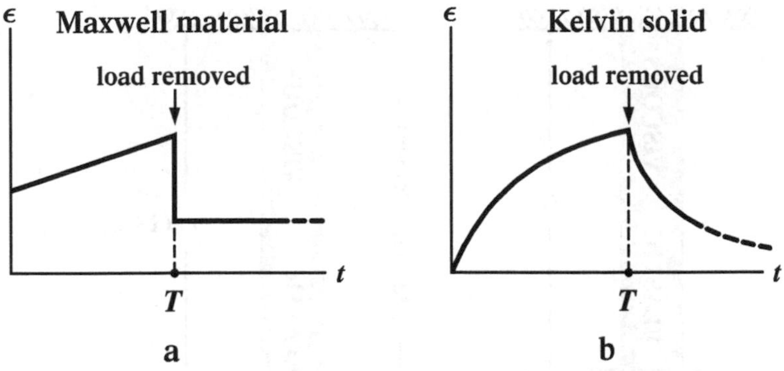

Figure 5.20 Responses of Maxwell and Kelvin materials

$$\frac{(E - i\mu\omega)}{(E^2 + \mu^2\omega^2)}. \tag{5.214}$$

Here $\delta(t)$ denotes the Dirac impulse at $t = 0$ and indicates that any attempt to enforce an instantaneous strain upon the Kelvin material requires a stress of unlimited magnitude.

Since the foregoing equations are linear, the strain caused by successive loadings can be obtained by superposition. In particular, the material can be subjected to the initial step $\tau = \tau_0 H(t)$ and, at a subsequent time $t = T$, the stress can be abruptly removed by the step

$$\tau = -\tau_0 H(t - T).$$

The strain $\epsilon = \epsilon(t)$ is obtained by superposing the solutions. The result for the Maxwell material is

$$\epsilon = \tau_0 \left[\left(\frac{1}{E} + \frac{t}{\mu} \right) H(t) - \left(\frac{1}{E} + \frac{t - T}{\mu} \right) H(t - T) \right], \tag{5.215}$$

and for the Kelvin material

$$\epsilon = \frac{\tau_0}{E} \left[\left(1 - e^{-(E/\mu)t} \right) H(t) - \left(1 - e^{-(E/\mu)(t-T)} \right) H(t - T) \right]. \tag{5.216}$$

Plots of (5.215) and (5.216) are shown in Figure 5.20.
Note: When the load is removed from the Maxwell body, there is an immediate recovery of the elastic strain, while the Kelvin body recovers gradually. On the other hand, the Maxwell body is permanently deformed, while the Kelvin body eventually recovers fully.

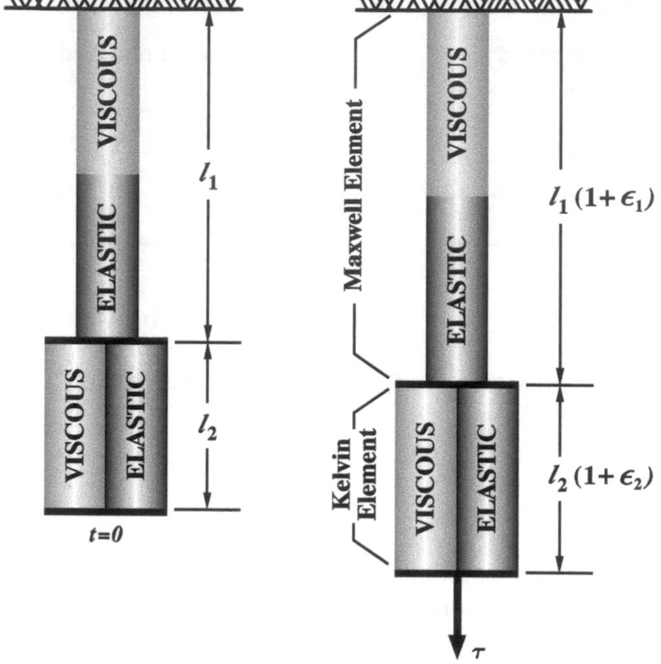

Figure 5.21 Model for a four-parameter viscoelastic material

The attributes of the Maxwell and Kelvin materials are incorporated in a material analogous to the mechanical assembly of Figure 5.21. Suppose that the material is subjected to the step stress

$$\tau = \tau_0 H(t).$$

It is evident from the analog of Figure 5.21 that the Maxwell and Kelvin elements transmit this stress. Then, in accordance with (5.203) and (5.212), the net strain is

$$\begin{aligned} \epsilon &= \epsilon_1 + \epsilon_2 \\ &= \tau_0 \left[\left(\frac{1}{E_1} + \frac{t}{\mu_1} \right) + \frac{1}{E_2} \left(1 - e^{-(E_2/\mu_2)t} \right) \right] H(t). \qquad (5.217) \end{aligned}$$

A plot of (5.217) is depicted in Figure 5.22.

In a creep experiment, the specimen is abruptly subjected to a constant load and deformations are subsequently measured during an extended pe-

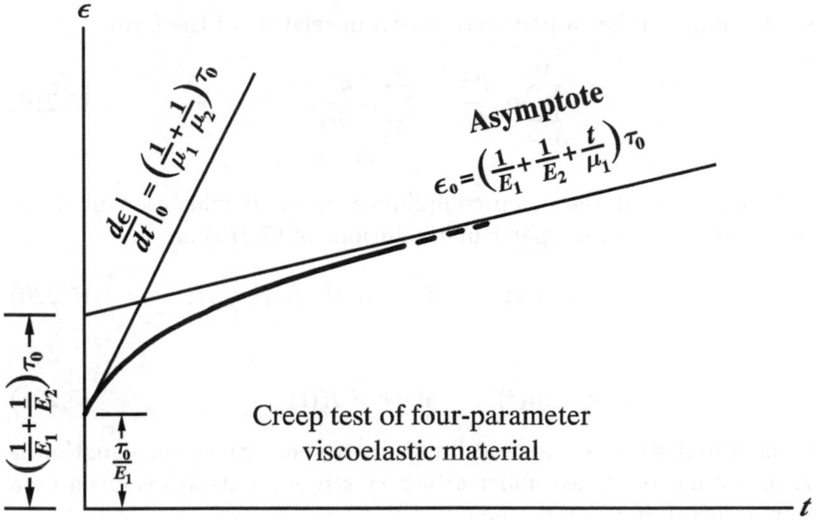

Figure 5.22 Creep of the four-parameter viscoelastic material

riod. The response according to (5.217) provides the plot of Figure 5.22. The initial strain and initial slope of the plot and the intercept and slope of the asymptote determine the four parameters E_1, E_2, μ_1, and μ_2.

By the analog of Figure 5.21,

$$\dot{\epsilon} = \dot{\epsilon}_1 + \dot{\epsilon}_2,$$

$$\dot{\epsilon}_1 = \frac{\tau}{\mu_1} + \frac{\dot{\tau}}{E_1},$$

$$\tau = \mu_2 \dot{\epsilon}_2 + E_2 \epsilon_2.$$

After eliminating ϵ_1 and ϵ_2 from these three equations, we obtain the differential equation

$$\mu_2 \ddot{\epsilon} + E_2 \dot{\epsilon} = \frac{\mu_2}{E_1}\ddot{\tau} + \left(\frac{\mu_2}{\mu_1} + \frac{E_2}{E_1} + 1\right)\dot{\tau} + \frac{E_2}{\mu_1}\tau. \tag{5.218}$$

Evidently, the conceptual model of Figure 5.21 serves to introduce the second derivatives of stress and strain into the constitutive equation. Further elaborations serve to introduce higher derivatives; in general, the linear viscoelastic material has a linear stress-strain relation of the form

$$\sum_{i=0}^{M} p_i \frac{d^i \tau}{dt^i} = \sum_{i=0}^{N} q_i \frac{d^i \epsilon}{dt^i}. \tag{5.219}$$

Let us assume that the creep compliance $k(t)$ and relaxation modulus $m(t)$ exist. In other words, particular solutions of (5.219) are

$$\epsilon = k(t), \qquad \text{if} \quad \tau = H(t), \tag{5.220}$$

and

$$\tau = m(t), \qquad \text{if} \quad \epsilon = H(t), \tag{5.221}$$

where the initial state is unstressed and unstrained. Since the equation is linear, the strain (or stress) under arbitrary stress (or strain) is given by a Duhamel integral; if $\tau = \tau(t)$, then

$$\epsilon(t) = \int_{-\infty}^{t} k(t-z) \frac{d\tau(z)}{dz}\, dz. \tag{5.222}$$

Alternatively, if $\epsilon = \epsilon(t)$, then

$$\tau(t) = \int_{-\infty}^{t} m(t-z) \frac{d\epsilon(z)}{dz}\, dz. \tag{5.223}$$

Equations (5.219), (5.222), and (5.223) are alternative forms of the constitutive equations for viscoelastic materials. Equations (5.222) and (5.223) are called *hereditary integrals* as they involve the entire past history of stress and strain.

If one adopts the Cauchy-Riemann definition of the integral, then equations (5.222) or (5.223) do not admit a discontinuity of stress or strain. Then, if an initial step is imposed upon a quiescent state:

$$\epsilon(t) = k(t)\, \tau_0 H(t) + \int_{0}^{t} k(t-z) \frac{d\tau}{dz}\, dz. \tag{5.224}$$

The first term on the right side of equation (5.224) describes the strains caused by the initial step stress $\tau_0 H(t)$, whereas the second term represents the strains resulting from subsequent changes in stress. Through integration by parts, the following, second version of the hereditary integral is obtained:

$$\epsilon(t) = \tau(t)\, k(0+) - \int_{0}^{t} \tau(z) \frac{dk(t-z)}{dz}\, dz. \tag{5.225}$$

5.40 Isotropic Linear Viscoelasticity

5.40.1 Differential Forms of the Stress-Strain Relations

To derive general constitutive equations of the Maxwell or Kelvin types, we recall the constitutive equations of the linear viscous fluid and linear elastic solid.

For simplicity, we cast the equations in a Cartesian/rectangular system. If the strains are small,[‡] then the equation (5.196) of the linear fluid takes the form:

$$s_D^{ij} = 2\mu\dot{\epsilon}_{ij} + \overset{*}{\lambda}\dot{\epsilon}_{kk}\delta_{ij}. \tag{5.226}$$

We note that the general theory of Section 5.37 admits a "reversible" component τ_R^{ij} [see equation (5.184a)]. That component is a hydrostatic pressure which can only cause a reversible dilatation of the isotropic medium. In the present circumstance of linear viscoelasticity, such dilatation is indistinguishable from any elastic contribution. The constitutive equations of isothermal linear elasticity are[†]

$$s^{ij} = 2G\epsilon_{ij} + \lambda\epsilon_{kk}\delta_{ij}. \tag{5.227}$$

The constitutive equations (5.226) and (5.227) can be expressed in terms of the stress deviator $'s^{ij}$, mean pressure p, strain deviator η_{ij}, and dilatation $'h \doteq \epsilon_{kk}$ of (3.74a–c). Then, the equations of the linear fluid take the forms

$$'s_D^{ij} = 2\mu\dot{\eta}_{ij}, \tag{5.228a}$$

$$p_D = \tfrac{1}{3}(2\mu + 3\overset{*}{\lambda})\,\dot{\epsilon}_{kk} \equiv K^*\,\dot{\epsilon}_{kk}. \tag{5.228b}$$

The linear elastic solid is governed by the equations

$$'s^{ij} = 2G\eta_{ij}, \tag{5.229a}$$

$$p = \tfrac{1}{3}(2G + 3\lambda)\,\epsilon_{kk} = K\,\epsilon_{kk}. \tag{5.229b}$$

[‡]The assumption of linear elasticity limits the theory to small strain, wherein $\tau^{ij} = s^{ij}$. Furthermore, if the strains are small, we have for the volumetric strains $'e$ and $'h$ of (3.73a–c) and (3.74a–c) the approximation $'h \doteq {}'e \doteq \epsilon_{kk}$.

[†]For simplicity, we limit our attention to isothermal deformations.

If the material is supposed to be a Maxwell medium, then

$$\dot{\eta}_{ij} = \dot{\eta}_{ij}^{E} + \dot{\eta}_{ij}^{P}, \tag{5.230a}$$

$$\dot{\epsilon}_{kk} = \dot{\epsilon}_{kk}^{E} + \dot{\epsilon}_{kk}^{P}, \tag{5.230b}$$

where suffixes E and P signify the recoverable (elastic) and irrecoverable (plastic) strains. The same stress is supposed to act upon the elastic and viscous constituents, i.e., $'s_D^{ij} = {}'s_E^{ij} = {}'s^{ij}$, $p_D = p_E = p$. In accordance with (5.228a, b) and (5.229a, b), the strains of (5.230a, b) are, respectively,

$$\dot{\eta}_{ij} = \frac{1}{2G}{}'\dot{s}^{ij} + \frac{1}{2\mu}{}'s^{ij}, \tag{5.231a}$$

$$\dot{\epsilon}_{kk} = \frac{1}{K}\dot{p} + \frac{1}{K^*}p. \tag{5.231b}$$

Here, any reversible dilatation of viscous constituents is also contained in the first term on the right side of (5.231b).

If the material is a Kelvin medium, then

$$'s^{ij} = {}'s_E^{ij} + {}'s_D^{ij}, \tag{5.232a}$$

$$p = p_E + p_D. \tag{5.232b}$$

It follows that

$$'s^{ij} = 2G\eta_{ij} + 2\mu\dot{\eta}_{ij}, \tag{5.233a}$$

$$s^{ii} = 3K\,\epsilon_{kk} + 3K^*\,\dot{\epsilon}_{kk}. \tag{5.233b}$$

Any isotropic, linear, viscoelastic material has constitutive equations in the forms

$$\underline{\boldsymbol{P}}\,{}'s^{ij} = \underline{\boldsymbol{Q}}\,\eta_{ij}, \tag{5.234a}$$

$$\underline{\boldsymbol{p}}\,p = \underline{\boldsymbol{q}}\,\epsilon_{kk}, \tag{5.234b}$$

where $\underline{\boldsymbol{P}}$, $\underline{\boldsymbol{Q}}$, $\underline{\boldsymbol{p}}$, and $\underline{\boldsymbol{q}}$ are linear differential operators. If the hereditary integrals are used, then $\underline{\boldsymbol{P}}$, $\underline{\boldsymbol{Q}}$, $\underline{\boldsymbol{p}}$, and $\underline{\boldsymbol{q}}$ are integral operators. The differential operators can also be written in the forms:

$$\underline{\boldsymbol{P}} = \sum_{k=0}^{\overline{m}} P_k \frac{\partial^k}{\partial t^k}, \qquad \underline{\boldsymbol{Q}} = \sum_{k=0}^{\overline{n}} Q_k \frac{\partial^k}{\partial t^k}, \tag{5.235a, b}$$

$$\underline{\boldsymbol{p}} = \sum_{k=0}^{\overline{\overline{m}}} p_k \frac{\partial^k}{\partial t^k}, \qquad \underline{\boldsymbol{q}} = \sum_{k=0}^{\overline{\overline{n}}} q_k \frac{\partial^k}{\partial t^k}. \tag{5.236a, b}$$

5.40.2 Integral Forms of the Stress-Strain Relations

If an isotropic, linear viscoelastic material is subjected to a deviatoric state of strain in the form of a step function, the consequent stress is given by a relaxation function‡ $2m_1(t)$, which is obtained by the solution of (5.234a). Likewise, the consequent hydrostatic pressure produced by a dilatational step is given by a relaxation function $m_2(t)$, which is obtained by the solution of (5.234b). Then the stress produced by a given history of strain is given by integrals in the manner of (5.223):

$$'s^{ij} = 2\int_{-\infty}^{t} m_1(t-z)\frac{\partial \eta_{ij}(z)}{\partial z}\,dz, \tag{5.237a}$$

$$s^{kk} = 3\int_{-\infty}^{t} m_2(t-z)\frac{\partial \epsilon_{kk}(z)}{\partial z}\,dz. \tag{5.237b}$$

Here, the variable z replaces t in the strain components of the integrand, while dependence on the spatial coordinate x^i is also understood; that is, $\epsilon_{ij}(z) = \epsilon_{ij}(x^1, x^2, x^3, z)$. The deviatoric and hydrostatic components of (5.237a) and (5.237b) can be combined as follows:

$$s^{ij} = 2\int_{-\infty}^{t} m_1(t-z)\frac{\partial \epsilon_{ij}(z)}{\partial z}\,dz + \delta_{ij}\int_{-\infty}^{t} \bar{\lambda}(t-z)\frac{\partial \epsilon_{kk}(z)}{\partial z}\,dz, \tag{5.238}$$

where m_1, m_2, and $\bar{\lambda}$ are analogous to the shear modulus, the bulk modulus, and the Lamé coefficient of elasticity; in particular,

$$\bar{\lambda}(t) = m_2(t) - \tfrac{2}{3}m_1(t). \tag{5.239}$$

‡The factor 2 is incorporated to achieve an analogy between the function m_1 and the shear modulus G.

As the stress components are expressed in terms of the strain components by (5.237a, b) or (5.238), the strain components can be expressed as integrals, like (5.222), which contain, as integrals, two creep compliances k_1 and k_2:

$$\eta_{ij} = \int_{-\infty}^{t} k_1(t-z)\frac{\partial\, {}'s^{ij}(z)}{\partial z}\, dz, \tag{5.240a}$$

$$\epsilon_{kk} = \int_{-\infty}^{t} k_2(t-z)\frac{\partial s^{kk}(z)}{\partial z}\, dz. \tag{5.240b}$$

Combining the deviatoric and dilatational parts, we obtain

$$\epsilon_{ij} = \int_{-\infty}^{t} k_1(t-z)\,\frac{\partial s^{ij}(z)}{\partial z}\, dz + \frac{1}{3}\delta_{ij}\int_{-\infty}^{t} [k_2(t-z) - k_1(t-z)]\,\frac{\partial s^{kk}(z)}{\partial z}\, dz. \tag{5.241}$$

5.40.3 Relations between Compliance and Modulus

We recall the Laplace transform $\bar{f}(s)$ of a function $f(t)$ (see, e.g., [97], [99]):

$$\bar{f}(s) = \int_{0}^{\infty} f(t)\, e^{-st}\, dt.$$

If the function f and all derivatives $\partial f^N/\partial t^N$ vanish at $t = 0$, then

$$\overline{\frac{\partial^N f}{\partial t^N}} = s^N \bar{f}(s). \tag{5.242}$$

We recall too the transform of the step function:

$$\overline{H(t)} = 1/s. \tag{5.243}$$

Now, we suppose that the body is quiescent prior to the application of stress or strain and we transform (5.234a, b):

$$\overline{{}'s^{ij}} \sum_{k=0}^{\overline{m}} P_k\, s^k = \overline{\eta_{ij}} \sum_{k=0}^{\overline{n}} Q_k\, s^k, \tag{5.244a}$$

$$\overline{p} \sum_{k=0}^{\overline{\overline{m}}} p_k\, s^k = \overline{\epsilon}_{kk} \sum_{k=0}^{\overline{\overline{n}}} q_k\, s^k. \tag{5.244b}$$

We recall that a creep compliance or a relaxation modulus is the response to a unit step of stress or strain, respectively. In the case of the isotropic material, we have two compliances, k_1 and k_2, and two moduli, $2m_1$ and $3m_2$, corresponding to the deviatoric and hydrostatic, or dilatational, parts. According to (5.243) and (5.244a, b), their transforms follow:

$$\overline{k}_1 = \frac{\sum_{k=0}^{\overline{m}} P_k\, s^k}{\sum_{k=0}^{\overline{n}} Q_k\, s^{k+1}}, \qquad \overline{k}_2 = \frac{\sum_{k=0}^{\overline{\overline{m}}} p_k\, s^k}{\sum_{k=0}^{\overline{\overline{n}}} q_k\, s^{k+1}}, \qquad (5.245\text{a, b})$$

$$2\overline{m}_1 = \frac{\sum_{k=0}^{\overline{n}} Q_k\, s^k}{\sum_{k=0}^{\overline{m}} P_k\, s^{k+1}}, \qquad 3\overline{m}_2 = \frac{\sum_{k=0}^{\overline{\overline{n}}} q_k\, s^k}{\sum_{k=0}^{\overline{\overline{m}}} p_k\, s^{k+1}}. \qquad (5.245\text{c, d})$$

Finally, we observe that

$$2\overline{k}_1(s)\,\overline{m}_1(s) = \frac{1}{s^2}, \qquad 3\overline{k}_2(s)\,\overline{m}_2(s) = \frac{1}{s^2}. \qquad (5.246\text{a, b})$$

Chapter 6

Principles of Work and Energy

6.1 Introduction

Theoretically, the problems of continuous solids and structures can be formulated, solved or approximated, by the mechanics of the preceding chapters. These provide the differential equations of equilibrium (or motion) governing stresses, the kinematical equations relating displacements and strains, and then certain constitutive equations relating the stresses and strains. That is the so-called "vectorial mechanics" (basic variables are vectors), also termed Newtonian. An alternative approach is based upon variations of work and energy. The latter is the so-called "analytical mechanics," or Lagrangean; both terms may be traced to J. L. Lagrange and his treatise, "Mécanique Analytique" [102]. In all instances, the alternative approach must provide the same description of the mechanical system. However, in many instances the energetic approach has advantages and frequently provides the most powerful means for effective approximations. This last attribute is especially important in the modern era of digital computation which facilitates the approximation of continuous bodies by discrete systems (e.g., finite elements).

Our treatment of the principles and methods is necessarily limited. We confine our attention to the concepts and procedures as applied to equilibrated systems. These are given in generality; principles, associated theorems, and necessary functionals apply to finite deformations of continuous bodies. In most instances, these can be readily extended to dynamic systems.

Although our primary concern is the analysis of continuous bodies, the salient features of the principles are most evident in simpler discrete systems. Then too, we recognize the important applications to discrete models of continuous bodies. Accordingly, the principles are initially formulated here for discrete mechanical systems and later reformulated for the continuous body. Specifically, the work or energy *functions* of the *discrete* variables are replaced by the *functionals* of the *continuous* functions; mathematical forms, but not concepts, differ. Those principal concepts follow our presentation of the essential terminology.

6.2 Historical Remarks

The basic concept of virtual work has been traced to Leonardo da Vinci (1452–1519). That notion was embodied in the numerous subsequent works of G. W. von Leibnitz (1646–1716), J. Bernoulli (1667–1748), and L. Euler (1707–1783); these and others embraced the concept of varying energy, both potential and kinetic. J. L. Lagrange (1736–1813) extended the concept of virtual work to dynamical systems via the D' Alembert's principle (1717–1783). Further generalization was formulated by W. R. Hamilton (1805–1865). An excellent historical account was given by G. Æ. Oravas and L. McLean [103], [104]. A most interesting philosophical/mathematical treatment and a historical survey are contained in the text of C. Lanczos [29]. Practical applications, as well as theoretical and historical background, are given in the book by H. L. Langhaar [10]. A rudimentary treatment may be found in the previous texts by G. A. Wempner [105], [106].

6.3 Terminology

Generalized Coordinates

The *configuration* of a mechanical system is determined by the positions of all the particles that comprise the system. The configuration of a discrete system consists of a finite number of particles or rigid bodies, and consequently, the configuration is defined by a finite number of real variables called *generalized coordinates*. An example is the angle which defines the configuration of a pendulum. A continuous body is conceived as an infinite collection of particles and, consequently, its configuration must be specified by an infinity of values, a continuous vector function.

Any change in the configuration of a system is referred to as a *displacement* of the system. To remove any ambiguity when the term "displacement" is used in such context, we define the magnitude of the displacement as the largest distance traveled by any of the particles comprising the system. In addition, we suppose that any motion occurs in a continuous fashion; that is, the distance traveled vanishes with the time of travel.

The generalized coordinates are assumed regular in the sense that any increment in the coordinate is accompanied by a displacement of the same order of magnitude and vice versa.

Degrees of Freedom

The number of *degrees of freedom* is the minimum number of generalized coordinates. For example, a particle lying on a table has two degrees of freedom; a ball resting on a table has five because three angles are also needed to establish its orientation.

Constraints

Any geometrical condition imposed upon the displacement of a system is called a *constraint*. A ball rolling upon a table has a particular kind of constraint imposed by the condition that no slip occurs at the point of contact. The latter is a nonintegrable differential relation between the five coordinates; such conditions are called *nonholonomic constraints*.

Virtual Displacements

At times, it is convenient to imagine small displacements of a mechanical system and to examine the work required to effect such movement. To be meaningful, such displacements are supposed to be consistent with all constraints imposed upon the system. Since these displacements are hypothetical, they are called *virtual displacements*.

Work

If $'\boldsymbol{f}$ denotes the resultant force upon a particle and $\boldsymbol{V}$ the displacement vector, then the work performed by $'\boldsymbol{f}$ during an infinitesimal displacement $\delta\boldsymbol{V}$ is

$$\delta w = '\boldsymbol{f} \cdot \delta\boldsymbol{V}. \tag{6.1a}$$

If $\boldsymbol{v} \equiv d\boldsymbol{V}/dt$, then the work done in time δt is

$$\delta w = '\boldsymbol{f} \cdot \boldsymbol{v}\,\delta t. \tag{6.1b}$$

In the time interval (t_0, t_1), the work done by $'\boldsymbol{f}$ is

$$\Delta w \equiv \int_{t_0}^{t_1} '\boldsymbol{f} \cdot \boldsymbol{v}\,dt. \tag{6.2}$$

Generalized Forces

In general, the particles comprising a system do not have complete freedom of movement but are constrained to move in certain ways. For example, a simple pendulum consists of a particle at the end of a massless link which constrains the particle to move on a circular path; the position of the particle is defined by one generalized coordinate, the angle of rotation of the link. In a discrete mechanical system, the displacements of all particles can be expressed in terms of generalized coordinates. If q_i $(i = 1, \ldots, n)$ denotes a regular generalized coordinate for a system with n degrees of freedom, then the position of a particle N is given by a vector $\boldsymbol{R}_N(q_1, \cdots, q_n)$ and the incremental displacement of the particle is

$$\delta \boldsymbol{V}_N = \sum_{i=1}^{n} \boldsymbol{a}_N^i \, \delta q_i, \qquad \boldsymbol{a}_N^i \equiv \frac{\partial \boldsymbol{R}_N}{\partial q_i} = \boldsymbol{R}_{N,i}. \tag{6.3a, b}$$

If $'\boldsymbol{f}^N$ denotes the force acting upon the *particle* N, then the work done during the displacement is

$$\delta w_N = {'\boldsymbol{f}^N} \cdot \delta \boldsymbol{V}_{\underline{N}} \qquad \text{(no summation on } N\text{)}.$$

The work performed by all forces acting upon *all particles* of the system is

$$\begin{aligned} \delta W &= \sum_N {'\boldsymbol{f}^N} \cdot \delta \boldsymbol{V}_N \\ &= \sum_{i=1}^{n} \left(\sum_N {'\boldsymbol{f}^N} \cdot \boldsymbol{a}_N^i \right) \delta q_i. \end{aligned}$$

If we define

$$Q^i \equiv \sum_N {'\boldsymbol{f}^N} \cdot \boldsymbol{a}_N^i,$$

then

$$\delta W = \sum_{i=1}^{n} Q^i \, \delta q_i. \tag{6.4}$$

The quantity Q^i is termed a *generalized force*. The modifier "generalized" is inserted because the quantity need not be a physical force; for example, if the angle of rotation is the generalized coordinate for a pendulum, then the moment about the pivot is the generalized force.

6.4 Work, Kinetic Energy, and Fourier's Inequality

Law of Kinetic Energy

According to Newton's law, the force on a particle of mass m moving with velocity $\boldsymbol{v}$ is

$$'\boldsymbol{f} = m\frac{d\boldsymbol{v}}{dt}. \tag{6.5}$$

It follows from (6.1b) and (6.5) that

$$\delta w = m\frac{d\boldsymbol{v}}{dt} \cdot \boldsymbol{v}\,\delta t = \frac{d}{dt}(\tfrac{1}{2}mv^2)\,\delta t.$$

The parenthetical term is the kinetic energy of the particle

$$\tau = \tfrac{1}{2}mv^2. \tag{6.6}$$

Therefore,

$$\delta w = \delta\tau. \tag{6.7}$$

Equation (6.7) states that the increment of work is equal to the increment of kinetic energy.

Viewing a mechanical system as a collection of particles and summing the work done upon all particles, we conclude that the work of all forces upon a mechanical system equals the increase of the kinetic energy of the system:

$$\delta W \equiv \sum \delta w = \sum \delta\tau \equiv \delta T, \tag{6.8}$$

where the summation extends over all particles of the system and T denotes the kinetic energy of the entire system.

If (6.8) is integrable, i.e., $\delta T = dT$, then over a time interval we obtain

$$\Delta W = \int_{t_0}^{t_1} dT \equiv \Delta T. \tag{6.9}$$

Equation (6.9) is called the *law of kinetic energy*. It is important to realize that the work W includes the work of internal forces as well as the work of forces exerted by external agencies.

Fourier's Inequality

According to the law of kinetic energy (6.9), the kinetic energy of a mechanical system cannot increase unless positive work is performed. If the system is at rest and there is no way that it can move such that the net work of all forces is positive, then the system must remain at rest. Stated otherwise, if the work performed during every virtual displacement is negative or zero, the system must remain at rest; that is, a sufficient condition for equilibrium is

$$\Delta W \leq 0 \tag{6.10}$$

for every small virtual displacement that is consistent with the constraints. The inequality (6.10) is known as *Fourier's inequality.*

6.5 The Principle of Virtual Work

According to Newton's first law, if the particle is at rest (or in steady motion $\dot{\boldsymbol{v}} = \mathbf{o}$) then the resultant force vanishes (${}'\boldsymbol{f} = \mathbf{o}$). During any infinitesimal movement of the particle, no work is done; that is,

$$\delta w = {}'\boldsymbol{f} \cdot \delta \boldsymbol{V} = 0.$$

Likewise, if we suppose that all particles of an equilibrated mechanical system are given infinitesimal displacements, then the work performed by all forces upon the system vanishes:

$$\delta W = 0. \tag{6.11}$$

Since the equality of (6.10) is a sufficient condition for equilibrium, equation (6.11) is both necessary and sufficient provided that the virtual displacements are consistent with any constraints imposed upon the system; the condition expresses the *principle of virtual work*. According to (6.11), the increment of virtual work vanishes if the system is in equilibrium and the movement is consistent with constraints. If the generalized coordinates q_i of (6.4) are independent, then the right side of (6.4) vanishes for *arbitrary* δq_i; it follows that

$$Q^i = 0. \tag{6.12}$$

The foregoing statement of the principle cannot be taken without qualification. Specifically, both forces and displacements are supposed to be continuous variables; abrupt discontinuities are inadmissible. The systems of Figure 6.1 illustrate the point. The roller of Figure 6.1a rests in a V-groove; any movement, as that from O to O', requires work. Note that the force of contact at A vanishes *abruptly*; the path of admissible displacement exhibits an *abrupt* corner. Moreover, work and displacement are of the same

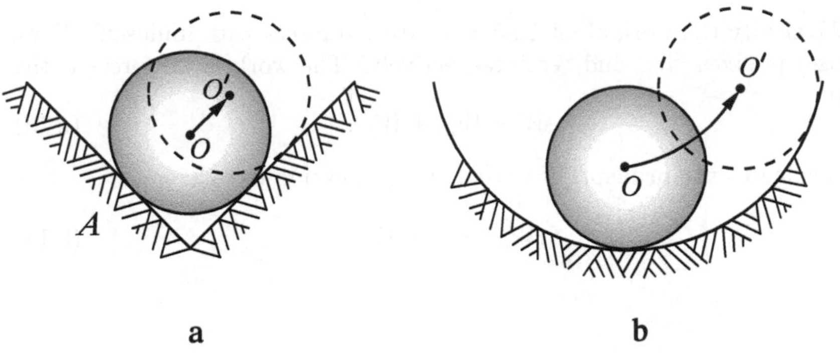

Figure 6.1 Inadmissible versus admissible paths

order of magnitude. The system of Figure 6.1b conforms to our criteria. A small motion along the smooth path requires work of higher order. Stated otherwise, the work ΔW and displacement Δs must not be of the same order of magnitude.[‡] To be more specific, let y denote the vertical position of the ball and s denote distance on the path OO' (see Figure 6.1). The work of *first order* performed by the weight f of the roller in Figures 6.1a, b follows:

$$\textbf{Case a:} \qquad \Delta W = -f\frac{dy}{ds}\,\Delta s, \qquad \frac{\Delta W}{\Delta s} = -f\frac{dy}{ds} \neq 0,$$

$$\textbf{Case b:} \qquad \Delta W = -f\frac{dy}{ds}\,\Delta s, \qquad \frac{\Delta W}{\Delta s} = 0.$$

The latter (**b**) follows since $dy/ds = 0$ at the equilibrium position, i.e., at the bottom of the trough. A small motion consistent with the constraint (rolls along the smooth path) requires work of higher order (*no* work of the *first* order in the displacement). The former (**a**) violates the requirement, since $dy/ds \neq 0$. It follows that the system of Figure 6.1b conforms to the criteria, the system of Figure 6.1a does not.

Internal and External Forces

In the analyses of structures, it is particularly convenient to classify the forces as internal or external forces. An internal force is an interaction between parts of the system while an external force is exerted by an external agency.

[‡]H. L. Langhaar [10] casts this statement in mathematical terms.

We signify the work of internal and external forces with minuscule (lowercase) prefixes "*in*" and "*ex*," respectively. The work of all forces is the sum

$$W = W_{in} + W_{ex}. \tag{6.13}$$

According to the principle of virtual work, equation (6.11),

$$\delta\,(W_{in}) = -\delta\,(W_{ex}). \tag{6.14}$$

6.6 Conservative Forces and Potential Energy

A force is said to be *conservative* if the work it performs upon a particle during transit from an initial position P_0 to another position P is independent of the path traveled. If the force is conservative, then it does no work during any motion which carries the particle along a closed path terminating at the point of origin P_0. The work performed depends upon the initial and current positions P_0 and P. Since we are concerned only with changes of the current position, the initial position is irrelevant and, therefore, we regard the work as a function of the current position, that is,

$$w = \int_{P_0}^{P} {}'\boldsymbol{f} \cdot d\boldsymbol{V} = w(P).$$

The potential of the force ${}'\boldsymbol{f}$ is denoted $v(P)$;

$$v(P) \equiv -w(P). \tag{6.15}$$

In view of the negative sign, it may be helpful to regard v as the work that you would do upon the particle if you held the particle (in equilibrium by exerting force $-{}'\boldsymbol{f}$) and transported it (slowly) to the position P. If P is defined by Cartesian coordinates X_i, then the differential of (6.15) is

$$dv = \frac{\partial v}{\partial X_i}\, dX_i \equiv -{}'f^i\, dX_i.$$

It follows that the conservative force derives from the potential

$${}'f^i = -\frac{\partial v}{\partial X_i}. \tag{6.16}$$

Two common examples of conservative forces are a gravitational force and a force exerted by an elastic spring. If one moves an object from the floor to a table, the work expended is independent of the path taken; it depends only on the change of elevation. If one grasps one end of the spring and changes its position, while the other end remains fixed, then the work performed is independent of the path; it depends only on the change of length.

If the configuration of a conservative mechanical system is defined by generalized coordinates q_i $(i = 1, \dots, n)$, and if W is the work of the (conservative) forces acting upon the entire system, then there exists a potential energy $\mathcal{V}$ such that

$$-W = \mathcal{V}(q_i).$$

The differential of work done by the conservative forces is

$$dW \equiv -d\,\mathcal{V}, \tag{6.17a}$$

$$= -\sum_{i=1}^{n} \frac{\partial \mathcal{V}}{\partial q_i}\, dq_i. \tag{6.17b}$$

In accordance with (6.4), the generalized forces are

$$Q^i = -\frac{\partial \mathcal{V}}{\partial q_i}. \tag{6.18}$$

The potential of internal and external forces are denoted by U and Π, that is,

$$W_{in} = -U, \qquad W_{ex} = -\Pi. \tag{6.19a, b}$$

The generalized internal and external forces are

$$'F^i = -\frac{\partial U}{\partial q_i}, \qquad P^i = -\frac{\partial \Pi}{\partial q_i}. \tag{6.20a, b}$$

The total potential energy of the system is the sum

$$\mathcal{V} = U + \Pi. \tag{6.21}$$

6.7 Principle of Stationary Potential Energy

A mechanical system is conservative if all forces, internal and external, are conservative. In this case, no mechanical energy is dissipated, that is, converted to another form such as heat. Then the principle of virtual work asserts that

$$d\mathcal{V} = \sum_{i=1}^{n} \frac{\partial \mathcal{V}}{\partial q_i} dq_i = 0. \tag{6.22}$$

If the coordinates are independent, then

$$Q^i = -\frac{\partial \mathcal{V}}{\partial q_i} = 0. \tag{6.23}$$

Since the equations (6.23) are the conditions which render the function $\mathcal{V}$ stationary, the principle of virtual work becomes the principle of stationary potential energy. In view of (6.21), (6.23) has the alternative form:

$$\frac{\partial U}{\partial q_i} = -\frac{\partial \Pi}{\partial q_i}. \tag{6.24}$$

In accordance with (6.20b) and (6.24), the external force P^i upon the *equilibrated system* satisfies the equation:

$$P^i = \frac{\partial U}{\partial q_i}. \tag{6.25}$$

6.8 Complementary Energy

Equation (6.25) expresses the force P^i in terms of the coordinates q_i and, if the second derivatives exist, then

$$dP^i = \sum_{j=1}^{n} \frac{\partial^2 U}{\partial q_i \partial q_j} dq_j.$$

If the determinant of the second derivatives does not vanish, then these equations can be inverted, such that

$$q_i = q_i(P^i). \tag{6.26}$$

In view of (6.26), we can define a complementary energy,

$$\mathcal{C}(P^i) \equiv \sum_{i=1}^{n} P^i q_i - U. \tag{6.27}$$

Then

$$\frac{\partial \mathcal{C}}{\partial P^j} = q_j + \sum_{i=1}^{n} \left(P^i - \frac{\partial U}{\partial q_i} \right) \frac{\partial q_i}{\partial P^j}.$$

Equation (6.25) holds for the *equilibrated system*; then in view of (6.25),

$$q_i = \frac{\partial \mathcal{C}}{\partial P^i}. \tag{6.28}$$

Note that the variables q_i are related to the complementary energy $\mathcal{C}(P^i)$ as the variables P^i are related to the potential $U(q_i)$. The transformation which reverses the roles of the complementary energies (U and $\mathcal{C}$) and the variables (P^i and q_i) is known as a *Legendre transformation.*

6.9 Principle of Minimum Potential Energy

The principle of virtual work asserts the vanishing of the virtual work in an infinitesimal virtual displacement, that is, $\delta W = 0$. Fourier's inequality asserts that a mechanical system does not move from rest (to an adjacent or distant configuration) unless it can move in some way that the active forces do positive work; if $\Delta W \leq 0$ the system must remain at rest.

Let us now distinguish between the inequality $\Delta W < 0$ and the equality $\Delta W = 0$, if the displacements are other then infinitesimals of the first order. The inequality implies that the system can be moved only if an outside agency does positive work upon the system. The equality means that the system can be slowly transported without doing work ("slowly," because no energy is supplied to increase the kinetic energy). The simple system of Figure 6.2 consisting of a ball resting (a) in a cup and (b) on a horizontal surface illustrates the difference. The active force on the ball is the gravitational force, the weight. The distances of the center from the reference point P is a suitable generalized coordinate if each ball is to roll upon its respective surface. Both are in equilibrium in position $P(\delta W = 0)$, but $\Delta W < 0$ in case (a) (from P to S) and $\Delta W = 0$ in case (b). In case (b), the position P is one of neutral equilibrium. In general, if $\Delta W < 0$ for sufficiently small displacements, then the system is stable. The

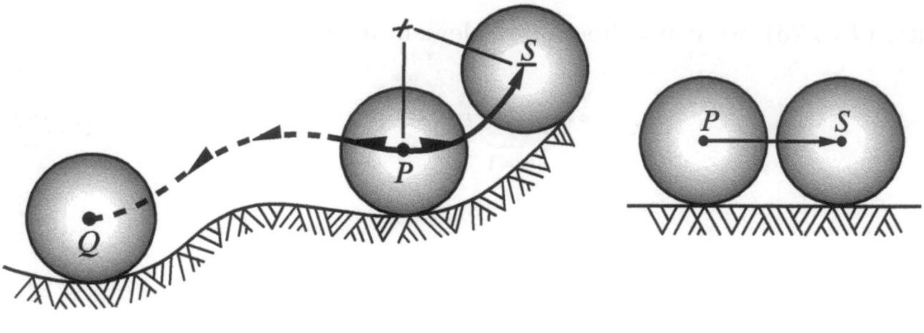

Figure 6.2 Stable, unstable, and neutral states of equilibrium

qualification of "sufficiently small" is needed to discount the circumstances in which negative work is performed initially but subsequently exceeded by a greater amount of positive work. For example, if the ball of Figure 6.2a is transported from P to Q the net work done by the gravitational force is positive, but some negative work is done as it is transported to the crest of the hill.

If the system is conservative, then the work done by all forces upon the system is equivalent to the decrease in the total potential energy, that is, $\Delta W = -\Delta \mathcal{V}$. Then the requirement for stable equilibrium is

$$\Delta \mathcal{V} > 0.$$

In words, an equilibrium configuration of a conservative mechanical system is regarded as stable if the potential energy is a proper minimum. However, the engineer must exercise great care in his application of this criterion, because a physical system may be more or less stable as demonstrated by the ball of Figure 6.2a; if the hill between the valley P and the lower valley Q is very low, then a small disturbance may render the weak stability at P insufficient to prevent the excursion to Q.

The principle of minimum potential is a cornerstone in the theory of elastic members and structures. Specifically, most criteria and analyses for the stability of equilibrium are founded upon this principle. The criterion of E. Trefftz [25] establishes the critical state for most structural systems. The criteria and analyses of W. T. Koiter [107] provide additional insights to the behavior at the critical state. Additionally, Koiter's work offers means to predict the abrupt, often catastrophic, "snap-through buckling" of certain shells and the related effects of imperfections. Here, we present the essential arguments and exhibit those criteria in the context of a discrete system. These are readily extended to continuous bodies; the algebraic functions of the discrete variables (e.g., generalized coordinates q_i) are supplanted by the functionals of the continuous functions [e.g., displacements $V_i(\theta^i)$].

Sources for further study are found in the work of J. W. Hutchinson and W. T. Koiter [108]. A clear presentation, practical applications and references, are contained in the book by H. L. Langhaar [10]. The monograph by Z. P. Băzant and L. Cedolin [109] offers a comprehensive treatment of the stability theory of structures. The book covers subjects relevant to many branches of engineering and the behavior of materials. It presents alternative methods of analysis, investigates the stability of various structural elements, and contains practical applications; the second part is devoted to the stability of inelastic systems. The monograph by H. Leipholz [110] gives an introduction to the stability of elastic systems, emphasizes the dynamic aspects of instability, and presents methods of solution; other topics include nonconservative loadings and stochastic aspects.

6.10 Structural Stability

The principle of minimum potential (Section 6.9) provides a basis for the analyses of structural stability. Some applications reveal certain practical implications and responses. To that end, we examine the simple linkage depicted in Figure 6.3. That system has but one degree of freedom; the lateral displacement w of the joint B. Although simple, the system exhibits important attributes of complex structures.

The two rigid bars, AB and BC, are joined by the frictionless pins at A, B, and C. Pin C inhibits translation; A constrains movement to the straight line AC. Extensional springs at B resist lateral movement by the force $F = kw$ and the torsional spring at B resists relative rotation (2θ) by a couple $C = \beta(2\theta)$. This simple system exhibits a behavior much as an elastic column (spring β simulates bending resistance) with lateral restraint (spring k simulates the support of a lateral beam). We now explore the response to the thrust force $P =$ constant (as a weight).

The potential energy of the displaced system of Figure 6.3b consists of the potential Π of the "dead" load P[‡] and the internal energy U of the springs (k and β):

$$\mathcal{V} = \Pi + U = 2Pl\cos\theta + \frac{kl^2}{2}\sin^2\theta + \frac{\beta}{2}(2\theta)^2. \tag{6.29}$$

[‡]Note that any constant can be added to the potential energy without affecting our analysis, since we are only concerned with variations in potential energy; we could as well set $\Pi = -2Pl(1 - \cos\theta)$.

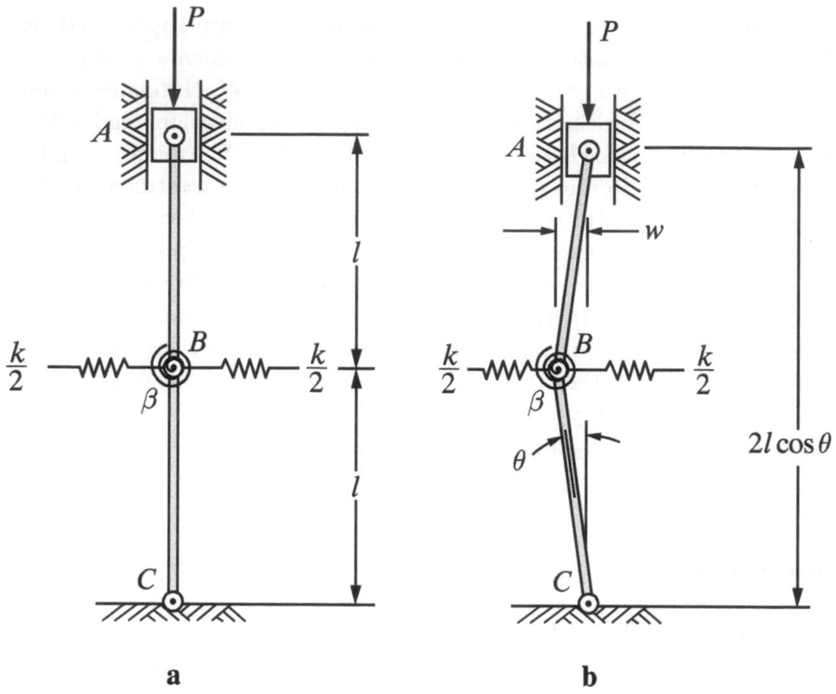

Figure 6.3 Concept of structural stability

By the principle of stationary potential, the system is in equilibrium if

$$\delta\mathcal{V} = (-2Pl\sin\theta + kl^2\sin\theta\cos\theta + 4\beta\theta)\,\delta\theta = 0. \tag{6.30}$$

The condition must hold for arbitrary $\delta\theta$; therefore, the equation of equilibrium follows:

$$-2Pl\sin\theta + kl^2\sin\theta\cos\theta + 4\beta\,\theta = 0.$$

Evidently, $\theta = 0$ is an equilibrium state for all choices of P, l, k, β. Additionally, the equation is satisfied for nonzero θ, if

$$P = \frac{kl}{2}\cos\theta + \frac{2\beta}{l}\frac{\theta}{\sin\theta}. \tag{6.31}$$

Note that $P(\theta)$ in (6.31) is an even function of the variable θ; it intersects the ordinate at

$$P \equiv P_{cr} = \frac{kl}{2} + \frac{2\beta}{l}. \tag{6.32}$$

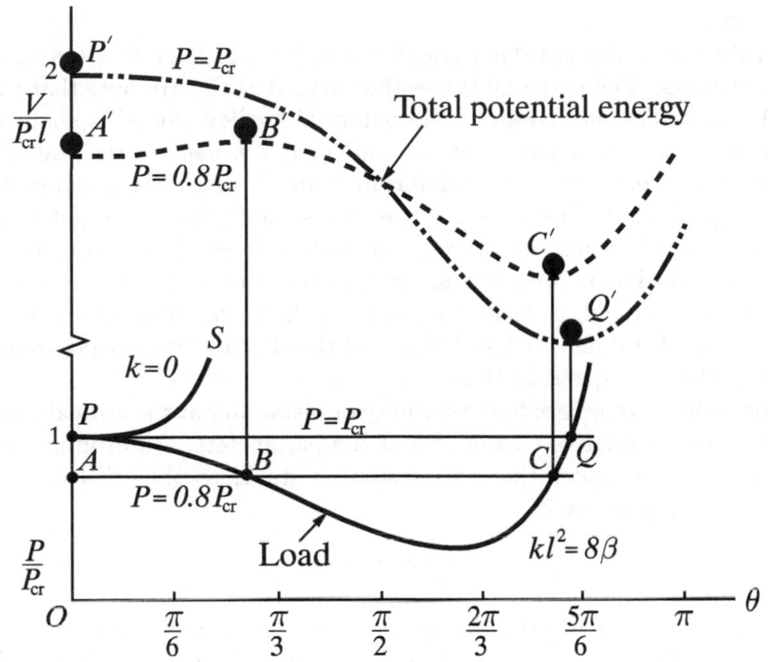

Figure 6.4 Stable versus unstable states of energy

Figure 6.4 displays plots of the dimensionless load (P/P_{cr}) for various values of the parameters k and β. The point P at the critical load $[(P/P_{cr}) = 1]$ is called a *bifurcation point*: The branches of equilibrium states (6.31) intersect the stem OP at point P.

First, let us consider the situation $k = 0$, the linkage without lateral support. The branch PS is initially normal to the stem, but henceforth has positive slope. The states on the stem OP are stable, unstable above P. States on the branch PS are stable, since an ever increasing load is required to cause additional deflection. Practically, the "structure" fails at $P \geq P_{cr}$, because slight increases in loading cause excessive deflection. A column, for example, is said to buckle at the critical load.

Now, consider the system when $kl^2 = 8\beta$. The perfect system is, strictly speaking, stable along the stem OP, unstable above P. The state at the critical load P_{cr} or slightly below, say $P = 0.8P_{cr}$, are especially interesting. As the unsupported linkage, the system does sustain a load $P < P_{cr}$ in the straight configuration to the critical load P_{cr}; at the critical load, the system *snaps-through* to the configuration of Q. The responses of this system are best explained by examining the potentials at the specific loads. At the critical load, the potential traces the curve $P'Q'$. Like the ball at the top of the hill (e.g., P'), the system moves to the valley (e.g., Q'), a state Q of lower energy.

The behavior of the latter system ($kl^2 = 8\beta$) under load $P = 0.8P_{cr}$ is more interesting. The potential traces the curve $A'B'C'$. We note that the state A is a stable one; however, the potential "valley" at A' is shallow. A slight disturbance can cause the system to *snap-through* to the state C. The system does not rest at the equilibrium state B, since it is unstable; B' is at the top of a hill. The system comes to rest at C, where the potential has a valley at C'. Some structures, especially thin shells, are subject to snap-through buckling. Note that slight imperfections or disturbances can cause the snap-through at loads far less than the theoretical critical load. The questions of stability, or instability, and the effects of imperfections are pursued in the subsequent sections.

The preceding example exhibits stability or instability at the critical load, depending on the relative magnitudes of the parameters. An examination of the potential indicates that this system is stable at the critical load, $\Delta V\,]_{P_{cr}} > 0$, if, and only if,

$$\frac{3kl^2}{4\beta} < 1.$$

This result can also be ascertained by examining the geometrical properties of the branch $P(\theta)$ at P_{cr}, $\theta = 0$. At the critical load, the trace $P(\theta)$ goes from concave to convex, from valley to hill.

6.11 Stability at the Critical Load

The simple example of Section 6.10 indicates that a mechanical system or structure can exhibit quite different responses (i.e, stable versus unstable) when subjected to critical loads. The energy criteria for such critical loads were established by E. Trefftz [25]. Criteria for stability and behavior *at* the critical load were developed by W. T. Koiter. Here, we present underlying concepts and some consequences of Koiter's criteria; these are set in the context of a discrete system, but apply as well to a continuous body via the alternative mathematics, i.e., functionals replace functions, integrals replace summations, etc.

All loads upon the system are assumed to increase in proportion and, therefore, the magnitude is given by a positive parameter λ. A configuration of the system is defined by N generalized coordinates q_i ($i = 1, \cdots, N$). As the loading parameter is increased from zero, the equilibrium states trace a path in a configuration-load space. For example, a system with two degrees of freedom (q_1, q_2) follows a path in the space ($q_1, q_2; \lambda$) of Figure 6.5a or 6.5b.

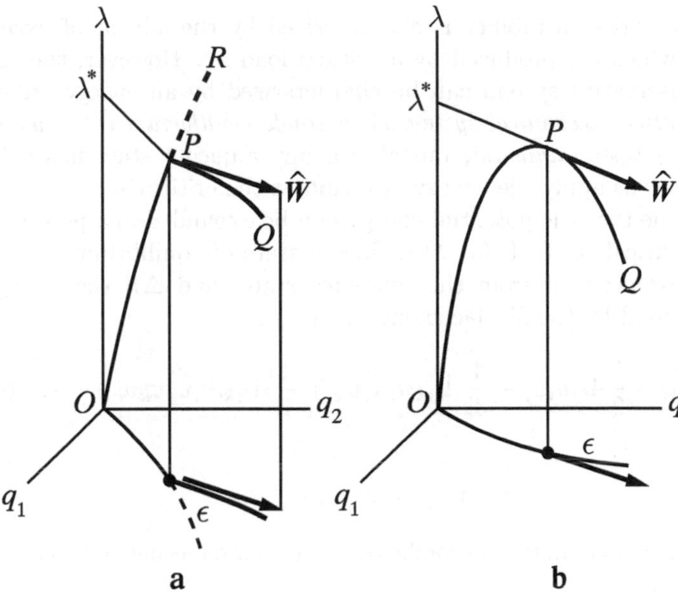

Figure 6.5 Configuration-load space

The point P of Figure 6.5a or 6.5b is a *critical state*, characterized by the existence of neighboring states which are not uniquely determined by an increment of the load. At the critical point P of Figure 6.5a the path OP forms two branches, PR and PQ. The branch PQ may ascend or descend, or the tangent $\hat{\boldsymbol{W}}$ may be normal to the λ axis. At the point P of Figure 6.5b, if the smooth curve has a tangent $\hat{\boldsymbol{W}}$ normal to the λ axis, and if S denotes arc length along the path, then the path PQ at P is characterized by the condition $d\lambda/dS = 0$; in words, the system tends to move from P with no increase of load.

The critical state of Figure 6.5a occurs at a *bifurcation point*; two paths of equilibrium emanate from P. However, the path PR represents unstable paths which cannot be realized. Actually, the system tends to move along PQ. If PQ is an ascending path, then additional loading is needed, and the system is said to be stable at the critical state. In actuality, a very slight increment is usually enough to cause an unacceptable deflection and the system is said to buckle. If the curve PQ is descending, the system collapses under the critical load λ^*.

The path of Figure 6.5b is entirely smooth, but reaches a so-called *limit point* P. The state of P is again critical in the sense that the tangent $\hat{\boldsymbol{W}}$ is normal to the λ axis. At P, the system tends to move under the critical load λ^*. Since the path descends from P to Q, the system exhibits a "snap-through buckling"; it essentially collapses to a much disfigured state.

By our remarks, instability is characterized by the advent of excessive deflections which are produced by a critical load λ^*. However, the stability of a conservative system can be characterized by an energy criterion: *the conservative mechanical system is in stable equilibrium if the potential energy is a proper minimum*, unstable if any adjacent state has a lower potential. Let us apply the energy criterion at the critical state.

We presume that the potential energy can be expanded in a power series about the critical state. If $(q_i; \lambda)$ defines a state of equilibrium, $u_i \equiv \Delta q_i$ defines a displacement from the reference state, and $\Delta \mathcal{V}$ the change of potential caused by the displacement, then

$$\Delta \mathcal{V} = \bar{A}_i u_i + \frac{1}{2} A_{ij} u_i u_j + \frac{1}{3!} A_{ijk} u_i u_j u_k + \frac{1}{4!} A_{ijkl} u_i u_j u_k u_l + \cdots, \quad (6.33)$$

where

$$A_{ij\cdots} = A_{ij\cdots}(\lambda). \quad (6.34)$$

With no loss of generality, the coefficients are treated as entirely symmetric in their indices.

Since the state is a state of equilibrium, in accordance with the principle of virtual work,

$$\bar{A}_i = 0. \quad (6.35)$$

If the quadratic term of (6.33) does not vanish identically, then it dominates for small enough displacement. It follows that the state is *stable if*

$$\mathcal{V}_2(u_i) \equiv \tfrac{1}{2} A_{ij} u_i u_j > 0. \quad (6.36)$$

The state is *critical if*

$$\tfrac{1}{2} A_{ij} u_i u_j \geq 0. \quad (6.37)$$

In words, the state is critical if there exists one (or more), nonzero displacement(s) u_i which causes the quadratic term to vanish, that is,

$$\mathcal{V}_2(\bar{u}_i) \equiv \tfrac{1}{2} A_{ij} \bar{u}_i \bar{u}_j = 0. \quad (6.38)$$

The displacement $\bar{u}_i$ is a buckling mode.

A minimum is characterized by a stationary condition. Here, the required minimum of $\mathcal{V}_2(u_i)$ is determined by the stationary criterion of E. Trefftz [25]. For an arbitrary variation δu_i,

$$\delta \mathcal{V}_2 = A_{ij} u_i \delta u_j = 0. \quad (6.39)$$

It follows that the buckling mode $\bar{u}_i$ is a nontrivial solution of the equations:

$$A_{ij} \bar{u}_i = 0. \quad (6.40)$$

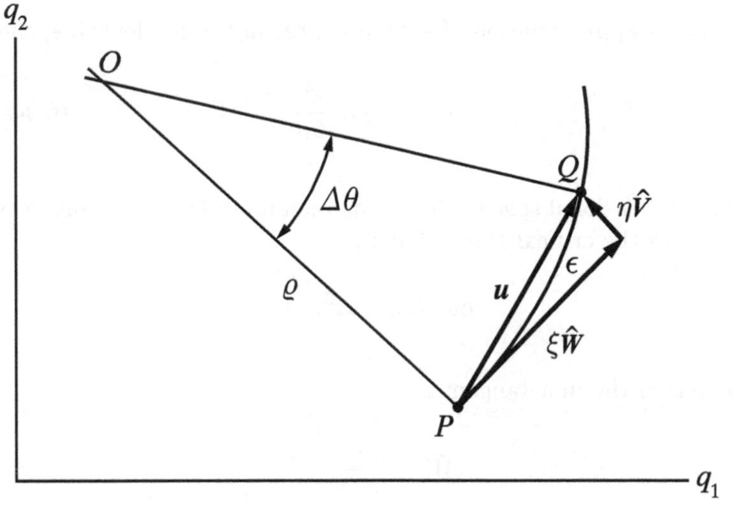

Figure 6.6 Displacement along the buckling path

The homogeneous system has a nontrivial solution if, and only if, the determinant of coefficients vanishes

$$\left|A_{ij}(\lambda)\right| = 0. \tag{6.41}$$

The least solution of (6.41) determines the critical load λ^*.

Let

$$\bar{u}_i = \epsilon\overline{W_i}, \tag{6.42}$$

where $\overline{W_i}$ are components of the unit vector in our N-dimensional space of q_i; that is,

$$\overline{W_i}\,\overline{W_i} = 1. \tag{6.43}$$

The parameter ϵ measures the magnitude of an excursion from the critical state and $\overline{W_i}$ defines the direction of the buckling.

Let us consider a movement along the path emanating from the critical point P. In the plane of (q_1, q_2), we see a path as shown in Figure 6.6. In Figure 6.6, ρ denotes the radius of the path PQ at P, $\hat{\boldsymbol{W}}$ is the unit tangent at P, and $\hat{\boldsymbol{V}}$ the unit normal. The displacement from P to Q can be expressed in the form $\boldsymbol{u} = \xi\hat{\boldsymbol{W}} + \eta\hat{\boldsymbol{V}}$ or, if ϵ denotes arc length along PQ,

$$\boldsymbol{u} = \epsilon\frac{d\boldsymbol{u}}{d\epsilon} + \frac{1}{2}\epsilon^2\frac{d^2\boldsymbol{u}}{d\epsilon^2} + \cdots = \epsilon\hat{\boldsymbol{W}} + \frac{\epsilon^2}{2\rho}\hat{\boldsymbol{V}} + \cdots.$$

If we accept an approximation of second degree in the arc length ϵ, then

$$\xi \doteq \epsilon, \qquad \eta \doteq \frac{\epsilon^2}{2\rho}. \tag{6.44a, b}$$

In the N-dimensional space, one can define an arc length ϵ along a path stemming from the critical state; that is,

$$du_i \, du_i = d\epsilon^2.$$

A component of the unit tangent is

$$\overline{W_i} = \frac{du_i}{d\epsilon}. \tag{6.45}$$

A component of the unit normal is

$$\kappa V_i = \frac{d^2 u_i}{d\epsilon^2}. \tag{6.46}$$

The displacement along a small segment is

$$\begin{aligned} u_i &= \frac{du_i}{d\epsilon}\epsilon + \frac{1}{2}\frac{d^2 u_i}{d\epsilon^2}\epsilon^2 + \cdots \\ &= \epsilon \overline{W_i} + \frac{\epsilon^2}{2}\kappa V_i + \cdots . \end{aligned} \tag{6.47}$$

Here, V_i is normalized in the manner of (6.43). In essence, we require the normal V_i and curvature κ which determine the *curved* path of minimum change $\Delta\mathcal{V}$. The change of potential follows from (6.33) and (6.47) and simplifies according to (6.35), (6.38), and (6.40):

$$\begin{aligned} \Delta\mathcal{V} = {} & \frac{\epsilon^3}{3!} A_{ijk} \overline{W_i}\,\overline{W_j}\,\overline{W_k} + \frac{\epsilon^4}{8}\kappa^2 A_{ij} V_i V_j \\ & + \frac{\epsilon^4}{4!} A_{ijkl} \overline{W_i}\,\overline{W_j}\,\overline{W_k}\,\overline{W_l} + \frac{\epsilon^4}{4}\kappa A_{ijk} \overline{W_i}\,\overline{W_j}\, V_k + O(\epsilon^5). \end{aligned} \tag{6.48}$$

If ϵ is sufficiently small, the initial term of (6.48) is dominant. Since the sign of the initial (cubic) term can be positive or negative, depending on the sense of the displacement $\overline{W}_i$, a *necessary condition for stability* follows:

$$A_3 \equiv \frac{1}{3!} A_{ijk} \overline{W}_i \overline{W}_j \overline{W}_k = 0. \tag{6.49}$$

If A_3 vanishes, as it usually does in the case of a symmetric structure, then the sign of $\Delta\mathcal{V}$ rests with the terms of higher degree. If $\Delta\mathcal{V}$ is negative for one displacement V_i, then the system is unstable. The minimum of (6.48) is stationary, that is, $\delta(\Delta\mathcal{V}) = 0$, for variations of V_i. The stationary conditions follow

$$A_{ij}\kappa V_j = -A_{jki}\overline{W}_j \overline{W}_k + \cdots + O(\epsilon). \tag{6.50}$$

If the terms of higher degree are neglected, then equation (6.50) constitutes a linear system in the displacement V_j. In accordance with (6.45) and (6.46), the solution $\overline{V}_i$ is to satisfy the orthogonality condition:

$$\overline{V}_i \overline{W}_i = 0. \tag{6.51}$$

It follows from (6.50) that

$$\kappa^2 A_{ij} \overline{V}_i \overline{V}_j = -\kappa A_{jki} \overline{W}_j \overline{W}_k \overline{V}_i + \cdots + O(\epsilon). \tag{6.52}$$

The potential change corresponding to the displacement

$$u_i = \epsilon \overline{W}_i + \frac{\epsilon^2}{2} \kappa \overline{V}_i$$

is obtained from (6.48) and simplified by means of (6.49) and (6.52):

$$\Delta\mathcal{V} = \epsilon^4 A_4, \tag{6.53}$$

where

$$A_4 \equiv \frac{1}{4!} A_{ijkl} \overline{W}_i \overline{W}_j \overline{W}_k \overline{W}_l - \frac{\kappa^2}{8} A_{ij} \overline{V}_i \overline{V}_j. \tag{6.54}$$

The system is *stable if*

$$A_4 > 0. \tag{6.55a}$$

The system is *unstable if*

$$A_4 < 0. \tag{6.55b}$$

In a system with one degree of freedom, $\overline{V}_i = 0$, and the final term of (6.54) vanishes.

6.12 Equilibrium States Near the Critical Load

In our preceding view of stability at the critical load λ^*, we examined the energy increment upon excursions from the critical state but assumed that the load remained constant. Such excursions follow the path of minimum potential on a hyperplane ($\lambda = \lambda^*$) in the configuration-load space $(q_i; \lambda)$. To trace a path of equilibrium from the critical state requires, in general, a change in the load. Let us now explore states of equilibrium near the reference state of equilibrium $(q_i; \lambda^*)$.[‡] To this end, we assume that the potential $\mathcal{V}(q_i; \lambda)$ can be expanded in a Taylor's series in the load λ, as well as the displacement u_i. Then, in place of (6.33), we have

$$\Delta\mathcal{V} = \left(\bar{A}_i u_i + \frac{1}{2}A_{ij}u_i u_j + \frac{1}{3!}A_{ijk}u_i u_j u_k + \cdots\right)$$

$$+ \left(\bar{A}'_i u_i + \frac{1}{2}A'_{ij}u_i u_j + \frac{1}{3!}A'_{ijk}u_i u_j u_k + \cdots\right)(\lambda - \lambda^*) + \cdots. \quad (6.56)$$

Here, the prime ($'$) signifies the derivative with respect to the parameter λ. Note that each of the coefficients ($\bar{A}_i, \bar{A}'_i$, etc.) is evaluated at the critical load.

Along a smooth path from the reference state in the configuration-load space, the "path" includes a step in the direction of λ, as well as the direction of q_i. In place of (6.47), we have

$$u_i = \epsilon u'_i + \frac{\epsilon^2}{2}\kappa V_i + \cdots, \quad (6.57a)$$

$$(\lambda - \lambda^*) = \epsilon\lambda' + \frac{\epsilon^2}{2}\kappa\mu + \cdots. \quad (6.57b)$$

Here, the vector $(u'_i; \lambda')$ is the unit tangent and $(V_i; \mu)$ is the principal normal at $(q_i; \lambda^*)$ of the path which traces equilibrium states in the space of configuration-load $(q_i; \lambda)$.

[‡]The logic follows the thesis of W. T. Koiter [107].

Upon substituting (6.57a, b) into (6.56), we obtain

$$\Delta\mathcal{V} = \epsilon(\bar{A}_i u_i') + \epsilon^2(\tfrac{1}{2}A_{ij}u_i'u_j' + \bar{A}_i' u_i'\lambda' + \tfrac{1}{2}\kappa\bar{A}_i V_i) + \cdots. \tag{6.58}$$

The principle of stationary potential energy gives the *equations of equilibrium at the reference state*:

$$\bar{A}_i = 0. \tag{6.59}$$

In view of (6.59), the quadratic terms (ϵ^2) dominate (6.58). The stationary principle, $\delta(\Delta\mathcal{V}) = 0$, gives the equilibrium equations for states very near the reference state:

$$A_{ij}u_j' = -\bar{A}_i'\lambda'. \tag{6.60}$$

Now, the *reference state is critical if*

$$\bar{A}_i'\lambda' = 0. \tag{6.61}$$

In words, either $\lambda' = 0$, which implies the existence of an adjacent state at the same level of loading, and/or $\bar{A}_i' = 0$; the latter holds if the reference configuration is an equilibrium configuration for $\lambda \neq \lambda^*$. Then, the *equilibrium equations of the neighboring state* follow:

$$A_{ij}u_i' = 0. \tag{6.62}$$

Equations (6.62) are the equations (6.40) of the Trefftz condition (6.39). The solution of (6.62) is the buckling mode

$$u_i' = \overline{W}_i. \tag{6.63}$$

Suppose that $\bar{A}_i' = 0$ in (6.61) and $\lambda' \neq 0$. Then, according to (6.59), (6.62), and (6.63), the potential of (6.56) and (6.58) takes the form:

$$\Delta\mathcal{V} = \epsilon^3(A_3 + A_2'\lambda' + \cdots) + O(\epsilon^4), \tag{6.64}$$

where

$$A_3 \equiv \frac{1}{3!}A_{ijk}\overline{W}_i\,\overline{W}_j\,\overline{W}_k, \tag{6.65}$$

$$A_2' \equiv \frac{1}{2}A_{ij}'\overline{W}_i\,\overline{W}_j. \tag{6.66}$$

We accept the indicated terms of (6.64) as our approximation and, in accordance with (6.57b), set

$$\epsilon\lambda' \doteq \lambda - \lambda^*. \tag{6.67}$$

Our approximation of (6.64) follows:

$$\Delta\mathcal{V} \doteq \epsilon^3 A_3 + \epsilon^2 A_2' \, (\lambda - \lambda^*). \tag{6.68}$$

The principle of stationary potential provides the *equation of equilibrium*:

$$\frac{d\,\Delta\mathcal{V}}{d\epsilon} = 3\epsilon^2 A_3 + 2\epsilon A_2' \, (\lambda - \lambda^*) = 0, \tag{6.69a}$$

or

$$\epsilon = -\frac{2A_2'}{3A_3} \, (\lambda - \lambda^*). \tag{6.69b}$$

The state is stable if the potential is a minimum, that is, if

$$\frac{d^2\Delta\mathcal{V}}{d\epsilon^2} = 6\epsilon A_3 + 2A_2' \, (\lambda - \lambda^*) > 0, \tag{6.70a}$$

or, in accordance with (6.69b), the *system is stable in the adjacent state if*

$$-A_2' \, (\lambda - \lambda^*) > 0, \qquad A_2' \, (\lambda - \lambda^*) < 0. \tag{6.70b, c}$$

In accordance with (6.56), (6.62), and (6.63), the quadratic terms of $\Delta\mathcal{V}$ in the buckled mode follow:

$$\mathcal{V}_2(\overline{W}_i) \equiv \tfrac{1}{2} \left[A_{ij} + A_{ij}' \, (\lambda - \lambda^*) + \tfrac{1}{2} A_{ij}'' \, (\lambda - \lambda^*)^2 \right] \overline{W}_i \, \overline{W}_j.$$

Since $\mathcal{V}_2(\overline{W}_i) = 0$ at the critical load, we expect that $\mathcal{V}_2(\overline{W}_i) > 0$ at loads slightly less than the critical value and that $\mathcal{V}_2(\overline{W}_i) < 0$ at loads slightly above the critical value. Therefore, we conclude that

$$A_2' \equiv \frac{1}{2} A_{ij}' \overline{W}_i \, \overline{W}_j < 0. \tag{6.71}$$

According to (6.71), the numerator of (6.69b) is always negative, but the denominator of (6.69b) is a homogeneous cubic in $\overline{W}_i$ and the sign is reversed by a reversal of the buckling mode. In this case, an adjacent state of equilibrium exists at loads above ($\lambda > \lambda^*$) or below ($\lambda < \lambda^*$) the critical value. In view of (6.70b) and (6.71), an equilibrium state above the critical load is stable and a state below is unstable.

Now, suppose that

$$\lambda' = A_3 = 0, \tag{6.72}$$

then, in view of (6.59), (6.62), (6.63), and (6.72), the potential of (6.56) and (6.58) takes the form:

$$\Delta\mathcal{V} = \epsilon^4 \left(\frac{\kappa^2}{8} A_{ij} V_i V_j + \frac{\kappa}{4} A_{ijk} V_i \overline{W}_j \overline{W}_k + \frac{1}{4!} A_{ijkl} \overline{W}_i \overline{W}_j \overline{W}_k \overline{W}_l + \cdots \right)$$

$$+ \epsilon^3 \mu \frac{\kappa}{2} \left(\underline{\bar{A}'_i \overline{W}_i} + \epsilon \frac{1}{2} A'_{ij} \overline{W}_i \overline{W}_j + \cdots \right) + O(\epsilon^5). \tag{6.73}$$

The underlined term of (6.73) dominates if $\mu \neq 0$ and if ϵ is sufficiently small. The term is odd in $\overline{W}_i$ and, therefore, always provides a negative potential change at any load $\lambda \neq \lambda^*$. A condition for the existence of stable states at noncritical values of load follows:

$$\bar{A}'_i \overline{W}_i = 0. \tag{6.74}$$

However, the buckling mode $\overline{W}_i$ is independent of the coefficients $\bar{A}'_i$. Therefore, equation (6.74) implies generally that

$$\bar{A}'_i = 0. \tag{6.75}$$

Now, we accept the remaining terms indicated in (6.73) as our approximation. Also, in view of (6.57b) and (6.72),

$$\epsilon^2 \frac{\kappa}{2} \mu \doteq \lambda - \lambda^*. \tag{6.76}$$

Our approximation of (6.73) follows:

$$\Delta\mathcal{V} \doteq \epsilon^4 \left(\frac{\kappa^2}{8} A_{ij} V_i V_j + \frac{\kappa}{4} A_{ijk} V_i \overline{W}_j \overline{W}_k + \frac{1}{4!} A_{ijkl} \overline{W}_i \overline{W}_j \overline{W}_k \overline{W}_l \right)$$

$$+ \left(\epsilon^2 \frac{1}{2} A'_{ij} \overline{W}_i \overline{W}_j \right) (\lambda - \lambda^*). \tag{6.77}$$

Again, we require a stationary potential for variations of the displacement V_i. The *equations of equilibrium* follow:

$$\epsilon^2 \kappa A_{ij} V_j = -\epsilon^2 A_{ijk} \overline{W}_j \overline{W}_k. \tag{6.78}$$

Let $\kappa \overline{V}_i$ denote the solution of (6.78). Then, it follows that

$$\epsilon^2 \kappa^2 A_{ij} \overline{V}_i \overline{V}_j = -\epsilon^2 \kappa A_{ijk} \overline{W}_j \overline{W}_k \overline{V}_i. \tag{6.79}$$

If the solution $\kappa \overline{V}_i$ and (6.79) are used in (6.77), then our approximation of the potential takes the form:

$$\Delta \mathcal{V} = \epsilon^4 A_4 + \epsilon^2 (\lambda - \lambda^*) A_2', \tag{6.80}$$

where A_2' is defined by (6.66) and A_4 by (6.54).

The solution of (6.78) determines the unit vector $\overline{V}_i$ which renders $\Delta\mathcal{V}$ stationary, but still dependent upon the *distance* ϵ. The principle of stationary potential gives the equilibrium condition:

$$\frac{d\,\Delta\mathcal{V}}{d\epsilon} = 4\epsilon^3 A_4 + 2\epsilon A_2' \,(\lambda - \lambda^*) = 0, \tag{6.81}$$

or

$$\epsilon^2 = -\frac{A_2'}{2A_4}\,(\lambda - \lambda^*). \tag{6.82}$$

According to (6.55a, b), (6.71), and (6.82), a stable adjacent state of equilibrium can exist only at loads above the critical value ($\lambda > \lambda^*$) and a state below the critical value is unstable.

W. T. Koiter [107] provides rigorous arguments for the conditions (6.71) and (6.75) if the critical configuration is a stable equilibrium configuration for loads less than the critical value. For example, the two-dimensional system has equilibrium configurations which trace a line along the λ axis, as shown in Figure 6.7. The portion OP represents stable states, the bifurcation point P represents the critical state, PR represents unstable states of the reference configuration, and PQ represents stable postbuckled equilibrium states. Here, the principle of stationary energy in the critical configuration at *any* load leads to equation (6.75), and the principle of minimum energy in the stable states of OP ($\lambda < \lambda^*$) leads to the inequality (6.71).

If the cubic term of $\Delta\mathcal{V}$ does not vanish, then equilibrium states trace paths with slope λ' at the critical load, as shown in Figure 6.7a. If the

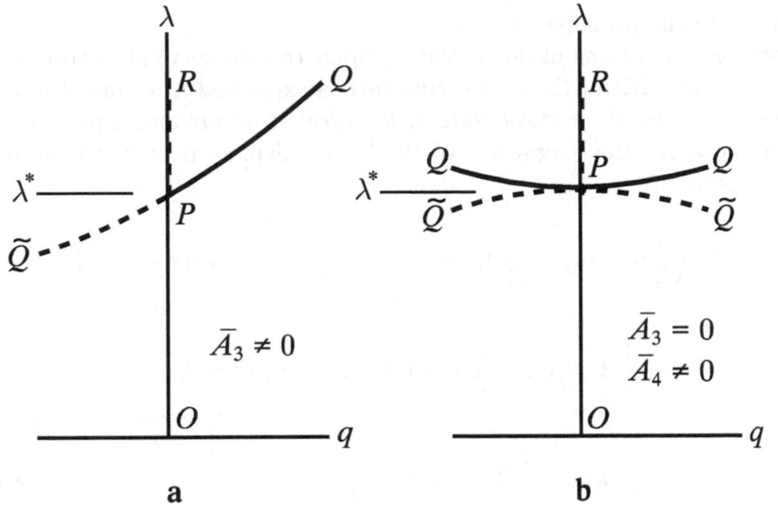

Figure 6.7 Critical state, stable and unstable states

cubic term vanishes, then $\lambda' = 0$, and the equilibrium states trace paths as shown in Figure 6.7b. In each figure, the solid lines are stable branches and the dotted lines are unstable.

Practically speaking, many structural systems display the instability patterns of Figure 6.7, that is, the prebuckled configuration of the ideal structure is an equilibrium state under all loads. Notable examples are the column under axial thrust, the spherical or cylindrical shell under external pressure, and the cylinder under uniform axial compression. Theoretically, each retains its form until the load reaches the critical value and then buckles. In the case of thin shells, initial imperfections cause pronounced departures from the initial form and often cause premature buckling ($\lambda \ll \lambda^*$).

Our analysis of stability at the critical load is limited. The reader should note, especially, that any of the various terms of the potential, for example, $\mathcal{V}_2, \mathcal{V}_4$, may vanish identically. Then, further investigation, involving terms of higher degree, is needed.

6.13 Effect of Small Imperfections upon the Buckling Load

In the monumental work of W. T. Koiter [107], an important practical achievement was his assessment of the effect of geometric imperfections upon the buckling load of an actual structure. Here, we outline the procedure and cite the principal results.

Under the conditions of dead loading upon the Hookean structure, the *energy potential* $\Delta\widetilde{\mathcal{V}}$ *of the actual structure* is expressed in terms of a *displacement* u_i *from the critical state of the ideal structure* and a parameter e which measures the magnitude of the initial displacement of the actual unloaded structure:

$$\Delta\widetilde{\mathcal{V}} = \left(\frac{1}{2}A_{ij}u_iu_j + \frac{1}{3!}A_{ijk}u_iu_ju_k + \frac{1}{4!}A_{ijkl}u_iu_ju_ku_l + \cdots\right)$$

$$+ \left(\frac{1}{2}A'_{ij}u_iu_j + \frac{1}{3!}A'_{ijk}u_iu_ju_k + \cdots\right)(\lambda - \lambda^*)$$

$$+ e\left(B_iu_i + \frac{1}{2}B_{ij}u_iu_j + \cdots\right) + \cdots. \tag{6.83}$$

Here, the linear terms in u_i, $\bar{A}_i$, and $\bar{A}'_i$, vanish because the reference configuration is an equilibrium configuration of the ideal structure ($e = 0$) at any load. *Note:* For all purposes, the coefficients $(A_{ij}\ldots, B_{ij}\ldots)$ are entirely symmetric with respect to the indices.

As before, the components u_i are expanded in powers of the arc length ϵ along the ideal curve of Figure 6.8. Here, we make an assumption that the initial deflection of the actual structure is nearly the buckling mode $\overline{W_i}$ of the ideal structure. Therefore, we have

$$u_i = \epsilon\overline{W_i} + \epsilon^2\frac{k}{2}V_i. \tag{6.84}$$

Here, the second-order term of (6.84) contains an unspecified parameter k, because this term does *not* represent a deviation from the tangent $(\epsilon\hat{\boldsymbol{W}})$ along the ideal path of Figure 6.8, but represents the displacement $\boldsymbol{d}$ which carries the system to the actual path as depicted in Figure 6.8.

Upon substituting (6.84) into (6.83) and acknowledging (6.38), (6.40), and (6.42), we obtain

$$\Delta\widetilde{\mathcal{V}} = \epsilon^3A_3 + \epsilon^2A'_2(\lambda - \lambda^*) + O(\epsilon^2)(\lambda - \lambda^*)^2 + O(\epsilon^5)$$

$$+ \epsilon^4\left[\frac{1}{4!}A_{ijkl}\,\overline{W_i}\,\overline{W_j}\,\overline{W_k}\,\overline{W_l} + \frac{k^2}{8}A_{ij}\,V_i\,V_j + \frac{k}{4}A_{ijk}\,\overline{W_i}\,\overline{W_j}V_k\right]$$

$$+ e\left(\epsilon B_i\,\overline{W_i} + \epsilon^2 B_i\frac{k}{2}V_i\right), \tag{6.85}$$

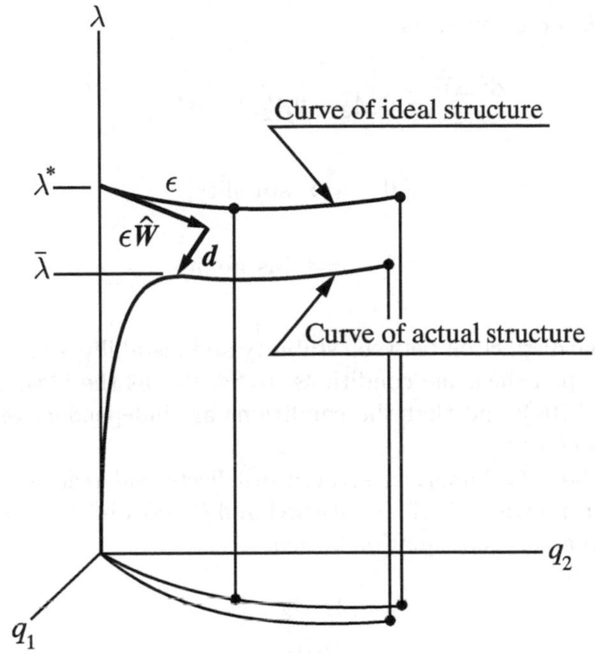

Figure 6.8 Effect of small imperfections upon the buckling load

where A_3 and A_2' are defined by (6.65) and (6.66), as before. Since the relative magnitudes of ϵ and e are unspecified, we must suppose that the terms $O(\epsilon^3)$ and $O(e\epsilon)$ dominate (6.85), if $A_3 \neq 0$ and $\lambda - \lambda^* \neq 0$. Then, we have the approximation

$$\Delta\widetilde{\mathcal{V}} \doteq \epsilon^3 A_3 + \epsilon^2 A_2' (\lambda - \lambda^*) + e\epsilon B_i \overline{W}_i. \qquad (6.86)$$

The stationary condition of equilibrium follows:

$$\frac{d\,\Delta\widetilde{\mathcal{V}}}{d\epsilon} = 3\epsilon^2 A_3 + 2\epsilon A_2'(\lambda - \lambda^*) + eB_i\overline{W}_i = 0. \qquad (6.87)$$

Now, recall that $\Delta\widetilde{\mathcal{V}}$ is *not* the potential increment from the critical configuration of the actual structure but the potential referred arbitrarily to the critical configuration of the ideal structure. An equilibrium configuration of the actual structure is stable or unstable, respectively, if the potential is a minimum or a maximum; therefore, the critical load $\bar{\lambda}$ of the actual structure satisfies the conditions

$$\frac{d^2\,\Delta\widetilde{\mathcal{V}}}{d\epsilon^2} = 6\epsilon A_3 + 2A_2'(\lambda - \lambda^*), \tag{6.88}$$

$$> 0 \implies \text{stability}, \tag{6.89a}$$

$$< 0 \implies \text{instability}. \tag{6.89b}$$

Observe that the distinction between stability and instability of a postbuckled state rests upon the same conditions, (6.89a, b), as the ideal structure [see equation (6.70a)] and that the conditions are independent of the imperfection parameter e.

If $A_3 \neq 0$, then the imperfect structure deflects and reaches a critical state of equilibrium when (6.87) is satisfied and (6.88) vanishes. If $(\lambda - \lambda^*)$ is eliminated from the two equations, then

$$e = \frac{3\epsilon^2 A_3}{B_i\overline{W_i}}. \tag{6.90}$$

The sign of the sum $B_i\overline{W_i}$ is arbitrary, since a change of sign is effected by redefining the parameter e. Therefore, we can choose e so that $B_i\overline{W_i}$ has the opposite sign of A_3. Then the condition (6.90) for a critical state is fulfilled only if $e < 0$. From our observations, we know that an imperfect structure tends to buckle in a preferred direction, depending upon the character of the geometric deviations. In the present case, if $A_3 < 0$, a critical state occurs only if $e < 0$. A plot of load versus deflection is depicted in Figure 6.9a; here, a negative value e produces buckling in a negative mode $(\epsilon\overline{W_i} < 0)$ according to the curve $O\widetilde{P}$, whereas the positive value e produces only stable states along the path OP.

A real structure which behaves in the manner of Figure 6.9a is the frame of Figure 6.9b. If the vertical strut is bent to the right or left, the imperfection parameter e is negative or positive, respectively. The rotation θ of the joint serves as a generalized coordinate $(\theta = q)$ and plots of load versus rotation take the forms of Figure 6.9a. The frame under eccentric loading has been studied experimentally by J. Roorda [111] and theoretically by W. T. Koiter [112], [113]. The latter computations show remarkable agreement with the former experimental results.

Let us turn to a structure that exhibits the behavior of Figure 6.7b, characterized by the condition

$$A_3 = 0. \tag{6.91}$$

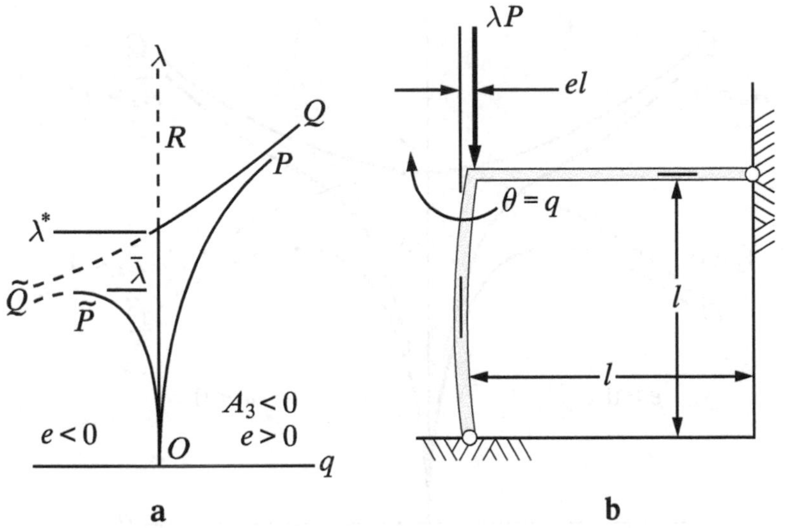

Figure 6.9 Effect of imperfections on an unsymmetric structure

At the critical load, terms $O(\epsilon^3)$ are absent from the potential of (6.85). The latter must be stationary with respect to the displacement V_i; for equilibrium,

$$\epsilon^2 k A_{ij} V_j = -\epsilon^2 A_{jki} \overline{W}_j \overline{W}_k - 2eB_i. \tag{6.92}$$

In this instance, we pursue a path (from P in Figure 6.7b) in the space (q_i, λ) with $\lambda = \lambda^*$. If $k\overline{V}_i$ denotes the solution of (6.92), then

$$\epsilon^2 k A_{jki} \overline{W}_j \overline{W}_k \overline{V}_i = -\epsilon^2 k^2 A_{ij} \overline{V}_i \overline{V}_j - 2ekB_i \overline{V}_i. \tag{6.93}$$

In accordance with (6.91) and (6.93), our approximation of (6.85) follows:

$$\Delta\widetilde{\mathcal{V}} \doteq \epsilon^4 \bar{A}_4 + \epsilon^2 A_2'(\lambda - \lambda^*) + e\epsilon B_i \overline{W}_i. \tag{6.94}$$

The coefficient A_2' is defined by (6.66) and $\bar{A}_4$ is defined, much as A_4 in (6.54):

$$\bar{A}_4 \equiv \frac{1}{4!} A_{ijkl} \overline{W}_i \overline{W}_j \overline{W}_k \overline{W}_l - \frac{k^2}{8} A_{ij} \overline{V}_i \overline{V}_j,$$

where k is the parameter in (6.84). As before, the potential $\Delta\widetilde{\mathcal{V}}$ is still dependent upon the distance ϵ. The stationary condition of equilibrium

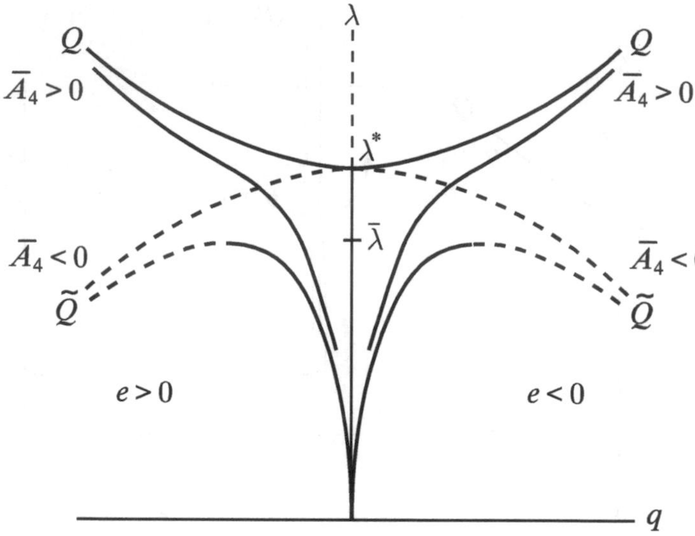

Figure 6.10 Effect of imperfections on a symmetrical system

follows:

$$\frac{d\,\Delta\widetilde{\mathcal{V}}}{d\epsilon} = 4\epsilon^3 \bar{A}_4 + 2\epsilon A_2' (\lambda - \lambda^*) + eB_i\,\overline{W_i} = 0. \tag{6.95}$$

The stability of equilibrium depends upon the second derivative as follows:

$$\frac{d^2\,\Delta\widetilde{\mathcal{V}}}{d\epsilon^2} = 12\epsilon^2 \bar{A}_4 + 2A_2' (\lambda - \lambda^*), \tag{6.96}$$

$$> 0 \quad \Longrightarrow \text{ stability,} \tag{6.97a}$$

$$< 0 \quad \Longrightarrow \text{ instability.} \tag{6.97b}$$

Again, the critical state of equilibrium is characterized by vanishing of the first derivative (6.95) and second derivative (6.96). The elimination of $(\lambda - \lambda^*)$ yields the result:

$$e = \frac{8\epsilon^3 \bar{A}_4}{B_i\,\overline{W_i}}. \tag{6.98}$$

Again, we note that the definition of e and the sign of $\overline{W_i}$ are arbitrary and, therefore, we assume that $\overline{W_i}$ renders the sum $B_i\overline{W_i} > 0$. Then, the

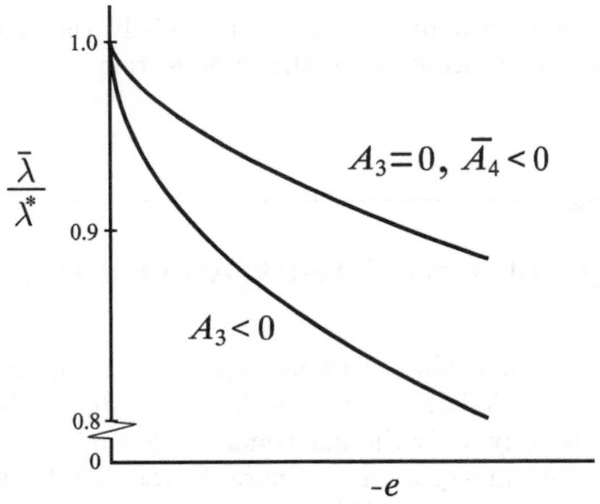

Figure 6.11 Effect of an imperfection on the buckling load

condition (6.98) for a critical load is attained if $\bar{A}_4 < 0$, $e < 0$, in keeping with (6.55b). Now, the structure also exhibits instability at the critical load if the sign of the parameter e and the buckling mode are both reversed. A plot of load versus deflection is depicted in Figure 6.10; it is characteristic of a symmetrical structure (e.g., a column) which could buckle in either direction depending upon the inherent imperfections. In either case, it is stable or unstable depending upon the sign of the constant $\bar{A}_4$. The critical load $\bar{\lambda}$ of the actual structure may be much less than the critical load λ^* of the ideal structure.

The instability of the imperfect systems are governed by equations (6.87) and (6.88), or by equations (6.95) and (6.96). The former prevail when $e < 0$ and $A_3 < 0$ (see Figure 6.9a). The latter govern when $A_3 = 0$ and $\bar{A}_4 < 0$ (see Figure 6.10). The effect of the imperfection e is assessed by eliminating the parameter ϵ from (6.87) and (6.88) in the first instance, or from (6.95) and (6.96) in the second. In the first instance ($A_3 < 0$), the critical load $\bar{\lambda} < \lambda^*$ is thereby expressed in terms of the imperfection $(-e)$:

$$\bar{\lambda} = \lambda^* - [\,3(-eB_i\overline{W}_i)(-A_3)\,]^{(1/2)}(-A_2')^{-1}. \tag{6.99}$$

In the second instance ($A_3 = 0$, $\bar{A}_4 < 0$), the corresponding relation between $\bar{\lambda}$ and λ^* assumes the following form:

$$\bar{\lambda} = \lambda^* - \tfrac{3}{2}[\,(-eB_i\overline{W}_i)^2(-\bar{A}_4)\,]^{(1/3)}(-A_2')^{-1}. \tag{6.100}$$

Plots of the actual buckling load versus the imperfection parameter for both cases are presented in Figure 6.11. Note that small imperfections can cause considerable reduction of the buckling load. Especially, the system $A_3 < 0$, as in Figure 6.9 is more susceptible to the imperfection.

6.14 Principle of Virtual Work Applied to a Continuous Body

The concepts of virtual displacement and work, and the principle as previously stated in Section 6.5, hold for any mechanical system. These apply as well to a continuous body; only the mathematical formulations differ. The continuum of particles undergoes a continuously variable displacement $\boldsymbol{V}$, and also the virtual displacement $\delta\boldsymbol{V}$. Hence, the finite number of discrete variables q_i are replaced by a continuous function $\boldsymbol{V}$. Continuous internal and external forces replace the discrete forces $'F^i$ and P^i. The former are expressed by a stress $\boldsymbol{s}^i$ (associated with the initial area), the latter by external body force $\boldsymbol{f}$ (per unit initial volume) and surface traction $\boldsymbol{T}$ (per unit initial area). Now, the virtual work is expressed by integrals extending over the initial volumes and surfaces. *Note:* The current equilibrium state *is* a deformed state; the stresses and forces act upon *deformed* surfaces and volumes. Our choices of measurement (initial volume and surfaces) are pragmatic: In practice, one knows only the initial-reference state, and those dimensions; one seeks the deformed-current state.

The virtual work of the stresses has any of the forms given previously in Section 4.5; there, increments are signified by the overdot $(\dot{\ })$. Here, the increments of displacement and corresponding strain are the arbitrarily imposed virtual increments signified by $\delta\boldsymbol{V}$ and $\delta\gamma_{ij}$. The virtual work expended by stresses (per unit volume) follows [see (4.19c–e) and (4.21b, c)]:

$$\delta w_s = \sqrt{g^{\underline{ii}}}\,\boldsymbol{\sigma}^i \cdot \delta\boldsymbol{V}_{,i} = \boldsymbol{s}^i \cdot \delta\boldsymbol{G}_i, \tag{6.101a}$$

$$= s^{ij}\boldsymbol{G}_j \cdot \delta\boldsymbol{G}_i, \tag{6.101b}$$

$$= t^{ij}\acute{\boldsymbol{g}}_j \cdot \delta\boldsymbol{G}_i. \tag{6.101c}$$

Notice that we give alternative representations: One incorporates tensorial components s^{ij} associated with the current/deformed vectors $\boldsymbol{G}_i$; the other embodies components t^{ij} associated with the rigidly rotated, but undeformed vectors $\acute{\boldsymbol{g}}_i$.

Let us note at the onset that our virtual displacements and, subsequently, variations of displacement are infinitesimals (small of first order); additionally, the displacement and derivatives are presumed continuous. From equations (3.114) and (3.115a), it follows that

$$\delta \boldsymbol{V} = \delta \boldsymbol{R}, \qquad \delta \boldsymbol{R}_{,i} = \delta \boldsymbol{G}_i = \delta \boldsymbol{V}_{,i}.$$

The work we must expend against the body force (per unit volume) is

$$\delta w_f = -\boldsymbol{f} \cdot \delta \boldsymbol{V} = -\boldsymbol{f} \cdot \delta \boldsymbol{R}. \tag{6.102}$$

We suppose that tractions $\overline{\boldsymbol{T}}$ act upon a part s_t of the bounding surface s. The virtual work that we do (per unit area) is

$$\delta w_t = -\overline{\boldsymbol{T}} \cdot \delta \boldsymbol{V} = -\overline{\boldsymbol{T}} \cdot \delta \boldsymbol{R}. \tag{6.103}$$

Finally, the virtual work δW upon the entire body is comprised of the integrals of δw_s and δw_f throughout the volume v, and δw_t over the surface s_t:

$$\delta W = \delta W_s + \delta W_f + \delta W_t, \tag{6.104a}$$

$$\delta W = \iiint_v (\boldsymbol{s}^i \cdot \delta \boldsymbol{G}_i - \boldsymbol{f} \cdot \delta \boldsymbol{V})\, dv - \iint_{s_t} \overline{\boldsymbol{T}} \cdot \delta \boldsymbol{V}\, ds. \tag{6.104b}$$

Here, according to (6.101b, c), we have the alternative forms of the vector $\boldsymbol{s}^i$:

$$\boldsymbol{s}^i = s^{ij} \boldsymbol{G}_j = t^{ij} \acute{\boldsymbol{g}}_j. \tag{6.105a, b}$$

We note that components of the increments $\delta \boldsymbol{G}_i$ can be treated, mathematically, as the rates of (3.139) and (3.147) (Section 3.18), viz.,

$$\delta \boldsymbol{G}_i = (\delta \gamma_{ji} + \delta \omega_{ji}) \boldsymbol{G}^j, \tag{6.106a}$$

$$= \delta h_{ij} \acute{\boldsymbol{g}}^j + C_{ij}\, \delta \boldsymbol{\omega} \times \acute{\boldsymbol{g}}^j = (\delta h_{ij} + C_i^l\, \delta \overline{\omega}^k\, e_{klj}) \acute{\boldsymbol{g}}^j. \tag{6.106b, c}$$

In each form, the first term comprises the deformation; moreover, both strains are symmetric,

$$\delta \gamma_{ij} = \delta \gamma_{ji}, \qquad \delta h_{ij} = \delta h_{ji}.$$

In each, the final term represents solely the *rigid rotation*. The virtual work of the stresses consists of the two parts, but only the symmetrical parts

($s^{(ij)} = s^{(ji)} = s^{ij} = s^{ji}$ and $t^{(ij)} = t^{(ji)}$) do work upon the respective strains. Therefore,

$$\delta w_s = s^{(ij)}\delta\gamma_{ji} + s^{ij}\delta\omega_{ji} = t^{(ij)}\delta h_{ij} + t^{ij}(C_i^l\,\delta\overline{\omega}^k\,e_{klj}). \qquad (6.107\text{a, b})$$

Equilibrium of an arbitrarily small element requires that the virtual work vanish for arbitrary virtual displacements. In this case, the rigid motions ($\delta\boldsymbol{V}$ and $\delta\boldsymbol{\omega}$) and the deformation ($\delta\gamma_{ij}$ or δh_{ij}) must be regarded as independent. Hence, equilibrium requires that the final term of (6.107a, b) vanish:

$$s^{ij}\delta\omega_{ji} = 0, \qquad t^{ij}(C_i^l\,\delta\overline{\omega}^k\,e_{klj}) = 0. \qquad (6.108\text{a, b})$$

These terms must vanish for *arbitrary* rotation; since $\delta\omega_{ji} = -\delta\omega_{ij}$ and $e_{klj} = -e_{kjl}$,

$$s^{ij} = s^{ji}, \qquad t^{ij}C_i^l = t^{il}C_i^j. \qquad (6.109\text{a, b})$$

Then the expression (6.104) incorporates only the virtual work of the symmetric tensor s^{ij} or the symmetric part $t^{(ij)}$ of the stress $\boldsymbol{s}^i$.

Having established the *equilibrium* conditions (6.109a, b), we return to the virtual work (6.104), but recall that $\delta\boldsymbol{V} = \delta\boldsymbol{R}$ and $\delta\boldsymbol{G}_i = \delta\boldsymbol{R}_{,i}$ with the requisite continuity of the vector $\boldsymbol{R}$, derivatives and variations:

$$\delta W = \iiint_v (\boldsymbol{s}^i\cdot\delta\boldsymbol{R}_{,i} - \boldsymbol{f}\cdot\delta\boldsymbol{R})\sqrt{g}\,d\theta_1\,d\theta_2\,d\theta_3 - \iint_{s_t}\overline{\boldsymbol{T}}\cdot\delta\boldsymbol{R}\,ds. \quad (6.110\text{a})$$

Applying Green's theorem to the first term [see equation (2.73)], we obtain

$$\delta W = -\iiint_v \left[\frac{1}{\sqrt{g}}(\sqrt{g}\,\boldsymbol{s}^i)_{,i} + \boldsymbol{f}\right]\cdot\delta\boldsymbol{R}\,dv$$

$$-\iint_{s_t}(\overline{\boldsymbol{T}} - \boldsymbol{s}^i n_i)\cdot\delta\boldsymbol{R}\,ds + \iint_{s_v}(\boldsymbol{s}^i n_i)\cdot\delta\boldsymbol{R}\,ds. \quad (6.110\text{b})$$

Here, $n_i = \hat{\boldsymbol{n}}\cdot\boldsymbol{g}_i$ is the component of the initial normal to the surface s; $s_v = s - s_t$ is that part of the bounding surface s where displacements are prescribed.

The principle of virtual work requires that the virtual work vanishes ($\delta W = 0$) for *arbitrary* admissible virtual displacements $\delta\boldsymbol{R}$, so-called variations (see e.g., [10], [29]). *Admissibility* implies that the variation $\delta\boldsymbol{R}$ has the requisite continuity and is consistent with the constraints ($\delta\boldsymbol{R}$ vanishes where displacements are prescribed, e.g., where the boundary surface is fixed). The condition is fulfilled if, and only if,

$$\frac{1}{\sqrt{g}}(\sqrt{g}\,\boldsymbol{s}^i)_{,i} + \boldsymbol{f} = \mathbf{o} \quad \text{in } v, \tag{6.111}$$

$$\boldsymbol{s}^i n_i = \overline{\boldsymbol{T}} \quad \text{on } s_t, \tag{6.112}$$

$$\delta\boldsymbol{R} = \mathbf{o} \quad \text{on } s_v. \tag{6.113}$$

The first equation (6.111) is the differential equation of equilibrium (4.44b) which can be expressed in terms of any of the stress tensors (s^{ij} or t^{ij}); see (4.45b, c). Equation (6.112) asserts the equilibrium of internal actions $\boldsymbol{s}^i n_i$ and surface traction $\overline{\boldsymbol{T}}$ upon an element at surface s_t. The last equation (6.113) expresses the fact that the displacements are prescribed on s_v.

6.15 Principle of Stationary Potential Applied to a Continuous Body

If the internal forces are conservative (practically speaking, if the body is elastic), then their incremental work derives from a potential:

$$\delta w_s = -\delta u_0 = -s^{ij}\,\delta\gamma_{ij}, \tag{6.114a}$$

$$= -\delta\bar{u}_0 = -t^{ij}\,\delta h_{ij}. \tag{6.114b}$$

The function $u_0 = u_0(\gamma_{ij})$ [or $\bar{u}_0(h_{ij})$] may be the internal energy or the free energy (see Section 5.11) when the deformation is adiabatic or isothermal, respectively. In either case,

$$s^{ij} = \frac{\partial u_0}{\partial \gamma_{ij}}, \qquad t^{ij} = \frac{\partial \bar{u}_0}{\partial h_{ij}}. \tag{6.115a, b}$$

The incremental work of all internal forces is the variation of an internal-energy potential U‡

$$\delta W_s = -\delta U \equiv -\iiint_v \delta u_0 \, dv. \tag{6.116}$$

If the body forces are conservative, the incremental work of body forces is the negative of the variation of a potential Π_f:

$$\delta W_f = -\delta \Pi_f \equiv -\iiint_v \delta \pi_f(X_i) \, dv, \tag{6.117}$$

where π_f is the potential (per unit initial volume) and X_i is the current Cartesian coordinate of the particle ($X_i \equiv \boldsymbol{R} \cdot \hat{\boldsymbol{\imath}}_i$). Since $X_i = \hat{\boldsymbol{\imath}}_i \cdot \boldsymbol{R}(\theta^i)$, the potential is also implicitly a function of the arbitrary coordinate θ^i. The body force has alternative forms:

$$\boldsymbol{f} = -\frac{\partial \pi_f}{\partial X_i} \hat{\boldsymbol{\imath}}_i = f^i \boldsymbol{g}_i = \overline{\overline{f}}{}^i \boldsymbol{G}_i. \tag{6.118}$$

Likewise, if the surface tractions are conservative, the incremental work of surface forces is the negative of the variation of a potential Π_t:

$$\delta W_t = -\delta \Pi_t \equiv -\iint_{s_t} \delta \pi_t(X_i) \, ds. \tag{6.119}$$

The traction also has alternative representations:

$$\boldsymbol{T} = -\frac{\partial \pi_t}{\partial X_i} \hat{\boldsymbol{\imath}}_i = T^i \boldsymbol{g}_i = \overline{\overline{T}}{}^i \boldsymbol{G}_i. \tag{6.120}$$

The principle of virtual work of (6.104a, b) applies to the conservative system, wherein

$$\delta W = -\delta \mathcal{V} \equiv -(\delta U + \delta \Pi_f + \delta \Pi_t) = 0. \tag{6.121}$$

In words, the system is in equilibrium if, and only if, the first-order variation $\delta\mathcal{V}$ of the total potential $\mathcal{V}$† vanishes; or stated otherwise, the potential is

‡Note that the integral $U = \iiint_v u \, dv$ has a value for each function $u(X_i)$; such quantities are called *functionals*. If u is assigned a variation $\delta u(X_i)$, the corresponding change in U is termed the variation of U and is denoted by δU. Note also that δW_s is not, in general, the variation of a functional.

†The *first-order variation*, or, simply, *first variation*, signifies a variation which includes only terms of first order in the varied quantity, in this instance, the displacement.

stationary. Again, as in (6.104a, b) and (6.110a, b), the variation is arbitrary but subject to constraints (e.g., $\boldsymbol{V} = \overline{\boldsymbol{V}} = \mathbf{o}$ on s_v) and requisite continuity. The stationary conditions are again the equilibrium conditions (6.109a, b) (arbitrary rotation), equation (6.111) (arbitrary displacement in v), and equation (6.112) (arbitrary displacement on s_t). Finally, we have also the constraint (6.113) ($\delta\boldsymbol{R} = \mathbf{o}$ on s_v).

The energy criteria for stability of equilibrium are presented in Sections 6.10 through 6.13. Those criteria rest upon the changes of potential which accompany small movements about a state of equilibrium. The analyses provide means to determine the critical loads which signal buckling of the perfect system, but also certain consequences of imperfections. The buckling can be the gradual, but excessive, deflection of a simple column or the abrupt severe disfiguration of a shell, so-called snap-buckling. The latter are also the most susceptible to imperfections and most prone to premature buckling. These effects in the continuous systems are analyzed as the similar phenomena of the discrete systems. Only the mathematical formalities are different; continuous functions supplant discrete variables and functionals replace the functions of those variables.

6.16 Generalization of the Principle of Stationary Potential

The potential of an elastic body is an *integral* $\mathcal{V}$ which *depends* on a *function*, the displacement $\boldsymbol{V}$ (or current position $\boldsymbol{R}$). As such, the integral is a *functional*. The principle of stationary potential asserts that the functional $\mathcal{V}$ is stationary with respect to admissible variations of the function $\boldsymbol{V}$. A modified and useful version of the potential and stationary conditions is credited to H. C. Hu [114] and K. Washizu [115]. The modified potential is also a functional with respect to strains and stresses. Here, we formulate alternative versions incorporating the different strains (γ_{ij} and h_{ij}) and the associated stresses (s^{ij} and t^{ij}).

The potential $\mathcal{V}$ consists of the internal energy U, and potentials Π_f and Π_t attributed to conservative external forces. We presume that the latter are explicitly dependent on the position $\boldsymbol{R}$. However, the internal energy is a functional of the strain: $U = U(\gamma_{ij})$ is an integral of the internal energy $u_0(\gamma_{ij})$ [or $\overline{U}(h_{ij})$ and $\bar{u}_0(h_{ij})$]. This integral is also implicitly a functional of $\boldsymbol{R}$ through either one of the following kinematic equations:

$$\gamma_{ij} = \tfrac{1}{2}(\boldsymbol{R}_{,i} \cdot \boldsymbol{R}_{,j} - g_{ij}), \tag{6.122}$$

$$h_{ij} = \tfrac{1}{2}(\acute{\boldsymbol{g}}_j \cdot \boldsymbol{R}_{,i} + \acute{\boldsymbol{g}}_i \cdot \boldsymbol{R}_{,j}) - g_{ij}. \tag{6.123}$$

The latter depends explicitly on the rigid rotation which carries the initial principal lines to their current orientation [see (3.87a, b)]:

$$\acute{\boldsymbol{g}}_k = r_k^{\cdot j} \boldsymbol{g}_j. \tag{6.124}$$

We digress to note that a distinction between the translation $\boldsymbol{R}$ and rotation $r_k^{\cdot j}$ is especially meaningful in certain structural problems: (1) The nonlinear behavior of *thin* bodies (rods, plates, and shells) is often characterized by large changes of rotation, though strains are usually small. (2) The behavior of one *small*, but finite, *element* is independent of the translation and rotation, but changes of rotation (a consequence of curvature) must be taken into account in an assembly ([116], [117]).

The modified potential $\mathcal{V}^*$ (or $\overline{\mathcal{V}}^*$) incorporates the internal energy $U(\gamma_{ij})$ [or $\overline{U}(h_{ij})$] as a functional of the strain γ_{ij} (or h_{ij}), wherein the strain is subject to the kinematical constraints (6.122) [or (6.123)]. The latter are imposed via Lagrange multipliers λ^{ij} (see, e.g., [118]). Also, the kinematical constraint ($\boldsymbol{R} = \overline{\boldsymbol{R}}$, the imposed position) on surface s_v is enforced by a multiplier $\widetilde{\boldsymbol{T}}$. The modified version of the potential follows:

$$\mathcal{V}^* = \iiint_v \left\{ u_0(\gamma_{ij}) - \lambda^{ij} \left[\gamma_{ij} - \tfrac{1}{2} (\boldsymbol{R}_{,i} \cdot \boldsymbol{R}_{,j} - g_{ij}) \right] + \pi_f(\boldsymbol{R}) \right\} dv$$

$$+ \iint_{s_t} \pi_t(\boldsymbol{R})\, ds - \iint_{s_v} \widetilde{\boldsymbol{T}} \cdot (\boldsymbol{R} - \overline{\boldsymbol{R}})\, ds. \tag{6.125}$$

The functional $\mathcal{V}^*(\gamma_{ij}, \lambda^{ij}, \widetilde{\boldsymbol{T}}$ and $\boldsymbol{R})$ is subject to variations of the displacement $\boldsymbol{R}$, strains γ_{ij} and multipliers λ^{ij} in v, $\boldsymbol{R}$ on s and multipliers $\widetilde{\boldsymbol{T}}$ on s_v. We note that $\mathcal{V}^* = \mathcal{V}$ when the kinematical constraints are satisfied; moreover, since $\gamma_{ij} = \gamma_{ji}$, there is no loss of generality by the restriction $\lambda^{ij} = \lambda^{ji}$. The first-order variation follows:

$$\delta\mathcal{V}^* = \iiint_v \left\{ \left[\frac{\partial u_0}{\partial \gamma_{ij}} - \lambda^{ij} \right] \delta\gamma_{ij} - \delta\lambda^{ij} \left[\gamma_{ij} - \frac{1}{2} (\boldsymbol{R}_{,i} \cdot \boldsymbol{R}_{,j} - g_{ij}) \right] \right.$$

$$\left. - \left[\frac{1}{\sqrt{g}} \left(\sqrt{g}\, \lambda^{ij} \boldsymbol{G}_j \right)_{,i} - \frac{\partial \pi_f}{\partial X_i} \hat{\boldsymbol{\imath}}_i \right] \cdot \delta\boldsymbol{R} \right\} dv$$

$$- \iint_{s_v} \delta\widetilde{\boldsymbol{T}} \cdot \left[\boldsymbol{R} - \overline{\boldsymbol{R}} \right] ds - \iint_{s_v} \left[\widetilde{\boldsymbol{T}} - \lambda^{ij} \boldsymbol{G}_j n_i \right] \cdot \delta\boldsymbol{R}\, ds$$

$$+ \iint_{s_t} \left[\lambda^{ij} \boldsymbol{G}_j n_i + \frac{\partial \pi_t}{\partial X_i} \hat{\boldsymbol{\imath}}_i \right] \cdot \delta\boldsymbol{R}\, ds. \tag{6.126}$$

Since the variations are arbitrary (γ_{ij}, λ^{ij}, $\boldsymbol{R}$ in v, $\boldsymbol{R}$ on s, and $\widetilde{\boldsymbol{T}}$ on s_v), each bracketed term must vanish; these are the Euler equations:

$$\lambda^{ij} = \frac{\partial u_0}{\partial \gamma_{ij}} \qquad \text{in } v, \tag{6.127}$$

$$\gamma_{ij} = \tfrac{1}{2}(\boldsymbol{R}_{,i} \cdot \boldsymbol{R}_{,j} - g_{ij}) \qquad \text{in } v, \tag{6.128}$$

$$\frac{1}{\sqrt{g}}(\sqrt{g}\,\lambda^{ij}\,\boldsymbol{G}_j)_{,i} - \frac{\partial \pi_f}{\partial X_i}\hat{\boldsymbol{\imath}}_i = \mathbf{o} \quad \text{in } v, \tag{6.129}$$

$$\lambda^{ij}\boldsymbol{G}_j n_i\,(\equiv \boldsymbol{T}) = -\frac{\partial \pi_t}{\partial X_i}\hat{\boldsymbol{\imath}}_i \qquad \text{on } s_t, \tag{6.130}$$

$$\boldsymbol{R} = \overline{\boldsymbol{R}} \qquad \text{on } s_v. \tag{6.131}$$

Additionally, if $\delta\boldsymbol{R} \neq \mathbf{o}$ on s_v, then

$$\widetilde{\boldsymbol{T}} = \lambda^{ij}\boldsymbol{G}_j n_i \qquad \text{on } s_v. \tag{6.132}$$

Equation (6.127) identifies the multiplier $\lambda^{ij} = s^{ij}$. Then (6.129) and (6.130) are recognized as the equilibrium equations (6.111) and (6.112). As anticipated, the variations of the multipliers (λ^{ij} in v and $\widetilde{\boldsymbol{T}}$ on s_v) provide the kinematical constraints (6.128) in v and (6.131) on s_v. According to (6.132), the multiplier $\widetilde{\boldsymbol{T}}$ emerges as the traction on s_v. *Note:* The latter (6.132) is *not* a condition *imposed* on s_v, since $\delta\boldsymbol{R} = \mathbf{o}$ ($\boldsymbol{R} = \overline{\boldsymbol{R}}$) on s_v. We have merely admitted such variation to reveal the identity of $\widetilde{\boldsymbol{T}}$.

Notice that the generalized potential $\mathcal{V}^*(\gamma_{ij}, s^{ij}, \boldsymbol{R})$ is a functional of all the basic functions, each subject to the required admissibility conditions in v and on s. Then the stationary conditions are the *constitutive* equations (6.127), all *kinematical* equations (6.128) and (6.131), and the *equilibrium* equations (6.129) and (6.130).

Alternative expressions can be derived if the multiplier $\widetilde{\boldsymbol{T}}$ is eliminated from (6.125) using (6.132), or if $\boldsymbol{R}$ is not variable ($\boldsymbol{R} = \overline{\boldsymbol{R}}$, $\delta\boldsymbol{R} = \mathbf{o}$) on s_v.

The foregoing functional (6.126) with the corresponding stationary criteria is cited by K. Washizu [115] and B. Fraeijs de Veubeke [119] as the *general principle* of stationary potential. The procedure is given by R. Courant and D. Hilbert [120].

To devise the modified potential incorporating the engineering strain h_{ij} and associated stress t^{ij}, we must enforce the kinematical conditions (3.96) or (3.99) which also entails the rotation of (3.87) (see Section 3.14), specifically, the variation of vector $\acute{\boldsymbol{g}}_i$. The latter is an infinitesimal rigid rotation; accordingly, it can be represented by a vector $\delta\boldsymbol{\omega}$:

$$\delta\acute{\boldsymbol{g}}_i = \delta\boldsymbol{\omega} \times \acute{\boldsymbol{g}}_i = \delta\bar{\omega}^k e_{kin}\acute{\boldsymbol{g}}^n \equiv \delta\bar{\omega}_{ni}\acute{\boldsymbol{g}}^n. \tag{6.133}$$

To implement the variational procedure, we also recall the relation between the rotated vector $\acute{\boldsymbol{g}}_i$ and the deformed version $\boldsymbol{G}_i$ [see (3.95)]:

$$\boldsymbol{G}_i \equiv \boldsymbol{R}_{,i} = C_i^j \acute{\boldsymbol{g}}_j = (h_i^j + \delta_i^j)\,\acute{\boldsymbol{g}}_j. \tag{6.134}$$

The alternative version of the potential takes the form:

$$\overline{\mathcal{V}}^* = \iiint_v \left[\overline{u}_0(h_{ij}) - \tilde{\lambda}^{ij}\left(h_{ij} - \acute{\boldsymbol{g}}_j \cdot \boldsymbol{R}_{,i} + g_{ij}\right) + \pi_f(\boldsymbol{R})\,dv\right]$$

$$+ \iint_{s_t} \pi_t(\boldsymbol{R})\,ds - \iint_{s_v} \widetilde{\boldsymbol{T}}\left(\boldsymbol{R} - \overline{\boldsymbol{R}}\right)\,ds. \tag{6.135}$$

This functional is dependent on position $\boldsymbol{R}$ (in v and s), strain h_{ij} (in v), multipliers $\tilde{\lambda}^{ij}$ (in v) and $\widetilde{\boldsymbol{T}}$ (on s_v), *and* also the rotation (in v) according to (3.87). Note that the vector $\delta\boldsymbol{\omega}$ effects the variation $\delta r_{ij} = r^n_{\bullet j} e_{kin}\,\delta\bar{\omega}^k$; see equation (3.105) or (6.124). Notice too that the rotation is most readily referred to the current state, i.e., $\delta\boldsymbol{\omega} = \delta\bar{\omega}^i\acute{\boldsymbol{g}}_i$.

The first-order variation assumes the form:

$$\delta\overline{\mathcal{V}}^* = \iiint_v \left\{\left[\frac{\partial\bar{u}_0}{\partial h_{ij}} - \tilde{\lambda}^{(ij)}\right]\delta h_{ij} - \delta\tilde{\lambda}^{(ij)}\left[h_{ij} - \frac{1}{2}(\acute{\boldsymbol{g}}_i \cdot \boldsymbol{R}_{,j} + \acute{\boldsymbol{g}}_j \cdot \boldsymbol{R}_{,i}) + g_{ij}\right]\right.$$

$$- \left[\frac{1}{\sqrt{g}}\left(\sqrt{g}\,\tilde{\lambda}^{ij}\,\acute{\boldsymbol{g}}_j\right)_{,i} - \frac{\partial\pi_f}{\partial X_i}\hat{\boldsymbol{\imath}}_i\right] \cdot \delta\boldsymbol{R}$$

$$\left. + \left[\tilde{\lambda}^{ij}(h_i^p + \delta_i^p)e_{kjp}\right]\delta\bar{\omega}^k + \delta\tilde{\lambda}^{[ij]}\frac{1}{2}\left[\acute{\boldsymbol{g}}_i \cdot \boldsymbol{R}_{,j} - \acute{\boldsymbol{g}}_j \cdot \boldsymbol{R}_{,i}\right]\right\} dv$$

$$- \iint_{s_v} \delta\widetilde{\boldsymbol{T}} \cdot \left[\boldsymbol{R} - \overline{\boldsymbol{R}}\right]\,ds - \iint_{s_v}\left[\widetilde{\boldsymbol{T}} - \tilde{\lambda}^{ij}\acute{\boldsymbol{g}}_j n_i\right] \cdot \delta\boldsymbol{R}\,ds$$

$$+ \iint_{s_t}\left[\tilde{\lambda}^{ij}\acute{\boldsymbol{g}}_j n_i + \frac{\partial\pi_t}{\partial X_i}\hat{\boldsymbol{\imath}}_i\right] \cdot \delta\boldsymbol{R}\,ds. \tag{6.136}$$

This is the counterpart of (6.126), but includes a term resulting from the rotation $\delta\boldsymbol{\omega}$ of vectors $\boldsymbol{\acute{g}}_i$. Note that the variation $\delta\tilde{\lambda}^{ij}$ has been split into symmetric $[\delta\tilde{\lambda}^{(ij)}]$ and antisymmetric $[\delta\tilde{\lambda}^{[ij]}]$ parts. Because the decomposition is such that $h_{ij} = h_{ji}$ and $\boldsymbol{\acute{g}}_i \cdot \boldsymbol{R}_{,j} = \boldsymbol{\acute{g}}_j \cdot \boldsymbol{R}_{,i}$ (see Subsection 3.14.1), the antisymmetric part makes no contribution to the variation $\delta\mathcal{V}^*$ (6.136).

The independence of the variations (δh_{ij}, $\delta\tilde{\lambda}^{ij}$, $\delta\boldsymbol{R}$, $\delta\widetilde{\boldsymbol{T}}$, $\delta\boldsymbol{\omega}$) requires that each bracketed term vanish; these are the Euler equations:

$$\tilde{\lambda}^{(ij)} = \frac{\partial \bar{u}_0}{\partial h_{ij}} \qquad \text{in } v, \tag{6.137}$$

$$h_{ij} = \tfrac{1}{2}(\boldsymbol{\acute{g}}_i \cdot \boldsymbol{R}_{,j} + \boldsymbol{\acute{g}}_j \cdot \boldsymbol{R}_{,i}) - g_{ij} \qquad \text{in } v, \tag{6.138}$$

$$\frac{1}{\sqrt{g}}(\sqrt{g}\,\tilde{\lambda}^{ij}\,\boldsymbol{\acute{g}}_j)_{,i} - \frac{\partial \pi_f}{\partial X_i}\hat{\boldsymbol{\imath}}_i = \mathbf{o} \qquad \text{in } v, \tag{6.139}$$

$$\tilde{\lambda}^{ji}(h_j^p + \delta_j^p) = \tilde{\lambda}^{jp}(h_j^i + \delta_j^i) \qquad \text{in } v, \tag{6.140}$$

$$\tilde{\lambda}^{ij}\boldsymbol{\acute{g}}_j n_i = -\frac{\partial \pi_t}{\partial X_i}\hat{\boldsymbol{\imath}}_i \qquad \text{on } s_t, \tag{6.141}$$

$$\boldsymbol{R} = \overline{\boldsymbol{R}} \qquad \text{on } s_v, \tag{6.142}$$

$$\widetilde{\boldsymbol{T}} = \tilde{\lambda}^{ij}\boldsymbol{\acute{g}}_j n_i \qquad \text{on } s_v. \tag{6.143}$$

The equations [(6.137) to (6.143)] differ from their counterparts [equations (6.127) to (6.132)] in two ways:

1. These are expressed in terms of the engineering strain h_{ij} and the associated stress $t^{ij} = \tilde{\lambda}^{ij}$.
2. The system includes the additional equations (6.140) that are the conditions of equilibrium associated with the rigid rotation $\delta\boldsymbol{\omega}$. In his memorable work, B. Fraeijs de Veubeke [121] calls this result [equations (6.140)] "a disguised form of the rotational equilibrium conditions."

The ultimate value of any theorem is the basis for greater understanding, further development, and useful applications. To those ends the general principle provides a powerful basis for examining and developing means of effective approximations by finite elements. Such applications are elucidated in Chapter 11.

6.17 General Functional and Complementary Parts

A primitive functional of stress $\boldsymbol{s}^i$ and position $\boldsymbol{R}$ has the form [122]:

$$\mathcal{P} = \iiint_v (\boldsymbol{s}^i \cdot \boldsymbol{R}_{,i} - \boldsymbol{f} \cdot \boldsymbol{R})\, dv - \iint_s \boldsymbol{T} \cdot \boldsymbol{R}\, ds - \iint_{s_v} (\boldsymbol{R} - \overline{\boldsymbol{R}}) \cdot \boldsymbol{T}\, ds. \tag{6.144}$$

As before, $\boldsymbol{s}^i$ may be represented in any of the alternative forms [e.g., (6.105a, b)]; $\boldsymbol{f}$ denotes the body force, $\boldsymbol{R}$ the current position, $\boldsymbol{T}$ the surface traction, and $\overline{\boldsymbol{R}}$ the imposed position on surface s_v. If the stress satisfies the equilibrium conditions $[(\sqrt{g}\,\boldsymbol{s}^i)_{,i} + \sqrt{g}\,\boldsymbol{f} = \mathbf{o}$ in v, $\boldsymbol{s}^i n_i = \boldsymbol{T}$ on $s_t]$ and the position is admissible [$\boldsymbol{R}$ and $\boldsymbol{R}_{,i}$ are continuous in v, $\boldsymbol{R} = \overline{\boldsymbol{R}}$ on s_v], then $\mathcal{P} = 0$.

Now, recall the relation between the density of internal energy (internal energy or free energy) and its complement (enthalpy or Gibbs potential) presented in Section 5.11; in the absence of thermal effects

$$s^{ij}\gamma_{ij} = u_0(\gamma_{ij}) + u_{c0}(s^{ij}), \tag{6.145a}$$

$$t^{ij}h_{ij} = \bar{u}_0(h_{ij}) + \bar{u}_{c0}(t^{ij}). \tag{6.145b}$$

Hereafter, we refer to these as internal density (u_0 or $\bar{u}_0$) and complementary density (u_{c0} or $\bar{u}_{c0}$).

With the requisite continuity

$$\boldsymbol{s}^i \cdot \boldsymbol{R}_{,i} = s^{ij}(\gamma_{ij} + \tfrac{1}{2}g_{ij}) + \tfrac{1}{2}\boldsymbol{s}^i \cdot \boldsymbol{R}_{,i} = t^{ij}(h_{ij} + g_{ij}).$$

For the elastic body [see (6.145a, b)],

$$\boldsymbol{s}^i \cdot \boldsymbol{R}_{,i} = u_0 + u_{c0} + \tfrac{1}{2}s^{ij}g_{ij} + \tfrac{1}{2}\boldsymbol{s}^i \cdot \boldsymbol{R}_{,i}, \tag{6.146a}$$

$$= \bar{u}_0 + \bar{u}_{c0} + t^i_i. \tag{6.146b}$$

In accordance with (4.26c), the components t^{ij} are associated via the initial metric, i.e., $t^{i}_{\cdot j} = g_{jk}t^{ik}$.

When (6.146a, b) are substituted into (6.144), one obtains

$$\mathcal{P} = \iiint_v \left(u_0 + u_{c0} + \tfrac{1}{2}s^{ij}g_{ij} + \tfrac{1}{2}\,\boldsymbol{s}^i \cdot \boldsymbol{R}_{,i} - \boldsymbol{f} \cdot \boldsymbol{R}\right) dv$$

$$- \iint_s \boldsymbol{T} \cdot \boldsymbol{R}\, ds - \iint_{s_v} \left(\boldsymbol{R} - \overline{\boldsymbol{R}}\right) \cdot \boldsymbol{T}\, ds,$$

or

$$\overline{\mathcal{P}} = \iiint_v \left(\bar{u}_0 + \bar{u}_{c0} + t^i_i - \boldsymbol{f} \cdot \boldsymbol{R}\right) dv$$

$$- \iint_s \boldsymbol{T} \cdot \boldsymbol{R}\, ds - \iint_{s_v} \left(\boldsymbol{R} - \overline{\boldsymbol{R}}\right) \cdot \boldsymbol{T}\, ds.$$

If the applied loads ($\boldsymbol{f}$ in v and $\overline{\boldsymbol{T}}$ on s_t) are "dead" loads (constant), then we recognize a part of each ($\mathcal{P}$ and $\overline{\mathcal{P}}$) which is the potential in one of the forms:

$$\mathcal{V} = \iiint_v (u_0 - \boldsymbol{f} \cdot \boldsymbol{R})\, dv - \iint_{s_t} \overline{\boldsymbol{T}} \cdot \boldsymbol{R}\, ds, \tag{6.147a}$$

or

$$\overline{\mathcal{V}} = \iiint_v (\bar{u}_0 - \boldsymbol{f} \cdot \boldsymbol{R})\, dv - \iint_{s_t} \overline{\boldsymbol{T}} \cdot \boldsymbol{R}\, ds. \tag{6.147b}$$

In the manner of (6.145a, b) we *define* total *complementary* potentials such that

$$\mathcal{P} \equiv \mathcal{V} + \mathcal{V}_c \equiv \overline{\mathcal{V}} + \overline{\mathcal{V}}_c. \tag{6.148a, b}$$

Then

$$\mathcal{V}_c \equiv \iiint_v \left(u_{c0} + \tfrac{1}{2}s^{ij}g_{ij} + \tfrac{1}{2}\,\boldsymbol{s}^i \cdot \boldsymbol{R}_{,i}\right) dv$$

$$- \iint_{s_v} \boldsymbol{T} \cdot \boldsymbol{R}\, ds - \iint_{s_v} \boldsymbol{T} \cdot \left(\boldsymbol{R} - \overline{\boldsymbol{R}}\right) ds, \tag{6.149a}$$

$$\overline{\mathcal{V}}_c \equiv \iiint_v \left(\bar{u}_{c0} + \boldsymbol{s}^i \cdot \acute{\boldsymbol{g}}_i\right) dv$$

$$- \iint_{s_v} \boldsymbol{T} \cdot \boldsymbol{R}\, ds - \iint_{s_v} \boldsymbol{T} \cdot \left(\boldsymbol{R} - \overline{\boldsymbol{R}}\right) ds. \tag{6.149b}$$

Apart from an irrelevant boundary integral, the form (6.149b) was derived by B. Fraeijs de Veubeke [121]; form (6.149a) was obtained by G. A. Wempner [122].

The reader is forewarned that B. Fraeijs de Veubeke's, and most earlier works, assign the opposite sign to the complementary potential. Our definition and our sign are chosen for the following reasons: First, we note the analogies [10] between the densities (u_0 and u_{c0}, $\bar{u}_0$ and $\bar{u}_{c0}$) of (6.145a, b) and the total potentials ($\mathcal{V}$ and $\mathcal{V}_c$, $\overline{\mathcal{V}}$ and $\overline{\mathcal{V}}_c$) in (6.148a, b). Moreover, in both cases, the left sides of these equations are *entirely independent* of the material properties. Most important however is the fact that the integral $\mathcal{P}$ *vanishes* for every admissible equilibrium state. Recall that the potential ($\mathcal{V}$ or $\overline{\mathcal{V}}$) is a relative minimum in a state of stable equilibrium. It follows that the complementary potential is a relative maximum. Of course, this holds only for *admissible variations*.

6.18 Principle of Stationary Complementary Potential

Two general forms of the complementary potential are $\mathcal{V}_c$ and $\overline{\mathcal{V}}_c$ defined by (6.149a and b), respectively. The former is expressed in terms of the Kirchhoff-Trefftz stress $[\mathcal{V}_c(s^{ij})]$; the latter in terms of the Jaumann stress $[\overline{\mathcal{V}}_c(t^{ij})]$. Neither is to be viewed as a functional of strain (γ_{ij} or h_{ij}). However, the stress components (s^{ij} or t^{ij}) are dependent upon a base vector ($\boldsymbol{G}_i$ or $\acute{\boldsymbol{g}}_i$) which is subject to rotation. Stated otherwise, a variation of the vector $\boldsymbol{s}^i$ is inherently effected by the basis, as well as the stress component. Specifically,

$$\delta \boldsymbol{s}^i = \delta s^{ij} \boldsymbol{G}_j + s^{ij}\, \delta\boldsymbol{\omega} \times \boldsymbol{G}_j = \delta t^{ij} \acute{\boldsymbol{g}}_j + t^{ij}\, \delta\boldsymbol{\omega} \times \acute{\boldsymbol{g}}_j,$$

$$\delta \boldsymbol{s}^i = \delta s^{ij} \boldsymbol{G}_j + s^{ij} \delta\omega^k E_{kjn} \boldsymbol{G}^n = \delta t^{ij} \acute{\boldsymbol{g}}_j + t^{ij} \delta\bar{\omega}^k e_{kjn} \acute{\boldsymbol{g}}^n. \tag{6.150a, b}$$

Note: The components of the rotation vector $\delta\boldsymbol{\omega}$ are $\delta\omega^k \equiv \boldsymbol{G}^k \cdot \delta\boldsymbol{\omega}$ or $\delta\bar{\omega}^k \equiv \acute{\boldsymbol{g}}^k \cdot \delta\boldsymbol{\omega}$ (see Subsection 3.14.3).

We seek the first-order variations of the complementary potentials. Variations of the stress are subject to admissibility, namely equilibrium conditions:

$$(\delta \boldsymbol{s}^i \sqrt{g}\,)_{,i} = \mathbf{o} \qquad \text{in } v, \tag{6.151a}$$

$$\delta \boldsymbol{s}^i n_i = \delta \boldsymbol{T} \quad \text{on } s_v, \tag{6.151b}$$

$$\delta \boldsymbol{T} = \mathbf{o} \qquad \text{on } s_t. \tag{6.151c}$$

In view of the aforestated constraints on $\delta \boldsymbol{s}^i$ (in v and on s_v) and on $\delta \boldsymbol{T}$ (on s_t), we take the variations of (6.149a and b); viz.,

$$\delta \mathcal{V}_c \equiv \iiint_v \left(\frac{\partial u_{c0}}{\partial s^{ij}}\, \delta s^{ij} + \frac{1}{2}\, \delta s^{ij} g_{ij} + \frac{1}{2}\, \delta \boldsymbol{s}^i \cdot \boldsymbol{R}_{,i} \right) dv$$

$$- \iint_{s_v} \delta \boldsymbol{s}^i \cdot \boldsymbol{R}\, n_i \, ds - \iint_{s_v} \delta \boldsymbol{T} \cdot \left(\boldsymbol{R} - \overline{\boldsymbol{R}} \right) ds, \tag{6.152a}$$

$$\delta \overline{\mathcal{V}}_c \equiv \iiint_v \left(\frac{\partial \bar{u}_{c0}}{\partial t^{ij}}\, \delta t^{ij} + \delta t^{ij} g_{ij} \right) dv$$

$$- \iint_{s_v} \delta \boldsymbol{s}^i \cdot \boldsymbol{R}\, n_i \, ds - \iint_{s_v} \delta \boldsymbol{T} \cdot \left(\boldsymbol{R} - \overline{\boldsymbol{R}} \right) ds. \tag{6.152b}$$

To the third term of the integral in v of (6.152a), we perform the *integration* [see equation (2.73)]:

$$\iiint_v \left(\delta \boldsymbol{s}^i \cdot \boldsymbol{R}_{,i} - \frac{1}{2} \delta \boldsymbol{s}^i \cdot \boldsymbol{G}_i \right) dv$$

$$= \iint_s \delta \boldsymbol{s}^i n_i \cdot \boldsymbol{R}\, ds - \iiint_v \left[\frac{1}{\sqrt{g}} (\sqrt{g}\, \delta \boldsymbol{s}^i)_{,i} \cdot \boldsymbol{R} + \frac{1}{2} \delta \boldsymbol{s}^i \cdot \boldsymbol{G}_i \right] dv. \tag{6.153}$$

When the constraints (6.151a–c) are enforced and (6.150a) is introduced into (6.153), we obtain from (6.152a)

$$\delta \mathcal{V}_c = \iiint_v \left[\left(\frac{\partial u_{c0}}{\partial s^{ij}} + \frac{1}{2} g_{ij} - \frac{1}{2} G_{ij} \right) \delta s^{ij} - \frac{1}{2} s^{ij} E_{kji} \delta \omega^k \right] dv$$

$$- \iint_{s_v} \delta \boldsymbol{T} \cdot (\boldsymbol{R} - \overline{\boldsymbol{R}})\, ds. \tag{6.154a}$$

By considering $\delta\boldsymbol{T} = \delta\boldsymbol{s}^i n_i = \mathbf{o}$ on s_t, the first integral on s_v in (6.152b) assumes the following form:

$$\iint_{s_v} \delta\boldsymbol{s}^i \cdot \boldsymbol{R} n_i \, ds = \iint_{s} \delta\boldsymbol{s}^i \cdot \boldsymbol{R} n_i \, ds$$

$$= \iiint_v \left[\frac{1}{\sqrt{g}} (\sqrt{g}\, \delta\boldsymbol{s}^i)_{,i} \cdot \boldsymbol{R} + \delta\boldsymbol{s}^i \cdot \boldsymbol{G}_i \right] dv.$$

Since $\delta\boldsymbol{s}^i$ is expressed by (6.150b) and also satisfies the equilibrium requirement (6.151a), we obtain

$$-\iint_{s_v} \delta\boldsymbol{s}^i \cdot \boldsymbol{R} n_i \, ds = -\iiint_v \left(\delta t^{ij} \acute{\boldsymbol{g}}_j \cdot \boldsymbol{G}_i + t^{ij}\, \delta\bar{\omega}^k e_{kjn}\, \acute{\boldsymbol{g}}^n \cdot \boldsymbol{G}_i \right) dv.$$

When this result is substituted into (6.152b) and $\boldsymbol{G}_i$ is expressed by (6.134), we obtain

$$\delta\overline{\mathcal{V}}_c = \iiint_v \left[\left(\frac{\partial \bar{u}_{c0}}{\partial t^{ij}} + g_{ij} - \acute{\boldsymbol{g}}_j \cdot \boldsymbol{G}_i \right) \delta t^{ij} - t^{ij} (h_i^n + \delta_i^n) \delta\bar{\omega}^k e_{kjn} \right] dv$$

$$- \iint_{s_v} \delta\boldsymbol{T} \cdot (\boldsymbol{R} - \overline{\boldsymbol{R}})\, ds. \tag{6.154b}$$

Since the variations of stress components and rotations of bases are independent in (6.154a, b), we have the respective stationary (Euler) equations:

$$\{\delta\mathcal{V}_c = 0\} \quad \Longleftrightarrow \quad \left\{ \begin{array}{lr} \frac{1}{2}(G_{ij} - g_{ij}) \equiv \gamma_{ij} = \frac{\partial u_{c0}}{\partial s^{ij}} & \text{in } v, \\ s^{ij} = s^{ji} & \text{in } v, \\ \boldsymbol{R} = \overline{\boldsymbol{R}} & \text{on } s_v. \end{array} \right\} \tag{6.155a}$$

$$\{\delta\overline{\mathcal{V}}_c = 0\} \quad \Longleftrightarrow \quad \left\{\begin{array}{ll} \acute{\boldsymbol{g}}_j \cdot \boldsymbol{G}_i - g_{ij} \equiv h_{ij} = \dfrac{\partial \bar{u}_{c0}}{\partial t^{(ij)}} & \text{in } v, \\[2ex] t^{ij}(h_i^n + \delta_i^n) = t^{in}(h_i^j + \delta_i^j) & \text{in } v, \\[2ex] \boldsymbol{R} = \overline{\boldsymbol{R}} & \text{on } s_v. \end{array}\right\} \tag{6.155b}$$

The latter [(6.154b), (6.155b)] were obtained by B. Fraeijs de Veubeke [121]; the former by G. A. Wempner [122]. We note that the complete system of equations is given by the stationary conditions of $\mathcal{V}$ (or $\overline{\mathcal{V}}$) and $\mathcal{V}_c$ (or $\overline{\mathcal{V}}_c$). The stationary conditions for the potential (variation of position $\boldsymbol{R}$), enforce equilibrium of stress $\boldsymbol{s}^i$ in v and on s_t. The stationary conditions for the complementary potential (variation of stress $\boldsymbol{s}^i$, both components and basis) enforce the kinematic conditions of (6.155a, b), but also the remaining conditions of equilibrium, viz., those which assert the equilibrium of moment ($s^{ij} = s^{ji}$ or $t^{ij}C_i^n = t^{in}C_i^j$).

6.19 Extremal Properties of the Complementary Potentials

To the initial integrands of $\mathcal{V}_c$ and $\overline{\mathcal{V}}_c$ of (6.149a, b) we add the terms:

$$+\frac{1}{2}\boldsymbol{s}^i \cdot \boldsymbol{R}_{,i} - \frac{1}{2}s^{ij}G_{ij} = 0, \tag{6.156a}$$

and

$$+\boldsymbol{s}^i \cdot \boldsymbol{R}_{,i} - t^{ij}C_{ij} = 0, \tag{6.156b}$$

respectively. The first term can be integrated by parts such that

$$\iiint_v \boldsymbol{s}^i \cdot \boldsymbol{R}_{,i}\, dv = \iint_s \boldsymbol{T} \cdot \boldsymbol{R}\, ds - \iiint_v \frac{1}{\sqrt{g}}(\sqrt{g}\, \boldsymbol{s}^i)_{,i} \cdot \boldsymbol{R}\, dv.$$

The final term vanishes for an equilibrated state in the absence of body force $\boldsymbol{f}$. Then, the introduction of (6.156a, b) into (6.149a, b) yields the alternative forms:

$$\mathcal{V}_c = \iiint_v [u_{c0} - s^{ij}\gamma_{ij}]\, dv + \iint_{s_t} \boldsymbol{T} \cdot \boldsymbol{R}\, ds, \tag{6.157a}$$

$$\overline{\mathcal{V}}_c = \iiint_v [\bar{u}_{c0} - t^{ij}h_{ij}]\, dv + \iint_{s_t} \boldsymbol{T} \cdot \boldsymbol{R}\, ds. \tag{6.157b}$$

Here, the final terms (integrals on s_v) have been omitted; those merely assert the forced condition $\boldsymbol{R} = \overline{\boldsymbol{R}}$ on s_v.

Observe that the first integrals of $\mathcal{V}_c$ and $\overline{\mathcal{V}}_c$ are the negatives of the internal energies, $-U$ and $-\overline{U}$. Note too that the stress $\boldsymbol{T}$ is not subject to variation on s_t. As the forms (6.149a, b), the forms (6.157a, b) are stationary with respect to the equilibrated stresses as the stresses appear *explicitly* in those functionals.

Recall the previous assessment of functionals $\mathcal{P}$ and $\overline{\mathcal{P}}$, the parts $\mathcal{V}$ and $\overline{\mathcal{V}}$, and parts $\mathcal{V}_c$ and $\overline{\mathcal{V}}_c$. The minimum of potential $\mathcal{V}$ (or $\overline{\mathcal{V}}$) implies a maximum of the complementary potential $\mathcal{V}_c$ (or $\overline{\mathcal{V}}_c$). However, those extremals are attained with the admissible variations of *all* variables. In the absence of body force $\boldsymbol{f}$, $-\mathcal{V}_c = \mathcal{V}$, and $-\overline{\mathcal{V}}_c = \overline{\mathcal{V}}$ [see (6.147a, b)]. As previously noted (see Section 6.17), the complementary potential is customarily the negative of $\mathcal{V}_c$ (or $\overline{\mathcal{V}}_c$). In accord with such custom, let

$$\mathcal{Q}_c \equiv -\mathcal{V}_c = \iiint_v u_0\, dv - \iint_{s_t} \boldsymbol{T} \cdot \boldsymbol{R}\, ds, \tag{6.158a}$$

$$\overline{\mathcal{Q}}_c \equiv -\overline{\mathcal{V}}_c = \iiint_v \bar{u}_0\, dv - \iint_{s_t} \boldsymbol{T} \cdot \boldsymbol{R}\, ds. \tag{6.158a}$$

In the equilibrium state, these functionals ($\mathcal{Q}_c$ and $\overline{\mathcal{Q}}_c$) are a relative *minimum* with respect to all *admissible* variations. It is the latter version (*minimum* complementary potential) which is most cited in earlier works.

6.20 Functionals and Stationary Theorem of Hellinger-Reissner

The functionals of Hellinger-Reissner [123], [124] are the complements of the modified potential of Hu-Washizu. Much as the potential entails the internal density (u_0 or $\bar{u}_0$), a complementary functional incorporates the complementary density (u_{c0} or $\bar{u}_{c0}$). Two forms may be deduced from (6.149a, b). It is only necessary to apply the Green/Gauss integration to the first integral on surface s_v:

$$\iint_{s_v} \boldsymbol{T} \cdot \boldsymbol{R}\, ds = \iiint_v \left[\frac{1}{\sqrt{g}} (\boldsymbol{s}^i \sqrt{g})_{,i} \cdot \boldsymbol{R} + \boldsymbol{s}^i \cdot \boldsymbol{R}_{,i} \right] dv - \iint_{s_t} \boldsymbol{T} \cdot \boldsymbol{R}\, ds.$$

Here, an admissible stress state satisfies equilibrium; $\boldsymbol{T} = \boldsymbol{s}^i n_i$ on s and $(\sqrt{g}\, \boldsymbol{s}^i)_{,i} = -\sqrt{g}\, \boldsymbol{f}$ in v. With these requirements, the complementary potentials ($\mathcal{V}_c$ and $\overline{\mathcal{V}}_c$) of (6.149a, b) assume the forms:

$$\mathcal{V}_c^* \equiv \iiint_v \left[u_{c0} - \tfrac{1}{2} s^{ij} (\boldsymbol{R}_{,j} \cdot \boldsymbol{R}_{,i} - g_{ij}) + \boldsymbol{f} \cdot \boldsymbol{R} \right] dv$$

$$+ \iint_{s_t} \boldsymbol{T} \cdot \boldsymbol{R}\, ds - \iint_{s_v} \left(\boldsymbol{R} - \overline{\boldsymbol{R}} \right) \cdot \boldsymbol{T}\, ds, \tag{6.159a}$$

$$\overline{\mathcal{V}}_c^* \equiv \iiint_v \left[\bar{u}_{c0} - t^{ij} (\acute{\boldsymbol{g}}_j \cdot \boldsymbol{R}_{,i} - g_{ij}) + \boldsymbol{f} \cdot \boldsymbol{R} \right] dv$$

$$+ \iint_{s_t} \boldsymbol{T} \cdot \boldsymbol{R}\, ds - \iint_{s_v} \left(\boldsymbol{R} - \overline{\boldsymbol{R}} \right) \cdot \boldsymbol{T}\, ds. \tag{6.159b}$$

The functionals $\mathcal{V}_c^*(s^{ij}, \boldsymbol{R})$ and $\overline{\mathcal{V}}_c^*(t^{ij}, \boldsymbol{R})$, dependent on stress *and* displacement, are stationary with respect to those functions if, but only if, the displacement-stress relations and equilibrium conditions are satisfied; these are the Euler equations:

$$\tfrac{1}{2}(\boldsymbol{R}_{,i} \cdot \boldsymbol{R}_{,j} - g_{ij}) (\equiv \gamma_{ij}) = \frac{\partial u_{c0}}{\partial s^{ij}} \quad \text{in } v, \tag{6.160a}$$

$$(\sqrt{g}\, s^{ij} \boldsymbol{G}_j)_{,i} + \sqrt{g}\, \boldsymbol{f} = \mathbf{o} \quad \text{in } v, \tag{6.160b}$$

$$s^{ij} \boldsymbol{G}_i n_i = \boldsymbol{T} \quad \text{on } s_t, \tag{6.160c}$$

$$\boldsymbol{R} = \overline{\boldsymbol{R}} \quad \text{on } s_v, \tag{6.160d}$$

or

$$(\acute{\boldsymbol{g}}_i \cdot \boldsymbol{R}_{,j} - g_{ij})(\equiv h_{ij}) = \frac{\partial \bar{u}_{c0}}{\partial t^{(ij)}} \quad \text{in } v, \tag{6.161a}$$

$$(\sqrt{g}\, t^{ij} \acute{\boldsymbol{g}}_j)_{,i} + \sqrt{g}\, \boldsymbol{f} = \mathbf{o} \quad \text{in } v, \tag{6.161b}$$

$$t^{ij} \acute{\boldsymbol{g}}_i n_i = \boldsymbol{T} \quad \text{on } s_t, \tag{6.161c}$$

$$\boldsymbol{R} = \overline{\boldsymbol{R}} \quad \text{on } s_v. \tag{6.161d}$$

In either case, one could accept the symmetry of the stress components ($s^{ij} = s^{ji}$ or $t^{(ij)} = t^{(ji)}$). Again, because $\acute{\boldsymbol{g}}_i \cdot \boldsymbol{R}_{,j} = \acute{\boldsymbol{g}}_j \cdot \boldsymbol{R}_{,i}$, an antisymmetric part of stress t^{ij} does not contribute to the functional. One could also admit the variation of base vectors $\boldsymbol{G}_i$ or $\acute{\boldsymbol{g}}_i$ which provides the equilibrium conditions presented in equations (6.155a, b), $s^{ij} = s^{ji}$ or $t^{ij}C_i^n = t^{in}C_i^j$.

Note: The modified potentials, forms (6.125) and (6.135), are directly related to their complementary counterparts, (6.159a, b), via (6.145a, b), as follows:

$$\mathcal{V}^* = -\mathcal{V}_c^*, \qquad \overline{\mathcal{V}}^* = -\overline{\mathcal{V}}_c^*. \tag{6.162}$$

A rudimentary form of the functional $\mathcal{V}_c^*$ was suggested by E. Hellinger [123]; the present form and stationarity were enunciated by E. Reissner [124]. The form $\overline{\mathcal{V}}_c^*$ was given by G. A. Wempner [122].

6.21 Functionals and Stationary Criteria for the Continuous Body; Summary

The functionals, potential $\mathcal{V}$, complementary potential $\mathcal{V}_c$, and modified potentials $\mathcal{V}^*$ and $\mathcal{V}_c^*$ (or $\overline{\mathcal{V}}$, $\overline{\mathcal{V}}_c$, $\overline{\mathcal{V}}^*$, $\overline{\mathcal{V}}_c^*$) are dependent on different functions. Here, the unbarred [or barred ($\overline{\ }$)] symbols signify the dependence upon the alternative strains and stresses, γ_{ij} and s^{ij} (or h_{ij} and t^{ij}), respectively.

In summary, the various functionals, variables, and properties follow:

- $\mathcal{V} = \mathcal{V}(\boldsymbol{R}) = \mathcal{V}[u_o(\gamma_{ij})] = \overline{\mathcal{V}}[\bar{u}_o(h_{ij})]$
  (see Sections 6.15 and 6.17)

  Admissible variations satisfy kinematic constraints.
  Stationary conditions enforce equilibrium.
  Minimum potential enforces stable equilibrium.

- $\mathcal{V}_c = \mathcal{V}_c\left(s^{ij}\right) = \overline{\mathcal{V}}_c\left(t^{ij}\right) = \mathcal{V}_c\left[u_{co}(s^{ij})\right] = \overline{\mathcal{V}}_c\left[\bar{u}_{co}(t^{ij})\right]$
(see Sections 6.17 through 6.19)

  Admissible variations satisfy equilibrium.
  Stationary conditions enforce the geometric constraints.

- $\mathcal{V}_c^* = \mathcal{V}_c^*\left(\boldsymbol{R}, s^{ij}\right) = \overline{\mathcal{V}}_c^*\left(\boldsymbol{R}, t^{ij}\right) = \mathcal{V}_c^*\left[\boldsymbol{R}, u_{co}(s^{ij})\right] = \overline{\mathcal{V}}_c^*\left[\boldsymbol{R}, \bar{u}_{co}(t^{ij})\right]$
(see Section 6.20)

  Stationarity with respect to displacements and stresses provides the equations of equilibrium and a form of constitutive equations (displacement-stress relations), respectively.

- $\mathcal{V}^* = \mathcal{V}^*\left(\boldsymbol{R}, \gamma^{ij}, s^{ij}\right) = \overline{\mathcal{V}}^*\left(\boldsymbol{R}, h^{ij}, t^{ij}\right)$
$= \mathcal{V}^*\left[\boldsymbol{R}, \gamma^{ij}, u_{co}(s^{ij})\right] = \overline{\mathcal{V}}^*\left[\boldsymbol{R}, h^{ij}, \bar{u}_{co}(t^{ij})\right]$
(see Section 6.16)

  Stationarity with respect to displacements, stresses, and strains provides *all* governing equations: equilibrium conditions, kinematic and constitutive relations.

Note: In the functionals $\overline{\mathcal{V}}_c$, $\overline{\mathcal{V}}_c^*$, and $\overline{\mathcal{V}}^*$, the stress expressed in the form $\boldsymbol{s}^i = t^{ij}\acute{\boldsymbol{g}}_j$ also admits a variation $\delta\acute{\boldsymbol{g}}_j = \delta\boldsymbol{\omega} \times \acute{\boldsymbol{g}}_j$ which provides the equilibrium criteria (6.140).

6.22 Generalization of Castigliano's Theorem on Displacement

As originally presented [125], Castigliano's theorem provided a means to treat the small displacements of Hookean bodies via a criterion of minimum complementary energy. H. L. Langhaar ([10], pp. 126–130) gives a formulation which applies to nonlinear elasticity, but is also limited to small displacements. In either case, that complementary energy is the integral of the complementary density. As such, that is not to be confused with the complementary energy of Sections 6.8, 6.17, and 6.18. Here, we offer modified forms of complementary energy. These are drawn from the integrals of (6.149a, b) and apply to any elastic body.

As before, we give forms in terms of the densities $u_{c0}(s^{ij})$ and $\bar{u}_{c0}(t^{ij})$ [see (6.145a, b)]. We define [see (6.149a, b)] an internal complementary energy by either of the following:

$$\mathcal{C}(s^{ij}) \equiv \iiint_v \left(u_{c0} + \tfrac{1}{2}s^{ij}g_{ij} + \tfrac{1}{2}s^{ij}G_{ij}\right) dv, \qquad (6.163a)$$

or

$$\overline{\mathcal{C}}(t^{ij}) \equiv \iiint_v \left(\bar{u}_{c0} + t^i_i\right) dv. \tag{6.163b}$$

We recall that

$$\frac{\partial u_{c0}}{\partial s^{ij}} = \gamma_{ij} = \tfrac{1}{2}(\boldsymbol{R}_{,i} \cdot \boldsymbol{R}_{,j} - g_{ij}), \tag{6.164a}$$

$$\frac{\partial \bar{u}_{c0}}{\partial t^{(ij)}} = h_{ij} = \acute{\boldsymbol{g}}_j \cdot \boldsymbol{R}_{,i} - g_{ij}. \tag{6.164b}$$

Therefore, the variations of the functionals $\mathcal{C}(s^{ij})$ and $\overline{\mathcal{C}}(t^{ij})$ assume the forms:

$$\delta\mathcal{C} = \iiint_v \left(\delta s^{ij} \boldsymbol{G}_j \cdot \boldsymbol{R}_{,i}\right) dv, \tag{6.165a}$$

$$\delta\overline{\mathcal{C}} = \iiint_v \left(\delta t^{ij} \acute{\boldsymbol{g}}_j \cdot \boldsymbol{R}_{,i}\right) dv. \tag{6.165b}$$

Note that $s^{ij} = s^{ji}$ and $t^{ij} = t^{(ij)} = t^{ji}$ in the foregoing equations.

By partial integration of each, we obtain

$$\delta\mathcal{C} = -\iiint_v \frac{1}{\sqrt{g}} \left(\sqrt{g}\, \delta s^{ij} \boldsymbol{G}_j\right)_{,i} \cdot \boldsymbol{R}\, dv + \iint_s \delta s^{ij} n_i \boldsymbol{G}_j \cdot \boldsymbol{R}\, ds, \tag{6.166a}$$

$$\delta\overline{\mathcal{C}} = -\iiint_v \frac{1}{\sqrt{g}} \left(\sqrt{g}\, \delta t^{ij} \acute{\boldsymbol{g}}_j\right)_{,i} \cdot \boldsymbol{R}\, dv + \iint_s \delta t^{ij} n_i \acute{\boldsymbol{g}}_j \cdot \boldsymbol{R}\, ds. \tag{6.166b}$$

Variations of stress are to satisfy equilibrium; these are the admissibility conditions:

$$\left(\sqrt{g}\, \delta s^{ij} \boldsymbol{G}_j\right)_{,i} = \left(\sqrt{g}\, \delta t^{ij} \acute{\boldsymbol{g}}_j\right)_{,i} = \mathbf{o} \quad \text{in } v, \tag{6.167a, b}$$

$$\delta s^{ij} n_i \boldsymbol{G}_j = \delta t^{ij} n_i \acute{\boldsymbol{g}}_j = \delta \boldsymbol{T} \qquad \text{on } s. \tag{6.168a, b}$$

Here $\delta\boldsymbol{T}$ denotes the variation of surface tractions on s. Our results follow:

$$\delta\mathcal{C} = \iint_s \delta\boldsymbol{T} \cdot \boldsymbol{R}\, ds, \qquad \delta\overline{\mathcal{C}} = \iint_s \delta\boldsymbol{T} \cdot \boldsymbol{R}\, ds. \tag{6.169a, b}$$

To obtain a general version of Castigliano's theorem, we follow the logic of H. L. Langhaar ([10], pp. 133–135). We suppose that the body is supported on some portion of the surface s_v so that rigid-body displacements of the entire system are impossible; there $\boldsymbol{R} = \boldsymbol{r}$, the initial position. The body may be subjected to certain concentrated tractions on the surface s_t. In the accepted spirit of engineering practice, we regard such concentrated force as distributed on a *small* spot. We are concerned only with the resultant, not the distribution, *but* understand that local stress and deformation are then beyond the scope of our theory. Let s_0 denote the very small surface which is subjected to the concentrated traction $\boldsymbol{T}_0$ and let $\boldsymbol{R}^0$ denote the current position of a particle within s_0. We now examine (6.169a, b) subject *only* to a variation of the traction on s_0. Then,

$$\delta \mathcal{C} = R_j^0 \iint_s n_i \delta s^{ij}\, ds = R_j^0\, \delta T_0^j,$$

$$\delta \overline{\mathcal{C}} = \overline{R}_j^0 \iint_s n_i \delta t^{ij}\, ds = \overline{R}_j^0\, \delta \overline{T}_0^j.$$

Note differences in the respective vectors: $R_j^0 \equiv \boldsymbol{G}_j \cdot \boldsymbol{R}^0$, $\overline{R}_j^0 \equiv \acute{\boldsymbol{g}}_j \cdot \boldsymbol{R}^0$, etc. The variations of stress (s^{ij} or t^{ij}) are a consequence of the variations of load (T_0^j or $\overline{T}_0^j$) and are in equilibrium with that variation [according to (6.167a, b) and (6.168a, b)]. It follows that

$$\delta \mathcal{C} = \frac{\partial \mathcal{C}}{\partial T_0^i} \delta T_0^i, \qquad \delta \overline{\mathcal{C}} = \frac{\partial \overline{\mathcal{C}}}{\partial \overline{T}_0^i} \delta \overline{T}_0^i,$$

or

$$R_i^0 = \frac{\partial \mathcal{C}}{\partial T_0^i}, \qquad \overline{R}_i^0 = \frac{\partial \overline{\mathcal{C}}}{\partial \overline{T}_0^i}. \tag{6.170a, b}$$

Practically, the result (6.170a, b) is useful if one can readily express the stress in terms of external loading. Engineers are familiar with elementary situations, particularly the so-called "statically determinate" problems, and also the method of a "dummy load." The forms (6.163a, b) are *not* limited to Hookean bodies, nor small deformations. However, we note that usage of form $\mathcal{C}$ of (6.163a) is more limited than the form $\overline{\mathcal{C}}$ of (6.163b) since the strain $[\gamma_{ij} = (G_{ij} - g_{ij})/2]$ appears explicitly. This points to an inherent disadvantage in the use of the components s^{ij}, as opposed to t^{ij}. The former components ($s^{ij} = \boldsymbol{G}^j \cdot \boldsymbol{s}^i$) are based upon the deformed basis ($\boldsymbol{G}_j$ and $\boldsymbol{G}^j$), whereas the latter ($t^{ij} = \acute{\boldsymbol{g}}^j \cdot \boldsymbol{s}^i$) are based on the rigidly rotated basis ($\acute{\boldsymbol{g}}_j$ and $\acute{\boldsymbol{g}}^j$). Hence, the integrand of $\overline{\mathcal{C}}$, like the components t^{ij}, is entirely divorced from the strain h_{ij}.

Let us now define alternative forms of complementary energy, viz.,

$$\mathcal{C}^* \equiv \iiint_v \left(u_{c0} + \tfrac{1}{2}s^{ij}g_{ij} + \tfrac{1}{2}s^{ij}G_{ij} - s^{ij}\boldsymbol{G}_j \cdot \boldsymbol{g}_i\right) dv, \qquad (6.171a)$$

$$\overline{\mathcal{C}}^* \equiv \iiint_v \left(\bar{u}_{c0} + t_i^i - t^{ij}\acute{\boldsymbol{g}}_j \cdot \boldsymbol{g}_i\right) dv. \qquad (6.171b)$$

In accordance with (6.164a, b) and (6.165a, b),

$$\delta\mathcal{C}^* = \iiint_v \left[\delta s^{ij}\boldsymbol{G}_j \cdot (\boldsymbol{R} - \boldsymbol{r})_{,i}\right] dv = \iint_s \delta\boldsymbol{T} \cdot (\boldsymbol{R} - \boldsymbol{r})\, ds,$$

$$\delta\overline{\mathcal{C}}^* = \iiint_v \left[\delta t^{ij}\acute{\boldsymbol{g}}_j \cdot (\boldsymbol{R} - \boldsymbol{r})_{,i}\right] dv = \iint_s \delta\boldsymbol{T} \cdot (\boldsymbol{R} - \boldsymbol{r})\, ds.$$

By the logic leading to (6.170a, b), we obtain similar results in terms of the displacement ($\boldsymbol{V} = \boldsymbol{R} - \boldsymbol{r}$) at the "point":

$$V_i^0 = \frac{\partial\mathcal{C}^*}{\partial T_0^i}, \qquad \overline{V}_i^0 = \frac{\partial\overline{\mathcal{C}}^*}{\partial \overline{T}_0^i}. \qquad (6.172a, b)$$

Again, we must question the limitations of the result and the practicality. To that end, let us rewrite the integrands of $\mathcal{C}^*$ and $\overline{\mathcal{C}}^*$ in (6.171a, b):

$$\mathcal{C}^* = \iiint_v \left[u_{c0} + \tfrac{1}{2}s^{ij}(\boldsymbol{G}_j - \boldsymbol{g}_j) \cdot (\boldsymbol{G}_i - \boldsymbol{g}_i)\right] dv, \qquad (6.173a)$$

$$\overline{\mathcal{C}}^* = \iiint_v \left[\bar{u}_{c0} + t^{ij}\acute{\boldsymbol{g}}_j \cdot (\acute{\boldsymbol{g}}_i - \boldsymbol{g}_i)\right] dv. \qquad (6.173b)$$

To grasp the significance of the final products, we recall the motion and deformation which carries the vector $\boldsymbol{g}_i$ to the vector $\acute{\boldsymbol{g}}_i$ and subsequently to $\boldsymbol{G}_i$. These are the rigid rotation of (3.87a, b) and then the deformation of (3.86a–c) and (3.95), viz.,

$$\acute{\boldsymbol{g}}_i = r_i^{\bullet j}\boldsymbol{g}_j, \qquad \boldsymbol{G}_k = (h_k^i + \delta_k^i)\acute{\boldsymbol{g}}_i.$$

In most practical problems (e.g., machines and structures), the strains are very small ($|h^i_j| \ll 1$) and rotations are small enough to treat them as vectors ($\boldsymbol{\omega}$, where $|\boldsymbol{\omega}| \ll 1$). Then, according to (3.164) and (3.165),

$$\acute{\boldsymbol{g}}_i \doteq \boldsymbol{g}_i + \Omega_{pi}\boldsymbol{g}^p, \qquad \boldsymbol{G}_i \doteq \boldsymbol{g}_i + \Omega_{pi}\boldsymbol{g}^p + h_{ip}\boldsymbol{g}^p \doteq \boldsymbol{g}_i + \Omega_{pi}\boldsymbol{g}^p + \gamma_{ip}\boldsymbol{g}^p,$$

where [see (3.163)], $\Omega_{ij} \equiv \Omega^k e_{kji}$. The requisite approximations follow:

$$\tfrac{1}{2}s^{ij}(\boldsymbol{G}_j - \boldsymbol{g}_j)\cdot(\boldsymbol{G}_i - \boldsymbol{g}_i) \doteq \tfrac{1}{2}s^{ij}\left[\Omega^k_{\cdot j}\Omega_{ki} + O(\gamma\Omega)\right], \tag{6.174}$$

$$t^{ij}\acute{\boldsymbol{g}}_j\cdot(\acute{\boldsymbol{g}}_i - \boldsymbol{g}_i) \doteq t^{ij}\Omega_{ji}. \tag{6.175}$$

Since we now consider small strains [see also (6.155b)], the relevant approximation is $t^{ij} = t^{ji}$; hence the right sides of (6.175) vanishes. The approximations of $\mathcal{C}^*$ and $\overline{\mathcal{C}}^*$ (6.173a, b) for small strain and *moderate* rotation follow:

$$\mathcal{C}^* \doteq \iiint_v \left(u_{c0} + \tfrac{1}{2}s^{ij}\Omega^k_{\cdot j}\Omega_{ki}\right) dv, \qquad \overline{\mathcal{C}}^* \doteq \iiint_v \bar{u}_{c0}\, dv. \tag{6.176a, b}$$

Once again, we detect merit in the decomposition of strain and rotation, the use of engineering strain h_{ij} and the associated stress t^{ij}, since $\overline{\mathcal{C}}^* = \overline{\mathcal{C}}^*(t^{ij})$ does not depend explicitly on strain or rotation. All of the foregoing complementary energies ($\mathcal{C}$, $\overline{\mathcal{C}}$, $\mathcal{C}^*$, or $\overline{\mathcal{C}}^*$) are independent of the elastic properties.

6.23 Variational Formulations of Inelasticity

Any process of inelastic deformation is nonconservative. One can form functionals, but not potentials, neither in a physical nor mathematical sense; the Gâtaux variations [126] of such functionals provide, as Euler equations, the equations that govern the inelastic deformation of the body. Such a formulation of an inelastic problem can be useful, if only as a consistent means to obtain a discrete model of the continuous body, e.g., a finite-element assembly.

In Section 5.12, one formulation of inelasticity presumes the existence of a free energy $\mathcal{F}$ (per unit mass ρ_0); here we employ the energy density *per unit volume* v, i.e.,

$$F(\gamma_{ij}, \gamma^N_{ij}, T) = \rho_0\,\mathcal{F}.$$

In general, the free energy F is a functional of the strain history; it is path-dependent and is also a function of the inelastic strains γ_{ij}^N.

Formally, one can write the Gâtaux variation (see, for example, [126]):

$$\delta W^* \equiv$$

$$\int_{t_0}^{t} \left\langle \iiint_v \left\{ - s^{ij}\,\delta\gamma_{ij} + S\,\delta T + \delta F + s_N^{ij}\,\delta\gamma_{ij}^N \right.\right.$$

$$- \left[\gamma_{ij} - \tfrac{1}{2}\left(\boldsymbol{R}_{,i}\boldsymbol{R}_{,j} - g_{ij}\right)\right]\,\delta s^{ij}$$

$$\left.\left. - \left[\frac{1}{\sqrt{g}}\left(\sqrt{g}\,s^{ij}\boldsymbol{G}_j\right)_{,i} + \boldsymbol{f}\right]\cdot\delta\boldsymbol{R}\right\} dv \right\rangle dt$$

$$- \int_{t_0}^{t} \left\{ \iint_{s_t} \left(\overline{\boldsymbol{T}} - n_i s^{ij}\boldsymbol{G}_j\right)\cdot\delta\boldsymbol{R}\,ds + \iint_{s_v} \left(\boldsymbol{R} - \overline{\boldsymbol{R}}\right)\cdot\delta\boldsymbol{T}\,ds \right\} dt. \quad (6.177)$$

Equation (6.177) is consistent with the thermodynamic relations of Chapter 5 [see Section 5.12, equations (5.44) to (5.47a, b)]. The variation (6.177) is but an elaboration of the Fréchet variation (6.126) of the potential $\mathcal{V}^*$ of (6.125). Here, to accommodate inelastic and time dependent strain, we acknowledge such dependence on time t; moreover, in all likelihood, a *potential* W^* does not exist. The variation vanishes (i.e., $\delta W^* = 0$) if, and only if, the constitutive equations (5.48a–c) of Section 5.12, kinematic equations and dynamic equations are satisfied in v and on s_v and s_t [see equations (6.128) to (6.131) of Section 6.16]. The stationary condition (6.177) is a version of Hamilton's principle [127]; cf. H. L. Langhaar [10]. If the behavior is time-independent, isothermal and elastic (mechanically conservative), then a potential W^* exists; specifically $W^* = \mathcal{V}^*$ of equation (6.125).

Let us turn now to inelasticity as described by the classical concepts of plasticity [see Chapter 5, Sections 5.24 to 5.32]: Yielding is initiated if the stress attains a yield condition, e.g., $\mathcal{Y}(s^{ij}) = \bar{\sigma}^2$. Inelastic strain ensues, if $\dot{\mathcal{Y}} \equiv (\partial\mathcal{Y}/\partial s^{ij})\dot{s}^{ij} \geq 0$. Note that the equality sign holds for ideally plastic material. The strain increment $\dot{\gamma}_{ij}$ consists of a recoverable (elastic) part $\dot{\gamma}_{ij}^E$ and a plastic (inelastic) part $\dot{\gamma}_{ij}^P$. The latter follows the normality condition, $\dot{\gamma}_{ij}^P = \dot{\lambda}\,(\partial\mathcal{Y}/\partial s^{ij})$. Strain hardening alters the parameter $\dot{\lambda}$ and also the yield condition $\mathcal{Y}$.

An incremental version of the Hu-Washizu functional [see (6.125)] has the form:

$$\dot{W}^* \equiv \iiint_v \left[\dot{u}_0 - s^{ij}\left(\dot{\gamma}^E_{ij} - \boldsymbol{R}_{,i}\cdot\dot{\boldsymbol{R}}_{,j}\right) - \boldsymbol{f}\cdot\dot{\boldsymbol{R}} - \dot{\lambda}\left(\mathcal{Y} - \bar{\sigma}^2\right)\right] dv$$

$$- \iint_{s_t} \overline{\boldsymbol{T}}\cdot\dot{\boldsymbol{R}}\, ds - \iint_{s_v} \boldsymbol{T}\cdot(\dot{\boldsymbol{R}} - \dot{\overline{\boldsymbol{R}}})\, ds. \qquad (6.178)$$

Here, $\dot{u}_0 = \dot{u}_0(\dot{\gamma}^E_{ij})$ denotes the internal energy; $\mathcal{Y} = \mathcal{Y}(s^{ij}) = \bar{\sigma}^2$, the yield condition; and $\dot{W}^* = \dot{W}^*(\dot{\gamma}^E_{ij},\ s^{ij},\ \dot{\boldsymbol{R}},\ \dot{\lambda})$. The usual integration by parts and enforcement of stationarity lead to the requisite governing equations:

$$s^{ij} = \frac{\partial \dot{u}_0}{\partial \dot{\gamma}^E_{ij}} \qquad \text{in } v, \qquad (6.179a)$$

$$\dot{\gamma}_{ij} \equiv \boldsymbol{R}_{,i}\cdot\dot{\boldsymbol{R}}_{,j} = \dot{\gamma}^E_{ij} + \dot{\lambda}\frac{\partial \mathcal{Y}}{\partial s^{ij}} \qquad \text{in } v, \qquad (6.179b)$$

$$\frac{1}{\sqrt{g}}\left(\sqrt{g}\, s^{ij}\boldsymbol{G}_i\right)_{,j} + \boldsymbol{f} = \mathbf{o} \qquad \text{in } v, \qquad (6.179c)$$

$$\mathcal{Y}(s^{ij}) = \bar{\sigma}^2 \qquad \text{in } v, \qquad (6.179d)$$

$$s^{ij}\boldsymbol{G}_i n_j = \overline{\boldsymbol{T}} \qquad \text{on } s_t, \qquad (6.179e)$$

$$\dot{\boldsymbol{R}} = \dot{\overline{\boldsymbol{R}}} \qquad \text{on } s_v. \qquad (6.179f)$$

An incremental version of the Hellinger-Reissner functional [see equation (6.159)a] has the form:

$$\dot{W}^*_c = \iiint_v \left[\dot{u}_{c0} - s^{ij}\boldsymbol{R}_{,j}\cdot\dot{\boldsymbol{R}}_{,i} + \boldsymbol{f}\cdot\dot{\boldsymbol{R}} + \dot{\lambda}(\mathcal{Y} - \bar{\sigma}^2)\right] dv$$

$$+ \iint_{s_t} \overline{\boldsymbol{T}}\cdot\dot{\boldsymbol{R}}\, ds - \iint_{s_v} \boldsymbol{T}\cdot\left(\dot{\boldsymbol{R}} - \dot{\overline{\boldsymbol{R}}}\right) ds. \qquad (6.180)$$

Here, $\dot{W}^*_c = \dot{W}^*_c(s^{ij},\ \dot{\boldsymbol{R}},\ \dot{\lambda})$; $\dot{u}_{c0} = \dot{u}_{c0}(s^{ij})$ denotes the complementary internal energy; and $\mathcal{Y} = \mathcal{Y}(s^{ij}) = \bar{\sigma}^2$, the yield condition. The stationary conditions follow:

$$\boldsymbol{R}_{,i} \cdot \dot{\boldsymbol{R}}_{,j} (\equiv \dot{\gamma}_{ij}) = \frac{\partial \dot{u}_{c0}}{\partial s^{ij}} + \dot{\lambda} \frac{\partial \mathcal{Y}}{\partial s^{ij}} \quad \text{in } v, \tag{6.181a}$$

$$\frac{1}{\sqrt{g}} \left(\sqrt{g}\, s^{ij} \boldsymbol{G}_i \right)_{,j} + \boldsymbol{f} = \mathbf{o} \quad \text{in } v, \tag{6.181b}$$

$$\mathcal{Y}(s^{ij}) = \bar{\sigma}^2 \quad \text{in } v, \tag{6.181c}$$

$$s^{ij} \boldsymbol{G}_i n_j = \overline{\boldsymbol{T}} \quad \text{on } s_t, \tag{6.181d}$$

$$\dot{\boldsymbol{R}} = \dot{\overline{\boldsymbol{R}}} \quad \text{on } s_v. \tag{6.181e}$$

The latter functional ($\dot{W}_c^*$) does not explicitly contain the elastic strain γ_{ij}^E.

In a practical application, the variational method provides a means to devise discrete approximations. The formulations and computations must proceed stepwise from the prevailing deformed state. Then, it might prove more expedient to cast the functionals entirely in terms of increments of displacement $\dot{\boldsymbol{R}}$, strain $\dot{\gamma}_{ij}$, and stress $\dot{s}^{ij}$. The latter are linearly related via the tangent modulus E_T^{ijkl} (see Section 5.32); that relation derives from a quadratic form:

$$\dot{u} = \tfrac{1}{2} E_T^{ijkl} \dot{\gamma}_{ij} \dot{\gamma}_{kl}, \qquad \dot{s}^{ij} = E_T^{ijkl} \dot{\gamma}_{kl}. \tag{6.182a, b}$$

These differ from the similar equations of Hookean elasticity: The expression (6.182a) for $\dot{u}$ includes dissipation; stress $\dot{s}^{ij}$ is nonconservative and $\dot{\gamma}_{ij}$ includes the inelastic strain ($\dot{\gamma}_{ij} = \dot{\gamma}_{ij}^E + \dot{\gamma}_{ij}^P$).

An incremental version of the functional $\dot{W}^*$ of (6.178) is a quadratic form in the increments ($\dot{\ }$). In that form, designated $\ddot{W}^*$, we signify quantities of the reference state by a tilde ($\tilde{\ }$):

$$\begin{aligned} \ddot{W}^* = \iiint_v \Big\{ & \tfrac{1}{2} E_T^{ijkl} \dot{\gamma}_{ij} \dot{\gamma}_{kl} - \dot{s}^{ij} \left[\dot{\gamma}_{ij} - \tfrac{1}{2} \left(\widetilde{\boldsymbol{R}}_{,i} \cdot \dot{\boldsymbol{R}}_{,j} + \widetilde{\boldsymbol{R}}_{,j} \cdot \dot{\boldsymbol{R}}_{,i} \right) \right] \\ & + \tfrac{1}{2} \tilde{s}^{ij} \dot{\boldsymbol{R}}_{,i} \cdot \dot{\boldsymbol{R}}_{,j} - \dot{\boldsymbol{f}} \cdot \dot{\boldsymbol{R}} \Big\} \, dv \\ & - \iint_{s_t} \dot{\overline{\boldsymbol{T}}} \cdot \dot{\boldsymbol{R}} \, ds - \iint_{s_v} \dot{\boldsymbol{T}} \cdot (\dot{\boldsymbol{R}} - \dot{\overline{\boldsymbol{R}}}) \, ds. \end{aligned} \tag{6.183}$$

We require that functional $\ddot{W}^*$ be stationary with respect to the functions $\dot{\gamma}_{ij}$, $\dot{s}^{ij}$, and $\dot{\boldsymbol{R}}$; the Euler equations follow:

$$\dot{s}^{ij} = E_T^{ijkl}\dot{\gamma}_{kl} \qquad \text{in } v, \tag{6.184a}$$

$$\dot{\gamma}_{ij} = \tfrac{1}{2}\left(\widetilde{\boldsymbol{R}}_{,i}\cdot\dot{\boldsymbol{R}}_{,j} + \widetilde{\boldsymbol{R}}_{,j}\cdot\dot{\boldsymbol{R}}_{,i}\right) \qquad \text{in } v, \tag{6.184b}$$

$$\frac{1}{\sqrt{g}}\left[\left(\sqrt{g}\,\dot{s}^{ij}\widetilde{\boldsymbol{R}}_{,i}\right)_{,j} + \left(\sqrt{g}\,\tilde{s}^{ij}\dot{\boldsymbol{R}}_{,i}\right)_{,j}\right] + \dot{\boldsymbol{f}} = \mathbf{o} \qquad \text{in } v, \tag{6.184c}$$

$$n_j\left(\dot{s}^{ij}\widetilde{\boldsymbol{R}}_{,i} + \tilde{s}^{ij}\dot{\boldsymbol{R}}_{,i}\right) = \dot{\overline{\boldsymbol{T}}} \qquad \text{on } s_t, \tag{6.184d}$$

$$\dot{\boldsymbol{R}} = \dot{\overline{\boldsymbol{R}}} \qquad \text{on } s_v. \tag{6.184e}$$

The moduli E_T^{ijkl} apply to loading, $\dot{s}^{ij}(\partial\mathcal{Y}/\partial s^{ij}) \geq 0$; in the event of unloading, the equations (6.184a–e) apply with E_T^{ijkl} replaced by the elastic moduli. The equations (6.184c, d) are merely the perturbations of (6.179c, e).

An alternative version of (6.183) expresses the strain as the sum of elastic and inelastic parts ($\dot{\gamma}_{ij} = \dot{\gamma}_{ij}^E + \dot{\gamma}_{ij}^P$), wherein [see equation (5.131)]

$$\dot{s}^{ij} = E^{ijkl}\dot{\gamma}_{ij}^E, \tag{6.185}$$

$$\frac{\partial\mathcal{Y}}{\partial s^{ij}}\dot{s}^{ij} = G_P\dot{\lambda}. \tag{6.186}$$

That alternative is a functional of $\dot{\gamma}_{ij}^E$, $\dot{s}^{ij}$, $\dot{\boldsymbol{R}}$, *and* $\dot{\lambda}$, as follows:

$$\begin{aligned}\ddot{W}^{**} = \iiint_v \Bigg\{ &\frac{1}{2}E^{ijkl}\,\dot{\gamma}_{ij}^E\,\dot{\gamma}_{kl}^E + \frac{1}{2}G_P(\dot{\lambda})^2 \\ &- \dot{s}^{ij}\left[\dot{\gamma}_{ij}^E + \dot{\lambda}\frac{\partial\mathcal{Y}}{\partial s^{ij}} - \frac{1}{2}\left(\widetilde{\boldsymbol{R}}_{,i}\cdot\dot{\boldsymbol{R}}_{,j} + \widetilde{\boldsymbol{R}}_{,j}\cdot\dot{\boldsymbol{R}}_{,i}\right)\right] \\ &+ \frac{1}{2}\tilde{s}^{ij}\dot{\boldsymbol{R}}_{,i}\cdot\dot{\boldsymbol{R}}_{,j} - \dot{\boldsymbol{f}}\cdot\dot{\boldsymbol{R}}\Bigg\}\,dv \\ &- \iint_{s_t}\dot{\overline{\boldsymbol{T}}}\cdot\dot{\boldsymbol{R}}\,ds - \iint_{s_v}\dot{\boldsymbol{T}}\cdot(\dot{\boldsymbol{R}} - \dot{\overline{\boldsymbol{R}}})\,ds. \end{aligned} \tag{6.187}$$

Now, the stationary conditions of $\ddot{W}^*$ are augmented by those derived by the variations of the functions $\dot{\gamma}_{ij}^E$ and $\dot{\lambda}$; instead of (6.184a), we obtain (6.185) and (6.186).

A complementary functional is achieved via the Legendre transformation (from $\dot{\gamma}_{ij}^E$ to $\dot{s}^{ij}$):

$$\dot{s}^{ij}\dot{\gamma}_{ij}^E = \tfrac{1}{2}E^{ijkl}\dot{\gamma}_{ij}^E\dot{\gamma}_{kl}^E + \tfrac{1}{2}D_{ijkl}\dot{s}^{ij}\dot{s}^{kl}. \tag{6.188}$$

The substitution of (6.188) into (6.187) provides an incremental version of the Hellinger-Reissner functional (6.180):

$$\begin{aligned}\ddot{W}_c^{**} = \iiint_v \Big\{ & -\frac{1}{2}D_{ijkl}\dot{s}^{ij}\dot{s}^{kl} + \frac{1}{2}G_P(\dot{\lambda})^2 \\ & - \dot{s}^{ij}\left[\dot{\lambda}\frac{\partial \mathcal{Y}}{\partial s^{ij}} - \frac{1}{2}\left(\widetilde{\boldsymbol{R}}_{,i}\cdot\dot{\boldsymbol{R}}_{,j} + \widetilde{\boldsymbol{R}}_{,j}\cdot\dot{\boldsymbol{R}}_{,i}\right)\right] \\ & + \frac{1}{2}\tilde{s}^{ij}\dot{\boldsymbol{R}}_{,i}\cdot\dot{\boldsymbol{R}}_{,j} - \dot{\boldsymbol{f}}\cdot\dot{\boldsymbol{R}}\Big\}\, dv \\ - \iint_{s_t} \dot{\overline{\boldsymbol{T}}}\cdot\dot{\boldsymbol{R}}\, ds - & \iint_{s_v} \dot{\boldsymbol{T}}\cdot(\dot{\boldsymbol{R}} - \dot{\overline{\boldsymbol{R}}})\, ds.\end{aligned} \tag{6.189}$$

Strains do not appear explicitly in $\ddot{W}_c^{**}$; it is a functional of the incremental stresses $\dot{s}^{ij}$, displacement $\dot{\boldsymbol{R}}$, and $\dot{\lambda}$. Variation of incremental stress $\dot{s}^{ij}$ provides a form of constitutive equations, viz.,

$$D_{ijkl}\,\dot{s}^{kl} + \dot{\lambda}\frac{\partial \mathcal{Y}}{\partial s^{ij}} = \frac{1}{2}\left(\widetilde{\boldsymbol{R}}_{,i}\cdot\dot{\boldsymbol{R}}_{,j} + \widetilde{\boldsymbol{R}}_{,j}\cdot\dot{\boldsymbol{R}}_{,i}\right). \tag{6.190}$$

The remaining equations, (6.184c–e) and (6.186), hold as before. In the enforcement of the foregoing elastic-plastic relations, it is essential that the yield condition prevails, viz.,

$$\frac{\partial \mathcal{Y}}{\partial s^{ij}}\dot{s}^{ij} = G_P\dot{\lambda} \geq 0.$$

In the event of unloading $[\dot{s}^{ij}(\partial\mathcal{Y}/\partial s^{ij}) < 0]$, the equations apply as well to the elastic response, wherein $\dot{\lambda} = 0$.

Chapter 7

Linear Theories of Isotropic Elasticity and Viscoelasticity

7.1 Introduction

The linear theory of elasticity originated with Hooke's law in 1678. However, the earliest formulation of a general theory for an isotropic body was given by C. L. M. H. Navier in 1821. Unfortunately, Navier's formulation suffered from one defect: his description contained but one elastic constant. Shortly thereafter, in 1822, A. L. Cauchy presented the formulation which we know today as the classical theory of elasticity (see Section 5.20 and the historical account presented by S. P. Timoshenko [46]).

In general, our ability to solve linear equations far exceeds our abilities to cope with nonlinear equations, whether the equations in question are differential or algebraic. Consequently, engineers and scientists have necessarily turned to linear approximations. As a result, the classical theory of elasticity has become a highly developed discipline during the years since its conception. Now, the structural engineer is expected to know the rudiments of the classical theory of elasticity. To this end, we present only the *basic* linear equations of isotropic elastic bodies. The reader seeking additional developments, special methods, or solutions, may consult one or more of the works by A. E. H. Love [1], N. I. Muskhelishvili [128], I. S. Sokolnikoff [129], C. E. Pearson [130], V. V. Novozhilov [131], M. Filonenko-Borodich [132], Y. C. Fung [54], S. P. Timoshenko and J. N. Goodier [133], and others.

The same kinematic equations govern the motions of all continuous media. In particular, the same linear equations are used to describe the kinematics and dynamics of a medium undergoing *small* displacements. Only the constitutive equations differ. Again, *linear* constitutive equations can be used to approximate viscoelastic media. Therefore, the linear problems of elasticity and viscoelasticity are, in many ways, similar. Indeed under certain conditions, the equations of the viscoelastic problem can be transformed into the form of a corresponding elastic problem. Because of the similarities and analogies, the linear theories of elasticity and viscoelasticity are brought together in the present chapter. The linear equations are drawn from the general equations of the preceding chapters.

The main purpose of this chapter is a presentation of the foundations and the most prominent formulations of the linear theories of isotropic elasticity and viscoelasticity. For the most part, we do not provide specific solutions; three notable exceptions are presented, because those solutions are particularly relevant to the engineer.

1. In Section 7.12, a solution via an Airy stress function describes the stress concentration caused by a circular hole in a plate. Such effect occurs at the riveted connections of buildings, bridges, and aircraft.

2. In Section 7.14, the solution for simple bending is examined. It is one of the few solutions which can be entirely anticipated as opposed to a formal solution of the governing differential equations. Such deduction has been termed an "inverse" solution. It is a result of fundamental importance in engineering and, moreover, it is applicable to circumstances of large rotation and displacement (certain kinematical nonlinearities).

3. In Section 7.15, solutions for torsion are described. These solutions are achieved by "semi-inverse" methods: Essential kinematical features are anticipated; complete solutions generally require the subsequent solution of the resulting differential equation and boundary conditions. Again, a specific problem of a notched shaft exhibits a type of stress concentration which confronts the engineer.

For simplicity, the equations are cast in a Cartesian/rectangular coordinate system. In this case, $\gamma_{ij} = \epsilon_{ij} = \gamma^i_j = \gamma^{ij}$. Since one can always conceive of a Cartesian/rectangular system and always transform to another, the equations can be readily generalized.

7.2 Uses and Limitations of the Linear Theories

In some practical situations, linear theory provides an adequate description of a deformable body. When applicable, linear theory can be very useful, because powerful methods are available for the solution of linear equations and modern computers can provide useful numerical approximations.

In a region of abrupt changes, at a concentrated load or at a corner, a numerical approximation is dubious, and an exact solution of the differential equations is needed. An important example is the so-called stress concentration at a hole or at a corner. When an exact solution is inaccessible, the linear theory can be used to formulate the basis of a discrete approximation which is governed by linear algebraic equations. The latter are usually amenable to numerical computation. In either case, we must appreciate the limitations of the linear formulation: linearity in the geometric relations between strains and displacements and linearity in the equations of motion imply that the *strains and rotations are very small*, or, strictly speaking, infinitesimal. Practically speaking, linearity in the constitutive equations implies that the *material is linearly viscoelastic*, or simply *linearly elastic.*

Numerous structural materials, for example, metals, are nearly Hookean within a range of small strains. Then, small deformations of the structural member may be treated by the theory of linear elasticity. If the member is too flexible, for example, a thin rod or plate, large rotations may invalidate the linear theory. Otherwise, the Hookean description is invalidated by the onset of plastic deformations or fracture. Other structural members display a viscoelastic behavior which can be approximated by linear equations, but only if the strains are very small.

7.3 Kinematic Equations of a Linear Theory

If strains and rotations are imperceptibly small, then the relationships between strains and displacements are given by (3.175a) as follows:

$$\epsilon_{ij} \doteq e_{ij} = \tfrac{1}{2}(V_{i,j} + V_{j,i}). \tag{7.1}$$

Also, recall the decomposition of the tangent vector according to (3.165):

$$\boldsymbol{G}_i = \hat{\boldsymbol{\imath}}_i + e_{ik}\hat{\boldsymbol{\imath}}_k + \Omega_{ki}\hat{\boldsymbol{\imath}}_k, \tag{7.2}$$

where the rotation vector is given by (see Section 3.22):

$$\boldsymbol{\Omega} \equiv \Omega_i\hat{\boldsymbol{\imath}}_i, \qquad \Omega_i = \tfrac{1}{2}\epsilon_{ijk}\Omega_{kj}, \tag{7.3a, b}$$

$$\Omega_{kj} = \Omega_n\epsilon_{njk} \doteq \tfrac{1}{2}(V_{k,j} - V_{j,k}). \tag{7.3c, d}$$

If the strain components are small compared to unity, then, in accord with (3.76), the dilatation $'e$ of (3.73a–c) [or $'h$ of (3.74a–c)] is given by the linear approximation:

$$'e \doteq {}'h \equiv \vartheta = \epsilon_{ii} = V_{i,i}. \tag{7.3e}$$

In theories of small strains and rotations, only the linear terms are retained in the integrand of (3.122c). Then, the displacement $\boldsymbol{V}\,]_Q$ at a cite $Q(\xi_i)$ is expressed in terms of the displacement and the rotation at another location $P(a_i)$ with coordinates a_i and an integral as follows:

$$\boldsymbol{V}\,]_Q = \boldsymbol{V}\,]_P + \hat{\boldsymbol{\imath}}_k(\xi_i - a_i)\Omega_{ki}\,]_P + \hat{\boldsymbol{\imath}}_k \int_P^Q [\epsilon_{km} + (\xi_i - x_i)(\epsilon_{mk,i} - \epsilon_{im,k})]\, dx_m. \tag{7.4}$$

The integral exists if, and only if, the strains satisfy the compatibility conditions; the linear versions of (3.130) are applicable, viz.,

$$^*R_{mipq} = \epsilon_{ip,qm} + \epsilon_{qm,ip} - \epsilon_{iq,pm} - \epsilon_{pm,iq} = 0, \tag{7.5}$$

$$^*R_{2112} = {}^*R_{3113} = {}^*R_{2332} = {}^*R_{2113} = {}^*R_{2331} = {}^*R_{1223} = 0. \tag{7.6}$$

7.4 Linear Equations of Motion

If the strain components are small enough, then the various tensorial and physical components of stress do not differ significantly:

$$\sigma^{ij} \doteq \overset{*}{\sigma}{}^{ij} \doteq \bar{\sigma}^{ij} \doteq \tau^{ij} \doteq t^{ij} \doteq s^{ij} = s^{ji}. \tag{7.7}$$

Therefore, we need only the symbol s^{ij} in our subsequent linear theory.

If the strains and rotations are small compared to unity, then the tangent $\boldsymbol{G}_i$ of (7.2), or the normal $\boldsymbol{G}^i$, is nearly the initial vector $\hat{\boldsymbol{\imath}}_i$. Consequently, the stress vector upon a surface with normal $\hat{\boldsymbol{n}} \doteq n_i\hat{\boldsymbol{\imath}}_i$ is given by a linear version of (4.33a, b); specifically,

$$\boldsymbol{T} = s^{ij}\hat{\boldsymbol{\imath}}_j n_i. \tag{7.8}$$

The linear version of the equilibrium equation (4.45a–c) implies that rotation and strain are neglected, i.e., $(s^{ij}\acute{\boldsymbol{g}}_j)_{,i} \doteq s^{ij}{}_{,i}\,\acute{\boldsymbol{g}}_j$ ($s^{ij}\acute{\boldsymbol{g}}_{j,i}$ is neglected). In the present case $\acute{\boldsymbol{g}}_i \doteq \hat{\boldsymbol{\imath}}_i$. It follows that the equations of equilibrium (or motion) assume the form:

$$\frac{\partial s^{ij}}{\partial x_i}\hat{\boldsymbol{\imath}}_j + \rho_0\,\tilde{\boldsymbol{f}} = \mathbf{o}. \tag{7.9}$$

If the body undergoes an acceleration, then the vector $\tilde{\boldsymbol{f}}$ includes the inertial term $(-\ddot{\boldsymbol{V}})$.

7.5 Linear Elasticity

The constitutive equations of the isotropic Hookean material are embodied in the free-energy potential of (5.93), the complementary (Gibbs) potential of (5.95), the stress-strain and the strain-stress equations of (5.94) and (5.96), in the order cited:

$$\rho_0\,\mathcal{F} = \frac{E}{2(1+\nu)}\left(\epsilon_{ij}\epsilon_{ij} + \frac{\nu}{1-2\nu}\epsilon_{ii}\epsilon_{jj}\right) - \frac{E\alpha}{1-2\nu}(T-T_0)\epsilon_{jj}, \tag{7.10}$$

$$\rho_0\,\mathcal{G} = -\frac{(1+\nu)}{2E}\left(s^{ij}s^{ij} - \frac{\nu}{1+\nu}s^{ii}s^{jj}\right) - \alpha(T-T_0)s^{ii}, \tag{7.11}$$

$$s^{ij} = \frac{\partial(\rho_0\,\mathcal{F})}{\partial\epsilon_{ij}}, \tag{7.12a}$$

$$= \frac{E}{1+\nu}\left(\epsilon_{ij} + \frac{\nu}{1-2\nu}\epsilon_{kk}\delta_{ij}\right) - \frac{E\alpha}{1-2\nu}(T-T_0)\delta_{ij}, \tag{7.12b}$$

$$\epsilon_{ij} = -\frac{\partial(\rho_0\,\mathcal{G})}{\partial s^{ij}}, \tag{7.13a}$$

$$= \frac{(1+\nu)}{E}\left(s^{ij} - \frac{\nu}{1+\nu}s^{kk}\delta_{ij}\right) + \alpha(T-T_0)\delta_{ij}. \tag{7.13b}$$

The foregoing equations describe the medium in terms of the most common coefficients, Young's modulus E and Poisson's ratio ν. Alternatively, the potentials and the stress-strain relations can be expressed in terms of the Lamé coefficient λ and shear modulus G. In accordance with (5.63), (5.64), and (5.66),

$$\rho_0\,\mathcal{F} = G\epsilon_{ij}\epsilon_{ij} + \frac{\lambda}{2}\epsilon_{ii}\epsilon_{jj} - \alpha(3\lambda+2G)(T-T_0)\epsilon_{kk}, \tag{7.14}$$

$$s^{ij} = 2G\epsilon_{ij} + \lambda\epsilon_{kk}\delta_{ij} - \alpha(3\lambda+2G)(T-T_0)\delta_{ij}, \tag{7.15}$$

$$2G\epsilon_{ij} = s^{ij} - \frac{\lambda}{3\lambda + 2G} s^{kk}\delta_{ij} + 2G\alpha(T - T_0)\delta_{ij}. \tag{7.16}$$

7.6 The Boundary-Value Problems of Linear Elasticity

The displacement, strain, and stress fields in the Hookean body are to satisfy the linear differential equations (7.1), (7.5), (7.9), and the linear algebraic equations (7.12a, b) and (7.13a, b), or (7.15) and (7.16) throughout the body. As usual, we assume that the body force and temperature distributions are prescribed.

In addition to the field equations cited, the stress components at the surface must equilibrate the applied traction. If $\overline{\boldsymbol{T}}$ is the traction prescribed upon a portion s_t of the surface, then the boundary condition has the form:

$$s^{ij}\hat{\boldsymbol{\imath}}_j n_i = \overline{\boldsymbol{T}} \quad \text{on } s_t. \tag{7.17}$$

It is likely that the displacement is prescribed upon some other portions s_v of the surface. If $\overline{\boldsymbol{V}}$ denotes the prescribed displacement, then the boundary condition asserts that

$$\boldsymbol{V} = \overline{\boldsymbol{V}} \quad \text{on } s_v. \tag{7.18}$$

From physical considerations, it is evident that other kinds of conditions could prevail upon portions of the body. For example, at the contact of two bodies, normal displacements may be equal but lubrication may preclude any significant shear traction. Yet another example is the elastically supported surface which requires a traction proportional to the displacements.

In the mathematical theory of elasticity, it is customary to distinguish three fundamental problems: the *first* problem arises if the displacements are prescribed upon the entire surface. The *second* problem occurs if the tractions are prescribed. The *third* problem is a *mixed* boundary-value problem, wherein tractions are prescribed upon a portion s_t and displacements upon the remaining portion s_v.

In a problem of the second kind, the distributions of body force and surface tractions cannot be prescribed in a completely arbitrary way since the resultant of all forces upon the body must satisfy equilibrium, that is, the resultant of the tractions must equilibrate the resultant of the body forces. In a problem of the second or third type, the support of the body must prohibit rigid-body motions under the prescribed loads.

For convenience, we record the basic field equations of the *isothermal* problem (first, second, or third kind). The results are taken from (7.1),

(7.5), (7.9), (7.15), or (7.16):

$$\epsilon_{ij} = \tfrac{1}{2}(V_{i,j} + V_{j,i}), \tag{7.19}$$

$$\epsilon_{ip,qm} + \epsilon_{qm,ip} - \epsilon_{iq,pm} - \epsilon_{pm,iq} = 0, \tag{7.20}$$

$$({s^{ij}}_{,i} + \rho_0 \tilde{f}^j)\hat{\boldsymbol{\imath}}_j = \mathbf{o}, \tag{7.21}$$

$$s^{ij} = 2G\epsilon_{ij} + \lambda\epsilon_{kk}\delta_{ij}, \tag{7.22a}$$

or

$$\epsilon_{ij} = \frac{1}{2G}\left(s^{ij} - \frac{\lambda}{3\lambda + 2G}s^{kk}\delta_{ij}\right). \tag{7.22b}$$

7.7 Kinematic Formulation

When the displacements are specifically required, it is natural to formulate the problem in terms of the displacements. For simplicity, we consider only the isothermal problem, that is, $T = T_0$. Then, the strain components can be eliminated from (7.22a) by means of (7.19) to obtain

$$s^{ij} = G(V_{i,j} + V_{j,i}) + \lambda V_{k,k}\delta_{ij}. \tag{7.23}$$

Upon substituting (7.23) into the equilibrium condition (7.21), we obtain

$$[\,GV_{j,ii} + (\lambda + G)V_{k,kj} + \rho_0 \tilde{f}^j\,]\hat{\boldsymbol{\imath}}_j = \mathbf{o}. \tag{7.24}$$

The equation (7.24) is referred to as the *Navier equation.*

In the problem of the first kind, the displacement must satisfy (7.24) in the region of the body and must have prescribed values on the boundary. In the second problem, the displacement must satisfy (7.24) but the stresses must satisfy (7.17) on the boundary. The latter can be phrased in terms of the displacements by means of (7.23), as follows:

$$[G(V_{i,j} + V_{j,i}) + \lambda V_{k,k}\delta_{ij}]n_i\hat{\boldsymbol{\imath}}_j = \overline{\boldsymbol{T}}. \tag{7.25}$$

Notice that the compatibility conditions of (7.20) are not relevant when the solution *is* the displacement.

The equilibrium equation (7.24) can be recast in terms of our approximations for the dilatation ϑ and rotation $\boldsymbol{\Omega}$ [see equations (7.3a–e)]. Since the derivatives are continuous,

$$\vartheta_{,j} = V_{i,ij}.$$

Rearranging the terms of (7.24), we have

$$\left[(\lambda + 2G)V_{k,ki} - G(V_{k,i} - V_{i,k})_{,k} + \rho_0 \tilde{f}^i\right] \hat{\boldsymbol{\imath}}_i = \mathbf{o},$$

or in the notations of (7.3a–e)

$$\left[(\lambda + 2G)\vartheta_{,i} - 2G\epsilon_{kji}\Omega_{j,k} + \rho_0 \tilde{f}^i\right] \hat{\boldsymbol{\imath}}_i = \mathbf{o}. \tag{7.26}$$

A differentiation and summation of the latter equation provide the result:

$$(\lambda + 2G)\vartheta_{,ii} + \rho_0 \tilde{f}^i{}_{,i} = 0. \tag{7.27}$$

Also from (7.24)

$$GV_{j,iikk} + (\lambda + G)\vartheta_{,jkk} + \rho_0 \tilde{f}^j{}_{,kk} = 0. \tag{7.28}$$

If the body force is absent ($\tilde{\boldsymbol{f}} = \mathbf{o}$) and if $\lambda + 2G \neq 0$, then according to (7.27), the dilatation is "harmonic," i.e., $\vartheta_{,ii} = 0$. Then, by (7.28) the displacement is "biharmonic," i.e., $V_{j,iikk} = 0$. Consequently, the homogeneous solution can be represented (in a finite simply connected domain) in terms of two harmonic functions, so-called displacement potentials. The implications and consequent representations are discussed in the text of I. S. Sokolnikoff ([129], Section 90).

7.8 Solutions via Displacements

In this section, we present various solutions to the equilibrium equations (7.24). Our attention is initially directed to solutions of the homogeneous version of equation (7.24), i.e., the body force is absent. Such homogeneous solutions must be augmented by the appropriate particular solution to account for any body force. We conclude with one particular solution for concentrated forces.

If we assume that the displacement components V_i can be derived from a scalar function F, where

$$V_i = F_{,i}, \tag{7.29}$$

then the homogeneous Navier equation (7.24) is satisfied, if F satisfies the Laplace equation

$$F_{,ii} = 0. \tag{7.30}$$

The scalar function F is known as a *displacement potential* or *Lamé's strain potential.* Using (7.29), we obtain from (7.24)

$$(\lambda + 2G)F_{,jii} = 0.$$

It follows from (7.3e) and (7.29) that

$$\vartheta \equiv V_{i,i} = F_{,ii} \equiv 0.$$

The solution implies no dilatation.

Another homogeneous solution of (7.24) takes the form

$$2GV_k = (x_i\psi_i + \phi)_{,k} - 4(1 - \nu)\psi_k, \tag{7.31}$$

where the scalar ϕ and vector ψ_k are both harmonic

$$\phi_{,ii} = 0, \qquad \psi_{k,ii} = 0.$$

The four harmonic functions [ψ_i $(i = 1, 2, 3)$ and ϕ] in equation (7.31) are known as the *Papkovich-Neuber displacement potentials.*‡ The deduction of the form (7.31) is given by I. S. Sokolnikoff [129]. A broad class of problems related to elastic equilibrium can be solved by determining the four previously mentioned functions (ψ_i, ϕ).

Another approach is provided by the so-called *Galerkin*† *vector* $\boldsymbol{F}$. The homogeneous version of (7.24) is satisfied if the displacement has the form:

$$2G\boldsymbol{V} = 2(1 - \nu)\boldsymbol{F}_{,ii} - F_{j,ji}\hat{\boldsymbol{\imath}}_i, \tag{7.32}$$

‡P. F. Papkovich, *Comptes Rendus Hebdomadaires des Séances de l'Académie des Sciences*, Paris, vol. 195, pp. 513–515 and pp. 754–756, 1932. H. Neuber, *Z. Angew. Math. Mech.*, vol. 14, p. 203, 1934.

†B. G. Galerkin, *Comptes Rendus Hebdomadaires des Séances de l'Académie des Sciences*, Paris, vol. 190, p. 1047, 1930.

where $\boldsymbol{F}$ is biharmonic; that is

$$\boldsymbol{F}_{,iijj} = \mathbf{o}. \tag{7.33}$$

By comparing (7.32) with (7.31), the Galerkin vector can be expressed in terms of the Papkovich-Neuber potentials, as follows:

$$F_{k,jj} = -2\psi_k,$$

$$F_{j,j} = -x_j\psi_j - \phi.$$

We end this section by presenting a useful particular solution of the nonhomogeneous Navier equations. If the force $(\rho_0 \tilde{\boldsymbol{f}})$ approaches a concentrated force $\boldsymbol{P}$,

$$\boldsymbol{P} = P_k \hat{\boldsymbol{\imath}}_k,$$

at (ξ_1, ξ_2, ξ_3), then a particular solution of (7.24) takes the form:

$$\boldsymbol{V} = \frac{\lambda + G}{8\pi G(\lambda + 2G)} \left[\left(\frac{\lambda + 3G}{\lambda + G} \right) \frac{P_i}{r} - \left(\frac{1}{r} \right)_{,i} (x_j - \xi_j) P_j \right] \hat{\boldsymbol{\imath}}_i, \tag{7.34}$$

where

$$r = \sqrt{(x_i - \xi_i)(x_i - \xi_i)}.$$

Observe that a particular solution for a distribution of body force can be generated by employing the notion of superposition and integrating (7.34) throughout the region of distribution. The solution is due to Lord Kelvin.‡

7.9 Formulation in Terms of Stresses

The problems of linear elasticity can be formulated in terms of stress components which must satisfy the equilibrium equation (7.21):

$$(s^{ij}{}_{,i} + \rho_0 \tilde{f}^j)\hat{\boldsymbol{\imath}}_j = \mathbf{o}. \tag{7.35}$$

‡Sir William Thomson (Lord Kelvin), *Cambridge and Dublin Math. J.*, 1848; reprinted in *Math. Phys. Papers*, vol. 1, p. 97.

The compatibility conditions (7.20) are expressed in terms of the stress components by means of (7.22b). The result follows:

$$s^{ip}{}_{,qn} + s^{qn}{}_{,ip} - s^{iq}{}_{,pn} - s^{pn}{}_{,iq}$$

$$-\frac{\lambda}{3\lambda+2G}(\delta_{ip}s^{kk}{}_{,qn} + \delta_{qn}s^{kk}{}_{,ip} - \delta_{iq}s^{kk}{}_{,pn} - \delta_{pn}s^{kk}{}_{,iq}) = 0. \quad (7.36)$$

This system (7.36) of linear equations is cast in another form by a summation implied if the index $q = n$:

$$s^{ip}{}_{,nn} - s^{in}{}_{,pn} - s^{pn}{}_{,in} - \frac{\lambda}{3\lambda+2G}\delta_{ip}s^{kk}{}_{,nn} + \frac{2(\lambda+G)}{3\lambda+2G}s^{nn}{}_{,ip} = 0. \quad (7.37a)$$

In view of (7.35), equation (7.37a) takes the form

$$s^{ip}{}_{,nn} - \frac{\lambda}{3\lambda+2G}\delta_{ip}s^{kk}{}_{,nn} + \frac{2(\lambda+G)}{3\lambda+2G}s^{nn}{}_{,ip} = -\rho_0\,(\tilde{f}^i{}_{,p} + \tilde{f}^p{}_{,i}). \quad (7.37b)$$

The second term of (7.37b) can be expressed in terms of the body force as follows: form a sum of (7.36) by setting $i = q$, $p = n$; that is,

$$s^{in}{}_{,in} - \frac{\lambda+2G}{3\lambda+2G}s^{ii}{}_{,nn} = 0. \quad (7.38)$$

The initial term of (7.38) is expressed in terms of the body force in accordance with the equilibrium equation (7.35); the result follows:

$$s^{ii}{}_{,nn} = -\frac{3\lambda+2G}{\lambda+2G}\rho_0\,\tilde{f}^j{}_{,j}. \quad (7.39)$$

By substituting (7.39) into the second term of (7.37b), we obtain

$$s^{ip}{}_{,nn} + \frac{2(\lambda+G)}{3\lambda+2G}s^{nn}{}_{,ip} = -\rho_0\left(\tilde{f}^i{}_{,p} + \tilde{f}^p{}_{,i} + \frac{\lambda}{\lambda+2G}\delta_{ip}\tilde{f}^j{}_{,j}\right). \quad (7.40a)$$

If the coefficient λ of (7.40a) is eliminated in favor of Poisson's ratio ν, then the compatibility conditions are given in the following form:

$$s^{ip}{}_{,nn} + \frac{1}{1+\nu}s^{nn}{}_{,ip} = -\rho_0\left(\tilde{f}^i{}_{,p} + \tilde{f}^p{}_{,i} + \frac{\nu}{1-\nu}\delta_{ip}\tilde{f}^j{}_{,j}\right). \quad (7.40b)$$

Equation (7.40b) is known as the *Beltrami-Michell equation.*‡

A solution of the problem of stresses must satisfy the equilibrium equations (7.35), the compatibility equations (7.40a, b), and boundary conditions (7.17). The displacements can only be obtained by an integration, for example, equation (7.4).

In the manner of (3.133) and (3.134), the compatibility equations (7.20) are recast into the form:

$$\epsilon^{ilk}\epsilon^{jmn}\epsilon_{lm,nk} = 0. \tag{7.41}$$

Then, employing (7.22b), we obtain:

$$\epsilon^{ilk}\epsilon^{jmn}\left(s^{lm}{}_{,nk} - \frac{\nu}{1+\nu}s^{pp}{}_{,nk}\delta_{lm}\right) = 0. \tag{7.42}$$

If Poisson's ratio vanishes, then it follows from (7.42):

$$\epsilon^{ilk}\epsilon^{jmn}s^{lm}{}_{,nk} = 0. \tag{7.43}$$

The solution of (7.43) is given by a stress function Φ as follows:

$$s^{ij} = \Phi_{,ij},$$

where

$$\Phi_{,pp} = \text{const.}$$

It can be readily shown that the stress function Φ also provides a solution of the equilibrium equations in the absence of body force.

If the body force is constant, then it follows from (7.39) that

$$s^{nn}{}_{,kk} = 0. \tag{7.44}$$

Furthermore, employing this result and differentiating equation (7.37b), we obtain

$$s^{ij}{}_{,kknn} = 0. \tag{7.45}$$

In words, (7.45) states that a component of stress is a *biharmonic* function.

Finally, we present formal solutions of the equilibrium equations in the absence of body force:

$$s^{ik}{}_{,k} = 0.$$

‡E. Beltrami, *Atti Reale Accad. Lincei*, Rome, ser. 5, vol. 1, 1892.
J. H. Michell, *Proc. London Math. Soc.*, vol. 31, pp. 100–124, 1900.

The homogeneous equations, also known as *Cauchy's equilibrium equations*, are satisfied by stress functions $T_{ij} = T_{ji}$, when

$$s^{ij} = \epsilon^{ipq}\epsilon^{jmn}T_{pm,qn}.$$

The stress functions T_{ij} include the stress functions of J. C. Maxwell‡ and G. Morera.† By setting $(T_{12} = T_{23} = T_{31} = 0)$ or $(T_{11} = T_{22} = T_{33} = 0)$, one obtains the solutions deduced by J. C. Maxwell or G. Morera (see also I. S. Sokolnikoff [129], p. 335 and H. Schäfer [134]). The functions T_{ij} are subject to the differential equations imposed by the compatibility conditions.

7.10 Plane Strain and Plane Stress

The plane boundary-value problems of linear isotropic elasticity have been studied extensively and powerful general methods have been established for their solution. It is not our purpose to propound these methods, nor is it our aim to present a multitude of known solutions. Rather, our present aim is to illustrate the salient features of the field theories and to exemplify their value with a solution of much importance in structural engineering (see Section 7.12).

Two physical conditions lead to elasticity problems of two dimensions in which the fields are dependent upon two coordinates, say x_1 and x_2. In such situations, we introduce Greek indices (e.g., x_α) which represent the numbers 1 or 2.

The simplest condition of two dimensions is that of *plane strain* which occurs if one displacement component (e.g., the component V_3 in the direction of x_3) vanishes and the others (V_1, V_2) are independent of that coordinate x_3; that is,

$$V_3 = 0, \qquad V_\alpha = V_\alpha(x_1, x_2). \tag{7.46a, b}$$

The condition of plane strain prevails in a prismatic body which extends uniformly along the x_3 axis, the ends are maintained in the plane condition (inhibited from axial displacement), and the loading is also independent

‡J. C. Maxwell, *Trans. Roy. Soc. Edinburgh*, vol. 26, p. 27, 1870.

†G. Morera, *Atti Reale Accad. Lincei*, Rome, ser. 5, vol. 1, pp. 137–141 and 223–234, 1892.

of x_3:

$$f^3 = 0, \qquad f^\alpha = f^\alpha(x_1, x_2), \qquad T = T(x_1, x_2).$$

It follows from (7.1) and (7.46a, b) that

$$\epsilon_{i3} = 0, \qquad \epsilon_{\alpha\beta} = \tfrac{1}{2}(V_{\alpha,\beta} + V_{\beta,\alpha}). \tag{7.47a, b}$$

All but one of the compatibility conditions vanish identically. The remaining compatibility condition follows:

$$R_{2112} = 0, \qquad \epsilon^{3\alpha\beta}\epsilon^{3\gamma\eta}\epsilon_{\alpha\gamma,\beta\eta} = 0, \tag{7.48a, b}$$

$$\epsilon_{11,22} - 2\epsilon_{12,12} + \epsilon_{22,11} = 0. \tag{7.48c}$$

The stress components follow from (7.12b) and (7.47a, b):

$$s^{3\alpha} = 0, \qquad s^{33} = \frac{E}{1-2\nu}\left[\frac{\nu}{1+\nu}\epsilon_{\alpha\alpha} - \alpha(T - T_0)\right], \tag{7.49a, b}$$

$$s^{\alpha\beta} = \frac{E}{1+\nu}\left(\epsilon_{\alpha\beta} + \frac{\nu}{1-2\nu}\epsilon_{\gamma\gamma}\delta_{\alpha\beta}\right) - \frac{E}{1-2\nu}\alpha(T - T_0)\delta_{\alpha\beta}. \tag{7.49c}$$

One component of the equilibrium equation vanishes identically, so that equation (7.9) assumes the form

$$\left(\frac{\partial s^{\alpha\beta}}{\partial x_\alpha} + \rho_0\,\tilde{f}^\beta\right)\hat{\boldsymbol{\imath}}_\beta = \mathbf{o}. \tag{7.50}$$

In view of (7.49a), the stress upon a lateral surface must satisfy a two-dimensional version of (7.17); namely,

$$s^{\alpha\beta} n_\alpha \hat{\boldsymbol{\imath}}_\beta = \overline{\boldsymbol{T}}(x_1, x_2). \tag{7.51}$$

Substitution of (7.47b) into (7.49c) yields an expression for the stresses in terms of the displacements. By means of this expression, the equilibrium equations (7.50) lead to the following Navier equations:

$$GV_{\beta,\alpha\alpha} + (\lambda + G)V_{\alpha,\alpha\beta} = -\rho_0\,\tilde{f}^\beta. \tag{7.52}$$

Using (7.52), it can be shown that the dilatation $\vartheta = \epsilon_{\alpha\alpha}$ and the rotation Ω_{12} of (7.3d) are both plane harmonic functions, i.e.,

$$\epsilon_{\alpha\alpha,\beta\beta} = 0, \qquad \Omega_{12,\alpha\alpha} = 0.$$

Inversion of (7.49c) yields an expression for the strains in terms of stresses:

$$\epsilon_{\alpha\beta} = \frac{1+\nu}{E}\left[s^{\alpha\beta} - \nu s^{\gamma\gamma}\delta_{\alpha\beta} + E\alpha(T-T_0)\delta_{\alpha\beta}\right]. \tag{7.53}$$

Employing the result (7.53), the compatibility equation (7.48c) takes the form:

$$s^{11}{}_{,22} - 2s^{12}{}_{,12} + s^{22}{}_{,11} - \nu s^{\gamma\gamma}{}_{,\beta\beta} = -E\alpha(T-T_0)_{,\gamma\gamma}. \tag{7.54}$$

A solution requires stresses $s^{\alpha\beta}$ which satisfy equilibrium (7.50), compatibility (7.54), and the requisite boundary conditions.

A state of *plane stress* exists with respect to an x_3 plane, if

$$s^{i3} = 0. \tag{7.55}$$

In reality, the situation is nearly realized in the case of a thin plate of thickness $(2h)$ with lateral surfaces $x_3 = \pm h$, provided that the plate and loading are uniform and symmetrical $(\tilde{f}^3 = 0)$ with respect to the middle x_3 plane. Here, we accept $s^{13} = 0$ as a valid assumption.

It follows from (7.13a, b) that

$$\epsilon_{\alpha 3} = 0, \qquad \epsilon_{33} = -\frac{\nu}{E}s^{\alpha\alpha} + \alpha(T-T_0), \tag{7.56a, b}$$

$$\epsilon_{\alpha\beta} = \frac{1+\nu}{E}\left(s^{\alpha\beta} - \frac{\nu}{1+\nu}s^{\gamma\gamma}\delta_{\alpha\beta}\right) + \alpha(T-T_0)\delta_{\alpha\beta}. \tag{7.56c}$$

The inverse of (7.56c) follows:

$$s^{\alpha\beta} = \frac{E}{1+\nu}\left(\epsilon_{\alpha\beta} + \frac{\nu}{1-\nu}\epsilon_{\gamma\gamma}\delta_{\alpha\beta}\right) - \frac{E\alpha}{1-\nu}(T-T_0)\delta_{\alpha\beta}. \tag{7.57}$$

Using (7.1) and (7.57), the equilibrium equations of plane stress can be expressed in terms of the displacements; the result follows:

$$G\left(V_{\beta,\alpha\alpha} + \frac{3\lambda+2G}{\lambda+2G}V_{\alpha,\alpha\beta}\right) = -\rho_0\,\tilde{f}^{\beta}. \tag{7.58}$$

The compatibility equation of plane stress follows:

$$(1+\nu)(s^{11}{}_{,22} - 2s^{12}{}_{,12} + s^{22}{}_{,11}) - \nu s^{\gamma\gamma}{}_{,\beta\beta} = -E\alpha(T-T_0)_{,\beta\beta}. \tag{7.59}$$

Again, a solution must satisfy equilibrium (7.50), compatibility (7.59), and the requisite boundary conditions.

Observe that the stress-strain relations of plane strain and plane stress, (7.49c) and (7.57), have the same form; only the elastic coefficients differ. In either case, we have

$$s^{\alpha\beta} = 2G\epsilon_{\alpha\beta} + \bar{\lambda}\epsilon_{\gamma\gamma}\delta_{\alpha\beta} - 3K\bar{\alpha}(T - T_0)\delta_{\alpha\beta}, \tag{7.60}$$

$$2G\epsilon_{\alpha\beta} = s^{\alpha\beta} - \frac{\bar{\lambda}}{2(\bar{\lambda} + G)}s^{\gamma\gamma}\delta_{\alpha\beta} + \frac{3KG}{G + \bar{\lambda}}\bar{\alpha}\,(T - T_0)\,\delta_{\alpha\beta}. \tag{7.61}$$

In the problem of plain strain, $\bar{\lambda} = \lambda$, $\bar{\alpha} = \alpha$, and in the problem of plane stress

$$\bar{\lambda} = \frac{2\lambda G}{\lambda + 2G} = \frac{\nu E}{1 - \nu^2}, \qquad \bar{\alpha} = \frac{\bar{\lambda}}{\lambda}\alpha = \left(\frac{1 - 2\nu}{1 - \nu}\right)\alpha. \tag{7.62a, b}$$

In either case, K is the bulk modulus of (5.84) and (5.86).

The problems of *plane strain* and *plane stress* are mathematically similar. Both are formulated in terms of the stress components by substituting the appropriate strain-stress relations into the compatibility condition (7.48a–c). The solutions can be expressed in terms of stresses which must satisfy the equilibrium equations (7.50) and the similar compatibility equations, (7.54) or (7.59). Either problem could be solved in terms of the displacements which must satisfy similar versions of the equilibrium equations, (7.52) or (7.58).

7.11 Airy Stress Function

In the absence of body force, a general solution of the equilibrium equation (7.50) is provided by the Airy function F, where

$$s^{\alpha\beta} = \epsilon^{\alpha\gamma 3}\epsilon^{\beta\eta 3}F_{,\gamma\eta}. \tag{7.63}$$

The substitution of (7.63) into (7.54) or (7.59) provides one differential equation governing the stress components

$$F_{,\alpha\alpha\beta\beta} = -kE\alpha\,\Delta T_{,\gamma\gamma}. \tag{7.64}$$

In equation (7.64), $\Delta T = (T - T_0)$, $k = 1/(1-\nu)$ in a problem of plane strain, and $k = 1$ in a problem of plane stress.

In the presence of body forces, the stresses of (7.63) must be augmented by a particular solution:

$$s^{\alpha\beta} = \epsilon^{\alpha\gamma 3}\epsilon^{\beta\eta 3}F_{,\gamma\eta} - P^{\alpha\beta}, \tag{7.65}$$

where

$$P^{\alpha\beta}{}_{,\alpha} = \rho_0\, \tilde{f}^{\beta}, \qquad P^{\alpha\beta} = P^{\beta\alpha}. \tag{7.66a, b}$$

Then, in place of (7.64), we obtain

$$F_{,\alpha\alpha\beta\beta} = k\left[-E\alpha\,\Delta T_{,\gamma\gamma} + \bar{k}\epsilon^{\alpha\beta 3}\epsilon^{\gamma\eta 3}P^{\alpha\gamma}{}_{,\beta\eta} - \nu P^{\gamma\gamma}{}_{,\beta\beta}\right], \tag{7.67}$$

where $\bar{k} = 1$ in a problem of plain strain and $\bar{k} = (1+\nu)$ in a problem of plane stress.

7.12 Stress Concentration at a Circular Hole in a Plate

Although our immediate goal is the formulation of the general theory, our broader purpose is a presentation which is useful and applicable. In Section 7.11 we have a tool, viz., the Airy stress function, which offers a means to demonstrate stress concentration, a matter of much importance in engineering. In the assembly of structural elements, it is often necessary to drill or punch holes in thin plates. The consequence is a so-called *stress concentration*, as demonstrated by the following example:

A thin Hookean isotropic plate contains a small circular hole as depicted in Figure 7.1. The plate is subjected only to normal stress $\sigma^{11} = T$, distributed uniformly along the *distant* edges $x_1 = \pm a_1$; no stress acts upon the *distant* edges $x_2 = \pm a_2$. The edge of the hole is also stress-free.

According to Saint-Venant's principle, the effects of the stress distribution near the hole are localized. Consequently, the actual size of the plate is unimportant, as long as the radius a is small in comparison with a_α. A practical solution is achieved if the stress field approaches the uniform stress ($\sigma^{11} = T$, $\sigma^{22} = 0$) at infinity.

Now, the circular hole calls for a formulation in cylindrical coordinates. Specifically, equations (7.63) and (7.64), wherein $\Delta T = 0$, $\theta^1 = r$, $\theta^2 = \theta$,

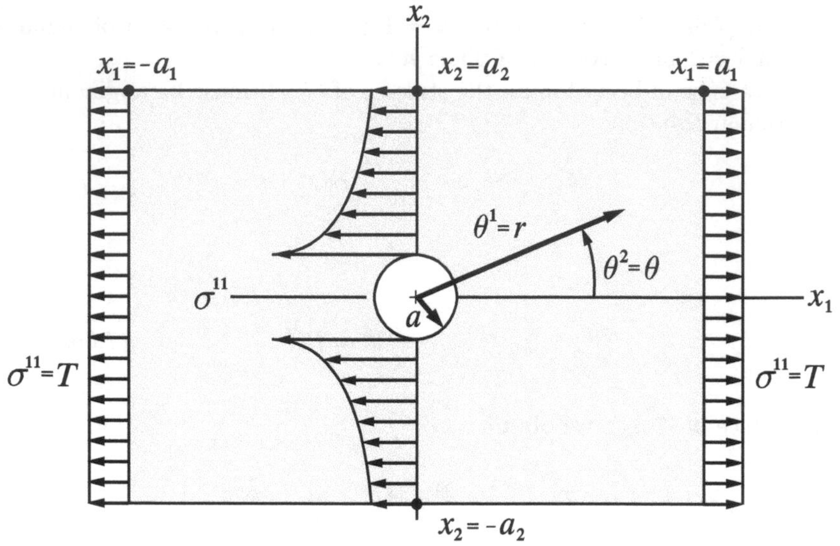

Figure 7.1 Stress concentration in a plate

assume the forms:

$$\sigma^{11} = s^{11} = \frac{1}{r^2}\frac{\partial^2 F}{\partial \theta^2} + \frac{1}{r}\frac{\partial F}{\partial r}, \tag{7.68a}$$

$$\sigma^{12} = rs^{12} = -\frac{1}{r}\frac{\partial^2 F}{\partial r \partial \theta} + \frac{1}{r^2}\frac{\partial F}{\partial \theta}, \tag{7.68b}$$

$$\sigma^{22} = r^2 s^{22} = \frac{\partial^2 F}{\partial r^2}, \tag{7.68c}$$

and

$$\left(\frac{\partial^2}{\partial r^2} + \frac{1}{r}\frac{\partial}{\partial r} + \frac{1}{r^2}\frac{\partial^2}{\partial \theta^2}\right)\left(\frac{\partial^2 F}{\partial r^2} + \frac{1}{r}\frac{\partial F}{\partial r} + \frac{1}{r^2}\frac{\partial^2 F}{\partial \theta^2}\right) = 0. \tag{7.69}$$

Here, s^{ij} denote the tensorial components and σ^{ij} the physical components.

The solution of our problem is given by the Airy function:

$$F = Mr^2 + N \ln \frac{r}{a} + (Ar^2 + Br^4 + Cr^{-2} + D) \cos 2\theta. \tag{7.70}$$

Here, the constants M, N, A, B, C, and D are determined by the edge conditions; details are given in many texts (cf. S. P. Timoshenko and J. N. Goodier [133], Section 35 and G. A. Wempner [105], pp. 505–507). The stress distribution assumes the following form:‡

$$\sigma^{11} = \frac{T}{2}\left(1 - \frac{a^2}{r^2}\right) + \frac{T}{2}\left[1 + 3\left(\frac{a}{r}\right)^4 - 4\left(\frac{a}{r}\right)^2\right]\cos 2\theta, \qquad (7.71a)$$

$$\sigma^{22} = \frac{T}{2}\left(1 + \frac{a^2}{r^2}\right) - \frac{T}{2}\left[1 + 3\left(\frac{a}{r}\right)^4\right]\cos 2\theta, \qquad (7.71b)$$

$$\sigma^{12} = -\frac{T}{2}\left[1 - 3\left(\frac{a}{r}\right)^4 + 2\left(\frac{a}{r}\right)^2\right]\sin 2\theta. \qquad (7.71c)$$

An important result is the stress at the edge of the hole, where

$$\sigma^{22} = T(1 - 2\cos 2\theta). \qquad (7.72)$$

The normal stress σ^{11} along the centerline $x_1 = 0$ is

$$\sigma^{11} = T\left[1 + \frac{1}{2}\left(\frac{a}{r}\right)^2 + \frac{3}{2}\left(\frac{a}{r}\right)^4\right]. \qquad (7.73)$$

A plot of the latter is shown in Figure 7.1. The maximum normal stress occurs at the edge of the hole ($x_1 = 0$, $x_2 = \pm a$), where

$$\sigma^{11}(0, \pm a) = 3\,T. \qquad (7.74)$$

From (7.73) it is evident that the stress diminishes rapidly with distance from the hole. For example, at $r = x_2 = 4a$

$$\sigma^{11}(0, \pm 4a) = 1.0307\,T.$$

The effect is termed *stress concentration* and occurs at abrupt reentrant corners, holes, or inclusions.

‡This solution was obtained by G. Kirsch (*VDI*, vol. 42, 1898).

7.13 General Solution by Complex Variables

An introduction to the plane problems of elasticity would be incomplete without mentioning a representation of the solution by analytic functions of the complex variable. The representation and numerous applications are contained in the treatise of N. I. Muskhelishvili [128]. As an introduction, we present some basic formulas:

First, we record a general representation of the biharmonic function F in terms of two analytic functions $\phi(z)$ and $\chi(z)$, where

$$z = x_1 + ix_2, \qquad \bar{z} = x_1 - i\chi_2, \qquad i^2 = -1.$$

The form given by É. Goursat‡ follows:

$$2F = \bar{z}\phi(z) + z\overline{\phi(z)} + \overline{\chi(z)} + \chi(z), \tag{7.75}$$

where a bar ($^-$) signifies the complex conjugate.

In accordance with (7.63), stress components are given in the forms:

$$s^{11} + is^{12} = \phi'(z) + \overline{\phi'(z)} - z\overline{\phi''(z)} - \overline{\chi''(z)}, \tag{7.76a}$$

$$s^{22} - is^{12} = \phi'(z) + \overline{\phi'(z)} + z\overline{\phi''(z)} + \overline{\chi''(z)}, \tag{7.76b}$$

where a prime ($'$) signifies the derivative.

By the integration of the strain-displacement equations, one can obtain the Kolosov representation [135] of the displacement:

$$2G(V_1 + iV_2) = \kappa\phi(z) - z\overline{\phi'(z)} - \overline{\chi'(z)}. \tag{7.77}$$

Herein κ is a constant, in the problem of plane strain, or plane stress, $\kappa = 3 - 4\nu$, or $\kappa = (3 - \nu)/(1 + \nu)$, respectively.

The solution of a plane boundary-value problem is provided by the functions ϕ and χ which satisfy the prescribed conditions upon the boundaries of the given region. A particular advantage of the complex representation lies in the properties, and consequent applicability, of conformal mappings. If, for example, the region in question can be mapped into the unit circle, then the boundary conditions can be transformed to the corresponding

‡É. Goursat. Sur l' équation $\Delta\Delta u = 0$. *Bull. Soc. Math. France*, vol. 26, p. 236, 1898.

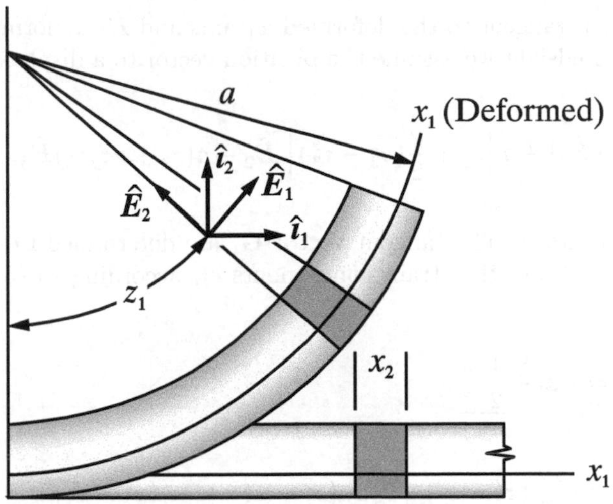

Figure 7.2 Simple bending of a slender road

conditions on the circle and, ultimately, the problem may be reduced to the solution of an integral equation.

The foregoing remarks serve only to indicate a well-established technique for the solution of plane problems. The interested reader can pursue the method in the works cited. An excellent introduction is given in English by I. S. Sokolnikoff [129]. Further developments and important applications are contained in the text of A. E. Green and W. Zerna [3].

7.14 Simple Bending of a Slender Rod

The prismatic body of Figure 7.2 is deformed so that the x_1 axis is bent to a circular arc of radius a; cross-sections (x_1 planes) remain plane and normal to the x_1 line.

Let

$$z_i = x_i/a, \tag{7.78}$$

$$\hat{\boldsymbol{E}}_1 = \hat{\boldsymbol{\imath}}_1 \cos z_1 + \hat{\boldsymbol{\imath}}_2 \sin z_1, \tag{7.79a}$$

$$\hat{\boldsymbol{E}}_2 = -\hat{\boldsymbol{\imath}}_1 \sin z_1 + \hat{\boldsymbol{\imath}}_2 \cos z_1. \tag{7.79b}$$

The vector $\hat{\boldsymbol{E}}_1$ is tangent to the deformed x_1 axis and $\hat{\boldsymbol{E}}_2$ is normal.

With great hindsight we assume the position vector to a displaced particle:

$$\boldsymbol{R} = a(\hat{\boldsymbol{\imath}}_2 - \hat{\boldsymbol{E}}_2) + a\left[z_2 + \frac{\nu}{2}(z_2^2 - z_3^2)\right]\hat{\boldsymbol{E}}_2 + a(z_3 + \nu z_2 z_3)\hat{\boldsymbol{E}}_3, \quad (7.80)$$

where ν is a constant. The tangent vector $\boldsymbol{G}_i$ are determined from (7.80) according to (3.3) and the strain components ϵ_{ij} according to (3.30) and (3.31):

$$\epsilon_{11} = -z_2 + \underline{\frac{1}{2}z_2^2}$$

$$\underline{-\frac{\nu}{2}\left(z_2^2 - z_3^2\right) + \frac{\nu}{2}z_2\left(z_2^2 - z_3^2\right) + \frac{\nu^2}{8}\left(z_2^2 - z_3^2\right)^2}, \quad (7.81a)$$

$$\epsilon_{22} = \nu z_2 + \underline{\frac{\nu^2}{2}\left(z_2^2 + z_3^2\right)}, \quad (7.81b)$$

$$\epsilon_{33} = \nu z_2 + \underline{\frac{\nu^2}{2}\left(z_2^2 + z_3^2\right)}, \quad (7.81c)$$

$$\epsilon_{12} = \epsilon_{13} = \epsilon_{23} = 0. \quad (7.81d\text{–}f)$$

Notice that the strain components are independent of z_1; each slice, as shown shaded in Figure 7.2, undergoes the same deformation. Also notice that the strains vanish at the x_1 axis, $\boldsymbol{G}_i(x_1, 0, 0) = \hat{\boldsymbol{E}}_i$, and an element at the axis experiences only rigid-body motion. Also, observe that the deformation is physically impossible if $z_2 > 1$ and unlikely unless $z_2 \ll 1$.

Since the rod is thin

$$z_2 \ll 1, \qquad z_3 \ll 1. \quad (7.82a, b)$$

Therefore, (7.81a–c) are approximated by

$$\epsilon_{11} = -z_2, \qquad \epsilon_{22} = \nu z_2, \qquad \epsilon_{33} = \nu z_2. \quad (7.83a\text{–}c)$$

If the material is linearly elastic and isotropic, the stress components are given by (7.12b). If ν is the Poisson ratio of (7.12b), then

$$s^{11} = -z_2 E, \qquad s^{22} = s^{33} = s^{12} = s^{13} = s^{23} = 0. \quad (7.84a\text{–}f)$$

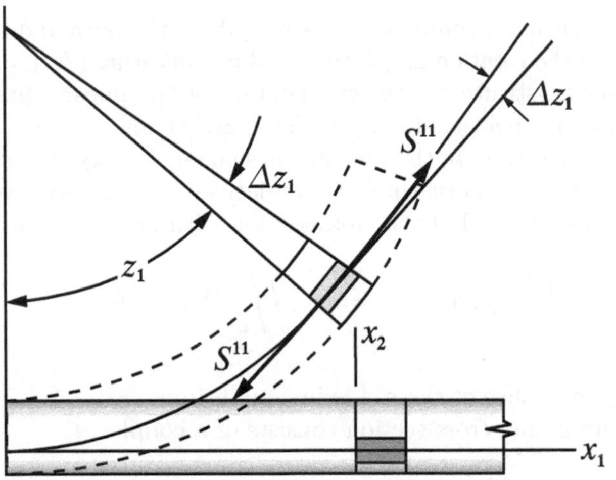

Figure 7.3 Bending of a rod: violation of equilibrium

It follows that the lateral surfaces of the bar are stress-free and the same distribution of normal stress acts on every cross-section. The deformation of (7.80) describes bending of a bar under the stresses of (7.84a–f) *if* the latter satisfy the conditions of equilibrium. Here, we return to the general version of Section 4.10, equation (4.45c), which admits finite rotation. With the omission of body force $\tilde{\boldsymbol{f}}$, that equation assumes the form:

$$\frac{\partial}{\partial \theta^i}\left(\sqrt{g}\, t^{ij} \acute{\boldsymbol{g}}_j\right) = \mathbf{o}. \tag{7.85a}$$

Note that our initial system is Cartesian/rectangular and the strains are small; therefore, $\sqrt{g} = 1$, $t^{ij} \doteq s^{ij}$, $\acute{\boldsymbol{g}}_j \doteq \hat{\boldsymbol{E}}_j$. Equation (7.85a) takes the simpler form:

$$\frac{\partial}{\partial x_i}\left(s^{ij} \hat{\boldsymbol{E}}_j\right) = \mathbf{o}. \tag{7.85b}$$

With the stress (7.84a) and derivative $\hat{\boldsymbol{E}}_{1,1} = \hat{\boldsymbol{E}}_2 / a$, we obtain

$$s^{11} \hat{\boldsymbol{E}}_{1,1} = -\frac{z_2 E}{a} \hat{\boldsymbol{E}}_2 \neq \mathbf{o}. \tag{7.86}$$

This mathematical statement asserts a violation of equilibrium. That violation is evident in the depiction of a fiber, as shown in Figure 7.3. It corresponds to the *small* component of stress s^{11} in the normal direction as a consequence of the rotation Δz_1 introduced by curvature ($\Delta z_1 = \Delta x_1/a$).

These results constitute an "exact" solution of the linear equations, in which the rotation is neglected. In practice, small strains can only occur in the elastic rod (or beam) *if* the member is slender; i.e., $z_2 \ll 1$ or, stated otherwise, the radius of curvature (a) is relatively large. Moreover, if the axis x_1 lies at the centroid of the cross-section with area A, then

$$\iint_A z_2\, dA = 0 \quad \Longrightarrow \quad \iint_A s^{11}\, dA = 0.$$

In words, any *finite* slice of the rod is in equilibrium.

The action upon any cross-section consists of a couple:

$$\begin{aligned}
\boldsymbol{M} &= \iint_A (x_2\hat{\boldsymbol{E}}_2 + x_3\hat{\boldsymbol{E}}_3) \times s^{11}\hat{\boldsymbol{E}}_1\, dA \\
&= -\hat{\boldsymbol{E}}_3 \iint_A x_2 s^{11}\, dA + \hat{\boldsymbol{E}}_2 \iint_A x_3 s^{11}\, dA \\
&= \frac{E}{a}\left(\hat{\boldsymbol{E}}_3 \iint_A x_2^2\, dA - \hat{\boldsymbol{E}}_2 \iint_A x_3 x_2\, dA\right) \\
&= \frac{E}{a}\left(I_{22}\hat{\boldsymbol{E}}_3 - I_{23}\hat{\boldsymbol{E}}_2\right).
\end{aligned} \tag{7.87}$$

The preceding result (7.87) is the familiar moment-curvature relationship of the Bernoulli-Euler theory. Although couched in the linear theory, it is applicable to circumstances of large rotation/displacement as demonstrated by Euler's "Theory of the Elastica" (see, for example, [105], pp. 198–201 and [109], pp. 39–45).

Finally, we should acknowledge that infinitesimal rotations are a mere abstraction, but negligible rotations are a reality. In the foregoing example, the rotations are negligible within a short segment Δx_1. Over an extended length, those small rotations are accumulative. This observation has practical implications when formulating finite-element approximations of thin beams, plates, or shells. One can utilize the linear description within an element, yet accommodate finite rotations by accounting for the *small* relative rotations (differences) between adjoining elements ([116], [117]).

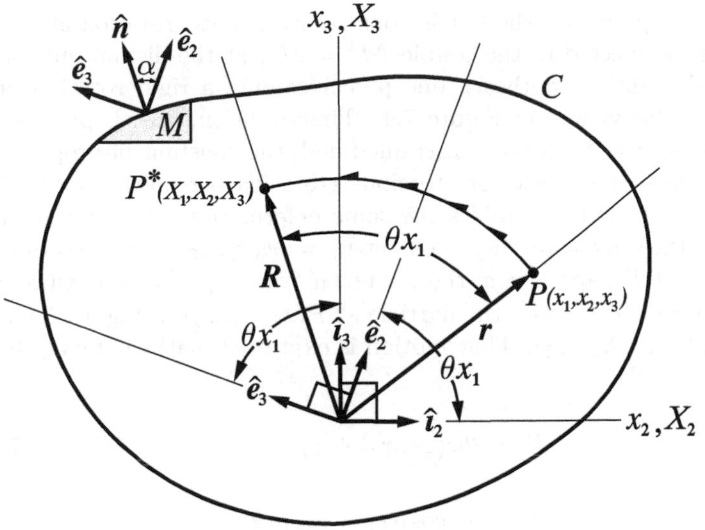

Figure 7.4 Rotated cross-section

7.15 Torsion of a Cylindrical Bar

7.15.1 Saint-Venant's Theory

The reader may recall the simple twisting of a uniform cylindrical bar with a circular cross-section: When the bar is subjected to the action of axial couples applied at the ends, cross-sections remain plane but rotate rigidly about the axis. If the cross-section of the bar is noncircular, then the deformation is more complex. To solve the linear problem of an elastic bar, Saint-Venant took a semi-inverse approach: He supposed that a cross-section does not deform in its plane (like the circular section), but he permitted the cross-sectional plane to warp (unlike the circular section).‡ Viewed from a position on the x_1 axis, a straight line of the cross-section appears to remain straight but, in fact, the line is warped because particles of the line displace *axially*. Here, we introduce such a rotation of unlimited magnitude, but restrict ourselves to small strains.

A slender, cylindrical bar has an arbitrary cross-section, which is shown rotated about the x_1 axis in Figure 7.4. The bar is composed of linearly elastic, isotropic, and homogeneous material. Since a rigid rotation is irrel-

‡This description of the torsion was employed by Saint-Venant to effect the solution of the problem of linear elasticity [27].

evant, we suppose that the bar is constrained against rotation at the end $x_1 = 0$ and subjected to the couple $\boldsymbol{M}^1 = M\hat{\boldsymbol{\imath}}_1$ at the distant end $x_1 = l$. By Saint-Venant's hypothesis, one perceives only a rigid rotation of the cross-section as viewed in Figure 7.4. That rotation, θx_1, is proportional to the distance x_1 from the constrained end; the constant of proportionality θ is termed the "twist" or "torsion" (rotation per unit length). Every similar slice of the bar exhibits the same deformation; every cross-section undergoes the same *axial* displacement $V_1 = \theta\psi(x_2, x_3)$. This describes the deformation fully, apart from the unknown "warping" function $\psi(x_2, x_3)$.

The deformation carries the particle at $P(x_1, x_2, x_3)$ in Figure 7.4 to the position $P^*(X_1, X_2, X_3)$. That motion is expressed mathematically by the equations:

$$X_1 = \theta\psi(x_2, x_3) + x_1, \tag{7.88a}$$

$$X_2 = x_2 \cos\theta x_1 - x_3 \sin\theta x_1, \tag{7.88b}$$

$$X_3 = x_2 \sin\theta x_1 + x_3 \cos\theta x_1. \tag{7.88c}$$

In order to appreciate the deformation of (7.88a–c), let us introduce the unit vectors $\hat{\mathbf{e}}_i$ obtained by rotating $\hat{\boldsymbol{\imath}}_i$ through the angle θx_1 about the axis:

$$\hat{\mathbf{e}}_1 \equiv \hat{\boldsymbol{\imath}}_1, \tag{7.89a}$$

$$\hat{\mathbf{e}}_2 \equiv \hat{\boldsymbol{\imath}}_2 \cos\theta x_1 + \hat{\boldsymbol{\imath}}_3 \sin\theta x_1, \tag{7.89b}$$

$$\hat{\mathbf{e}}_3 \equiv -\hat{\boldsymbol{\imath}}_2 \sin\theta x_1 + \hat{\boldsymbol{\imath}}_3 \cos\theta x_1. \tag{7.89c}$$

Now, the position vector to a particle of the deformed body takes the form:

$$\boldsymbol{R} = (x_1 + \theta\psi)\hat{\mathbf{e}}_1 + x_2\hat{\mathbf{e}}_2 + x_3\hat{\mathbf{e}}_3. \tag{7.90}$$

The tangent vectors follow from (3.3):

$$\boldsymbol{G}_1 = \theta(x_2\hat{\mathbf{e}}_3 - x_3\hat{\mathbf{e}}_2) + \hat{\mathbf{e}}_1, \tag{7.91a}$$

$$\boldsymbol{G}_2 = \theta\psi_{,2}\hat{\mathbf{e}}_1 + \hat{\mathbf{e}}_2, \tag{7.91b}$$

$$\boldsymbol{G}_3 = \theta\psi_{,3}\hat{\mathbf{e}}_1 + \hat{\mathbf{e}}_3. \tag{7.91c}$$

In accordance with (3.30) and (3.31), the strain components take the forms:

$$\epsilon_{11} = \frac{\theta^2}{2}(x_2^2 + x_3^2), \qquad \epsilon_{22} = \frac{\theta^2}{2}(\psi_{,2})^2, \qquad \epsilon_{33} = \frac{\theta^2}{2}(\psi_{,3})^2, \quad \text{(7.92a–c)}$$

$$\epsilon_{12} = \frac{\theta}{2}(\psi_{,2} - x_3), \qquad \epsilon_{13} = \frac{\theta}{2}(\psi_{,3} + x_2), \qquad \epsilon_{23} = \frac{\theta}{2}(\psi_{,2}\,\psi_{,3}). \quad \text{(7.92d–f)}$$

Notice that the strains are independent of x_1; *every* slice along the x_1 axis deforms similarly.

If the strain components throughout the bar are small compared to unity, then it follows from (7.92a–f) that

$$\theta x_2 \ll 1, \qquad \theta x_3 \ll 1, \qquad \theta \psi_{,2} \ll 1, \qquad \theta \psi_{,3} \ll 1. \quad \text{(7.93a–d)}$$

Consequently, the strain components ϵ_{11}, ϵ_{22}, ϵ_{33}, and ϵ_{23} are negligible in comparison with ϵ_{12} and ϵ_{13}.

Since the material is linearly elastic and isotropic, the stresses are obtained by means of (7.12b). In view of (7.92a–f) and (7.93a–d), we have the close approximation

$$\epsilon_{11} \doteq \epsilon_{22} \doteq \epsilon_{33} \doteq \epsilon_{23} \doteq 0, \quad \text{(7.94a–d)}$$

$$\epsilon_{12} = \frac{\theta}{2}(\psi_{,2} - x_3), \qquad \epsilon_{13} = \frac{\theta}{2}(\psi_{,3} + x_2), \quad \text{(7.94e, f)}$$

$$s^{11} \doteq s^{22} \doteq s^{33} \doteq s^{23} \doteq 0, \quad \text{(7.95a–d)}$$

$$s^{12} = G\theta(\psi_{,2} - x_3), \qquad s^{13} = G\theta(\psi_{,3} + x_2). \quad \text{(7.95e, f)}$$

We note that the foregoing approximations presume small strains but the rotation (θx_1) is not limited.

Since the state described by the equations (7.94a–f) and (7.95a–f) is derived from a continuous displacement, it remains only to determine the warping function ψ such that the stresses $(s^{12},\ s^{13})$ satisfy equilibrium. Since the strains are small $(\boldsymbol{G}_i \doteq \acute{\boldsymbol{g}}_i)$, the relevant equations are (4.45c); viz.,

$$\frac{\partial}{\partial x_i}\left(s^{ij}\acute{\boldsymbol{g}}_j\right) \doteq \frac{\partial}{\partial x_i}\left(s^{ij}\boldsymbol{G}_j\right) = \mathbf{o}. \quad \text{(7.96)}$$

According to (7.95a–f), we obtain

$$\frac{\partial}{\partial x_1}\left[(\psi_{,2} - x_3)\boldsymbol{G}_2 + (\psi_{,3} + x_2)\boldsymbol{G}_3\right]$$

$$+ \frac{\partial}{\partial x_2}\left[(\psi_{,2} - x_3)\boldsymbol{G}_1\right]$$

$$+ \frac{\partial}{\partial x_3}\left[(\psi_{,3} + x_2)\boldsymbol{G}_1\right] = \mathbf{o}. \tag{7.97}$$

By the preceding arguments [see the inequalities (7.93a–d)], the terms of order $(\theta x_2,\ \theta x_3,\ \theta\psi_{,2},\ \theta\psi_{,3})$ are all small compared to unity, negligible as a consequence of small strain. In that case, a consistent approximation of the equilibrium equation (7.96) assumes the following form:

$$\psi_{,22} + \psi_{,33} = 0. \tag{7.98}$$

The result (7.98) infers that the rotation θx_1, even though finite, does not preclude the linear solution. This might be anticipated because a large rotation (θx_1) occurs only at a large distance (x_1) from the constrained end. Within a small segment (i.e., $\Delta x_1 \ll l$), the relative rotation $(\theta\,\Delta x_1)$ is imperceptible.

In addition to the differential equation of equilibrium, the function ψ must meet the conditions for equilibrium of an element M at the lateral surface (on boundary C of Figure 7.4). The traction $\boldsymbol{s}^n$ must vanish:

$$\boldsymbol{s}^n = s^{ij}\hat{\boldsymbol{e}}_j n_i. \tag{7.99}$$

Here, $n_i = \hat{\boldsymbol{e}}_i \cdot \hat{\boldsymbol{n}}$, where $\hat{\boldsymbol{n}}$ is the unit normal to C. In particular,

$$n_1 = 0, \qquad n_2 = \cos\alpha, \qquad n_3 = \sin\alpha.$$

It follows from (7.99) and (7.95a–f) that

$$\left.(s^{12}\cos\alpha + s^{13}\sin\alpha)\right]_C = 0, \tag{7.100}$$

$$\left[(\psi_{,2} - x_3)\cos\alpha + (\psi_{,3} + x_2)\sin\alpha\right]_C = 0. \tag{7.101}$$

Finally, given a solution to the boundary-value problem, i.e., a function ψ which satisfies the Laplace equation (7.98) and boundary condition (7.101), the couple-twist relation follows:

$$M = \iint_A (x_2 s^{13} - x_3 s^{12})\, dA$$

$$= G\theta \iint_A (x_2 \psi_{,3} - x_3 \psi_{,2} + x_2^2 + x_3^2)\, dA. \qquad (7.102)$$

7.15.2 Prandtl Stress Function

From the preceding description of the torsion problem, as given by the Saint-Venant's theory, two nonvanishing components of stress (s^{12}, s^{13}) need only satisfy the linear differential equation of equilibrium; viz.,

$$\frac{\partial s^{21}}{\partial x_2} + \frac{\partial s^{31}}{\partial x_3} = 0. \qquad (7.103)$$

A solution is given by the Prandtl stress function‡ $\Phi(x_2, x_3)$, if

$$s^{21} \equiv G\theta \Phi_{,3}, \qquad s^{31} \equiv -G\theta \Phi_{,2}. \qquad (7.104\text{a, b})$$

It follows immediately that

$$\epsilon_{21} = \frac{\theta}{2}\Phi_{,3}, \qquad \epsilon_{31} = -\frac{\theta}{2}\Phi_{,2}. \qquad (7.105\text{a, b})$$

Any function Φ with the requisite continuity provides a solution of (7.103), but need not admit a continuous displacement. If we consider, for the present, that the cross-section is simply connected (has no holes), then the existence of the displacement is assured if, and only if, the compatibility conditions are fulfilled. In the present case, the strains of (7.94a–f) and (7.105a, b) must satisfy the linear version of the compatibility equations (7.5); specifically, the nonvanishing equations

$$^*R_{2331} = (\Phi_{,22} + \Phi_{,33})_{,3} = 0,$$

$$^*R_{1223} = -(\Phi_{,22} + \Phi_{,33})_{,2} = 0.$$

It follows that

$$\Phi_{,22} + \Phi_{,33} = B, \qquad (7.106)$$

‡L. Prandtl, *Physik Z.*, vol. 4, pp. 758–770, 1903.

where B denotes a constant. However, by Saint-Venant's hypothesis, the strain components ϵ_{21} and ϵ_{31} are given by (7.94e, f) as well as (7.105a, b). Equating the right sides of (7.94e, f) and (7.105a, b), we obtain

$$\Phi_{,3} = \psi_{,2} - x_3, \qquad \Phi_{,2} = -\psi_{,3} - x_2. \qquad (7.107\text{a, b})$$

It follows from (7.106) and (7.107a, b) that Φ must satisfy Poisson's equation

$$\Phi_{,22} + \Phi_{,33} = -2. \qquad (7.108)$$

The boundary condition (7.101) assumes the form

$$(-\Phi_{,3}\cos\alpha + \Phi_{,2}\sin\alpha)\big]_C = 0.$$

If we denote the arc-length on C by c progressing counterclockwise, then

$$\cos\alpha = \frac{dx_3}{dc}, \qquad \sin\alpha = -\frac{dx_2}{dc}.$$

Therefore, it follows that

$$\frac{d\Phi}{dc} = 0, \qquad \Phi]_C = K, \qquad (7.109\text{a, b})$$

where K denotes a constant.

Observe that equations (7.100) and (7.109a, b) assert that the stress vector at the boundary is tangent to the curve C; that is, $\boldsymbol{s}^1 \cdot \hat{\boldsymbol{n}} = 0$. Likewise, at any point in the cross-section, the stress vector is tangent to a contour $\overline{C}$, $\Phi = \overline{K}$ (constant), through the point as depicted in Figure 7.5. Also, in accordance with

$$\boldsymbol{s}^i \doteq s^{ij}\hat{\mathbf{e}}_j,$$

the (shear) stress follows:

$$\boldsymbol{s}^1 \equiv s^{12}\hat{\mathbf{e}}_2 + s^{13}\hat{\mathbf{e}}_3. \qquad (7.110)$$

The magnitude of $\boldsymbol{s}^1$ is

$$s^1 = -s^{12}\sin\alpha + s^{13}\cos\alpha. \qquad (7.111\text{a})$$

If n denotes the distance along a normal to the contour $\overline{C}$, then

$$\cos\alpha = \frac{dx_2}{dn}, \qquad \sin\alpha = \frac{dx_3}{dn}. \qquad (7.111\text{b})$$

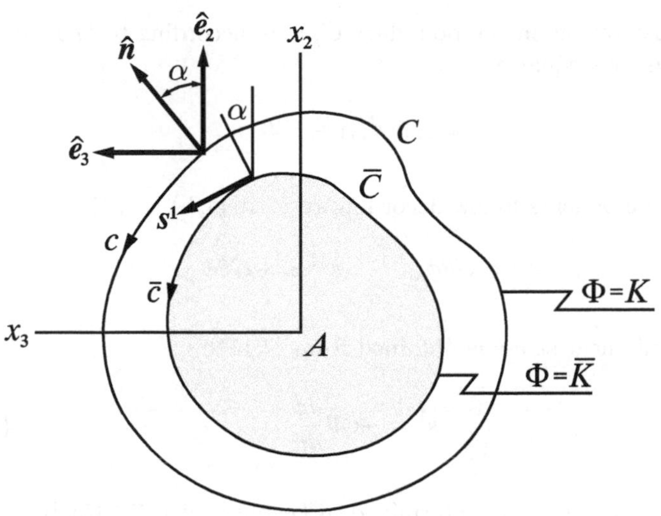

Figure 7.5 Prandtl stress function: boundary conditions

Consequently, it follows from (7.104a, b) and (7.111a, b) that

$$s^1 = -G\theta \frac{d\Phi}{dn}. \tag{7.111c}$$

The resultant on a cross-section is an axial couple:

$$M^1 = \iint_A (x_2 s^{13} - x_3 s^{12})\, dA. \tag{7.112}$$

Substituting (7.104a, b) into (7.112) and integrating by parts, we obtain

$$M^1 = -G\theta \iint_A (x_2 \Phi_{,2} + x_3 \Phi_{,3})\, dA, \tag{7.113a}$$

$$= 2G\theta \iint_A (\Phi - K)\, dA, \tag{7.113b}$$

where K is the constant boundary value.

Equation (7.113b) suggests that we define the function

$$\phi = \Phi - K. \tag{7.114}$$

Then, ϕ must vanish on the boundary C, and, according to (7.108), ϕ also satisfies Poisson's equation:

$$\phi_{,22} + \phi_{,33} = -2. \tag{7.115}$$

The stress components follow according to (7.104a, b) and (7.114):

$$s^{21} = G\theta\phi_{,3}, \qquad s^{31} = -G\theta\phi_{,2}. \tag{7.116}$$

The resultant shear stress is obtained from (7.111c):

$$s^1 = -G\theta\frac{d\phi}{dn}. \tag{7.117}$$

The warping function ψ is determined in keeping with (7.107a, b):

$$\psi_{,2} = x_3 + \phi_{,3}, \qquad \psi_{,3} = -x_2 - \phi_{,2}. \tag{7.118}$$

The twisting couple is given by the integral

$$M^1 = 2G\theta \iint_A \phi\, dA. \tag{7.119}$$

The function ϕ is known as the *Prandtl stress function.*

A geometrical interpretation of the Prandtl function is helpful. A plot of ϕ versus x_2 and x_3 is depicted in Figure 7.6. Here, the stress vector $\boldsymbol{s}^1$ is drawn tangent to the contour line. The magnitude of the shear stress $\boldsymbol{s}^1$ is proportional to the slope $(-\tan\beta)$ in accordance with (7.117). The magnitude of the twisting couple M^1 is proportional to the volume under the dome-like surface in accordance with (7.119).

It is a fact of practical importance that the stress s^1 attains a maximum at the boundary C. Therefore, according to most theories, yielding or fracture would be initiated at a point on the surface of the rod.

7.15.3 Alternative Formulation

The solution of the torsion problem can be expressed in terms of another function ϕ_1, where

$$\phi_1 \equiv \phi + \tfrac{1}{2}(x_2^2 + x_3^2). \tag{7.120}$$

Instead of the differential equation (7.115), we have

$$\phi_{1,22} + \phi_{1,33} = 0. \tag{7.121}$$

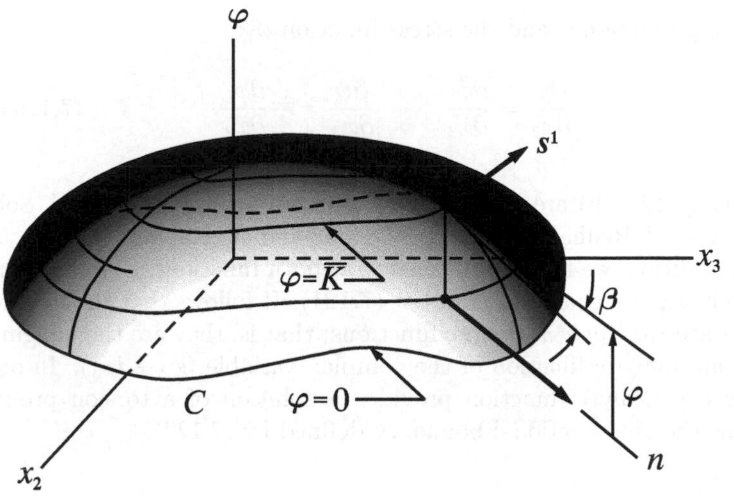

Figure 7.6 Geometrical interpretation of the Prandtl stress function

On the boundary C,

$$\left[\phi_1 - \tfrac{1}{2}(x_2^2 + x_3^2)\right]_C = 0. \tag{7.122}$$

Instead of (7.116), we have

$$s^{21} = G\theta(\phi_{1,3} - x_3), \tag{7.123a}$$

$$s^{31} = -G\theta(\phi_{1,2} - x_2). \tag{7.123b}$$

The twisting couple of (7.119) is given by the integral

$$M^1 = 2G\theta \iint_A \phi_1 \, dA - G\theta I, \tag{7.124}$$

where I is the so-called polar moment of the cross-section:

$$I = \iint_A (x_2^2 + x_3^2) \, dA. \tag{7.125}$$

By means of (7.118) and (7.120), we obtain the following relation between the warping function ψ and the stress function ϕ_1:

$$\frac{\partial\psi}{\partial x_2} = \frac{\partial\phi_1}{\partial x_3}, \qquad \frac{\partial\psi}{\partial x_3} = -\frac{\partial\phi_1}{\partial x_2}. \tag{7.126a, b}$$

Equations (7.126a, b) are the Cauchy-Riemann equations (cf. I. S. Sokolnikoff and R. M. Redheffer [136], pp. 538–543 and W. Kaplan [137], pp. 510, 545). In addition, we have established that both functions, ψ and ϕ_1, must satisfy the Laplace equation, (7.98) or (7.121). It follows that the functions ψ and ϕ_1 are *conjugate harmonic* functions; that is, they are the imaginary parts of an analytic function of the complex variable $(x_2 + ix_3)$. In other words, every analytic function provides a solution of a torsion problem which has the cross-sectional boundary defined by (7.122).

7.15.4 An Example of Stress Concentration

To illustrate the preceding formulation, consider the function

$$F \equiv ai\left(z - \frac{b^2}{z}\right) + i\frac{b^2}{2},$$

where i is the imaginary number and z the complex number:

$$z = x_2 + ix_3.$$

Alternatively,

$$F = -a\left(x_3 + \frac{b^2 x_3}{x_2^2 + x_3^2}\right) + i\left(ax_2 - \frac{ab^2 x_2}{x_2^2 + x_3^2} + \frac{b^2}{2}\right).$$

The imaginary part of F provides the stress function:

$$\phi_1 = ax_2 - \frac{ab^2 x_2}{x_2^2 + x_3^2} + \frac{b^2}{2}. \tag{7.127}$$

The function must satisfy (7.122) on some curve C which determines the shape of the cross-section. Accordingly, we seek the curve defined by the equation:

$$ax_2 - \frac{ab^2 x_2}{x_2^2 + x_3^2} + \frac{b^2}{2} - \frac{1}{2}(x_2^2 + x_3^2) = 0. \tag{7.128a}$$

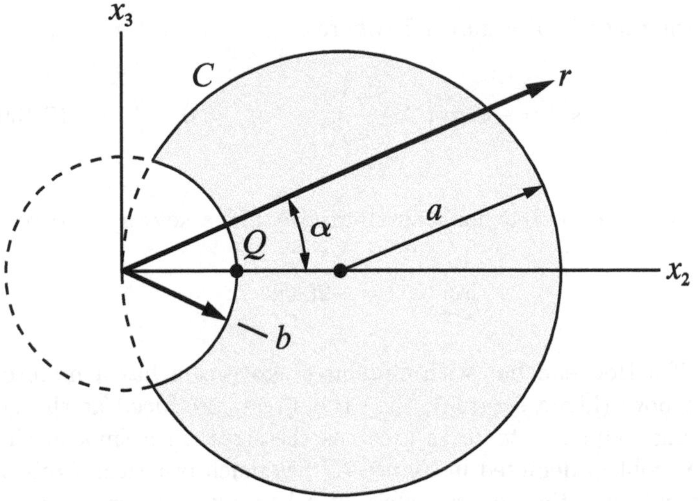

Figure 7.7 Stress concentration: cross-section with a groove

The curve is recognized more readily in the polar coordinates of Figure 7.7, namely,

$$r^2 = x_2^2 + x_3^2, \qquad \alpha = \arctan \frac{x_3}{x_2}.$$

Equation (7.128a) takes the form:

$$\left(r^2 - b^2\right)\left(\frac{a}{r}\cos\alpha - \frac{1}{2}\right) = 0. \tag{7.128b}$$

The left side vanishes on the curves

$$r = b, \qquad a\cos\alpha = \frac{r}{2}. \tag{7.129a, b}$$

Evidently, (7.129a) is a circle of radius b and center at the origin. Equation (7.129b) has an alternative form, namely,

$$(x_2 - a)^2 + x_3^2 = a^2.$$

The latter is evidently a circle of radius a and center at ($x_2 = a,\ x_3 = 0$). Together, the two circles form the boundary C as shown in Figure 7.7.

The stress components are given by (7.123a, b). The maximum stress occurs at the point Q of Figure 7.7, where

$$s^{13} = -G\theta a\left(2 - \frac{b}{a}\right), \qquad s^{12} = 0. \tag{7.130a, b}$$

Now, suppose that the bar has a small groove, like a keyway, and consider the limit:

$$\lim_{b/a \to 0} s^{13} = -2G\theta a.$$

In words, if a Hookean bar with circular cross-section has a minute longitudinal groove (like a scratch), then the stress, produced at the groove by a twisting couple, is *twice* as great as the stress in a smooth circular shaft. The problem depicted in Figure 7.7 has much practical significance, because it indicates the adverse effect of a keyway or a scratch in shafts intended to transmit torque.

7.16 Linear Viscoelasticity

As noted in our introductory remarks, the kinematic equations (7.1) to (7.5) and the equations of motion (7.8) and (7.9) are applicable to a theory of very small strains and rotations, elastic or inelastic. These need only be augmented by the constitutive equations of linear viscoelasticity in order to have the basis of a linear theory. The constitutive equations of a linear theory may take the form of (5.234a, b)

$$\underline{\boldsymbol{P}}\,{}'s^{ij} = \underline{\boldsymbol{Q}}\,\eta_{ij}, \qquad \underline{\boldsymbol{p}}\,p = \underline{\boldsymbol{q}}\,\epsilon_{kk}. \tag{7.131a, b}$$

Of course, the linear equations (7.131a, b), which govern the deviatoric and dilatational parts, can be combined in a single form relating the stress and strain components. Alternatively, the relations can be given in the form of integrals, such as (5.238) and (5.241); namely,

$$s^{ij} = 2\int_{-\infty}^{t} m_1(t-z)\frac{\partial \epsilon_{ij}(z)}{\partial z}\,dz + \delta_{ij}\int_{-\infty}^{t} \bar{\lambda}(t-z)\frac{\partial \epsilon_{kk}(z)}{\partial z}\,dz, \tag{7.132}$$

$$\epsilon_{ij} = \int_{-\infty}^{t} k_1(t-z)\,\frac{\partial s^{ij}(z)}{\partial z}\,dz$$

$$+\frac{1}{3}\delta_{ij}\int_{-\infty}^{t} [k_2(t-z) - k_1(t-z)]\,\frac{\partial s^{kk}(z)}{\partial z}\,dz. \qquad (7.133)$$

7.17 Kinematic Formulation

A problem of linear viscoelasticity can be formulated in terms of displacements by substituting the strain-displacement equations (7.1) into (7.132) and the latter into the equations of motion (7.9) and boundary conditions (7.8). The equation of motion (7.9) takes the form

$$\hat{\imath}_j \int_{-\infty}^{t} m_1(t-z)\frac{\partial V_{j,kk}(z)}{\partial z}\,dz$$

$$+\,\hat{\imath}_j \int_{-\infty}^{t} [\bar{\lambda}(t-z) + m_1(t-z)]\,\frac{\partial V_{k,kj}(z)}{\partial z}\,dz = \rho_0\left(\frac{\partial^2 V_j}{\partial t^2} - \tilde{f}^j\right)\hat{\imath}_j. \quad (7.134)$$

The reader should compare (7.134) with the Navier equation (7.24).

7.18 Quasistatic Problems and Separation of Variables

A problem of viscoelasticity is termed *quasistatic* if the inertial term is omitted from the equations of motion. In this and the following sections, we focus our attention on such problems.

It is natural to attempt a solution of partial differential equations by a separation of variables, for example, a displacement in the form:

$$V_i = \overline{V}_i(x_1, x_2, x_3)g(t). \qquad (7.135)$$

It follows from (7.1) and (7.135) that

$$\epsilon_{ij} = \bar{\epsilon}_{ij}(x_1, x_2, x_3)g(t), \qquad (7.136)$$

where $\bar{\epsilon}_{ij}$ and $\overline{V}_i$ are related as ϵ_{ij} and V_i.

If we consider the equilibrium equation (7.134) in the absence of inertial and body force, we see that a solution of the form (7.135) requires that

$$\bar{\lambda}(t) = C\, m_1(t),$$

where C is a constant. Moreover, equation (5.239) then requires that

$$m_2(t) = (C + \tfrac{2}{3}) m_1(t). \tag{7.137}$$

From the Laplace transformation of (7.137) and from (5.246a, b), we conclude that

$$(1 + \tfrac{3}{2}C) k_2(t) = k_1(t). \tag{7.138}$$

We observe that the moduli m_1 and m_2 play roles as the shear and bulk moduli in elasticity and, therefore, we set

$$C = \frac{2\nu}{1-2\nu}, \qquad \frac{m_2}{m_1} = \frac{2(1+\nu)}{3(1-2\nu)}, \qquad \frac{k_2}{k_1} = \frac{1-2\nu}{1+\nu}. \tag{7.139a–c}$$

According to (7.139), the Poisson's ratio is to be constant.

7.19 Quasistatic Problems in Terms of Displacements

Consider the fundamental problem, wherein body forces are absent and the displacement is prescribed upon the entire surface in the form

$$V_i(x_1, x_2, x_3, t) = \overline{V_i}(x_1, x_2, x_3) g(t) \quad \text{on } s. \tag{7.140}$$

A solution in the form (7.135) and (7.136) together with the relations (7.132) and (7.139a–c) leads to a spatial distribution of stress, similar to the distribution in the elastic body, but varying with time in a manner dependent upon the viscoelastic behavior. The separation of variables in (7.134) provides the equation governing $\overline{V_i}$

$$\overline{V}_{i,jj} + \frac{1}{1-2\nu} \overline{V}_{j,ji} = 0. \tag{7.141}$$

Observe that the displacements $\overline{V_i}$ and, consequently, the strains $\bar{\epsilon}_{ij}$ correspond to the solution of an elastic problem. However, the actual displacement V_i and strains ϵ_{ij} change with time in accordance with (7.135) and (7.136) and the stress components are determined by (7.132).

7.20 Quasistatic Problems in Terms of Stresses

Consider a second fundamental problem in which body forces and tractions are prescribed everywhere in the forms:

$$\widetilde{f}^i(x_1, x_2, x_3, t) = \overline{\overline{f}}^i(x_1, x_2, x_3)F(t), \tag{7.142}$$

$$s^{ij}(x_1, x_2, x_3, t) = \overline{s}^{ij}(x_1, x_2, x_3)F(t). \tag{7.143}$$

In view of (7.133) and (7.139a–c), the spatial distribution of strain is similar to the distribution in an elastic body under these loadings, but varying in time in a way that depends upon the viscoelastic properties, specifically, the compliances k_1 and k_2 of (7.133). The governing equations are the compatibility conditions expressed in terms of the stress components by means of (7.133). In view of (7.139a–c), the compatibility conditions provide the following equations, similar to the Beltrami-Michell equations, governing the field $\overline{s}^{ij}$:

$$\overline{s}^{ip}{}_{,nn} = \frac{1}{1+\nu}\overline{s}^{nn}{}_{,ip} = -\rho_0\left(\overline{\overline{f}}^i{}_{,p} + \overline{\overline{f}}^p{}_{,i} + \frac{\nu}{1-\nu}\overline{\overline{f}}^n{}_{,n}\delta_{ip}\right). \tag{7.144}$$

When the stress components are determined, the strain components may be assumed in the form (7.136). Then, it follows from (7.133) and (7.139a–c) that

$$\overline{\epsilon}_{ij} = \overline{s}^{ij} - \frac{\nu}{1+\nu}\overline{s}^{kk}\delta_{ij}, \tag{7.145}$$

$$\widetilde{f}(t) = \int_{-\infty}^{t} k_1(t-z)\frac{\partial F(z)}{\partial z}\, dz. \tag{7.146}$$

The displacement must be determined in the form of (7.140) by the integration of (7.145).

7.21 Laplace Transforms and Correspondence with Elastic Problems

Let us denote the Laplace transform of a function by a bar ($^{-}$); thus, the transform of a function $g(t)$ is given by

$$\overline{g}(s) = \int_0^\infty g(t) e^{-st}\, dt. \tag{7.147}$$

If the function and derivatives vanish initially ($t = 0$), then

$$\overline{\frac{d^N g}{dt^N}} = s^N \overline{g}(s). \tag{7.148}$$

If a viscoelastic body is initially in a quiescent state, then the transforms of the kinematical (7.1) and equilibrium (7.9) equations take the forms:

$$\overline{\epsilon}_{ij} = \tfrac{1}{2}\left(\overline{V}_{i,j} + \overline{V}_{j,i}\right), \tag{7.149}$$

$$\left(\overline{s}^{ij}{}_{,i} + \rho_0 \overline{\overline{f}}^j\right)\hat{\boldsymbol{\imath}}_j = \mathbf{o}. \tag{7.150}$$

The dynamic or kinematic conditions in (7.17) and (7.18) are similarly transformed:

$$\overline{s}^{ij}\hat{\boldsymbol{\imath}}_j n_i = \overline{\overline{T}}^j \hat{\boldsymbol{\imath}}_j \quad \text{on } s_t, \tag{7.151}$$

$$\overline{V}_i = \overline{\overline{V}}_i \quad \text{on } s_v. \tag{7.152}$$

The Laplace transforms of the stress-strain equations in the form of relations (5.234a, b) or (5.237a, b) follow in accordance with (5.244a, b) and (5.245a, b)

$$\overline{{}'s^{ij}} = 2s\overline{m}_1\overline{\eta}_{ij}, \tag{7.153}$$

$$\overline{s^{kk}} = 3s\overline{m}_2\overline{\epsilon}_{kk}. \tag{7.154}$$

The system of linear equations (7.149) to (7.154) are entirely similar to the basic system of the linear elastic problem; see equations (7.17) to (7.19),

(7.21), and (7.22a, b). Here, the coefficients $s\overline{m}_1$ and $s\overline{m}_2$ replace the shear and bulk moduli, G and K. It follows that the transformed variables $\overline{V}_i(x_1, x_2, x_3)$, $\overline{\epsilon}_{ij}(x_1, x_2, x_3)$, and $\overline{s}^{ij}(x_1, x_2, x_3)$ are given by the solution of an elasticity problem corresponding to the system of equations (7.149) to (7.154). The solution of the viscoelastic problem is accomplished by inversion of the Laplace transforms.

The correspondence between the transformed variables of the viscoelastic problem and the variables of an elastic problem provides a general method of solving quasistatic problems. Of course, the method works only if the transformation and inversion are possible. An exceptional problem arises when the surfaces s_t and s_v are changing in time, for example, the contacting surfaces of viscoelastic bodies growing as the bodies are pressed together. The method is applicable to problems which admit the separable solutions in the forms (7.135), (7.136), (7.142), and (7.143).

Our introduction to linear viscoelasticity is cursory, intended only to draw the analogies with linear elasticity. The interested reader can consult numerous texts on the topic (e.g., [96] to [99]).

Chapter 8

Differential Geometry of a Surface

8.1 Introduction

A shell is a *thin layer* of material. Consequently, the kinematics of a shell are intimately related to the geometry of surfaces, the boundary surfaces, or some intermediate surface. In most practical theories, the differential geometry of one reference surface completely determines the strain components throughout the thickness. Our study is limited to those aspects of the geometry which are essential to a full understanding of these theories of shells.

8.2 Base Vectors and Metric Tensors of the Surface

Let us direct our attention to the surface $\theta^3 = 0$ of Figure 8.1. The position vector to a point P on that surface is

$$\boldsymbol{r}(\theta^1, \theta^2, 0) \equiv {}_0\boldsymbol{r}(\theta^1, \theta^2). \tag{8.1}$$

Let θ^3 be the distance along the normal to our reference surface ($\theta^3 = 0$) and let $\hat{\boldsymbol{a}}_3$ denote the unit *normal* vector at the point P. Then the position vector to an arbitrary point Q is

$$\boldsymbol{r}(\theta^1, \theta^2, \theta^3) = {}_0\boldsymbol{r}(\theta^1, \theta^2) + \theta^3 \hat{\boldsymbol{a}}_3(\theta^1, \theta^2). \tag{8.2}$$

Observe that θ^α ($\alpha = 1, 2$) are arbitrary coordinates of the surface, while θ^3 is a special coordinate, the *distance* along the normal.

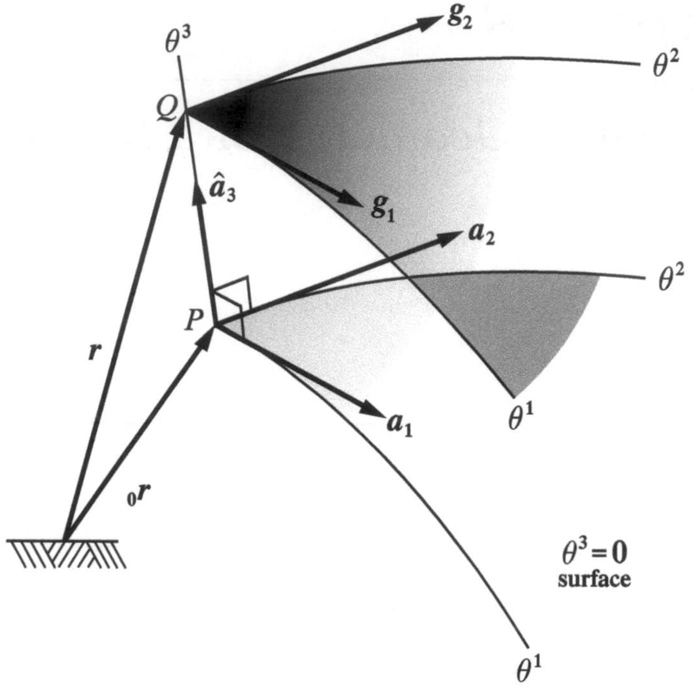

Figure 8.1 Position and coordinates of a surface

The tangent base vectors of our coordinate system follow from (2.5) and (8.2):

$$\boldsymbol{g}_\alpha = {}_0\boldsymbol{r}_{,\alpha} + \theta^3 \hat{\boldsymbol{a}}_{3,\alpha}, \qquad \boldsymbol{g}_3 = \hat{\boldsymbol{a}}_3, \tag{8.3a, b}$$

where the Greek indices have the range 1, 2. Let us denote the tangent vectors at the reference surface as follows:

$$\boldsymbol{a}_i(\theta^1, \theta^2) \equiv \boldsymbol{g}_i(\theta^1, \theta^2, 0). \tag{8.4}$$

It follows from (8.3a, b) and (8.4) that

$$\boldsymbol{g}_\alpha = \boldsymbol{a}_\alpha + \theta^3 \hat{\boldsymbol{a}}_{3,\alpha}, \tag{8.5}$$

$$\boldsymbol{a}_\alpha = {}_0\boldsymbol{r}_{,\alpha}, \qquad \boldsymbol{g}_3 = \hat{\boldsymbol{a}}_3. \tag{8.6a, b}$$

In the manner of (2.6), we define the reciprocal base vector $\boldsymbol{a}^i$ which is

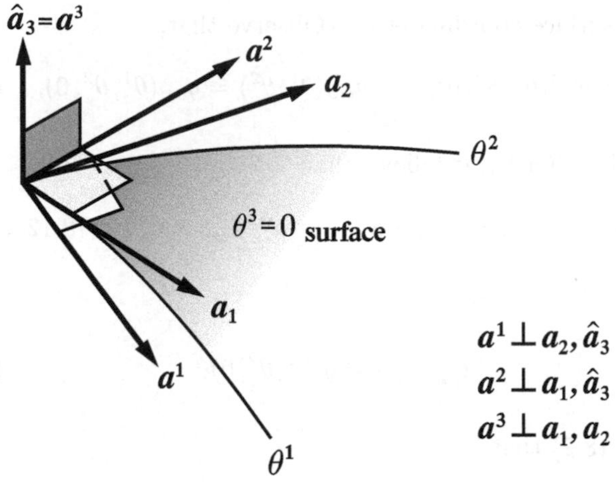

Figure 8.2 Tangent and normal base vectors at the reference surface ($\theta^3 = 0$)

normal to the θ^i surface:

$$\boldsymbol{a}^i(\theta^1, \theta^2) \equiv \boldsymbol{g}^i(\theta^1, \theta^2, 0), \tag{8.7}$$

$$\boldsymbol{a}^i \cdot \boldsymbol{a}_j = \delta^i_j. \tag{8.8}$$

Since the θ^3 is *distance* along the *normal* to the reference surface $\theta^3 = 0$,

$$\boldsymbol{a}^3 = \hat{\boldsymbol{a}}^3 = \boldsymbol{g}^3 = \hat{\boldsymbol{a}}_3. \tag{8.9a, b}$$

At each point of the surface we have two basic triads, $\boldsymbol{a}_i$ and $\boldsymbol{a}^i$, where $\boldsymbol{a}_\alpha$ and $\boldsymbol{a}^\alpha$ are tangent to the surface and $\boldsymbol{a}_3 = \boldsymbol{a}^3 = \hat{\boldsymbol{a}}_3$ is the unit normal. The base vector $\boldsymbol{a}_\alpha$ is tangent to the θ^α line, and the vector $\boldsymbol{a}^\alpha$ is normal to the θ^α line as shown in Figure 8.2.

The vectors $\boldsymbol{a}^\alpha$ can be expressed as a linear combination of $\boldsymbol{a}_\alpha$ and vice versa:

$$\boldsymbol{a}^\alpha = a^{\alpha\beta}\boldsymbol{a}_\beta, \qquad \boldsymbol{a}_\alpha = a_{\alpha\beta}\boldsymbol{a}^\beta. \tag{8.10a, b}$$

These linear relations are a subcase of (2.7) and (2.8), and the coefficients $a^{\alpha\beta}$ and $a_{\alpha\beta}$ are the components of the contravariant and covariant *metric tensors* of our surface coordinates θ^α. Observe that,

$$a^{\alpha\beta}(\theta^1,\theta^2) = g^{\alpha\beta}(\theta^1,\theta^2,0), \qquad a_{\alpha\beta}(\theta^1,\theta^2) = g_{\alpha\beta}(\theta^1,\theta^2,0). \quad (8.11\text{a, b})$$

From (8.8) and (8.10a, b), it follows that

$$a^{\alpha\beta} = \boldsymbol{a}^\alpha \cdot \boldsymbol{a}^\beta, \qquad a_{\alpha\beta} = \boldsymbol{a}_\alpha \cdot \boldsymbol{a}_\beta, \qquad a^{\alpha\beta} a_{\beta\gamma} = \delta^\alpha_\gamma. \quad (8.12\text{a, b}), (8.13)$$

Let

$$|a_{\alpha\beta}| \equiv a = g(\theta^1,\theta^2,0). \quad (8.14\text{a, b})$$

It follows from (8.2) that

$$a^{11} = \frac{a_{22}}{a}, \qquad a^{22} = \frac{a_{11}}{a}, \qquad a^{12} = -\frac{a_{12}}{a}, \quad (8.15\text{a, c})$$

$$|a^{\alpha\beta}| = \frac{1}{a}. \quad (8.16)$$

The vectors $\boldsymbol{a}^a$, $\boldsymbol{a}_a$ and the tensors $a^{\alpha\beta}$, $a_{\alpha\beta}$ play a role in the two-dimensional subspace (surface) as the vectors $\boldsymbol{g}^a$, $\boldsymbol{g}_a$ and the tensors $g^{\alpha\beta}$, $g_{\alpha\beta}$ in the three-dimensional space.

8.3 Products of the Base Vectors

In accordance with (2.24a, b) and (8.14a, b), we define

$$\overline{e}_{\alpha\beta}(\theta^1,\theta^2) \equiv e_{\alpha\beta 3}(\theta^1,\theta^2,0) = \sqrt{a}\,\epsilon_{\alpha\beta 3}, \quad (8.17\text{a, b})$$

$$\overline{e}^{\alpha\beta}(\theta^1,\theta^2) \equiv e^{\alpha\beta 3}(\theta^1,\theta^2,0) = \frac{\epsilon_{\alpha\beta 3}}{\sqrt{a}}. \quad (8.17\text{c, d})$$

Following (2.26a, b), we have

$$\boldsymbol{a}_\alpha \times \boldsymbol{a}_\beta = \overline{e}_{\alpha\beta}\hat{\boldsymbol{a}}_3, \qquad \boldsymbol{a}^\alpha \times \boldsymbol{a}^\beta = \overline{e}^{\alpha\beta}\hat{\boldsymbol{a}}_3, \quad (8.18\text{a, b})$$

$$\hat{\boldsymbol{a}}_3 \times \boldsymbol{a}_\alpha = \overline{e}_{\alpha\beta}\boldsymbol{a}^\beta, \qquad \hat{\boldsymbol{a}}_3 \times \boldsymbol{a}^\alpha = \overline{e}^{\alpha\beta}\boldsymbol{a}_\beta. \quad (8.18\text{c, d})$$

8.4 Derivatives of the Base Vectors

In accordance with (2.46a, b), we define the Christoffel symbols for the surface coordinates:

$$\overline{\Gamma}_{\alpha\beta\gamma}(\theta^1,\theta^2) \equiv \boldsymbol{a}_\gamma \cdot \boldsymbol{a}_{\alpha,\beta} = \Gamma_{\alpha\beta\gamma}(\theta^1,\theta^2,0), \qquad (8.19\text{a, b})$$

$$\overline{\Gamma}^\gamma_{\alpha\beta}(\theta^1,\theta^2) \equiv \boldsymbol{a}^\gamma \cdot \boldsymbol{a}_{\alpha,\beta} = \Gamma^\gamma_{\alpha\beta}(\theta^1,\theta^2,0). \qquad (8.20\text{a, b})$$

Here, the overbar ($\overline{\ }$) signifies that the symbol is evaluated at the reference surface $\theta^3 = 0$. If the derivatives are continuous (the surface is smooth), then it follows from (8.6a) and the definitions (8.19a) and (8.20a) that

$$\overline{\Gamma}_{\alpha\beta\gamma} = \overline{\Gamma}_{\beta\alpha\gamma}, \qquad \overline{\Gamma}^\gamma_{\alpha\beta} = \overline{\Gamma}^\gamma_{\beta\alpha}. \qquad (8.21\text{a, b})$$

Also, from (8.2b) and (8.19a) we have

$$\overline{\Gamma}_{\alpha\beta\gamma} = \tfrac{1}{2}(a_{\alpha\gamma,\beta} + a_{\beta\gamma,\alpha} - a_{\alpha\beta,\gamma}). \qquad (8.22)$$

Recall (8.8), (8.9a, b), and (8.10a, b): viz.,

$$\boldsymbol{a}^\alpha \cdot \boldsymbol{a}_\beta = \delta^\alpha_\beta, \qquad \boldsymbol{a}^\alpha = a^{\alpha\beta}\boldsymbol{a}_\beta, \qquad \boldsymbol{a}_\alpha = a_{\alpha\beta}\boldsymbol{a}^\beta, \qquad (8.23\text{a–c})$$

$$\hat{\boldsymbol{a}}_3 \cdot \boldsymbol{a}_\alpha = \hat{\boldsymbol{a}}_3 \cdot \boldsymbol{a}^\alpha = 0, \qquad \hat{\boldsymbol{a}}_3 \cdot \hat{\boldsymbol{a}}_3 = 1. \qquad (8.24\text{a–c})$$

From (8.23a) and the definition (8.20a), we obtain

$$\overline{\Gamma}^\gamma_{\alpha\beta} = -\boldsymbol{a}_\alpha \cdot \boldsymbol{a}^\gamma{}_{,\beta}. \qquad (8.25)$$

In view of (8.23b, c) and the definitions (8.19a) and (8.20a), we see that

$$\overline{\Gamma}_{\alpha\beta\gamma} = a_{\gamma\eta}\overline{\Gamma}^\eta_{\alpha\beta}, \qquad \overline{\Gamma}^\eta_{\alpha\beta} = a^{\eta\gamma}\overline{\Gamma}_{\alpha\beta\gamma}. \qquad (8.26\text{a, b})$$

Normal components of the derivatives $\boldsymbol{a}_{\alpha,\beta}$ are denoted as follows:

$$b_{\alpha\beta}(\theta^1,\theta^2) \equiv \hat{\boldsymbol{a}}_3 \cdot \boldsymbol{a}_{\alpha,\beta} = \Gamma_{\alpha\beta 3}(\theta^1\theta^2,0), \qquad (8.27\text{a, b})$$

$$b^\alpha_\beta(\theta^1,\theta^2) \equiv \hat{\boldsymbol{a}}_3 \cdot \boldsymbol{a}^\alpha{}_{,\beta} = -\Gamma^\alpha_{3\beta}(\theta^1\theta^2,0). \qquad (8.28\text{a, b})$$

From (8.27a, b), (8.28a, b), and (8.24a–c) we obtain

$$b_{\alpha\beta} = b_{\beta\alpha} = -\boldsymbol{a}_\alpha \cdot \hat{\boldsymbol{a}}_{3,\beta}, \tag{8.29a, b}$$

$$b^\alpha_\beta = -\boldsymbol{a}^\alpha \cdot \hat{\boldsymbol{a}}_{3,\beta}. \tag{8.30}$$

According to (8.23b, c), (8.29b), and (8.30), we have

$$b_{\alpha\beta} = a_{\alpha\gamma} b^\gamma_\beta, \qquad b^\alpha_\beta = a^{\alpha\gamma} b_{\gamma\beta}. \tag{8.31a, b}$$

The components $b_{\alpha\beta}$ and b^a_β constitute associated surface tensors which we call *curvature tensors*. Finally, it follows from (8.24c) that

$$\hat{\boldsymbol{a}}_3 \cdot \hat{\boldsymbol{a}}_{3,a} = 0, \tag{8.32a}$$

and, consequently,

$$\overline{\Gamma}_{3a3} = \overline{\Gamma}^3_{3a} = 0. \tag{8.32b, c}$$

In accordance with (8.19a), (8.20a), (8.23b), (8.26b), and (8.27a), we have

$$\boldsymbol{a}_{\alpha,\beta} = \overline{\Gamma}^\gamma_{\alpha\beta} \boldsymbol{a}_\gamma + b_{\alpha\beta} \hat{\boldsymbol{a}}_3, \tag{8.33a}$$

$$= \overline{\Gamma}_{\alpha\beta\gamma} \boldsymbol{a}^\gamma + b_{\alpha\beta} \hat{\boldsymbol{a}}_3. \tag{8.33b}$$

Likewise, according to (8.25) and (8.28a)

$$\boldsymbol{a}^\alpha{}_{,\beta} = -\overline{\Gamma}^\alpha_{\beta\gamma} \boldsymbol{a}^\gamma + b^\alpha_\beta \hat{\boldsymbol{a}}_3. \tag{8.34}$$

Equations (8.33a, b) and (8.34) can also be derived by setting $\theta^3 = 0$ in (2.45a, b) and (2.49) and employing the aforementioned formulas. Equations (8.33a, b) and (8.34) are *Gauss' formulas* for the derivatives of the tangent vectors. Observe that the Christoffel symbols $\overline{\Gamma}^\gamma_{\alpha\beta}$ or $\overline{\Gamma}_{\alpha\beta\gamma}$ determine the components (of $\boldsymbol{a}_{\alpha,\beta}$ and $\boldsymbol{a}^\alpha{}_{,\beta}$) tangent to the surface while the components $b_{\alpha\beta}$ or b^α_β are normal.

The derivatives of the normal $\hat{\boldsymbol{a}}_3$ follow from (8.29a, b) and (8.30)

$$\hat{\boldsymbol{a}}_{3,\beta} = -b_{\alpha\beta} \boldsymbol{a}^\alpha = -b^\alpha_\beta \boldsymbol{a}_\alpha. \tag{8.35a, b}$$

Equations (8.35a, b) are known as *Weingarten's formulas.*

8.5 Metric Tensor of the Three-Dimensional Space

Returning to the base vector $\boldsymbol{g}_\alpha$ of (8.5) and employing (8.29b) and (8.30), we have

$$\boldsymbol{g}_a = \boldsymbol{a}_\alpha - \theta^3 b^\beta_\alpha \boldsymbol{a}_\beta, \tag{8.36a}$$

$$= \boldsymbol{a}_\alpha - \theta^3 b_{\alpha\beta} \boldsymbol{a}^\beta. \tag{8.36b}$$

The component $g_{\alpha\beta}$ is obtained according to (2.9) with the aid of (8.2b), (8.31a), and (8.36a, b):

$$g_{\alpha\beta} = a_{\alpha\beta} - 2\theta^3 b_{\alpha\beta} + (\theta^3)^2 b_{\alpha\gamma} b^\gamma_\beta. \tag{8.37a}$$

From (8.6b), (8.24a–c), and (8.36a), we have

$$g_{\alpha 3} = 0, \qquad g_{33} = 1. \tag{8.37b, c}$$

According to (2.12) and (8.37a–c), the components $g^{\alpha\beta}$ are rational functions of θ^3 while $g^{a3} = 0$ and $g^{33} = 1$.

8.6 Fundamental Forms

An incremental change of position on the surface is accompanied by a change in the position vector ${}_0\boldsymbol{r}$ and a change in the normal vector $\hat{\boldsymbol{a}}_3$; the first-order differentials are:

$$d\,{}_0\boldsymbol{r} = {}_0\boldsymbol{r}_{,\alpha}\, d\theta^\alpha = \boldsymbol{a}_a\, d\theta^\alpha, \tag{8.38}$$

$$d\hat{\boldsymbol{a}}_3 = \hat{\boldsymbol{a}}_{3,a}\, d\theta^a = -b_{\alpha\beta} \boldsymbol{a}^\beta\, d\theta^a. \tag{8.39}$$

From (8.38) and (8.39), we can form three scalar products, namely

$$d\,{}_0\boldsymbol{r} \cdot d\,{}_0\boldsymbol{r} \equiv d\,{}_0\ell^2 = a_{\alpha\beta}\, d\theta^\alpha\, d\theta^\beta, \tag{8.40}$$

$$d\,{}_0\boldsymbol{r} \cdot d\boldsymbol{a}_3 = -b_{\alpha\beta}\, d\theta^\alpha\, d\theta^\beta, \tag{8.41}$$

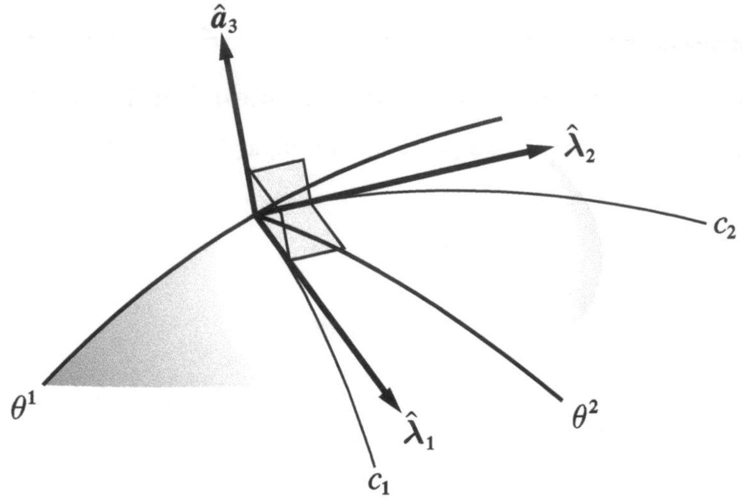

Figure 8.3 Curvature and torsion: orthogonal curves c_1 and c_2 on a surface

$$d\hat{\boldsymbol{a}}_3 \cdot d\hat{\boldsymbol{a}}_3 = a_{\gamma\eta} b^\eta_\alpha b^\gamma_\beta \, d\theta^\alpha \, d\theta^\beta. \tag{8.42}$$

Equations (8.40) to (8.42) are known as the *first, second*, and *third fundamental forms*, respectively. The coefficients of the first quadratic form are components of the metric tensor of the surface coordinates; these components serve to measure distances on the surface. Now, let us examine the role of the coefficients $b_{\alpha\beta}$ appearing in the second quadratic form.

8.7 Curvature and Torsion

Let c_1 and c_2 denote orthogonal curves on the surface of Figure 8.3 and let ℓ_1 and ℓ_2 denote arc lengths along curves c_1 and c_2, respectively.

At each point of curve c_1, we may construct a triad of orthonormal vectors as shown in Figure 8.3; $\hat{\boldsymbol{a}}_3$ is normal to the surface, $\hat{\boldsymbol{\lambda}}_1$ is tangent to c_1, and $\hat{\boldsymbol{\lambda}}_2$ is tangent to c_2.

Now, consider the motion of this triad as it advances along c_1, $\hat{\boldsymbol{a}}_3$ remaining normal to the surface, and $\hat{\boldsymbol{\lambda}}_1$ tangent to c_1. The tangent vector $\hat{\boldsymbol{\lambda}}_1$ turns about $\hat{\boldsymbol{a}}_3$ and about $\hat{\boldsymbol{\lambda}}_2$; that is,

$$\frac{d\hat{\boldsymbol{\lambda}}_1}{d\ell_1} = \kappa_{11}\hat{\boldsymbol{a}}_3 + \sigma_{12}\hat{\boldsymbol{\lambda}}_2. \tag{8.43}$$

This derivative has no component in the direction $\hat{\boldsymbol{\lambda}}_1$ because

$$\frac{d}{d\ell_1}(\hat{\boldsymbol{\lambda}}_1 \cdot \hat{\boldsymbol{\lambda}}_1) = 0.$$

The normal curvature of the surface in the direction of $\hat{\boldsymbol{\lambda}}_1$ is determined by the rate at which $\hat{\boldsymbol{\lambda}}_1$ turns about $\hat{\boldsymbol{\lambda}}_2$, toward the direction of $\hat{\boldsymbol{a}}_3$; the *normal curvature* in the direction of $\hat{\boldsymbol{\lambda}}_1$ is

$$\kappa_{11} \equiv \hat{\boldsymbol{a}}_3 \cdot \frac{d\hat{\boldsymbol{\lambda}}_1}{d\ell_1} = -\frac{d\hat{\boldsymbol{a}}_3}{d\ell_1} \cdot \hat{\boldsymbol{\lambda}}_1. \tag{8.44a, b}$$

The component σ_{12} is known as the *geodesic curvature*:

$$\sigma_{12} = \hat{\boldsymbol{\lambda}}_2 \cdot \frac{d\hat{\boldsymbol{\lambda}}_1}{d\ell_1}. \tag{8.45}$$

As the triad advances along c_1, the tangent vector $\hat{\boldsymbol{\lambda}}_2$ turns about $\hat{\boldsymbol{a}}_3$ and about $\hat{\boldsymbol{\lambda}}_1$

$$\frac{d\hat{\boldsymbol{\lambda}}_2}{d\ell_1} = -\sigma_{21}\hat{\boldsymbol{\lambda}}_1 + \tau_{21}\hat{\boldsymbol{a}}_3. \tag{8.46}$$

Multiplying (8.43) and (8.46) by $\hat{\boldsymbol{\lambda}}_2$ and $\hat{\boldsymbol{\lambda}}_1$, respectively, and noting the orthogonality $(\hat{\boldsymbol{\lambda}}_1 \cdot \hat{\boldsymbol{\lambda}}_2 = 0)$, we have

$$\hat{\boldsymbol{\lambda}}_1 \cdot \frac{d\hat{\boldsymbol{\lambda}}_2}{d\ell_1} = -\hat{\boldsymbol{\lambda}}_2 \cdot \frac{d\hat{\boldsymbol{\lambda}}_1}{d\ell_1}, \tag{8.47a}$$

$$\sigma_{12} = \sigma_{21}. \tag{8.47b}$$

The *torsion* of the surface in the direction of $\hat{\boldsymbol{\lambda}}_1$ is determined by the rate at which $\hat{\boldsymbol{\lambda}}_2$ turns about $\hat{\boldsymbol{\lambda}}_1$, toward the direction of $\hat{\boldsymbol{a}}_3$; the torsion in the direction of $\hat{\boldsymbol{\lambda}}_1$ is

$$\tau_{21} \equiv \hat{\boldsymbol{a}}_3 \cdot \frac{d\hat{\boldsymbol{\lambda}}_2}{d\ell_1} = -\frac{d\hat{\boldsymbol{a}}_3}{d\ell_1} \cdot \hat{\boldsymbol{\lambda}}_2. \tag{8.48a, b}$$

The *torsion* τ_{21} is also called the *geodesic torsion* in the direction of $\hat{\boldsymbol{\lambda}}_1$ (see, for example, A. J. McConnell [138] and I. S. Sokolnikoff [139]).

To express the *normal curvature* and *torsion* in other terms we note that

$$\hat{\boldsymbol{\lambda}}_1 = \frac{\partial \theta^\alpha}{\partial \ell_1}\, \boldsymbol{a}_\alpha \equiv \lambda_1^\alpha \boldsymbol{a}_\alpha, \tag{8.49a, b}$$

$$\hat{\boldsymbol{\lambda}}_2 = \frac{\partial \theta^\alpha}{\partial \ell_2}\, \boldsymbol{a}_\alpha \equiv \lambda_2^\alpha \boldsymbol{a}_\alpha. \tag{8.50a, b}$$

With the aid of (8.33a), we obtain from (8.49a)

$$\frac{d\hat{\boldsymbol{\lambda}}_1}{d\ell_1} = (\lambda^\mu_{1,\beta}\lambda_1^\beta + \lambda_1^\alpha \lambda_1^\beta \overline{\Gamma}^\mu_{\alpha\beta})\boldsymbol{a}_\mu + \lambda_1^\alpha \lambda_1^\beta b_{\alpha\beta}\hat{\boldsymbol{a}}_3.$$

Then, according to (8.44a)

$$\kappa_{11} = \lambda_1^\alpha \lambda_1^\beta b_{\alpha\beta}. \tag{8.51a}$$

Likewise, the normal curvature in the direction of $\hat{\boldsymbol{\lambda}}_2$ is

$$\kappa_{22} = \lambda_2^\alpha \lambda_2^\beta b_{\alpha\beta}. \tag{8.51b}$$

To obtain an expression for the torsion τ_{21} from (8.48a), we require the derivative of (8.50b). By means of (8.33a), we obtain

$$\frac{d\hat{\boldsymbol{\lambda}}_2}{d\ell_1} = (\lambda^\mu_{2,\beta}\lambda_1^\beta + \lambda_2^\alpha \lambda_1^\beta \overline{\Gamma}^\mu_{\alpha\beta})\boldsymbol{a}_\mu + \lambda_2^\alpha \lambda_1^\beta b_{\alpha\beta}\hat{\boldsymbol{a}}_3.$$

Then, according to (8.48a)

$$\tau_{21} = \lambda_2^\alpha \lambda_1^\beta b_{\alpha\beta}. \tag{8.52}$$

Observe that

$$\tau_{12} = \tau_{21};$$

that is, the torsion in the orthogonal directions of $\hat{\boldsymbol{\lambda}}_1$ and $\hat{\boldsymbol{\lambda}}_2$ is the same.

In accordance with (8.18c),

$$\hat{\boldsymbol{\lambda}}_2 = \hat{\boldsymbol{a}}_3 \times \hat{\boldsymbol{\lambda}}_1 = \lambda_1^\alpha\, \overline{e}_{\alpha\beta}\boldsymbol{a}^\beta,$$

$$\hat{\boldsymbol{\lambda}}_1 = \hat{\boldsymbol{\lambda}}_2 \times \hat{\boldsymbol{a}}_3 = \lambda_2^\alpha\, \overline{e}_{\beta\alpha}\boldsymbol{a}^\beta,$$

or, according to (8.49a, b) and (8.50),

$$\lambda_2^\gamma = \lambda_1^\alpha \bar{e}_{\alpha\beta} a^{\beta\gamma}, \tag{8.53a}$$

$$\lambda_1^\gamma = \lambda_2^\alpha \bar{e}_{\beta\alpha} a^{\beta\gamma}. \tag{8.53b}$$

In view of (8.53a, b), equation (8.52) has the alternative forms

$$\tau_{12} = \lambda_1^\alpha \lambda_2^\beta b_{\alpha\beta}, \tag{8.54a}$$

$$= \lambda_1^\alpha \lambda_1^\mu a^{\beta\eta} \bar{e}_{\mu\eta} b_{\alpha\beta}, \tag{8.54b}$$

$$= \lambda_2^\beta \lambda_2^\mu a^{\alpha\eta} \bar{e}_{\eta\mu} b_{\alpha\beta}. \tag{8.54c}$$

The normal curvature κ in the direction of $\hat{\boldsymbol{\lambda}} = \lambda^\alpha \boldsymbol{a}_\alpha$ is given by a form like (8.51a), namely,

$$\kappa = \lambda^\alpha \lambda^\beta b_{\alpha\beta}. \tag{8.55}$$

Now, let us seek the directions in which the normal curvature κ has extremal values. The conditions for an extremal value are

$$\frac{\partial \kappa}{\partial \lambda^\alpha} = 0.$$

Additionally, λ^α must satisfy the auxiliary (normality) condition,

$$\lambda^\alpha \lambda^\beta a_{\alpha\beta} = 1. \tag{8.56}$$

The directions are given by the solution of the linear equations:

$$(b_{\alpha\beta} - \kappa a_{\alpha\beta})\lambda^\alpha = 0. \tag{8.57}$$

However, equations (8.57) have a nontrivial solution if, and only if, the determinant of the coefficients vanishes; that is,

$$|b_{\alpha\beta} - \kappa a_{\alpha\beta}| = 0, \tag{8.58a}$$

or, in expanded form,

$$\frac{a}{2} \bar{e}^{\alpha\mu} \bar{e}^{\beta\eta} (b_{\alpha\beta} - \kappa a_{\alpha\beta})(b_{\mu\eta} - \kappa a_{\mu\eta}) = 0, \tag{8.58b}$$

or

$$\kappa^2 - b^\alpha_\alpha \kappa + \tfrac{1}{2}\bar{e}_{\alpha\gamma}\bar{e}^{\beta\eta} b^\alpha_\beta b^\gamma_\eta = 0. \tag{8.58c}$$

The coefficients in the quadratic equation (8.58c) play an important role in the differential geometry of surfaces. One is known as the *Gaussian curvature*:

$$\widetilde{k} \equiv |b^\alpha_\beta| = \tfrac{1}{2}\,\bar{e}_{\alpha\gamma}\bar{e}^{\beta\eta} b^\alpha_\beta b^\gamma_\eta = b^1_1 b^2_2 - b^1_2 b^2_1. \tag{8.59}$$

The *mean curvature* is defined as follows:

$$\widetilde{h} \equiv \tfrac{1}{2} b^\alpha_\alpha. \tag{8.60}$$

With the notations of (8.59) and (8.60), equation (8.58c) has the form:

$$\kappa^2 - 2\widetilde{h}\kappa + \widetilde{k} = 0. \tag{8.61}$$

Let $\tilde{\kappa}_{11}$ and $\tilde{\kappa}_{22}$ be the two real roots of (8.61) and $\tilde{\lambda}^\alpha_1$ and $\tilde{\lambda}^\beta_2$ the corresponding solutions of (8.57). Then, multiplying (8.57) by $\tilde{\lambda}^\beta_1$, likewise by $\tilde{\lambda}^\beta_2$, and recalling equation (8.56), we obtain

$$\tilde{\kappa}_{11} = \tilde{\kappa}_{11} a_{\alpha\beta}\tilde{\lambda}^\alpha_1\tilde{\lambda}^\beta_1 = b_{\alpha\beta}\tilde{\lambda}^\alpha_1\tilde{\lambda}^\beta_1,$$

$$\tilde{\kappa}_{22} = \tilde{\kappa}_{22} a_{\alpha\beta}\tilde{\lambda}^\alpha_2\tilde{\lambda}^\beta_2 = b_{\alpha\beta}\tilde{\lambda}^\alpha_2\tilde{\lambda}^\beta_2.$$

The two real roots $\tilde{\kappa}_{11}$ and $\tilde{\kappa}_{22}$ are the extremal values of the normal curvature; they are called the *principal curvatures*. The corresponding directions are called the *principal directions*.

Also, from (8.57), we have

$$b_{\alpha\beta}\tilde{\lambda}^\alpha_1\tilde{\lambda}^\beta_2 - \tilde{\kappa}_{11} a_{\alpha\beta}\tilde{\lambda}^\alpha_1\tilde{\lambda}^\beta_2 = 0,$$

$$b_{\alpha\beta}\tilde{\lambda}^\alpha_2\tilde{\lambda}^\beta_1 - \tilde{\kappa}_{22} a_{\alpha\beta}\tilde{\lambda}^\alpha_2\tilde{\lambda}^\beta_1 = 0.$$

However, $b_{\alpha\beta}$ and $a_{\alpha\beta}$ are symmetric; consequently, the difference of the last two equations is

$$(\tilde{\kappa}_{11} - \tilde{\kappa}_{22}) a_{\alpha\beta}\tilde{\lambda}^\alpha_1\tilde{\lambda}^\beta_2 = 0.$$

If the principal curvatures are distinct, i.e, $\tilde{\kappa}_{11} \neq \tilde{\kappa}_{22}$, then $a_{\alpha\beta}\tilde{\lambda}^\alpha_1\tilde{\lambda}^\beta_2 = 0$; that is, the principal directions are orthogonal.

Since $\tilde{\kappa}_{11}$ and $\tilde{\kappa}_{22}$ are the roots of (8.61),

$$(\kappa - \tilde{\kappa}_{11})(\kappa - \tilde{\kappa}_{22}) = \kappa^2 - 2\widetilde{h}\kappa + \widetilde{k} = 0.$$

It follows that

$$\widetilde{k} = \tilde{\kappa}_{11}\tilde{\kappa}_{22}, \tag{8.62}$$

$$\widetilde{h} = \tfrac{1}{2}(\tilde{\kappa}_{11} + \tilde{\kappa}_{22}). \tag{8.63}$$

Consider the directions in which the *torsion* vanishes: according to equation (8.54a), the condition of vanishing torsion is

$$b_{\alpha\beta}\lambda_1^\alpha\lambda_2^\beta = 0,$$

with the supplementary conditions

$$a_{\alpha\beta}\lambda_1^\alpha\lambda_1^\beta = a_{\alpha\beta}\lambda_2^\alpha\lambda_2^\beta = 1,$$

$$a_{\alpha\beta}\lambda_1^\alpha\lambda_2^\beta = 0.$$

In view of (8.57), and the orthogonality of the principal directions, the principal directions fulfill the foregoing conditions when $\tilde{\kappa}_{11} \neq \tilde{\kappa}_{22}$. If $\tilde{\kappa}_{11} = \tilde{\kappa}_{22}$ at a point of a surface, then the normal curvature is the same in all directions and the torsion vanishes in all directions; such a point is known as an *umbilic* of the surface. At an umbilic $b_{\alpha\beta} = \kappa a_{\alpha\beta}$, where κ is the normal curvature in all directions.

If c_1 and c_2 are (orthogonal) coordinate lines and if the coordinates are arc lengths ℓ_1 and ℓ_2, then $\lambda_1^1 = \lambda_2^2 = 1$, $\lambda_2^1 = \lambda_1^2 = 0$, and, according to (8.51a, b) and (8.54a),

$$\kappa_{11} \equiv \frac{1}{r_1} = b_{11} = b_1^1, \qquad \kappa_{22} \equiv \frac{1}{r_2} = b_{22} = b_2^2,$$

$$\tau_{21} = \tau_{12} = b_{12} = b_2^1 = b_1^2.$$

Here, r_1 and r_2 are the principal radii of curvature; these are the radii of the arcs traced by the intersections of the surface with the planes of $(\hat{\mathbf{a}}_3, \hat{\boldsymbol{\lambda}}_1)$ and $(\hat{\mathbf{a}}_3, \hat{\boldsymbol{\lambda}}_2)$.

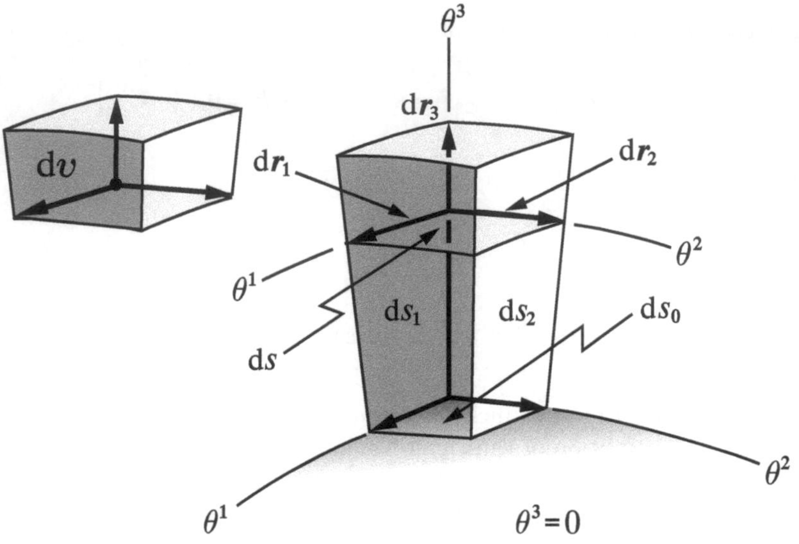

Figure 8.4 Volume and area differentials

8.8 Volume and Area Differentials

An elemental area ds of an arbitrary θ^3 surface is delineated by θ^α lines as shown in Figure 8.4. The corresponding θ^α surfaces contain faces of the elemental volume dv. The edges of the elemental area ds_0 of the reference surface are contained in the same θ^α surfaces.

An edge of the elemental volume approaches the vector

$$d\boldsymbol{r}_i = \boldsymbol{r}_{,i}\, d\theta^{\underline{i}} = \boldsymbol{g}_i\, d\theta^{\underline{i}}. \tag{8.64a, b}$$

The volume dv is expressed as follows:

$$dv = d\boldsymbol{r}_3 \cdot (d\boldsymbol{r}_1 \times d\boldsymbol{r}_2) = \boldsymbol{g}_3 \cdot (\boldsymbol{g}_1 \times \boldsymbol{g}_2)\, d\theta^1\, d\theta^2\, d\theta^3. \tag{8.65}$$

The area ds is given by the expression:

$$\begin{aligned} ds &= \hat{\boldsymbol{a}}_3 \cdot (d\boldsymbol{r}_1 \times d\boldsymbol{r}_2) \\ &= \hat{\boldsymbol{a}}_3 \cdot (\boldsymbol{g}_1 \times \boldsymbol{g}_2)\, d\theta^1\, d\theta^2. \end{aligned} \tag{8.66}$$

The corresponding area ds_0 on the reference surface has the form:

$$ds_0 = \hat{\boldsymbol{a}}_3 \cdot (\boldsymbol{a}_1 \times \boldsymbol{a}_2)\, d\theta^1\, d\theta^2. \tag{8.67}$$

In accordance with (2.23a) and (8.3b), we recall that

$$\boldsymbol{g}_3 \cdot (\boldsymbol{g}_1 \times \boldsymbol{g}_2) = \hat{\boldsymbol{a}}_3 \cdot (\boldsymbol{g}_1 \times \boldsymbol{g}_2) = \sqrt{g}\,. \tag{8.68a, b}$$

Likewise, according to (8.17a) and (8.18a),

$$\hat{\boldsymbol{a}}_3 \cdot (\boldsymbol{a}_1 \times \boldsymbol{a}_2) = \sqrt{a}\,. \tag{8.69}$$

In view of (8.68a, b) and (8.69), equations (8.65) to (8.67) take the forms:

$$dv = \sqrt{g}\, d\theta^1\, d\theta^2\, d\theta^3, \tag{8.70}$$

$$ds = \sqrt{g}\, d\theta^1\, d\theta^2, \tag{8.71}$$

$$ds_0 = \sqrt{a}\, d\theta^1\, d\theta^2. \tag{8.72}$$

In accordance with (8.17c, d), equation (8.68b) has the alternative form

$$\sqrt{g} = \sqrt{a}\, \tfrac{1}{2}\, \bar{e}^{\alpha\beta} \hat{\boldsymbol{a}}_3 \cdot (\boldsymbol{g}_\alpha \times \boldsymbol{g}_\beta). \tag{8.73a}$$

Substituting (8.36a) into (8.73a) and employing (8.18a), we obtain

$$\sqrt{g} = \sqrt{a}\, \tfrac{1}{2}\, \bar{e}^{\alpha\beta}\, \bar{e}_{\mu\eta} (\delta^\mu_\alpha - \theta^3 b^\mu_\alpha)(\delta^\eta_\beta - \theta^3 b^\eta_\beta). \tag{8.73b}$$

With the expressions (8.59) and (8.60) for the Gaussian and mean curvatures, $\widetilde{k}$ and $\widetilde{h}$, equation (8.73b) is reduced to the form:

$$\sqrt{g} = \sqrt{a}\, [1 - 2\widetilde{h}\theta^3 + \widetilde{k}(\theta^3)^2\,]. \tag{8.73c}$$

It follows from (8.70) to (8.72) and (8.73b) that

$$\frac{ds}{ds_0} = \frac{dv}{ds_0\, d\theta^3} = \sqrt{\frac{g}{a}} = 1 - 2\widetilde{h}\theta^3 + \widetilde{k}(\theta^3)^2. \tag{8.74a–c}$$

The area ds_1 of the θ^1 face of the elemental volume is obtained from the expression:

$$ds_1 = \hat{\boldsymbol{e}}^1 \cdot (d\boldsymbol{r}_2 \times d\boldsymbol{r}_3),$$

where $\hat{\boldsymbol{e}}^1$ denotes the unit normal to ds_1. Since $\hat{\boldsymbol{e}}^1 = \boldsymbol{g}^1/\sqrt{g^{11}}$, we have

$$ds_1 = \frac{g^{1j}}{\sqrt{g^{11}}}\boldsymbol{g}_j \cdot (\boldsymbol{g}_2 \times \boldsymbol{g}_3)\, d\theta^2\, d\theta^3$$

$$= \sqrt{g^{11}\, g}\, d\theta^2\, d\theta^3. \tag{8.75a}$$

Likewise,

$$ds_2 = \sqrt{g^{22}\, g}\, d\theta^1\, d\theta^3. \tag{8.75b}$$

8.9 Vectors, Derivatives, and Covariant Derivatives

A vector $\boldsymbol{V}$ can be expressed as a linear combination of the base vectors $\boldsymbol{a}_i$ or $\boldsymbol{a}^i$:

$$\boldsymbol{V} = V^\alpha \boldsymbol{a}_\alpha + V^3 \hat{\boldsymbol{a}}_3$$

$$= V_\alpha \boldsymbol{a}^\alpha + V^3 \hat{\boldsymbol{a}}_3.$$

It follows from (8.10a, b) that the components V_α and V^α are associated as follows:

$$V^\alpha = a^{\alpha\beta} V_\beta, \qquad V_\alpha = a_{\alpha\beta} V^\beta.$$

With the aid of (8.33a, b) to (8.35a, b) we can express the partial derivatives $\boldsymbol{V}_{,\alpha}$ in the alternative forms:

$$\boldsymbol{V}_{,\alpha} = (V^\mu{}_{,\alpha} + V^\beta \overline{\Gamma}^\mu_{\beta\alpha} - V^3 b^\mu_\alpha)\boldsymbol{a}_\mu + (V^3{}_{,\alpha} + V^\beta b_{\alpha\beta})\hat{\boldsymbol{a}}_3, \tag{8.76a}$$

$$\boldsymbol{V}_{,\alpha} = (V_{\mu,\alpha} - V_\beta \overline{\Gamma}^\beta_{\mu\alpha} - V^3 b_{\mu\alpha})\boldsymbol{a}^\mu + (V^3{}_{,\alpha} + V_\beta b^\beta_\alpha)\hat{\boldsymbol{a}}_3. \tag{8.76b}$$

The covariant derivatives of the components V^α and V_α are defined in the manner of (2.50a, b) and (2.51a, b); namely,

$$V^\alpha\|_\beta \equiv V^\alpha{}_{,\beta} + V^\mu \overline{\Gamma}^\alpha_{\mu\beta}, \tag{8.77a}$$

$$V_\alpha\|_\beta \equiv V_{\alpha,\beta} - V_\mu \overline{\Gamma}^\mu_{\alpha\beta}. \tag{8.77b}$$

The double bar signifies covariant differentiation with respect to the surface rather than the three-dimensional space; note the difference:

$$V^{\alpha}|_{\beta} \equiv V^{\alpha}{}_{,\beta} + V^{i}\,\Gamma^{\alpha}_{i\beta}.$$

In the latter, the repeated Latin index implies a sum of three terms; moreover, the components need not be evaluated at a particular surface. With the notations of (8.77a, b), equations (8.76a, b) assume the forms

$$\boldsymbol{V}_{,\alpha} = (V^{\mu}\|_{\alpha} - V^{3}b^{\mu}_{\alpha})\boldsymbol{a}_{\mu} + (V^{3}{}_{,\alpha} + V^{\beta}b_{\beta\alpha})\hat{\boldsymbol{a}}_{3}, \tag{8.78a}$$

$$\boldsymbol{V}_{,\alpha} = (V_{\mu}\|_{\alpha} - V^{3}b_{\mu\alpha})\boldsymbol{a}^{\mu} + (V^{3}{}_{,\alpha} + V_{\beta}b^{\beta}_{\alpha})\hat{\boldsymbol{a}}_{3}. \tag{8.78b}$$

From equation (2.53), we infer that

$$\frac{\partial\sqrt{a}}{\partial\theta^{\alpha}} = \sqrt{a}\,\overline{\Gamma}^{\lambda}_{\lambda\alpha}. \tag{8.79a}$$

Finally, by means of the result (8.79a), we obtain

$$\frac{\partial}{\partial\theta^{\alpha}}\left(\sqrt{a}\,V^{\alpha}\right) = \sqrt{a}\,V^{\alpha}\|_{\alpha}. \tag{8.79b}$$

8.10 Surface Tensors

A quantity such as

$$F^{\alpha\beta\cdots}(\theta^{1},\theta^{2}), \qquad P_{\alpha\beta\cdots}(\theta^{1},\theta^{2}), \quad \text{or} \quad T^{\alpha\cdots}_{\beta\cdots}(\theta^{1},\theta^{2}), \tag{8.80a–c}$$

is the component of a contravariant, covariant, or mixed surface tensor, respectively, if the component

$$\overline{F}^{\alpha\beta\cdots}(\bar{\theta}^{1},\bar{\theta}^{2}), \qquad \overline{P}_{\alpha\beta\cdots}(\bar{\theta}^{1},\bar{\theta}^{2}), \quad \text{or} \quad \overline{T}^{\alpha\cdots}_{\beta\cdots}(\bar{\theta}^{1},\bar{\theta}^{2}), \tag{8.81a–c}$$

in another coordinate system $\bar{\theta}^{\alpha}$ is obtained from (8.80a–c) by a linear transformation of the types (2.55), (2.57), and (2.58), with Greek indices in place of Latin indices.

A surface invariant has no free indices; for example,

$$A = T^{\alpha\beta} S_{\alpha\beta}.$$

From the definitions, it follows that the base vectors $\boldsymbol{a}_\alpha$ and $\boldsymbol{a}^\alpha$ transform as components of covariant and contravariant tensors and, therefore, from (8.2a, b) it follows that $a_{\alpha\beta}$ and $a^{\alpha\beta}$ are components of surface tensors, covariant and contravariant, respectively. Likewise, from (8.27a) and (8.28a), it follows that $b_{\alpha\beta}$ and b^α_β are components of covariant and mixed surface tensors. Furthermore, $\overline{e}_{\alpha\beta}$ and $\overline{e}^{\alpha\beta}$ are components of associated surface tensors, covariant and contravariant,

$$\overline{e}^{\alpha\beta} = a^{\alpha\gamma} a^{\beta\eta} \, \overline{e}_{\gamma\eta}, \qquad \overline{e}_{\alpha\beta} = a_{\alpha\gamma} a_{\beta\eta} \, \overline{e}^{\gamma\eta}.$$

With the aid of (8.17c, d) and (8.79a), we obtain for the covariant derivative:

$$\overline{e}^{\alpha\beta}\|_\gamma = 0.$$

Two surface tensors are said to be associated tensors if each component of one is a *particular* linear combination of the components of the other. The combination is formed by multiplying components of the tensor with components of the metric tensor and summing. For example, $T_\alpha^{\bullet\gamma\beta}$ and $T_{\alpha\mu\eta}$ are components of associated tensors related as follows:

$$T_\alpha^{\bullet\gamma\beta} = a^{\gamma\mu} a^{\beta\eta} T_{\alpha\mu\eta} \quad \Longleftrightarrow \quad T_{\alpha\mu\eta} = a_{\mu\gamma} a_{\eta\beta} T_\alpha^{\bullet\gamma\beta}.$$

Observe that these components of the associated tensors are referred to the same surface coordinates. Also notice that a vacant position is marked by a dot. If the tensor is completely symmetric; that is

$$T_{\alpha\beta\gamma} = T_{\beta\alpha\gamma} = T_{\alpha\gamma\beta} = T_{\gamma\beta\alpha},$$

then the positions need not be marked. The quantities $b_{\alpha\beta}$ and b^α_β of (8.31a, b) are components of associated symmetric tensors, covariant and mixed. Forming an associated tensor is often termed *raising*, or *lowering*, indices.

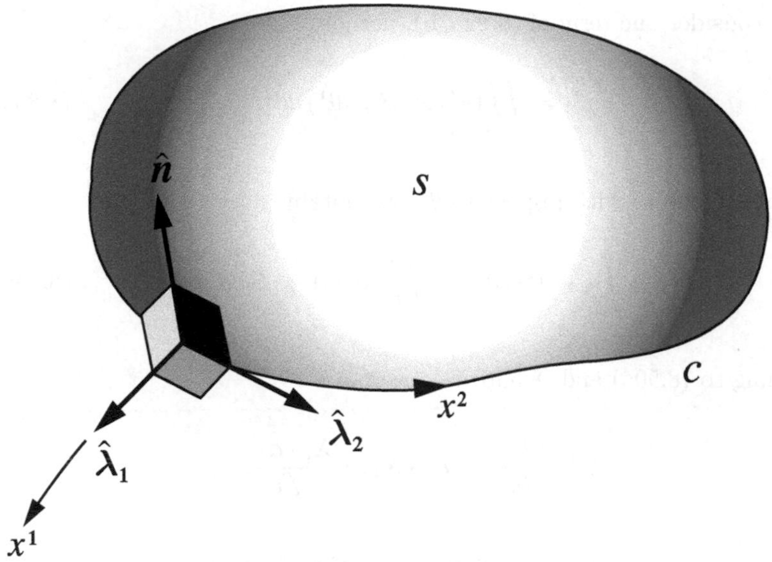

Figure 8.5 Green's theorem: surface s bounded by a curve c

8.11 Green's Theorem (Partial Integration) for a Surface

The analysis of shells by energy principles often involves integrals of the form

$$I \equiv \iint_s A^\alpha B_{,\alpha}\, ds = \iint_s A^\alpha B_{,\alpha} \sqrt{a}\, d\theta^1\, d\theta^2, \qquad (8.82\text{a, b})$$

wherein integration extends over a surface s (e.g., the midsurface or a reference surface of the shell) bounded by a curve c; A^α and B are functions of the surface coordinates. We assume that the surface s and curve c are smooth and that A^α, B, and derivatives are continuous. If, in subsequent developments, the surface in question may be the deformed surface of a shell, then the minuscules (s), (c), and (a) may be replaced by (S), (C), and (A), respectively.

Let $\hat{\boldsymbol{\lambda}}_2$ denote the unit tangent to c, and $\hat{\boldsymbol{\lambda}}_1$ the normal to c tangent to s. Let x^2 denote arc length along c and x^1 the arc length along a curve normal to c, as shown in Figure 8.5.

Now consider one term of (8.82a, b), namely,

$$I_1 = \iint_s (\sqrt{a}\, A^1 B_{,1}\, d\theta^1)\, d\theta^2. \tag{8.83a}$$

Integrating (8.83a) with respect to θ^1, we obtain

$$I_1 \equiv \int_c \sqrt{a}\, A^1 B\, d\theta^2 - \iint_s (\sqrt{a}\, A^1)_{,1} B\, d\theta^1\, d\theta^2. \tag{8.83b}$$

According to (8.50a) and (8.53b),

$$\lambda_2^2 \equiv \frac{\partial \theta^2}{\partial x^2} = \boldsymbol{a}^2 \cdot \hat{\boldsymbol{\lambda}}_2 = \frac{\hat{\boldsymbol{\lambda}}_1 \cdot \boldsymbol{a}_1}{\sqrt{a}}.$$

On the curve c, we have $\theta^\alpha = \theta^\alpha(x^2)$, and, therefore,

$$d\theta^2 = \frac{\hat{\boldsymbol{\lambda}}_1 \cdot \boldsymbol{a}_1}{\sqrt{a}}\, dx^2.$$

Consequently, (8.83b) takes the form:

$$I_1 \equiv \int_c A^1 B(\hat{\boldsymbol{\lambda}}_1 \cdot \boldsymbol{a}_1)\, dx^2 - \iint_s (\sqrt{a}\, A^1)_{,1} B\, d\theta^1\, d\theta^2. \tag{8.83c}$$

Likewise,

$$\begin{aligned} I_2 &\equiv \iint_s (\sqrt{a}\, A^2 B_{,2}\, d\theta^2)\, d\theta^1 \\ &= \int_c A^2 B(\hat{\boldsymbol{\lambda}}_1 \cdot \boldsymbol{a}_2)\, dx^2 - \iint_s (\sqrt{a}\, A^2)_{,2} B\, d\theta^1\, d\theta^2. \end{aligned} \tag{8.84}$$

It follows that

$$I = \int_c A^\alpha B(\hat{\boldsymbol{\lambda}}_1 \cdot \boldsymbol{a}_\alpha)\, dx^2 - \iint_s \frac{1}{\sqrt{a}} (\sqrt{a}\, A^\alpha)_{,\alpha} B\, ds. \tag{8.85}$$

In physical problems I is an invariant, A^α transforms as the component of a contravariant surface tensor, and B is invariant.

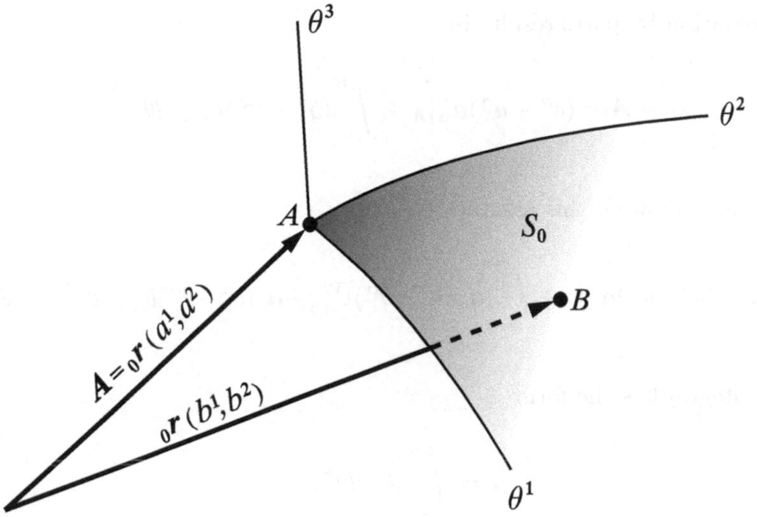

Figure 8.6 Position of particles A and B on the surface s_0

8.12 Equations of Gauss and Codazzi

Because a shell is thin, the position of any particle is often identified with a neighboring point on a reference surface. Indeed, in most theories[‡] an approximation serves to express the three-dimensional field $\boldsymbol{r}(\theta^1, \theta^2, \theta^3)$ explicitly in terms of the two-dimensional field ${}_0\boldsymbol{r}(\theta^1, \theta^2)$, the position vector of particles on the reference surface. We are concerned about the existence of the vector ${}_0\boldsymbol{r}(\theta^1, \theta^2)$.

Since a rigid-body motion is irrelevant to a discussion of deformation, let us assume that the position vector $\boldsymbol{A}$ of one particle at $A(a^1, a^2)$ of a surface s_0 is known. Now, let us set out to compute the position vector ${}_0\boldsymbol{r}(b^1, b^2)$ of an arbitrary particle at $B(b^1, b^2)$:

The relative position of particles A and B in Figure 8.6 is

$$
\begin{aligned}
{}_0\boldsymbol{r} - \boldsymbol{A} &= \int_A^B d\boldsymbol{r} = \int_A^B \boldsymbol{r}_{,\alpha}\, d\theta^\alpha \\
&= \int_A^B \boldsymbol{a}_\alpha\, d\theta^\alpha = \int_A^B \boldsymbol{a}_\alpha\, d(\theta^\alpha - b^\alpha).
\end{aligned}
$$

[‡]A prominent example is the Kirchhoff-Love theory (see Chapter 10).

An integration by parts results in

$$ {}_0\boldsymbol{r} - \boldsymbol{A} = (b^\alpha - a^\alpha)\boldsymbol{a}_\alpha]_A + \int_A^B (b^\alpha - \theta^\alpha)\boldsymbol{a}_{\alpha,\beta}\, d\theta^\beta . $$

By means of (8.33a), this assumes the form:

$$ {}_0\boldsymbol{r} - \boldsymbol{A} = (b^\alpha - a^\alpha)\boldsymbol{a}_\alpha]_A + \int_A^B [\boldsymbol{a}_\gamma (b^\alpha - \theta^\alpha)\overline{\Gamma}^\gamma_{\alpha\beta} + \hat{\boldsymbol{a}}_3 (b^\alpha - \theta^\alpha) b_{\alpha\beta}]\, d\theta^\beta . \quad (8.86) $$

The integral has the form

$$ I = \int_A^B \mathcal{F}_\beta \, d\theta^\beta . $$

The integral exists in a simply connected region (no holes in the surface) independently of the path of integration if, and only if,‡

$$ \frac{\partial \mathcal{F}_\beta}{\partial \theta^\mu} = \frac{\partial \mathcal{F}_\mu}{\partial \theta^\beta} . \quad (8.87) $$

In view of (8.33a), (8.34), and the symmetry of $\overline{\Gamma}^\gamma_{\alpha\beta}$ and $b_{\alpha\beta}$, condition (8.87) takes the form

$$ (b^\alpha - \theta^\alpha)\Big[\big(\overline{\Gamma}^\eta_{\alpha\beta,\mu} - \overline{\Gamma}^\eta_{\alpha\mu,\beta} + \overline{\Gamma}^\eta_{\gamma\mu}\overline{\Gamma}^\gamma_{\alpha\beta} - \overline{\Gamma}^\eta_{\gamma\beta}\overline{\Gamma}^\gamma_{\alpha\mu} - b^\eta_\mu b_{\alpha\beta} + b^\eta_\beta b_{\alpha\mu}\big)\boldsymbol{a}_\eta $$

$$ + \big(b_{\gamma\mu}\overline{\Gamma}^\gamma_{\alpha\beta} - b_{\gamma\beta}\overline{\Gamma}^\gamma_{\alpha\mu} + b_{\alpha\beta,\mu} - b_{\alpha\mu,\beta}\big)\hat{\boldsymbol{a}}_3\Big] = \mathbf{o}. $$

However, the condition must be satisfied at every point of the surface. Consequently, the vector in brackets must vanish; this requires that each component vanish, that is

$$ \overline{R}^\eta_{\boldsymbol{\cdot}\alpha\mu\beta} \equiv \overline{\Gamma}^\eta_{\alpha\beta,\mu} - \overline{\Gamma}^\eta_{\alpha\mu,\beta} + \overline{\Gamma}^\eta_{\gamma\mu}\overline{\Gamma}^\gamma_{\alpha\beta} - \overline{\Gamma}^\eta_{\gamma\beta}\overline{\Gamma}^\gamma_{\alpha\mu} $$

$$ = b^\eta_\mu b_{\alpha\beta} - b^\eta_\beta b_{\alpha\mu}, \quad (8.88) $$

‡If the surface contain holes, then an additional condition is needed to ensure the continuity about each hole.

and

$$b_{\alpha\beta,\mu} - b_{\gamma\beta}\overline{\Gamma}^{\gamma}_{\alpha\mu} = b_{\alpha\mu,\beta} - b_{\gamma\mu}\overline{\Gamma}^{\gamma}_{\alpha\beta}. \tag{8.89}$$

Equation (8.88) is equivalent to

$$a_{\lambda\eta}\overline{R}^{\eta}_{\boldsymbol{\cdot}\alpha\mu\beta} \equiv \overline{R}_{\lambda\alpha\mu\beta} = b_{\lambda\mu}b_{\alpha\beta} - b_{\lambda\beta}b_{\alpha\mu}. \tag{8.90a}$$

Moreover, in view of the symmetry of $\overline{\Gamma}^{\gamma}_{\alpha\beta}$ and $b_{\alpha\beta}$, equation (8.90a) is an identity except in the case:

$$\overline{R}_{1212} = \overline{R}_{2121} = -\overline{R}_{2112} = -\overline{R}_{1221}. \tag{8.90b}$$

In other words, equation (8.90a) represents one condition for the existence of ${}_0\boldsymbol{r} - \boldsymbol{A}$; that condition can be restated as follows:

$$\tfrac{1}{4}\,\overline{e}^{\lambda\alpha}\,\overline{e}^{\mu\beta}\,\overline{R}_{\lambda\alpha\mu\beta} = \tfrac{1}{4}\,\overline{e}^{\lambda\alpha}\,\overline{e}^{\mu\beta}(b_{\lambda\mu}b_{\alpha\beta} - b_{\lambda\beta}b_{\alpha\mu}).$$

According to (8.59), the right side is the Gaussian curvature. Therefore, this condition takes the form

$$\tfrac{1}{4}\,\overline{e}^{\lambda\alpha}\,\overline{e}^{\mu\beta}\,\overline{R}_{\lambda\alpha\mu\beta} = \widetilde{k}. \tag{8.91}$$

Equation (8.91) is the *Gauss equation* of the surface.

If we introduce the covariant derivative of the tensor $b_{\alpha\beta}$, namely,

$$b_{\alpha\beta}\|_{\mu} = b_{\alpha\beta,\mu} - b_{\gamma\beta}\overline{\Gamma}^{\gamma}_{\alpha\mu} - b_{\alpha\gamma}\overline{\Gamma}^{\gamma}_{\beta\mu},$$

then (8.89) takes the form

$$b_{\alpha\beta}\|_{\mu} = b_{\alpha\mu}\|_{\beta}. \tag{8.92}$$

Equation (8.92) represents either of the two equations, namely,

$$b_{11}\|_{2} = b_{12}\|_{1}, \qquad b_{21}\|_{2} = b_{22}\|_{1}. \tag{8.93a, b}$$

These are the *Codazzi equations* of the surface.

The Gauss and Codazzi equations (8.91) and (8.93a, b) are the necessary and sufficient conditions for the existence of the position vector ${}_0\boldsymbol{r} - \boldsymbol{A}$. They play an important role in the subsequent discussion of compatibility in the deformation of shells.

Using the expressions for the covariant derivative [equation (8.77a, b)] and the previously derived equations (8.88) and (8.91), we obtain for the second covariant derivatives of the components V_α (or V^α) of any vector $\boldsymbol{V}(\theta^1, \theta^2)$:

$$V_\alpha\|_{\beta\gamma} - V_\alpha\|_{\gamma\beta} = \overline{R}_{\mu\alpha\beta\gamma} V^\mu, \tag{8.94}$$

$$= (b_{\mu\beta} b_{\alpha\gamma} - b_{\mu\gamma} b_{\alpha\beta}) V^\mu, \tag{8.95}$$

$$V_1\|_{12} - V_1\|_{21} = -\widetilde{k}\, V^2, \tag{8.96}$$

$$V_2\|_{12} - V_2\|_{21} = \widetilde{k}\, V^1. \tag{8.97}$$

It follows from equations (8.96) and (8.97) that the order of covariant differentiation is immaterial, i.e.,

$$V_\alpha\|_{\beta\gamma} \doteq V_\alpha\|_{\gamma\beta},$$

if $\widetilde{k} \ll 1$, i.e, as one approaches a plane surface.

Chapter 9

Theory of Shells

9.1 Introduction

Shells are both practical and aesthetically appealing in various mechanical and structural applications. Consequently, much has been written about their attributes and the theories which are intended to describe and predict their mechanical behavior. Quite often the theoretical writings are entirely shrouded in mathematical language, sometimes devoid of mechanical bases or interpretations. Our intent is to employ the essential mathematics for a concise presentation, but also to offer some insights into the mechanics which serve to explain and to predict the unique behavior of shells. Towards these goals, we begin by citing some simple structural elements and their distinctive behaviors.

At the risk of appearing simplistic, we mention first the case of a thin, straight, and homogeneous rod. When supported at both ends, transverse loads cause perceptible bending and deflection. The same rod can sustain much greater *axial* load with imperceptible deformation. Similar observations apply to a thin flat plate. Transverse loadings are accompanied by bending and resisted by bending couples upon a cross-section. In-plane loadings cause only stretching, compressing and/or shearing, resisted by in-plane forces on a cross-section; these are termed *membrane* forces. Simultaneous bending and stretching of rods or plates usually occur only under circumstances of combined loadings. An exception is noteworthy: If a plate is fixed or constrained along its entire perimeter and subjected to transverse loads, then the bending necessarily deforms the plane to a curved surface; such deformation is accompanied by stretching *and* the attendant membrane forces. The latter are small enough to be neglected unless deflections are large—the interactions are included in nonlinear theories. The simultaneous occurrence of bending *and* stretching, resisting couples *and* membrane forces, provides a key to the mechanical behavior of shells.

a

b

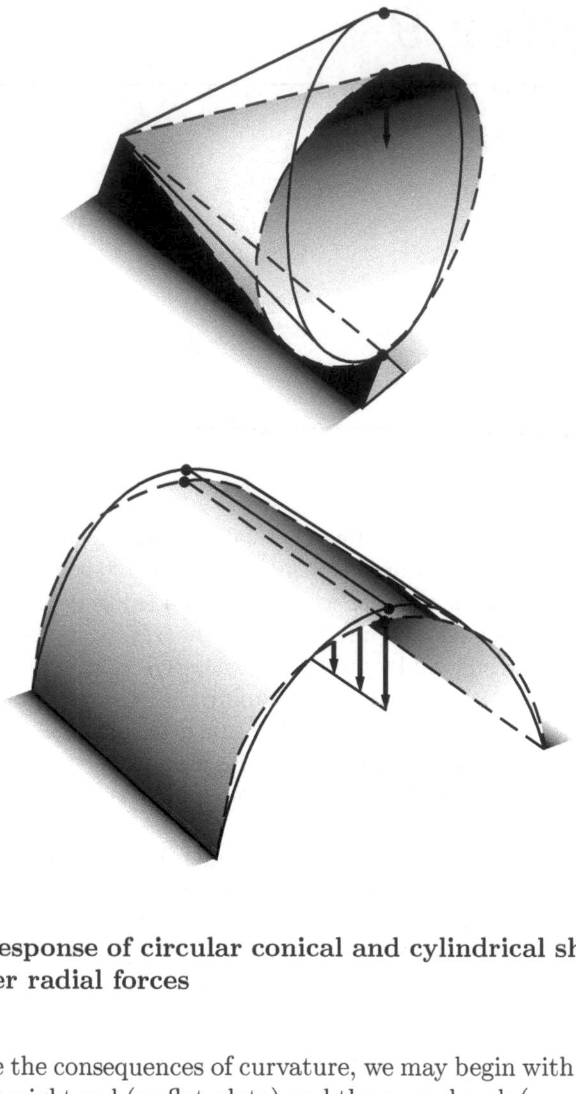

Figure 9.1 Response of circular conical and cylindrical shells under radial forces

To appreciate the consequences of curvature, we may begin with the comparison of the straight rod (or flat plate) and the curved arch (or cylindrical shell). Any transverse load upon the curved arch is accompanied by forces (membrane forces) upon a cross-section. Indeed, it is this feature of the arch (or shell) which provides the stiffness. The most striking example is a thin circular cylindrical shell (a tube) which can sustain extreme normal pressure with imperceptible deformation, little bending and inconsequential bending couples. The compressed tube also exhibits another characteristic, namely, the susceptibility to buckling under external pressure. Typically, the thin tube supports an external pressure by virtue of the membrane action; because it is thin, it offers little resistance to bending. Consequently,

Figure 9.2 Buckled beverage cans

the initiation of bending (near a critical load) precipitates abrupt collapse. Unlike the straight column, which can sustain the critical load but bends excessively, the shell cannot, but collapses. It is a price one pays for the inherent stiffness of such thin shells.

To carry our prelude further, we observe the responses of circular conical and cylindrical shells. If these are without supporting diaphragms at their open ends, then opposing radial forces cause pronounced deformations as depicted in Figure 9.1a, b. Such deformations ensue because these shells can readily deform to noncircular conical and cylindrical forms, respectively, without stretching and without the attendant membrane forces. These are very special surfaces generated by straight lines; they readily deflect to the noncircular forms which possess the same straight generators. Expressed in geometrical terms, the Gaussian curvature vanishes at all points of the conical and cylindrical surfaces. As in the case of a simple rod, both cylinder and cone can support much greater forces applied in axial and symmetrical manner. Such loading is resisted by the membrane action, especially along the straight generators but coupled also with some circumferential actions. In both instances the advent of bending may precipitate an abrupt collapse, or snap-through buckling. The latter may be demonstrated by a simple experiment upon common beverage cans. Those depicted in Figure 9.2 are capable of supporting an axial load of 1112 Newton or 250 pounds. A slight disturbance will cause the buckling; then the

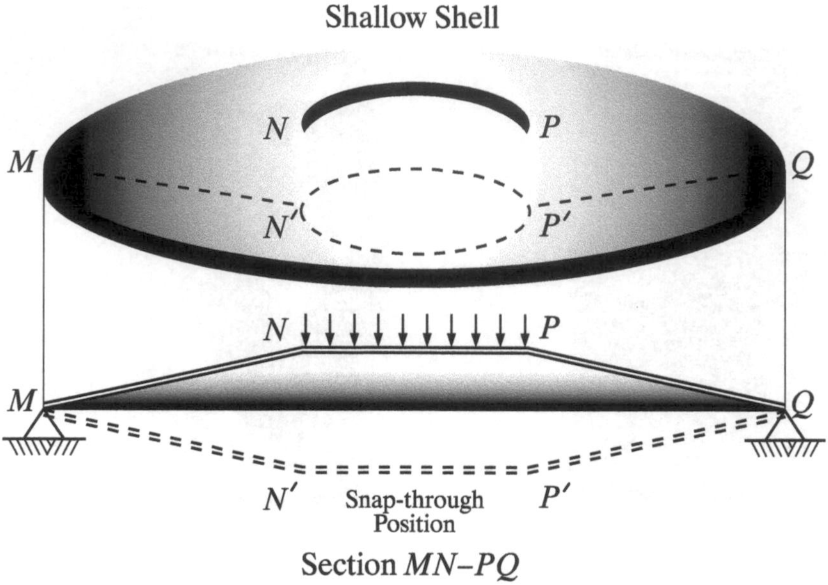

Figure 9.3 Snap-through buckling of a shallow conical shell

resistance to the axial load drops abruptly to 667 Newton (150 pounds). Figure 9.2 shows the postbuckled (deformed) cans; the cited loads are typical results from numerous tests (cf. [106], p. 578, Figure 9.7).

The shallow conical shell of Figure 9.3 provides a simple example of a snap-buckling which is a characteristic phenomenon of thin shells. This shell resists the symmetrical axial loading by virtue of the membrane forces. As the load increases, the shallow shell is compressed to ever flatter form; the stiffness diminishes. A plot of load versus deflection traces a nonlinear path as that shown in the Figure 9.4. We note that a conical surface can assume another conical form *without* stretching: It is an inversion of the initial surface. Indeed, the shallow shell can snap-through to another state which is near that inverted form. Thus, at some load the shell snaps from state A to B. Because the shell possesses some bending resistance, the postbuckled form differs slightly from the conical form. Moreover, the shell at state B is subjected to the same load of state A. One can appreciate the behavior if one supposes that the conical shell has *no* bending resistance, i.e., it is very thin. Then, an inverted state can exist with no externally applied force, no internal forces, and no energy. Our hypothetical shell (no flexural stiffness) must trace the symmetrical path, dotted in Figure 9.4. This hypothetical shell is effectively a membrane; but one with hypothetical constraints which maintain the symmetrical conical form as deflection pursues the path from O to O'. The inverted form at O' possesses the

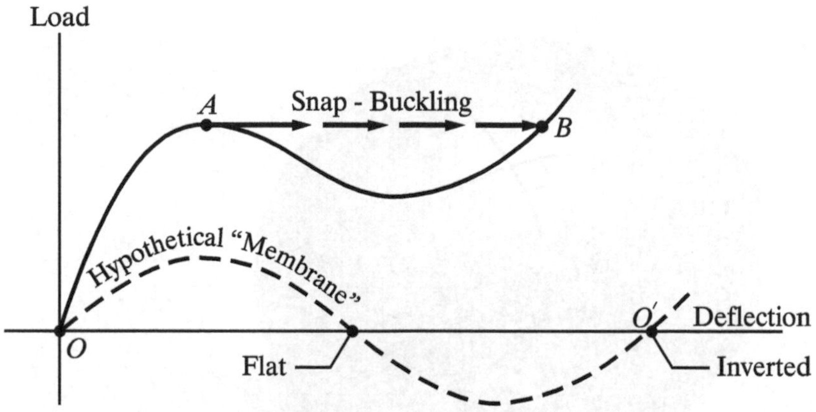

Figure 9.4 Snap-through buckling behavior

identical attributes of the initial state O; the load must be reversed to retrace the path O' to O.

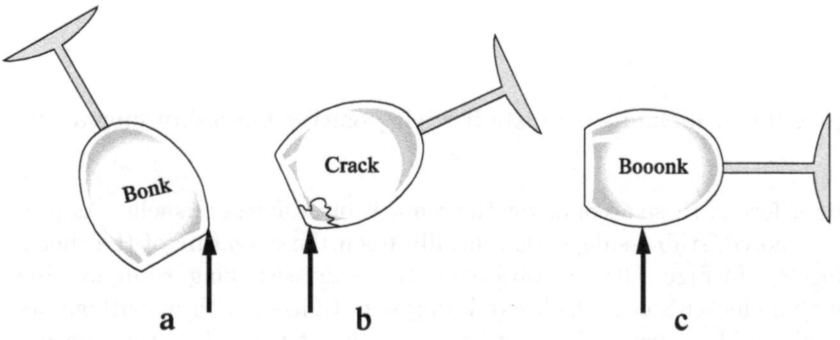

Figure 9.5 Role of membrane forces at the edges of a shell with positive Gaussian curvature

The role of membrane forces is often evident at the edges: This is illustrated in Figure 9.5. The impulsive load of Figure 9.5a is almost tangent to the surface, hence resisted by membrane forces. The load of Figure 9.5b acts transversely, causes bending and consequent fracture. The load of Figure 9.5c acts at an interior site; since the shell has curvature (positive Gaussian curvature), inextensional deformation is not possible and membrane forces resist the impact. The reader can conduct such an experiment; the authors do recommend that the wine be removed prior to aforesaid experimentation.

Finally, we reiterate the dominant role of membrane action and the at-

Figure 9.6 Chicken's egg: shell with positive Gaussian curvature

tendant forces; these account for the remarkable stiffness of shells. As previously noted, stiffness depends crucially upon the curvature of the shell's surface(s). In Figure 9.6, we observe nature's most striking example: The very thin chicken's egg which exhibits great stiffness, with or without its natural liquid contents. Such stiffness can be attributed to the nonzero positive Gaussian curvature at every point:[‡]

$$\frac{1}{r_1} \cdot \frac{1}{r_2} > 0.$$

Consequently, some stretching must accompany a deformation; that, in turn, is resisted by membrane forces.

[‡]Here, r_1 and r_2 denote the "principal" radii of curvature.

9.2 Historical Perspective

Shells, because of their distinct physical attributes (thinness and form), exhibit distinctly different behaviors, as illustrated by the examples of Section 9.1. Not surprisingly, practical analyses of shells call for special mathematical descriptions: These are the so-called theories of shells. Precise origins of specific theories are matters of some conjecture: V. V. Novozhilov [140] attributes the earliest formulations to the works of A. L. Cauchy [141] and S. D. Poisson [42]. The former was "based on an expansion of the displacements and stresses ... in a power series of z" (the distance from a midsurface). Similar approaches have been employed by numerous scholars (cf. P. M. Naghdi [142], [143]). The higher-moment theories of E. R. A. Oliveira ([144], [145]) and G. A. Wempner [146] are examples. Legendre polynomials serve as effective alternatives to the power series (cf. A. I. Soler [147]). A variant is the method of asymptotic expansion which serves to reduce the three-dimensional theory to a progression of two-dimensional systems; each system is identified with a power of a thickness parameter. Such expansions were employed by M. W. Johnson and E. Reissner [148] and most recently by F. I. Niordson ([149], [150]). Any of these methods is intended to progress from the three-dimensional to the two-dimensional theory; as such, they provide bridges between thick and thin shells and, in particular, they offer means to assess the adequacies of the simpler approximations for thin shells. For brevity, let us call these various theories (via expansions) "progressive" theories.

The most prevalent contemporary theories are descendants of that contained in A. E. H. Love's treatise of 1892 [1]. Love's formulation was founded on the prior works of H. Aron [151], G. R. Kirchhoff [152], and his paper of 1888 [153]. The theory of Love is commonly called the "first-approximation"; because it embodies the hypothesis of Kirchhoff (as applied to plates), it is also termed the Kirchhoff-Love theory. As applied to thin Hookean shells, the Kirchhoff-Love theory has been refined by the works of A. I. Lur'e [154], J. L. Synge and W. Z. Chien [155], J. L. Sanders ([156], [157]), W. T. Koiter [158], R. W. Leonard [159], P. M. Naghdi ([142], [143]), and F. I. Niordson [160]. The theory has been extended to accommodate nonlinearities associated with finite rotations (cf. J. L. Sanders [157], W. T. Koiter [161], J. G. Simmonds and D. A. Danielson [162], and W. Pietraszkiewicz [163]). The foundation of the Kirchhoff-Love theory is the geometrical hypothesis: Normals to the reference surface remain *straight* and *normal* to the deformed reference surface. *Note:* The hypothesis precludes any transverse-shear strain, i.e., no change in the right angle between the normal and any line in the surface. This does impose limitations; most notably, it is strictly applicable to *thin* shells. Additionally, it is not descriptive of the behavior near localized loads or junctures.

Between the "progressive" theories and the "first-approximations" lie many higher-order theories. These are mainly theories which elaborate upon the first-approximation. The most relevant refinements account for a transverse shear and some include also an extension of the normal. Here, we must mention the theories of E. Reissner [164] and R. D. Mindlin [165]. The interested reader will find an extensive historical review in the treatise by P. M. Naghdi [143].

Although our intent is a historical background, our readers deserve an entree to the literature and to the monographs of technical importance. Numerous references are given in the reviews by G. A. Wempner ([166], [167]) and in the following monographs:

Applied theories that are directed toward engineering applications are contained in [168] to [180]. Many of these monographs also contain practical methods of analysis and solutions for various shell problems. Linear and nonlinear theories for shells of arbitrary geometry using tensor analysis are also presented in [160], [176], [177], and [180] to [183]. The interested reader will find useful analyses of anisotropic shells in the monographs [184] and [185]. Other works address the vibrations of thin shells ([186] to [188]). In a subsequent section (Section 10.8), we turn to the inelastic deformations of thin shells (see, e.g., [189] to [191]) and cite some relevant works.

Our citations are decidedly limited, but, hopefully, those noted, and their bibliographies, will serve to reveal the vast literature on the subject.

9.3 The Essence of Shell Theory

We define a shell as a thin body, one bounded by two nearby surfaces, a top surface s_+ and a bottom surface s_-, separated by a small distance h which is much less than a radius of curvature r; $h/r \ll 1$. If these surfaces are not closed, then the shell has edges and the distance between such edges is normally large in comparison with the thickness. All of these geometrical features are implicit in the examples and remarks of Section 9.1.

A theory of shells acknowledges the overriding geometrical feature, *thinness*: Customarily, we identify one (or more) surface(s), the top, bottom, or an intermediate surface, as the reference surface s_0. Two coordinates, θ^1 and θ^2, locate particles on that surface; then a third coordinate θ^3 along the normal to the undeformed reference surface locates an arbitrary particle of the shell (see Figure 8.1). Thinness means that distance z along the normal line θ^3 is everywhere small compared to lengths along surface coordinates θ^1 and θ^2. The mechanical behavior of a thin shell is embodied in the underlying approximations with respect to the distributions (or integrals) of displacement, strain, and stress through the thickness—i.e., dependence on coordinate θ^3 is eliminated. Whatever approximations these may be, they render the theory of the shell as one of two dimensions; the ultimate dependent variables are functions only of the surface coordinates θ^α ($\alpha = 1, 2$). From all previous observations, one notes that the relative displacements of particles along the normal are small compared to the displacement at the reference surface. Accordingly, most theories of thin shells are concerned only with deflections of a surface. Here, we limit our attention to the theory of thin shells which is based upon a simple, yet effective, approximation: *The displacement is assumed to vary linearly through the thickness*;

$$\boldsymbol{V} \doteq {}_0\boldsymbol{V}(\theta^1, \theta^2) + \theta^3 \boldsymbol{u}(\theta^1, \theta^2), \tag{9.1a}$$

$$\boldsymbol{u}(\theta^1, \theta^2) \equiv \boldsymbol{A}_3(\theta^1, \theta^2) - \hat{\boldsymbol{a}}_3(\theta^1, \theta^2). \tag{9.1b}$$

Here, $\boldsymbol{A}_3$ denotes the tangent vector to the deformed θ^3 line and $\hat{\boldsymbol{a}}_3$ is the normal to the undeformed midsurface. With this assumption, the displacement throughout the shell is expressed by two vectors, ${}_0\boldsymbol{V}$ and $\boldsymbol{u}$, functions of the surface coordinates (θ^1, θ^2).

9.4 Scope of the Current Treatment

Our treatment of shells [hypothesis: $\boldsymbol{V} \doteq {}_0\boldsymbol{V} + \theta^3(\boldsymbol{A}_3 - \hat{\boldsymbol{a}}_3)$] encompasses the accepted first-approximation, but also accommodates *finite rotations* and *finite strains*. The kinematical and dynamical equations hold for *any* continuous cohesive media. Those fundamental equations are applicable to finite elastic deformations (e.g., rubber) or inelastic deformations (e.g., plastic). Strains at the reference surface ($\theta^3 = z = 0$) include transverse extension and shear; in other words, all six components ϵ_{ij} are included. From a practical viewpoint, such theory provides the means to progress, via finite elements, to a three-dimensional description. From the standpoint of contemporary digital computation the theory provides the basis of a bridge between the classical (first-approximation) and the three-dimensional approximations (see Section 10.13).

Throughout our development, we provide the mechanical and geometrical interpretation of the variables. Specifically, the two-dimensional tensors of strain and stress, the invariant energies, and the governing equations are correlated with their three-dimensional counterparts. The corresponding principles of work and energy are presented together with the functionals which provide the governing equations. Those functionals also provide the basis for subsequent approximations via finite elements.

9.5 Kinematics

Apart from the basic assumption (9.1) that the displacement varies linearly through the thickness, the shell theory presented here is without restrictions upon the deformations: The equations admit large deformations, displacements, rotations, and strains. The deformations of thin bodies are often accompanied by large rotations, though strains may remain small. Consequently, it is most appropriate to decompose the motion into rotation and strain (see Section 3.14). Accordingly, a rotation, a stretch, and an engineering strain are the appropriate choices. In choosing the most convenient variables, we anticipate the role of the reference surface s_0 and the merits of using the differential geometry (Chapter 8) of that surface, before and after the deformation. Towards these goals, we observe that the θ^α lines of the initial surface s_0 have tangents $\boldsymbol{a}_\alpha$ which are determined by the position ${}_0\boldsymbol{r}$:

$$\boldsymbol{a}_\alpha \equiv {}_0\boldsymbol{r}_{,\alpha}. \tag{9.2}$$

The deformed lines have tangents $\boldsymbol{A}_\alpha$ which are determined by the position ${}_0\boldsymbol{R}$ to the deformed surface S_0:

$$\boldsymbol{A}_\alpha \equiv {}_0\boldsymbol{R}_{,\alpha}. \tag{9.3}$$

The vectors $\boldsymbol{A}_\alpha$ play the same roles in the geometry of the deformed surface, as $\boldsymbol{a}_\alpha$ of the undeformed surface. In particular,

$$\boldsymbol{A}_\alpha \cdot \boldsymbol{A}^\beta = \delta_\alpha^\beta, \qquad \boldsymbol{A}_\alpha \cdot \boldsymbol{A}_\beta = A_{\alpha\beta}.$$

In the initial state, an arbitrary particle of the shell is located at the initial distance θ^3 along the normal to the surface s_0. If $\hat{\boldsymbol{a}}_3$ denotes the unit normal, then

$$\boldsymbol{r} = {}_0\boldsymbol{r} + \theta^3 \hat{\boldsymbol{a}}_3. \tag{9.4}$$

The same particle of the deformed shell is located by position $\boldsymbol{R}$:

$$\boldsymbol{R} = {}_0\boldsymbol{R} + \theta^3 \boldsymbol{A}_3. \tag{9.5}$$

Because we do *not* preclude stretching of the normal, *nor* transverse shear, the vector $\boldsymbol{A}_3$ is *not* a unit vector; *nor* is $\boldsymbol{A}_3$ normal to the surface S_0. The initial and deformed (current) states are depicted in Figure 9.7.

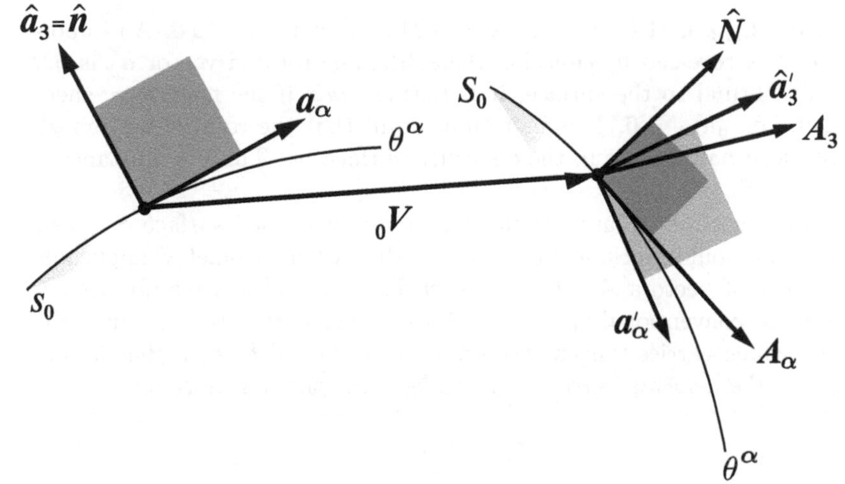

Figure 9.7 Rotated base triad a'_α, $\hat{a}'_3$ (rotation of principal axes)

We recall that the rigid rotation employed in the earlier decomposition (see Section 3.14) is the rotation of the orthogonal *principal* axes. This rotation is represented by the tensor $r_n^{\cdot j}$ of (3.87a, b):

$$\acute{\boldsymbol{g}}_n = r_n^{\cdot j}\boldsymbol{g}_j, \qquad \acute{\boldsymbol{g}}^n = r^n_{\cdot j}\boldsymbol{g}^j.$$

At the reference surface,

$$\acute{\boldsymbol{g}}_n(\theta^1,\theta^2,0) \equiv \boldsymbol{a}'_n = r_n^{\cdot j}(\theta^1,\theta^2,0)\,\boldsymbol{a}_j.$$

From (3.95), we have the tangent vector:

$$\boldsymbol{A}_n \equiv \boldsymbol{G}_n(\theta^1,\theta^2,0) = (\delta^j_n + {}_0h^j_n)\boldsymbol{a}'_j, \qquad {}_0h^j_n \equiv h^j_n(\theta^1,\theta^2,0),$$

and

$$\boldsymbol{a}'_i \cdot \boldsymbol{A}_n = a_{ni} + {}_0h_{ni},$$

where

$$a_{ni} \equiv g_{ni}(\theta^1,\theta^2,0).$$

In particular, since $a_{3\alpha} = a_{\alpha 3} = 0$,

$$\boldsymbol{a}'_3 \cdot \boldsymbol{A}_\alpha = \boldsymbol{a}'_\alpha \cdot \boldsymbol{A}_3 = {}_0h_{\alpha 3}, \tag{9.6a, b}$$

$$\boldsymbol{a}'_\alpha \cdot \boldsymbol{A}_\beta = a_{\alpha\beta} + {}_0h_{\alpha\beta} \equiv C_{\alpha\beta}. \tag{9.7a, b}$$

The latter, $C_{\alpha\beta}$, is the "stretch" [see (3.94a, b)] at the surface. An important point is revealed by equation (9.6a, b): The rotated vector $\boldsymbol{a}'_3$ is *not* generally normal to the surface; it is normal *only* if the transverse shear vanishes, i.e., ${}_0h_{\alpha 3} = 0$. It is more important that the rotated vectors $\boldsymbol{a}'_\alpha$ are therefore *not* tangent to the deformed surface S_0. This circumstance is depicted in Figure 9.7.

Since the vectors $\boldsymbol{a}'_\alpha$ are *not* tangent to the deformed surface S_0, their employment would preclude our use of the differential geometry, applicable to the tangent vectors $\boldsymbol{A}_i$. Indeed, such bases would not readily accommodate the conventional theories. More conveniently, we introduce the rotation which carries tangent vectors $\boldsymbol{a}_i$ to the triad $\boldsymbol{b}_i$ such that $\boldsymbol{b}_\alpha$ are *tangent to the deformed surface* S_0 and $\boldsymbol{b}_3 \equiv \hat{\boldsymbol{N}}$, the unit normal:

$$\boldsymbol{b}_i = \bar{r}_i^{\bullet j} \boldsymbol{a}_j. \tag{9.8}$$

Note: The rotation $\bar{r}_i^{\bullet j}$ of (9.8) is *not* that which carries $\boldsymbol{a}_j$ to $\boldsymbol{a}'_j$! The orientation of tangents $\boldsymbol{b}_\alpha$ is such that the stretch $C_{\alpha\beta}$ *of the reference* surface is symmetric:

$$C_{\alpha\beta} \equiv \boldsymbol{b}_\alpha \cdot \boldsymbol{A}_\beta = \boldsymbol{b}_\beta \cdot \boldsymbol{A}_\alpha. \tag{9.9}$$

Stretch *of the normal* θ^3 line is

$$C_{3i} \equiv \boldsymbol{b}_i \cdot \boldsymbol{A}_3. \tag{9.10}$$

Notice that $C_{3\alpha}$ is the transverse shear and that

$$\boldsymbol{b}_3 \cdot \boldsymbol{A}_\alpha \equiv \hat{\boldsymbol{N}} \cdot \boldsymbol{A}_\alpha = 0. \tag{9.11}$$

Note: C_{ij} in (9.9) to (9.11) is *not* the stretch $(\boldsymbol{a}'_i \cdot \boldsymbol{A}_j)$ as employed in Chapter 3! Notice that the triad $\boldsymbol{b}_i$ is a rigidly rotated version of the triad $\boldsymbol{a}_i$. Therefore, the expressions for the products and derivatives of the base vectors $\boldsymbol{b}_i$ (see Sections 8.3 and 8.4) are similar to those for the vectors $\boldsymbol{a}_i$; for example,

$$\boldsymbol{b}_\alpha \cdot \boldsymbol{b}_\beta = a_{\alpha\beta}, \qquad \boldsymbol{b}^\alpha \cdot \boldsymbol{b}^\beta = a^{\alpha\beta}, \qquad \boldsymbol{b}^\alpha \cdot \boldsymbol{b}_\beta = \delta^\alpha_\beta.$$

$$\boldsymbol{b}_\alpha \times \boldsymbol{b}_\beta = \bar{e}_{\alpha\beta} \boldsymbol{b}_3, \qquad \boldsymbol{b}_3 \times \boldsymbol{b}_\alpha = \bar{e}_{\alpha\beta} \boldsymbol{b}^\beta, \qquad \bar{e}_{\alpha\beta} = \sqrt{a}\,\epsilon_{\alpha\beta 3}.$$

The rotated triad $\boldsymbol{b}_i$ and the triad $\boldsymbol{A}_i$ are depicted in Figure 9.8. In accor-

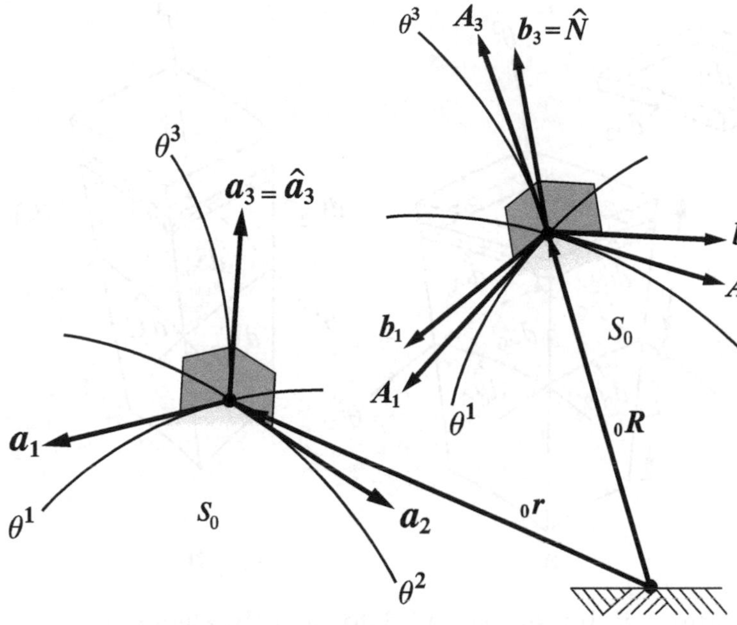

Figure 9.8 Motion of base triad

dance with the foregoing relations:

$$A_\beta = C^\gamma_\beta b_\gamma = C_{\beta\gamma} b^\gamma. \qquad (9.12a, b)$$

Note that the covariant and mixed components $C_{\alpha\beta}$ and C^β_α, respectively, are thereby associated via the initial metric $a_{\alpha\beta}$ or $a^{\alpha\beta}$. For our subsequent convenience, we also define the inverse stretch c^β_α:

$$b_\alpha \equiv c^\mu_\alpha A_\mu = c_{\alpha\mu} A^\mu. \qquad (9.13a, b)$$

Also notice that this is an exceptional instance, wherein the tensors C^α_β and c^α_β are mixed:

$$C^\alpha_\beta = a^{\alpha\gamma} C_{\beta\gamma}, \qquad c^\alpha_\beta = A^{\alpha\gamma} c_{\beta\gamma}.$$

From equations (9.12a, b) and (9.13a, b), it follows that

$$C^\gamma_\alpha c^\beta_\gamma = c^\gamma_\alpha C^\beta_\gamma = \delta^\beta_\alpha. \qquad (9.14)$$

Moreover,

$$C_{\alpha\beta} = C^\gamma_\alpha a_{\gamma\beta} = c_{\alpha\beta} = c^\gamma_\alpha A_{\gamma\beta} = b_\alpha \cdot A_\beta. \qquad (9.15)$$

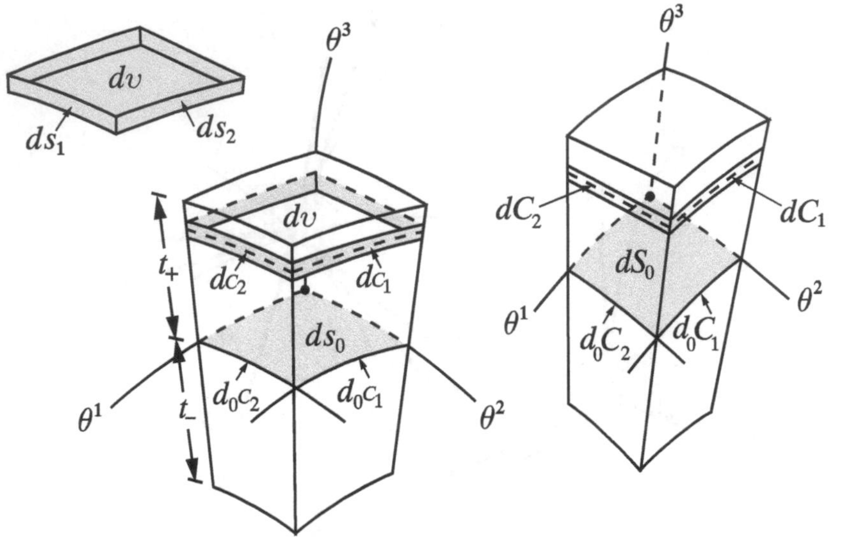

Figure 9.9 Initial (a) and deformed (b) elements

Thin shells of common solid materials are most frequently analyzed via theory based upon the hypothesis of H. Aron [151], G. R. Kirchhoff [152], and A. E. H. Love [153]: Briefly, they presumed that normal lines to surface s_0 remain straight and normal to the surface S_0; then $\hat{\boldsymbol{N}} \cdot \boldsymbol{A}_\alpha = 0$. That is a special case; the transverse shear $C_{3\alpha}$ of equation (9.10) vanishes. In that case the rotation of equation (9.8) is also the rotation of equation (3.87a, b).

Our study of the kinematics remains incomplete, since the six stretches ($C_{\alpha\beta}$, C_{3i}) characterize only the deformation *at* the reference surface. We require more to adequately describe the deformation of an element of the shell. Such an element contains the differential surfaces, ds_0 and dS_0 of Figure 9.9a and b, respectively; the element has the finite thickness $h = t_- + t_+$.

Reduction from the theory of three dimensions necessitates expressions for the deformed triad $\boldsymbol{G}_i$ in accord with the underlying assumption (9.5):

$$\boldsymbol{G}_\alpha = \boldsymbol{A}_\alpha + \theta^3 \boldsymbol{A}_{3,\alpha}, \qquad \boldsymbol{G}_3 = \boldsymbol{A}_3. \tag{9.16a, b}$$

Since our basis is the triad $\boldsymbol{b}_i$, we require the components

$$D_{\alpha\beta} \equiv \boldsymbol{A}_{3,\alpha} \cdot \boldsymbol{b}_\beta, \qquad D_{\alpha 3} \equiv \boldsymbol{A}_{3,\alpha} \cdot \hat{\boldsymbol{N}}. \tag{9.17a, b}$$

We recall the expressions for the curvatures of a surface [see Section 8.4, equations (8.27a, b) and (8.35a, b)] as applied to our deformed surface S_0:

$$B_{\alpha\beta} = \boldsymbol{A}_{\alpha,\beta} \cdot \hat{\boldsymbol{N}}, \tag{9.18a}$$

$$\hat{\boldsymbol{N}}_{,\alpha} = -B^{\beta}_{\alpha}\boldsymbol{A}_{\beta}. \tag{9.18b}$$

Note that these components (covariant $B_{\alpha\beta}$ and mixed B^{β}_{α}) are strictly associated via the metric $A_{\alpha\beta}$ of the deformed surface S_0.

The requisite components $D_{\alpha i}$ are obtained from expression (9.10) by using the inverse (9.13a, b) and equations (9.18a, b) and (8.78a), as follows:

$$\boldsymbol{A}_3 = C^{\eta}_{3}\boldsymbol{b}_{\eta} + C^{3}_{3}\hat{\boldsymbol{N}} = C^{\eta}_{3}c^{\gamma}_{\eta}\boldsymbol{A}_{\gamma} + C^{3}_{3}\hat{\boldsymbol{N}}, \tag{9.19a, b}$$

$$\boldsymbol{A}_{3,\alpha} = (C^{\eta}_{3}c^{\gamma}_{\eta})\|^{*}_{\alpha}\boldsymbol{A}_{\gamma} + (C^{\eta}_{3}c^{\gamma}_{\eta})B_{\gamma\alpha}\hat{\boldsymbol{N}} + C^{3}_{3,\alpha}\hat{\boldsymbol{N}} - C^{3}_{3}B^{\gamma}_{\alpha}\boldsymbol{A}_{\gamma}, \tag{9.20a}$$

$$\equiv -K^{\gamma}_{\alpha}\boldsymbol{A}_{\gamma} + D_{\alpha 3}\hat{\boldsymbol{N}}, \tag{9.20b}$$

where

$$K^{\gamma}_{\alpha} \equiv -(C^{\eta}_{3}c^{\gamma}_{\eta})\|^{*}_{\alpha} + C^{3}_{3}B^{\gamma}_{\alpha}, \tag{9.21a}$$

$$D_{\alpha 3} \equiv (C^{\eta}_{3}c^{\gamma}_{\eta})B_{\gamma\alpha} + C^{3}_{3,\alpha}. \tag{9.21b}$$

Note that ($\|^{*}$) signifies the covariant derivative with respect to the metric of the deformed surface. Again, using (9.12a, b) in (9.20a, b) and considering (9.17a, b), we obtain

$$\boldsymbol{b}_{\beta} \cdot \boldsymbol{A}_{3,\alpha} = -K^{\gamma}_{\alpha}C_{\gamma\beta} = D_{\alpha\beta}, \tag{9.22a}$$

$$\boldsymbol{b}_{3} \cdot \boldsymbol{A}_{3,\alpha} \equiv \hat{\boldsymbol{N}} \cdot \boldsymbol{A}_{3,\alpha} = D_{\alpha 3}. \tag{9.22b}$$

With the aid of (9.22a, b) and (9.12b), expression (9.20b) assumes the form:

$$\boldsymbol{A}_{3,\alpha} = D_{\alpha\beta}\boldsymbol{b}^{\beta} + D_{\alpha 3}\hat{\boldsymbol{N}}. \tag{9.22c}$$

In the next section, we require the power of the stress which calls for the rate (or increment) $\dot{\boldsymbol{G}}_i$; this, in turn, requires the rates $\dot{\boldsymbol{A}}_i$ and $\dot{\boldsymbol{A}}_{3,\alpha}$. As before, we begin with expressions (9.12a, b), (9.19a), and (9.22c) and recall that the rate of $\boldsymbol{b}_i$ is attributed only to a spin $\dot{\boldsymbol{\Omega}}$ (rigid rotation). The steps follow:

$$\dot{\boldsymbol{A}}_\alpha = \dot{C}_\alpha^\beta \boldsymbol{b}_\beta + C_\alpha^\beta \dot{\boldsymbol{b}}_\beta, \tag{9.23a}$$

$$\dot{\boldsymbol{A}}_3 = \dot{C}_3^\beta \boldsymbol{b}_\beta + C_3^\beta \dot{\boldsymbol{b}}_\beta + \dot{C}_3^3 \hat{\boldsymbol{N}} + C_3^3 (\hat{\boldsymbol{N}})^{\boldsymbol{\cdot}}, \tag{9.23b}$$

$$\dot{\boldsymbol{A}}_{3,\alpha} = \dot{D}_{\alpha\beta} \boldsymbol{b}^\beta + D_{\alpha\beta} \dot{\boldsymbol{b}}^\beta + \dot{D}_{\alpha 3} \hat{\boldsymbol{N}} + D_{\alpha 3} (\hat{\boldsymbol{N}})^{\boldsymbol{\cdot}}. \tag{9.23c}$$

In general,

$$\dot{\boldsymbol{b}}_i = \dot{\boldsymbol{\Omega}} \times \boldsymbol{b}_i = \dot{\varpi}^k e_{kij} \boldsymbol{b}^j = \dot{\varpi}_{ji} \boldsymbol{b}^j, \tag{9.24a–c}$$

where

$$\dot{\boldsymbol{\Omega}} = \dot{\varpi}^k \boldsymbol{b}_k, \qquad \dot{\varpi}_{ji} \equiv \dot{\varpi}^k e_{kij} = \tfrac{1}{2}(\boldsymbol{b}_j \cdot \dot{\boldsymbol{b}}_i - \boldsymbol{b}_i \cdot \dot{\boldsymbol{b}}_j). \tag{9.25a–c}$$

Substituting (9.24c) into (9.23a–c), we obtain

$$\dot{\boldsymbol{A}}_\alpha = \dot{C}_\alpha^\beta \boldsymbol{b}_\beta + C_\alpha^\beta \dot{\varpi}_{\mu\beta} \boldsymbol{b}^\mu + C_\alpha^\beta \dot{\varpi}_{3\beta} \hat{\boldsymbol{N}}, \tag{9.26}$$

$$\begin{aligned}\dot{\boldsymbol{A}}_3 &= \dot{C}_{3\alpha} \boldsymbol{b}^\alpha + C_{3\alpha} \dot{\varpi}_\mu{}^{\cdot\alpha} \boldsymbol{b}^\mu + C_{3\alpha} \dot{\varpi}_3{}^{\cdot\alpha} \hat{\boldsymbol{N}} \\ &\quad + \dot{C}_3^3 \hat{\boldsymbol{N}} + C_3^3 \dot{\varpi}_\mu{}^{\cdot 3} \boldsymbol{b}^\mu,\end{aligned} \tag{9.27}$$

$$\begin{aligned}\dot{\boldsymbol{A}}_{3,\alpha} &= \dot{D}_{\alpha\beta} \boldsymbol{b}^\beta + D_{\alpha\beta} \dot{\varpi}_\mu{}^{\cdot\beta} \boldsymbol{b}^\mu + D_{\alpha\beta} \dot{\varpi}_3{}^{\cdot\beta} \hat{\boldsymbol{N}} \\ &\quad + \dot{D}_{\alpha 3} \hat{\boldsymbol{N}} + D_{\alpha 3} \dot{\varpi}_\mu{}^{\cdot 3} \boldsymbol{b}^\mu.\end{aligned} \tag{9.28}$$

Our further analysis of the element requires expressions for the elemental area of a face ds_α consistent with volume dv as depicted in Figure 9.9a; here, dc_α is the differential length of the slice:

$$dc_\alpha = \sqrt{g_{\underline{\alpha\alpha}}}\, d\theta^\alpha, \qquad d\,{}_0c_\alpha = \sqrt{a_{\underline{\alpha\alpha}}}\, d\theta^\alpha, \tag{9.29a, b}$$

$$ds_\alpha = \sqrt{g^{\underline{\alpha\alpha}}\, g}\; d\theta^\beta\, d\theta^3 \quad \alpha \neq \beta, \tag{9.29c}$$

$$dv = \sqrt{g}\; d\theta^1\, d\theta^2\, d\theta^3. \tag{9.30}$$

In our system (see Sections 8.5 and 8.8)

$$\boldsymbol{g}_\alpha = \boldsymbol{a}_\alpha + \theta^3 \boldsymbol{a}_{3,\alpha} = \boldsymbol{a}_\alpha - \theta^3 b^\beta_\alpha \boldsymbol{a}_\beta, \tag{9.31}$$

$$\sqrt{g} = \sqrt{a}\, \mu, \tag{9.32}$$

where

$$\mu \equiv [1 - 2\,\widetilde{h}\theta^3 + \widetilde{k}(\theta^3)^2]. \tag{9.33}$$

Here, the properties are those of the reference state; b^α_β is the curvature tensor, $\sqrt{a}$ the area metric, $\widetilde{h}$ the mean curvature, and $\widetilde{k}$ the Gaussian curvature.

Final Remark: All discussion of displacement is deferred to Sections 10.9 and 10.10.

9.6 Strains and Stresses

Here, as in our description of the three-dimensional theory, the choices of strains and stresses must be consistent. Specifically, we require an invariant expression of internal power or incremental work (see Section 4.5). Presently, that invariant expression is work per unit of *surface* ($\sqrt{a}\, d\theta^1\, d\theta^2$). The work of stress [see Section 4.5, equation (4.16) and Section 6.14, equation (6.101)] is

$$\dot{W}_s = \boldsymbol{s}^i \cdot \dot{\boldsymbol{G}}_i\, dv. \tag{9.34}$$

As implied in the previous studies (e.g., Section 6.14) the vector $\boldsymbol{s}^i$ can be expressed via alternative bases. Here, it is most appropriate to employ the triad $\boldsymbol{b}_i$ (and associated triad $\boldsymbol{b}^i$). Together with equations (9.16a, b), (9.26) to (9.28), and (9.32), expression (9.34) provides the power per unit *area*, as the integral:

$$\dot{u}_s \equiv \int_{-t_-}^{t_+} \dot{w}_s\, \mu\, d\theta^3, \tag{9.35a}$$

$$= \dot{\boldsymbol{A}}_\alpha \cdot \int_{-t_-}^{t_+} \boldsymbol{s}^\alpha \mu\, d\theta^3 + \dot{\boldsymbol{A}}_{3,\alpha} \cdot \int_{-t_-}^{t_+} \boldsymbol{s}^\alpha \mu \theta^3\, d\theta^3$$

$$+ \dot{\boldsymbol{A}}_3 \cdot \int_{-t_-}^{t_+} \boldsymbol{s}^3 \mu\, d\theta^3. \tag{9.35b}$$

Let us introduce notations for the integrals:

$$\boldsymbol{N}^{\alpha} \equiv \int_{-t_-}^{t_+} \boldsymbol{s}^{\alpha} \mu \, d\theta^3, \tag{9.36a}$$

$$\boldsymbol{M}^{\alpha} \equiv \int_{-t_-}^{t_+} \boldsymbol{s}^{\alpha} \mu \theta^3 \, d\theta^3, \tag{9.36b}$$

$$\boldsymbol{T} \equiv \int_{-t_-}^{t_+} \boldsymbol{s}^{3} \mu \, d\theta^3. \tag{9.37}$$

Each vector transforms as a surface tensor in accordance with the indicial notation ($\boldsymbol{T}$ as invariant; $\boldsymbol{N}^{\alpha}$ and $\boldsymbol{M}^{\alpha}$ as first-order contravariant). These have the tensorial components:

$$N^{\alpha\beta} \equiv \boldsymbol{b}^{\beta} \cdot \boldsymbol{N}^{\alpha}, \qquad N^{\alpha 3} \equiv \hat{\boldsymbol{N}} \cdot \boldsymbol{N}^{\alpha}, \tag{9.38a, b}$$

$$M^{\alpha\beta} \equiv \boldsymbol{b}^{\beta} \cdot \boldsymbol{M}^{\alpha}, \qquad M^{\alpha 3} \equiv \hat{\boldsymbol{N}} \cdot \boldsymbol{M}^{\alpha}, \tag{9.39a, b}$$

$$T^{\alpha} \equiv \boldsymbol{b}^{\alpha} \cdot \boldsymbol{T}, \qquad T^{3} \equiv \hat{\boldsymbol{N}} \cdot \boldsymbol{T}. \tag{9.40a, b}$$

The corresponding physical components are forces and moments (see Section 9.9).

In view of (9.22a), (9.26) to (9.28), and with the notations (9.38a, b) to (9.40a, b), the integral $\dot{u}_s$ of (9.35) gives the rate of work:

$$\begin{aligned}\dot{u}_s = {} & N^{\alpha\beta}\dot{C}_{\alpha\beta} + M^{\alpha\beta}\dot{D}_{\alpha\beta} + T^{\alpha}\dot{C}_{3\alpha} + T^{3}\dot{C}_{33} + M^{\alpha 3}\dot{D}_{\alpha 3} \\ & + [(N^{\gamma 3} - K^{\gamma}_{\alpha} M^{\alpha 3})C^{\mu}_{\gamma} - T^{\mu}C_{33} + T^{3}C^{\mu}_{3} - M^{\alpha\mu}D_{\alpha 3}]\,\dot{\bar{\omega}}_{3\mu} \\ & + [(N^{\alpha\beta} - K^{\alpha}_{\mu} M^{\mu\beta})C^{\gamma}_{\alpha} + T^{\beta}C^{\gamma}_{3}]\,\dot{\bar{\omega}}_{\beta\gamma}.\end{aligned} \tag{9.41}$$

The definitions of stresses and strains are always matter of convenience, as dictated by practice or theory. It is important that these are consistent; the power $\dot{u}_s$ must be invariant. Here, the apparent choices are the "stresses" ($N^{\alpha\beta}$, $M^{\alpha\beta}$, T^{α}, T^{3}, and $M^{\alpha 3}$) which are associated with the "stretch rates" ($\dot{C}_{\alpha\beta}$, $\dot{D}_{\alpha\beta}$, $\dot{C}_{\alpha 3}$, $\dot{C}_{33}$, and $\dot{D}_{\alpha 3}$). Then, the associated strains are defined as to vanish in the reference state:

$$h_{\alpha\beta} \equiv C_{\alpha\beta} - a_{\alpha\beta}, \qquad \kappa_{\alpha\beta} \equiv D_{\alpha\beta} + b_{\alpha\beta}, \tag{9.42a, b}$$

$$h_{3\alpha} \equiv C_{3\alpha}, \qquad h_{33} \equiv C_{33} - 1, \tag{9.42c, d}$$

$$\kappa_{\alpha 3} \equiv D_{\alpha 3}. \tag{9.42e}$$

The strain $h_{\alpha\beta}$ is the engineering strain of the surface s_0; $h_{3\alpha}$ is the transverse shear strain and h_{33} the transverse extensional strain. Again, these are not precisely the engineering strains of the three-dimensional theory (see Section 3.14). The most significant differences are in the transverse shear; here $h_{3\alpha} = \boldsymbol{b}_\alpha \cdot \boldsymbol{A}_3 \neq \boldsymbol{b}_3 \cdot \boldsymbol{A}_\alpha$. Here, that strain is entirely determined by the vectors $\boldsymbol{b}_\alpha$ and $\boldsymbol{A}_3$ (see Figure 9.8). Furthermore, $\kappa_{\alpha\beta}$ is the flexural strain and $\kappa_{\alpha 3}$ a gradient of transverse shear. Since the spin $\dot{\boldsymbol{\Omega}}$ constitutes rigid motion, the bracketed terms of (9.41) must vanish.

Consistent with the previous definitions of the stretch C_i^j and inverse c_i^j, we can introduce the strain h_i^j and inverse H_i^j:

$$\boldsymbol{A}_i = C_i^j \boldsymbol{b}_j = (\delta_i^j + h_i^j)\boldsymbol{b}_j, \tag{9.43a}$$

$$\boldsymbol{b}_k = c_k^l \boldsymbol{A}_l = (\delta_k^l + H_k^l)\boldsymbol{A}_l. \tag{9.43b}$$

As noted previously, C_i^j, c_i^j, h_i^j, and H_i^j are mixed tensors:

$$h_{ij} = a_{im} h_j^m, \qquad H_{ij} = A_{im} H_j^m,$$

etc. By means of expressions (9.12a, b) to (9.15), we obtain

$$C_\beta^\alpha = \delta_\beta^\alpha + h_\beta^\alpha, \qquad c_\beta^\alpha = \delta_\beta^\alpha + H_\beta^\alpha. \tag{9.44a, b}$$

In a simple form, this shows the nature of the stretch C_α^β and inverse c_α^β, and of the strain h_α^β and its inverse H_α^β [see equations (9.71a, b)].

9.7 Equilibrium

Our most expedient and precise route to the equations of equilibrium and edge (boundary) conditions is the principle of virtual work (see Section 6.14). Here, our body has three "boundaries": the top and bottom surfaces (s_+ and s_-) with applied tractions $\boldsymbol{t}^+$ and $\boldsymbol{t}^-$, and the edge defined by the curve c_t on the reference surface s_0 with tractions $\boldsymbol{t}^c$. The general form of (6.104a, b) follows [virtual increments are designated by the dot $(\dot{\ })$]:

$$\dot{W} = \iint_{s_0} d\theta^1\, d\theta^2 \left[\int_{-t_-}^{t_+} (\boldsymbol{s}^i \cdot \dot{\boldsymbol{G}}_i - \boldsymbol{f} \cdot \dot{\boldsymbol{R}}) \sqrt{g}\, d\theta^3 \right]$$

$$- \iint_{s_-} \boldsymbol{t}^- \cdot \dot{\boldsymbol{R}} \sqrt{a_-}\, d\theta^1\, d\theta^2 - \iint_{s_+} \boldsymbol{t}^+ \cdot \dot{\boldsymbol{R}} \sqrt{a_+}\, d\theta^1\, d\theta^2$$

$$- \int_{c_t} dc \int_{-t_-}^{t_+} \boldsymbol{t}^c \cdot \dot{\boldsymbol{R}}\, d\theta^3. \tag{9.45}$$

In the integral we employ $\dot{\boldsymbol{R}}$ and $\dot{\boldsymbol{G}}_i$ as dictated by (9.5) and (9.16a, b) and the integrals of $\boldsymbol{s}^i$ as defined by (9.36a, b) and (9.37). Also, we note from (9.32) that

$$\sqrt{a_-} = \sqrt{a}\, \mu(-t_-) \equiv \sqrt{a}\, \mu_-,$$

$$\sqrt{a_+} = \sqrt{a}\, \mu(t_+) \equiv \sqrt{a}\, \mu_+.$$

Our integrations include the following:

$$\boldsymbol{t}^+ \mu_+ + \boldsymbol{t}^- \mu_- \equiv \boldsymbol{F}_s, \qquad (\boldsymbol{t}^+ \mu_+)\, t_+ - (\boldsymbol{t}^- \mu_-)\, t_- \equiv \boldsymbol{C}_s, \tag{9.46a, b}$$

$$\int_{-t_-}^{t_+} \boldsymbol{f} \mu\, d\theta^3 \equiv \boldsymbol{F}_b, \qquad \int_{-t_-}^{t_+} \boldsymbol{f} \theta^3 \mu\, d\theta^3 \equiv \boldsymbol{C}_b. \tag{9.46c, d}$$

These variables ($\boldsymbol{F}$ signifies force, $\boldsymbol{C}$ signifies couple) constitute the loadings upon the breadth of the shell (s signifies the contributions from top and bottom surfaces, b from the body forces). Integrals across the edge contribute to the edge-loading:

$$\int_{-t_-}^{t_+} \boldsymbol{t}^c\, d\theta^3 \equiv \overline{\boldsymbol{N}}, \tag{9.47a}$$

$$\int_{-t_-}^{t_+} \boldsymbol{t}^c \theta^3\, d\theta^3 = \overline{\boldsymbol{M}}. \tag{9.47b}$$

These are the force $\overline{\boldsymbol{N}}$ and couple $\overline{\boldsymbol{M}}$ per unit length on curve c_t. In these notations, the virtual work takes the form:

$$\dot{W} = \iint_{s_0} (\boldsymbol{N}^\alpha \cdot {}_0\dot{\boldsymbol{R}}_{,\alpha} + \boldsymbol{M}^\alpha \cdot \dot{\boldsymbol{A}}_{3,\alpha} + \boldsymbol{T} \cdot \dot{\boldsymbol{A}}_3 - \boldsymbol{F} \cdot {}_0\dot{\boldsymbol{R}} - \boldsymbol{C} \cdot \dot{\boldsymbol{A}}_3) \sqrt{a}\, d\theta^1\, d\theta^2$$

$$- \int_{c_t} (\overline{\boldsymbol{N}} \cdot {}_0\dot{\boldsymbol{R}} + \overline{\boldsymbol{M}} \cdot \dot{\boldsymbol{A}}_3)\, dc. \tag{9.48}$$

Here,

$$\boldsymbol{F} \equiv \boldsymbol{F}_s + \boldsymbol{F}_b, \qquad \boldsymbol{C} \equiv \boldsymbol{C}_s + \boldsymbol{C}_b. \tag{9.49a, b}$$

The first two terms of (9.48) can be integrated by parts; then

$$\dot{W} = \iint_{s_0} \left[-\frac{1}{\sqrt{a}} (\boldsymbol{N}^\alpha \sqrt{a})_{,\alpha} - \boldsymbol{F} \right] \cdot {}_0\dot{\boldsymbol{R}} \sqrt{a}\, d\theta^1\, d\theta^2$$

$$+ \iint_{s_0} \left[-\frac{1}{\sqrt{a}} (\boldsymbol{M}^\alpha \sqrt{a})_{,\alpha} + \boldsymbol{T} - \boldsymbol{C} \right] \cdot \dot{\boldsymbol{A}}_3 \sqrt{a}\, d\theta^1\, d\theta^2$$

$$+ \iint_{c_t} [\boldsymbol{N}^\alpha n_\alpha - \overline{\boldsymbol{N}}] \cdot {}_0\dot{\boldsymbol{R}}\, dc + \iint_{c_t} [\boldsymbol{M}^\alpha n_\alpha - \overline{\boldsymbol{M}}] \cdot \dot{\boldsymbol{A}}_3\, dc.$$

(9.49c)

The virtual work vanishes for arbitrary variations ${}_0\dot{\boldsymbol{R}}$ and $\dot{\boldsymbol{A}}_3$ on s_0 and on c if, and only if, the bracketed terms vanish. These are the equilibrium equations consistent with the assumption (9.5):

$$\frac{1}{\sqrt{a}} (\boldsymbol{N}^\alpha \sqrt{a})_{,\alpha} + \boldsymbol{F} = \mathbf{o} \qquad \text{on } s_0, \tag{9.50a}$$

$$\frac{1}{\sqrt{a}} (\boldsymbol{M}^\alpha \sqrt{a})_{,\alpha} - \boldsymbol{T} + \boldsymbol{C} = \mathbf{o} \qquad \text{on } s_0, \tag{9.50b}$$

$$\boldsymbol{N}^\alpha n_\alpha = \overline{\boldsymbol{N}}, \qquad \boldsymbol{M}^\alpha n_\alpha = \overline{\boldsymbol{M}} \qquad \text{on } c_t. \tag{9.51a, b}$$

A portion of the boundary may be constrained such that ${}_0\dot{\boldsymbol{R}}$ and/or $\dot{\boldsymbol{A}}_3$ are given; then the conditions (9.51) are supplanted by those constraints.

Equation (9.50a) represents the requirement that the force upon an element vanishes. Equation (9.50b) requires the vanishing of moment.

In a strict sense, our theory lacks only the constitutive equations. We proceed to examine alternative formulations for an elastic shell. Specifically, we now obtain the complementary potentials, counterparts of $\mathcal{P}$, $\overline{\mathcal{V}}$, $\overline{\mathcal{V}}_c$ in the three-dimensional forms: (6.144), (6.147b), and (6.149b).

9.8 Complementary Potentials

The foremost tools for the approximation of elastic shells are the principles of stationary and minimum potential. Others are the theorems for stationarity of the modified potential and complementary potential. We formulate these two-dimensional versions as their three-dimensional counterparts (see Sections 6.16 to 6.18). First, we set forth the two-dimensional counterpart of the primitive functional (6.144): That is obtained by integrating (6.144) through the thickness, across the breadth and the edge. We obtain in the notations of the preceding Sections 9.5 and 9.7:

$$\mathcal{P} = \iint_{s_0} \left(\boldsymbol{N}^\alpha \cdot {}_0\boldsymbol{R}_{,\alpha} - \boldsymbol{F} \cdot {}_0\boldsymbol{R} + \boldsymbol{M}^\alpha \cdot \boldsymbol{A}_{3,\alpha} + \boldsymbol{T} \cdot \boldsymbol{A}_3 - \boldsymbol{C} \cdot \boldsymbol{A}_3 \right) ds_0$$

$$- \int_c \left(\boldsymbol{N} \cdot {}_0\boldsymbol{R} + \boldsymbol{M} \cdot \boldsymbol{A}_3 \right) dc$$

$$- \int_{c_v} \left[\boldsymbol{N} \cdot \left({}_0\boldsymbol{R} - {}_0\overline{\boldsymbol{R}} \right) + \boldsymbol{M} \cdot \left(\boldsymbol{A}_3 - \bar{\boldsymbol{A}}_3 \right) \right] dc. \tag{9.52}$$

Here, the notations follow the previous form: c_v denotes a part of the edge where the displacement is constrained; the overbar ($\overline{}$) signifies the imposed quantity and $\boldsymbol{N}$, $\boldsymbol{M}$ are the edge tractions.

Note that the variation of the displacement ${}_0\dot{\boldsymbol{R}}$ and $\dot{\boldsymbol{A}}_3$ (consistent with the constraints), enforces the equilibrium equations (9.50a, b) and (9.51a, b).

If the shell is elastic, then the stresses are presumed to derive from an internal energy $\bar{u}_0(h_{ij}, \kappa_{\alpha i})$ (per unit area); conversely, the strains are to derive from a complementary energy $\bar{u}_{c0}(N^{\alpha\beta}, M^{\alpha i}, T^i)$. These are related as their three-dimensional counterparts [see equation (6.145b)]:

$$N^{\alpha\beta} h_{\alpha\beta} + M^{\alpha i} \kappa_{\alpha i} + T^i h_{3i} = \bar{u}_0 + \bar{u}_{c0}. \tag{9.53}$$

The two-dimensional counterparts of equations (6.147b) and (6.149b) are the potential and the complementary potential of the elastic shell:

$$\overline{\mathcal{V}} = \iint_{s_0} (\bar{u}_0 - \boldsymbol{F} \cdot {}_0\boldsymbol{R} - \boldsymbol{C} \cdot \boldsymbol{A}_3)\, ds_0 - \int_{c_t} (\overline{\boldsymbol{N}} \cdot {}_0\boldsymbol{R} + \overline{\boldsymbol{M}} \cdot \boldsymbol{A}_3)\, dc, \quad (9.54)$$

$$\overline{\mathcal{V}}_c = \iint_{s_0} (\bar{u}_{c0} + N^{\alpha\beta} a_{\alpha\beta} - M^{\alpha\beta} b_{\alpha\beta} + T^3)\, ds_0$$

$$- \int_{c_v} (\boldsymbol{N} \cdot {}_0\boldsymbol{R} + \boldsymbol{M} \cdot \boldsymbol{A}_3)\, dc$$

$$- \int_{c_v} [\boldsymbol{N} \cdot ({}_0\boldsymbol{R} - {}_0\overline{\boldsymbol{R}}) + \boldsymbol{M} \cdot (\boldsymbol{A}_3 - \bar{\boldsymbol{A}}_3)]\, dc. \quad (9.55)$$

The sum of the two potentials (9.54) and (9.55) is the primitive functional:

$$\mathcal{P} = \overline{\mathcal{V}} + \overline{\mathcal{V}}_c. \quad (9.56)$$

This is analogous to the previous result [see equation (6.148b)]. The complementary potential $\overline{\mathcal{V}}_c$ is to be stationary with respect to admissible (equilibrated) variations of the stresses. The stationary conditions are then the kinematic requirements.

Section 6.20 describes the stationary theorem of Hellinger-Reissner. We recall that the functional is a modified complementary potential. In analogous manner, we now derive the two-dimensional version for the elastic shell. We perform a partial integration upon the first boundary integral of $\overline{\mathcal{V}}_c$ [see Section 8.11, equation (8.85)] and enforce equilibrium as required:

$$\int_{c_v} (\boldsymbol{N} \cdot {}_0\boldsymbol{R} + \boldsymbol{M} \cdot \boldsymbol{A}_3)\, dc$$

$$= \int_{c} (\boldsymbol{N} \cdot {}_0\boldsymbol{R} + \boldsymbol{M} \cdot \boldsymbol{A}_3)\, dc - \int_{c_t} (\overline{\boldsymbol{N}} \cdot {}_0\boldsymbol{R} + \overline{\boldsymbol{M}} \cdot \boldsymbol{A}_3)\, dc$$

$$= \iint_{s_0} (N^{\alpha i} \boldsymbol{b}_i \cdot {}_0\boldsymbol{R}_{,\alpha} + M^{\alpha i} \boldsymbol{b}_i \cdot \boldsymbol{A}_{3,\alpha} + \boldsymbol{T} \cdot \boldsymbol{A}_3 - \boldsymbol{F} \cdot {}_0\boldsymbol{R} - \boldsymbol{C} \cdot \boldsymbol{A}_3)\, ds_0$$

$$- \int_{c_t} (\overline{\boldsymbol{N}} \cdot {}_0\boldsymbol{R} + \overline{\boldsymbol{M}} \cdot \boldsymbol{A}_3)\, dc. \quad (9.57)$$

When equation (9.57) is substituted into (9.55), the modified complementary potential follows:

$$
\begin{aligned}
\overline{\mathcal{V}}_c^* = \iint_{s_0} & \left[\bar{u}_{c0} - N^{\alpha\beta}(\boldsymbol{b}_\beta \cdot {}_0\boldsymbol{R}_{,\alpha} - a_{\alpha\beta}) - N^{\alpha 3}\boldsymbol{b}_3 \cdot {}_0\boldsymbol{R}_{,\alpha} \right. \\
& - M^{\alpha\beta}(\boldsymbol{b}_\beta \cdot \boldsymbol{A}_{3,\alpha} + b_{\alpha\beta}) - M^{\alpha 3}\boldsymbol{b}_3 \cdot \boldsymbol{A}_{3,\alpha} \\
& \left. - T^\alpha \boldsymbol{b}_\alpha \cdot \boldsymbol{A}_3 - T^3(\boldsymbol{b}_3 \cdot \boldsymbol{A}_3 - 1) + \boldsymbol{F} \cdot {}_0\boldsymbol{R} + \boldsymbol{C} \cdot \boldsymbol{A}_3 \right] ds_0 \\
& + \int_{c_t} \left(\overline{\boldsymbol{N}} \cdot {}_0\boldsymbol{R} + \overline{\boldsymbol{M}} \cdot \boldsymbol{A}_3\right) dc \\
& - \int_{c_v} \left[\boldsymbol{N} \cdot ({}_0\boldsymbol{R} - {}_0\overline{\boldsymbol{R}}) + \boldsymbol{M} \cdot (\boldsymbol{A}_3 - \bar{\boldsymbol{A}}_3)\right] dc. \qquad (9.58)
\end{aligned}
$$

Finally, we complete the picture with the generalization of the potential (see Section 6.16). We obtain the modified potential $\overline{\mathcal{V}}^*$ from $\overline{\mathcal{V}}_c^*$ of equation (9.58) via (9.53) in the manner of (6.162):

$$
\begin{aligned}
\overline{\mathcal{V}}^* &= (-\overline{\mathcal{V}}_c^*) \\
&= \iint_{s_0} \left[\bar{u}_0 - N^{\alpha\beta}(h_{\alpha\beta} - \boldsymbol{b}_\beta \cdot {}_0\boldsymbol{R}_{,\alpha} + a_{\alpha\beta}) - M^{\alpha\beta}(\kappa_{\alpha\beta} - \boldsymbol{b}_\beta \cdot \boldsymbol{A}_{3,\alpha} - b_{\alpha\beta}) \right. \\
&\quad - T^\alpha(h_{3\alpha} - \boldsymbol{b}_\alpha \cdot \boldsymbol{A}_3) - T^3(h_{33} - \boldsymbol{b}_3 \cdot \boldsymbol{A}_3 + 1) \\
&\quad \left. + N^{\alpha 3}\boldsymbol{b}_3 \cdot {}_0\boldsymbol{R}_{,\alpha} - M^{\alpha 3}(\kappa_{\alpha 3} - \boldsymbol{b}_3 \cdot \boldsymbol{A}_{3,\alpha}) - \boldsymbol{F} \cdot {}_0\boldsymbol{R} - \boldsymbol{C} \cdot \boldsymbol{A}_3 \right] ds_0 \\
&\quad - \int_{c_t} \left[\overline{\boldsymbol{N}} \cdot {}_0\boldsymbol{R} + \overline{\boldsymbol{M}} \cdot \boldsymbol{A}_3\right] dc \\
&\quad + \int_{c_v} \left[\boldsymbol{N} \cdot ({}_0\boldsymbol{R} - {}_0\overline{\boldsymbol{R}}) + \boldsymbol{M} \cdot (\boldsymbol{A}_3 - \bar{\boldsymbol{A}}_3)\right] dc. \qquad (9.59)
\end{aligned}
$$

In *summary*, the various functionals, variables, and properties follow:

- $\overline{\mathcal{V}} = \overline{\mathcal{V}}({}_0\boldsymbol{R}, \boldsymbol{A}_3)$

  Admissible variations satisfy kinematic constraints.
  Stationary conditions enforce equilibrium.
  Minimum potential enforces stable equilibrium.

- $\overline{\mathcal{V}}_c = \overline{\mathcal{V}}_c(\boldsymbol{N}, \boldsymbol{M}, \boldsymbol{T})$

  Admissible variations satisfy equilibrium.
  Stationary conditions enforce the kinematic constraints.

- $\overline{\mathcal{V}}_c^* = \overline{\mathcal{V}}_c^*({}_0\boldsymbol{R}, \boldsymbol{A}_3, N^{\alpha i}, M^{\alpha i}, T^i)$

  Stationarity with respect to displacements and stresses provides the equations of equilibrium and constitutive equations, respectively.

- $\overline{\mathcal{V}}^* = \overline{\mathcal{V}}^*({}_0\boldsymbol{R}, \boldsymbol{A}_3, N^{\alpha i}, M^{\alpha i}, T^i, h_{ij}, \kappa_{\alpha i})$

  Stationarity with respect to displacements, stresses and stains provides all governing equations: equilibrium conditions, kinematic and constitutive relations.

Note that the functionals are dependent on the rotation of the triad $\boldsymbol{b}_i$. Variations of the rotation provide the conditions of equilibrium (of moments); these are the conditions that the bracketed terms of (9.41) vanish. Because the last functional admits variations of all variables, it provides a powerful basis for approximation via finite elements. With appropriate limitations, attention to mechanical and mathematical implications, one has possibilities to introduce various approximations for each variable. This provides additional capability and admits limited incompatibilities which are described in Sections 11.11 and 11.13.3.

9.9 Physical Interpretations

The stress resultants acting upon an edge of the element are obtained by considering the force acting upon the differential area ds_α of the θ^α-face in Figure 9.9. According to (9.29c), (4.19c), and (9.32), this force is given by

$$\boldsymbol{\sigma}^\alpha\, ds_{\underline{\alpha}} = \boldsymbol{\sigma}^\alpha \sqrt{g^{\underline{\alpha\alpha}}\, g}\, d\theta^\beta\, d\theta^3 = \boldsymbol{s}^\alpha \mu \sqrt{a}\, d\theta^\beta\, d\theta^3 \quad (\alpha \neq \beta). \qquad (9.60)$$

The force per unit of θ^β (not necessarily length) is obtained by dividing (9.60) by $d\theta^\beta$ and integrating through the thickness. In accordance with the notations (9.36a, b), the force and moment per unit of θ^β assume the

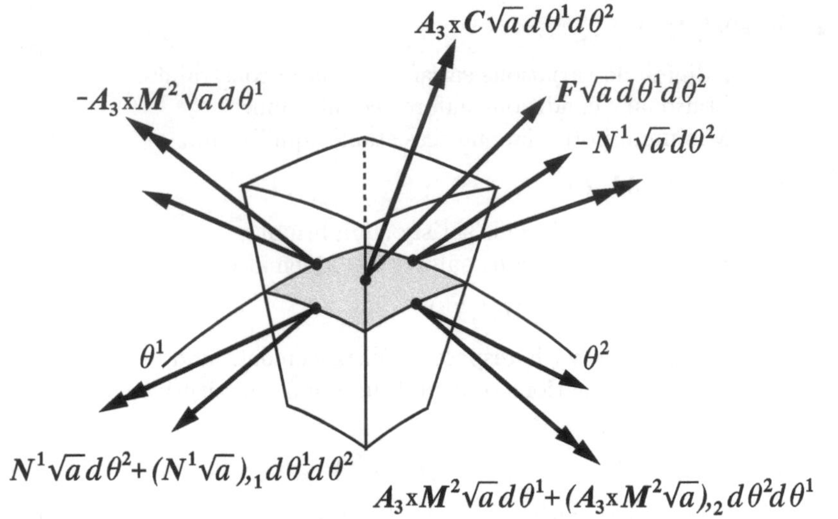

Figure 9.10 Physical actions upon an element

forms $\sqrt{a}\,\boldsymbol{N}^\alpha$ and $(\boldsymbol{A}_3 \times \boldsymbol{M}^\alpha)\sqrt{a}$, respectively. The force and moment *per unit length* ${}_0c_\alpha$ of an edge are [see equations (9.29b) and (8.15)]:

$$\boldsymbol{\mathcal{N}}^\alpha \equiv \frac{\boldsymbol{N}^\alpha}{\sqrt{a^{\underline{\alpha\alpha}}}}, \qquad \boldsymbol{\mathcal{M}}^\alpha \equiv \frac{\boldsymbol{A}_3 \times \boldsymbol{M}^\alpha}{\sqrt{a^{\underline{\alpha\alpha}}}}. \tag{9.61a, b}$$

Additionally, we have the net external force $\boldsymbol{F}$ of (9.46a) and (9.46c), and the net physical couple $\boldsymbol{\mathcal{C}}$ from (9.46b) and (9.46d):

$$\boldsymbol{\mathcal{C}} = \boldsymbol{A}_3 \times (\boldsymbol{C}_s + \boldsymbol{C}_b).$$

The actions upon an element are shown in Figure 9.10. Note that the external loads are multiplied by the differential area $\sqrt{a}\,d\theta^1\,d\theta^2$ and the internal actions $(\boldsymbol{\mathcal{N}}^\alpha, \boldsymbol{\mathcal{M}}^\alpha)$ are multiplied by the lengths of edges $\sqrt{a\,a^{\underline{\alpha\alpha}}}$.

Equilibrium of forces (from the free-body diagram) provides the equation (9.50a) of Section 9.7:

$$(\boldsymbol{N}^\alpha\sqrt{a}\,)_{,\alpha} + \sqrt{a}\,\boldsymbol{F} = \mathbf{o}.$$

Equilibrium of moment includes the moment of edge forces, viz.,

$$\boldsymbol{A}_\alpha \times \boldsymbol{N}^\alpha\sqrt{a}\,d\theta^1\,d\theta^2;$$

that equation has the form:

$$(\boldsymbol{A}_3 \times \boldsymbol{M}^\alpha \sqrt{a}\,)_{,\alpha} + \boldsymbol{A}_\alpha \times \boldsymbol{N}^\alpha \sqrt{a} + \boldsymbol{A}_3 \times \boldsymbol{C}\sqrt{a} = \mathbf{o}. \tag{9.62a}$$

This equation appears different than the previous version (9.50b); however, by means of (9.50b) and (9.62a), we obtain

$$\boldsymbol{A}_{3,\alpha} \times \boldsymbol{M}^\alpha + \boldsymbol{A}_\alpha \times \boldsymbol{N}^\alpha = -\boldsymbol{A}_3 \times \boldsymbol{T}. \tag{9.62b}$$

Now, by considering the equations (9.12a), (9.19a), (9.20b), (8.18a), and (8.18c), the above expression (9.62b) is another form of the condition that the work upon the spin $\dot{\boldsymbol{\Omega}}$ vanishes in (9.41), viz.,

$$\left[(N^{\gamma 3} - K^\gamma_\alpha M^{\alpha 3})C^\beta_\gamma - T^\beta C^3_3 + T^3 C^\beta_3 - M^{\alpha\beta} D_{\alpha 3}\right] \bar{e}_{\mu\beta}\, \boldsymbol{b}^\mu$$

$$- \left[(N^{\alpha\mu} - K^\alpha_\gamma M^{\gamma\mu})C^\beta_\alpha + T^\mu C^\beta_3\right] \bar{e}_{\mu\beta}\, \hat{\boldsymbol{N}} = \mathbf{o}. \tag{9.62c}$$

These are the same bracketed terms of (9.41).

We observe that the transverse shear stress $N^{\alpha 3}$ is not associated with a strain, nor does that stress appear in the equilibrium equation (9.50b). It is *interactive* through the equilibrium equation (9.62a–c). In cases of small strain the initial bracket of (9.62c) provides the approximation $T^\alpha \doteq N^{\alpha 3}$ [see equation (9.76b)].

9.10 Theory of Membranes

Apart from the initial assumption (9.5) the preceding theory is without restrictions upon the deformations. Specifically, the equations admit large deformations, displacements, rotations and strains. Also, the kinematic and equilibrium equations hold for any continuous cohesive medium. Only the potentials presume elasticity.

Now, let us suppose that the body is so thin that it has no resistance to bending and transverse shear. Then $M^{\alpha i} = N^{\alpha 3} = T^\alpha = 0$; only surface strains $h_{\alpha\beta}$ and associated stresses $N^{\alpha\beta}$ are relevant. Equilibrium is governed by the differential equation (9.50a) and edge condition (9.51a). The rate of (9.41) reduces to

$$\dot{u}_s = N^{\alpha\beta} \dot{h}_{\alpha\beta} + T^3 \dot{C}_{33} + N^{\alpha\beta} \dot{\bar{\omega}}_{\beta\alpha}. \tag{9.63}$$

The second term might be retained to account for circumstances of large external pressure; in most instances one would anticipate a state of "plane" stress ($T^3 = 0$). The final term reasserts the symmetry of the stress tensor ($N^{\alpha\beta} = N^{\beta\alpha}$).

The various functionals/potentials of Section 9.8 apply with the indicated simplifications, viz., the suppression of terms attributed to bending and transverse shear. Note that the deformed membrane resists transverse loading by virtue of its curvature. Specifically, no restrictions are placed on the magnitudes of displacement $\boldsymbol{R}$, nor rotation $\bar{r}_{i}^{\bullet j}$ of equation (9.8) which carries the initial triad $\boldsymbol{a}_i$ to the convected triad $\boldsymbol{b}_i$. In the event of large rotations, the resulting nonlinear equations are most effectively treated by a method of incremental loading and a succession of linear equations governing such increments [192] (see Section 11.15).

9.11 Approximations of Small Strain

Most shells in structural and mechanical applications are intended to sustain small strains (physical strains $\epsilon_{ij} \ll 1$). Indeed, most structures or machines have failed when yielding or fracture occurs. Accordingly, it is most appropriate to adopt those approximations which offer simpler, yet adequate, means of analysis.

Let us re-examine the shear stress defined by (9.39b), viz.,

$$M^{\alpha 3} \equiv \boldsymbol{b}^3 \cdot \boldsymbol{M}^\alpha = \int_{-t_-}^{t_+} \boldsymbol{b}^3 \cdot \boldsymbol{s}^\alpha \mu \theta^3 \, d\theta^3. \tag{9.64}$$

First, we note that our distinctions between vectors $\boldsymbol{a}'_i$ and $\boldsymbol{b}_i$ are negligible in the circumstances of *small* strain, i.e., $\boldsymbol{a}'_3 = \acute{\boldsymbol{g}}_3 \doteq \boldsymbol{b}_3 \equiv \hat{\boldsymbol{N}}$. Under most conditions of loading, the component $t^{\alpha 3} = \acute{\boldsymbol{g}}^3 \cdot \boldsymbol{s}^\alpha$ is negligible at the outer surfaces. Hence, $(\boldsymbol{b}^3 \cdot \boldsymbol{s}^\alpha)$ is also negligible at the top and bottom surfaces. If the distribution through the thickness is assumed parabolic:

$$t^{\alpha 3} \doteq {}_0t^{\alpha 3} \left[1 - \frac{t_- - t_+}{t_+ t_-} \theta^3 - \frac{1}{t_+ t_-} (\theta^3)^2 \right],$$

or, if $t_+ = t_- = h/2$, then

$$t^{\alpha 3} \doteq {}_0t^{\alpha 3} \left[1 - (2\theta^3/h)^2 \right].$$

We note that $\mu \doteq 1$ in a thin shell. Also, we note that the integrand of (9.64) vanishes at the reference surface ($\theta^3 = 0$). By all accounts we can neglect $M^{\alpha 3}$ ($\doteq 0$) in thin shells.

In most practical situations, the transverse normal stress $t^{33} = \acute{\boldsymbol{g}}^3 \cdot \boldsymbol{s}^3$ is negligible. Then, we can reasonably assume that T^3 of (9.40b) is also negligible.

With the foregoing assumptions, the internal power of (9.41) assumes the simpler form:

$$\begin{aligned}\dot{u}_s = {} & N^{\alpha\beta}\dot{C}_{\alpha\beta} + M^{\alpha\beta}\dot{D}_{\alpha\beta} + T^{\alpha}\dot{C}_{3\alpha} \\ & + [N^{\alpha 3}C^{\mu}_{\alpha} - T^{\mu}C_{33} - M^{\alpha\mu}D_{\alpha 3}]\dot{\bar{\omega}}_{3\mu} \\ & + [(N^{\alpha\beta} - K^{\alpha}_{\mu}M^{\mu\beta})C^{\gamma}_{\alpha} + T^{\beta}C^{\gamma}_{3}]\dot{\bar{\omega}}_{\beta\gamma}. \qquad (9.65)\end{aligned}$$

It remains to identify geometrical simplifications which are justified by small strain ($\epsilon_{ij} \ll 1$).

We recall the definitions of the strains [equations (9.9), (9.10), (9.17a, b), (9.20a, b) to (9.22a–c), and (9.42a–c)]:

$$h_{\alpha\beta} \equiv \boldsymbol{b}_{\beta} \cdot \boldsymbol{A}_{\alpha} - a_{\alpha\beta} = C_{\alpha\beta} - a_{\alpha\beta}, \qquad (9.66\text{a, b})$$

$$h_{3\alpha} \equiv \boldsymbol{b}_{\alpha} \cdot \boldsymbol{A}_3 = C_{3\alpha}, \qquad (9.66\text{c, d})$$

$$h_{33} \equiv \hat{\boldsymbol{N}} \cdot \boldsymbol{A}_3 - 1 = C_{33} - 1, \qquad (9.66\text{e})$$

$$\kappa_{\alpha 3} \equiv D_{\alpha 3} \equiv \hat{\boldsymbol{N}} \cdot \boldsymbol{A}_{3,\alpha}, \qquad (9.67\text{a, b})$$

$$\equiv C^{3}_{3,\alpha} + (C^{\eta}_{3}c^{\gamma}_{\eta})B_{\gamma\alpha}, \qquad (9.67\text{c})$$

$$\kappa_{\alpha\beta} \equiv D_{\alpha\beta} + b_{\alpha\beta}, \qquad (9.68\text{a})$$

$$\equiv \boldsymbol{b}_{\beta} \cdot \boldsymbol{A}_{3,\alpha} + b_{\alpha\beta}, \qquad (9.68\text{b})$$

$$D_{\alpha\beta} \equiv -C^{3}_{3}C_{\gamma\beta}B^{\gamma}_{\alpha} + (C^{\eta}_{3}c^{\gamma}_{\eta})\|^{*}_{\alpha}C_{\gamma\beta}. \qquad (9.69)$$

Let us first address the approximations of the bracketed terms of (9.65); these are equilibrium equations. Strains appear in those terms as a consequence of the deformed size and shape. For example, C^{α}_{β} appears in place of δ^{α}_{β} as a consequence of stretching of surface s_0, C^3_3 in place of unity, K^{α}_{β} in place of B^{α}_{β} [see (9.21a)], etc. If these strains are small compared to unity, then we have the approximations

$$(N^{\mu 3} - T^{\mu})\dot{\bar{\omega}}_{3\mu} + (N^{\gamma\beta} - M^{\mu\beta}B^{\gamma}_{\mu})\dot{\bar{\omega}}_{\beta\gamma} = 0. \tag{9.70}$$

This version (9.70) still admits the deformed curvature B^{γ}_{μ}; in most practical situations, one can also set $B^{\gamma}_{\mu} \doteq b^{\gamma}_{\mu}$. Note that the approximation (9.70) merely neglects the deformation of the element in Figure 9.10 when enforcing equilibrium of moment.

The remaining question concerns the approximation of the flexural strain $\kappa_{\alpha\beta}$ of (9.42b) and (9.22a), viz.,

$$\kappa_{\alpha\beta} = \boldsymbol{b}_{\beta} \cdot \boldsymbol{A}_{3,\alpha} + b_{\alpha\beta}.$$

It follows from (9.14) and (9.44a, b) that

$$\delta^{\beta}_{\alpha} = (\delta^{\mu}_{\alpha} + h^{\mu}_{\alpha})(\delta^{\beta}_{\mu} + H^{\beta}_{\mu}) = \delta^{\beta}_{\alpha} + h^{\beta}_{\alpha} + H^{\beta}_{\alpha} + h^{\mu}_{\alpha}H^{\beta}_{\mu}. \tag{9.71a}$$

For small strains ($h^{\alpha}_{\beta} \ll 1$, $H^{\alpha}_{\beta} \ll 1$); consequently, we can neglect the quadratic term in (9.71a) and use the approximation:

$$h^{\beta}_{\alpha} \doteq -H^{\beta}_{\alpha}. \tag{9.71b}$$

Then, by considering (9.12b), (9.19b), (9.20a), (9.43b), (9.66a), (9.68a), and (9.71b), we obtain

$$\boldsymbol{A}_{\alpha} = (h_{\alpha\beta} + a_{\alpha\beta})\boldsymbol{b}^{\beta}, \tag{9.72a}$$

$$\boldsymbol{A}_3 \doteq \hat{\boldsymbol{N}} + h^{\gamma}_3 \boldsymbol{b}_{\gamma}, \tag{9.72b}$$

$$\boldsymbol{A}_{3,\alpha} \doteq -B^{\gamma}_{\alpha}\boldsymbol{A}_{\gamma} + h^{\gamma}_3\|_{\alpha}\boldsymbol{b}_{\gamma} + h^{\gamma}_3 B_{\gamma\alpha}\hat{\boldsymbol{N}}, \tag{9.72c}$$

$$\boldsymbol{b}_{\beta} \cdot \boldsymbol{A}_{3,\alpha} \doteq -B_{\alpha\beta} - B^{\gamma}_{\alpha}H_{\beta\gamma} + h_{3\beta}\|_{\alpha}, \tag{9.73a}$$

$$\doteq -B_{\alpha\beta} + B^{\gamma}_{\alpha}h_{\beta\gamma} + h_{3\beta}\|_{\alpha}. \tag{9.73b}$$

Our approximation of the flexural strain follows:

$$\kappa_{\alpha\beta} \doteq -(B_{\alpha\beta} - b_{\alpha\beta}) + B^{\gamma}_{\alpha}h_{\beta\gamma} + h_{3\beta}\|_{\alpha}. \tag{9.74}$$

The virtual work of (9.35a, b) is simplified by the aforementioned approximations ($M^{\alpha 3} \doteq 0,\ T^3 \doteq 0$). Equation (9.35b) assumes the form:

$$\dot{u}_s \doteq (\dot{\boldsymbol{A}}_\alpha \cdot \boldsymbol{b}_i) N^{\alpha i} + (\dot{\boldsymbol{A}}_{3,\alpha} \cdot \boldsymbol{b}_\gamma) M^{\alpha\gamma} + (\dot{\boldsymbol{A}}_3 \cdot \boldsymbol{b}_\mu) T^\mu. \tag{9.75a}$$

Again, we neglect higher-order terms in (9.72a–c) to obtain

$$\dot{\boldsymbol{A}}_\alpha \doteq \dot{h}_{\alpha\beta} \boldsymbol{b}^\beta + (\dot{\boldsymbol{\Omega}} \times \boldsymbol{b}_\alpha),$$

$$\dot{\boldsymbol{A}}_3 \doteq \dot{h}_{3\gamma} \boldsymbol{b}^\gamma + (\dot{\boldsymbol{\Omega}} \times \hat{\boldsymbol{N}}),$$

$$\begin{aligned}\dot{\boldsymbol{A}}_{3,\alpha} &\doteq (-\dot{B}_{\alpha\beta} + B^\gamma_\alpha \dot{h}_{\beta\gamma} + \dot{h}_{3\beta}\|_\alpha) \boldsymbol{b}^\beta + B_{\gamma\alpha} \dot{h}^\gamma_3 \hat{\boldsymbol{N}} - B^\gamma_\alpha (\dot{\boldsymbol{\Omega}} \times \boldsymbol{b}_\gamma) \\ &\doteq \dot{\kappa}_{\alpha\gamma} \boldsymbol{b}^\gamma + B_{\gamma\alpha} \dot{h}^\gamma_3 \hat{\boldsymbol{N}} - B^\gamma_\alpha (\dot{\boldsymbol{\Omega}} \times \boldsymbol{b}_\gamma).\end{aligned}$$

Here, the strains are neglected in each of the terms associated with the spin $\dot{\boldsymbol{\Omega}}$ [see equations (9.25a–c)]. Therefore,

$$\begin{aligned}\dot{u}_s &= N^{\alpha\beta} \dot{h}_{\alpha\beta} + (\dot{\boldsymbol{\Omega}} \times \boldsymbol{b}_\alpha) \cdot (\boldsymbol{b}_i N^{\alpha i}) \\ &\quad + M^{\alpha\beta} \dot{\kappa}_{\alpha\beta} - (\dot{\boldsymbol{\Omega}} \times \boldsymbol{b}_\gamma) \cdot (\boldsymbol{b}_\eta M^{\alpha\eta}) B^\gamma_\alpha \\ &\quad + T^\alpha \dot{h}_{3\alpha} + (\dot{\boldsymbol{\Omega}} \times \hat{\boldsymbol{N}}) \cdot (\boldsymbol{b}_\alpha T^\alpha), \qquad (9.75\text{b}) \\ &= N^{\alpha\beta} \dot{h}_{\alpha\beta} + M^{\alpha\beta} \dot{\kappa}_{\alpha\beta} + T^\alpha \dot{h}_{3\alpha} \\ &\quad + (N^{\alpha\beta} + B^\beta_\phi M^{\phi\alpha}) \dot{\bar{\omega}}^3 \bar{e}_{\alpha\beta} + (N^{\alpha 3} - T^\alpha) \dot{\bar{\omega}}^\beta \bar{e}_{\beta\alpha}. \qquad (9.75\text{c})\end{aligned}$$

Since no work is expended in the rigid spin $\dot{\boldsymbol{\Omega}}$,

$$N^{\alpha\beta} + B^\beta_\phi M^{\phi\alpha} = N^{\beta\alpha} + B^\alpha_\phi M^{\phi\beta}, \tag{9.76a}$$

$$N^{\alpha 3} = T^\alpha. \tag{9.76b}$$

Note that these equations follow also from (9.70). Note too that the "strain" $\kappa_{\alpha\beta}$ in (9.74) and rate $\dot{\kappa}_{\alpha\beta}$ include a linear term $B^\gamma_\alpha \dot{h}_{\beta\gamma}$. This suggests an alternative flexural strain:

$$\rho_{\alpha\beta} \equiv -(B_{\alpha\beta} - b_{\alpha\beta}) + h_{3\beta}\|_\alpha. \tag{9.77}$$

Then, the power of the stresses takes the form:

$$\dot{u}_s = (N^{\alpha\beta} + B^{\beta}_{\gamma} M^{\gamma\alpha})\dot{h}_{\alpha\beta} + M^{\alpha\beta}\dot{\rho}_{\alpha\beta} + T^{\mu}\dot{h}_{3\mu}. \tag{9.78}$$

Now, the stress associated with the strain $h_{\alpha\beta}$ is the *symmetric* stress:

$$n^{\alpha\beta} \equiv N^{\alpha\beta} + B^{\beta}_{\gamma} M^{\gamma\alpha}. \tag{9.79}$$

The simplified forms of the equations (9.50a, b) follow:

$$\frac{1}{\sqrt{a}}(\sqrt{a}\, N^{\alpha\beta}\boldsymbol{b}_{\beta})_{,\alpha} + \frac{1}{\sqrt{a}}(\sqrt{a}\, T^{\alpha}\boldsymbol{b}_{3})_{,\alpha} + \boldsymbol{F} = \mathbf{o}, \tag{9.80a}$$

$$\frac{1}{\sqrt{a}}(\sqrt{a}\, M^{\alpha\beta}\boldsymbol{b}_{\beta})_{,\alpha} - T^{\alpha}\boldsymbol{b}_{\alpha} + \boldsymbol{C} = \mathbf{o}. \tag{9.80b}$$

The latter (9.80b) can be solved for components T^{α} (transverse shear) and substituted into the former (9.80a). The result is a second-order differential equation (vectorial) or three components. That equation and (9.76a) comprise the equilibrium conditions. *Note:* Consistency requires that (see Section 8.4)

$$\boldsymbol{b}_{\alpha,\beta} = \overline{\Gamma}_{\alpha\beta\gamma}\boldsymbol{b}^{\gamma} + B_{\alpha\beta}\boldsymbol{b}_{3}, \tag{9.81a}$$

$$\boldsymbol{b}_{3,\alpha} = -B^{\beta}_{\alpha}\boldsymbol{b}_{\beta}. \tag{9.81b}$$

In words, the triad $\boldsymbol{b}_i$ are associated with the deformed (bent) surface S_0, where we are neglecting stretch (${}^{*}\overline{\Gamma}_{\alpha\beta\gamma} \doteq \overline{\Gamma}_{\alpha\beta\gamma}$).

9.12 The Meaning of Thin

Since we are concerned with thin shells, it is appropriate to define thinness. From our previous discussion, it is evident that the components b^{α}_{β} are proportional to the curvature and/or torsion. Indeed, a component b^{α}_{β} can be expressed as a linear combination of the extremal values κ_1 and κ_2 by a transformation of coordinates. Now, suppose that κ denotes the greatest curvature or torsion of the reference surface in the undeformed shell. *The shell is thin if*

$$\kappa h \ll 1,$$

where h is the thickness. This means that quantities of order κh are to be neglected in comparison with unity: unfortunately, the error committed depends on many factors besides the magnitude of κh; for example, the nature of the loads and boundary conditions. There is no magic number ($\kappa h = ?$) which separates thin from thick. A rule-of-thumb says that the shell qualifies as thin if $\kappa h < 1/20$; it must be questioned when the circumstances are unusual or great accuracy is required.

Local effects of transverse shear are often present in the vicinity of a concentrated transverse load or in a zone adjacent to an edge, but the overall effects of flexural and extensional deformations dominate the behavior of thin shells. Therefore, most analyses of thin shells neglect the effect of transverse shear and follow the theory of Kirchhoff-Love (Chapter 10).

9.13 Theory of Hookean Shells with Transverse Shear Strain

Our initial approximation [see Section 9.5, equation (9.5)] admits finite strain and transverse shear $h_{3\alpha}$. We now seek the linear constitutive equations, simplifications for small strain ($\epsilon_{ij} \ll 1$), and "plane" stress ($t^{33} = s^{33} = 0$). To avail ourselves of the inherent symmetry, we employ the Green strain:

$$\gamma_{ij} = (\boldsymbol{G}_i \cdot \boldsymbol{G}_j - g_{ij}). \tag{9.82}$$

From the approximation (9.5),

$$\boldsymbol{R} = {}_0\boldsymbol{R} + \theta^3 \boldsymbol{A}_3 = {}_0\boldsymbol{R} + \theta^3(\hat{\boldsymbol{N}} + 2\gamma_\mu \boldsymbol{A}^\mu), \tag{9.83}$$

$$\boldsymbol{G}_\alpha \doteq \boldsymbol{A}_\alpha + \theta^3(2\gamma_\mu\|^*_\alpha - B_{\mu\alpha})\boldsymbol{A}^\mu + 2\theta^3\gamma_\mu B^\mu_\alpha \hat{\boldsymbol{N}}, \tag{9.84a}$$

$$\boldsymbol{A}_3 \equiv \boldsymbol{G}_3 = \hat{\boldsymbol{N}} + 2\gamma_\mu \boldsymbol{A}^\mu, \tag{9.84b}$$

where

$$\gamma_\mu \equiv \tfrac{1}{2}\boldsymbol{A}_3 \cdot \boldsymbol{A}_\mu.$$

We note [see equations (9.43a, b) and (9.71b)] the following:

$$\boldsymbol{A}_i = \boldsymbol{b}_i + h^j_i \boldsymbol{b}_j, \qquad \boldsymbol{b}_i = \boldsymbol{A}_i + H^j_i \boldsymbol{A}_j.$$

The first-order approximation follows:

$$\gamma_{ij} \doteq h_{ij} \doteq -H_{ij}.$$

Recall that if the material of the shell is elastically symmetric with respect to the θ^3 surface (see Section 5.14), i.e., the properties of the shell are unaltered by the transformation ($\theta^3 = -\bar{\theta}^3$), then the free energy per unit volume (5.52) for a linearly elastic material takes the form

$$\rho_0 \mathcal{F} = \tfrac{1}{2} E^{\alpha\beta\gamma\eta}\gamma_{\alpha\beta}\gamma_{\gamma\eta} + E^{33\gamma\eta}\gamma_{33}\gamma_{\gamma\eta} + 2E^{\alpha 3\beta 3}\gamma_{\alpha 3}\gamma_{\beta 3} + \tfrac{1}{2}E^{3333}\gamma_{33}^2$$
$$-\alpha^{\alpha\beta}\gamma_{\alpha\beta}(T - T_0) - \alpha^{33}\gamma_{33}(T - T_0). \tag{9.85}$$

In previous discussions (see Sections 6.16 and 9.8), we extolled the merits of the generalized potential $\mathcal{V}^*$ and the stationary criteria as the basis of approximation. As a functional of position $\boldsymbol{R}$, strain γ_{ij}, and stress s^{ij} [see equations (6.125) and (6.126)], we can introduce the most practical approximations and derive the consistent equations. For the shell with $t_+ = t_- = h/2$, we have the generalized version:

$$\mathcal{V}^* = \iint_{s_0}\int_{-h/2}^{h/2} \{u_0(\gamma_{ij}) - s^{ij}\left[\gamma_{ij} - \tfrac{1}{2}(\boldsymbol{G}_i \cdot \boldsymbol{G}_j - g_{ij})\right] - \boldsymbol{f}\cdot\boldsymbol{R}\}\,\mu\, d\theta^3\, ds_0$$
$$- \iint_{s_0} (\boldsymbol{t}^+ \cdot \boldsymbol{R}_+ \mu_+ + \boldsymbol{t}^- \cdot \boldsymbol{R}_- \mu_-)\, ds_0 - \int_c \int_{-h/2}^{h/2} \boldsymbol{t}^c \cdot \boldsymbol{R}\, d\theta^3\, dc. \tag{9.86}$$

With respect to the variation of position $\dot{\boldsymbol{R}}$, the variation $\dot{\mathcal{V}}^*$ appears like $\dot{W}$ of (9.45) (here $s^{ij}\boldsymbol{G}_i = \boldsymbol{s}^j$ and $ds_0 = \sqrt{a}\, d\theta^1\, d\theta^2$). The resulting stationary conditions are the equilibrium equations (9.50a, b) and boundary conditions (9.51a, b). Our present concerns are the strain-displacement and the constitutive equations which are implicit in the approximations of (9.84a, b) and those chosen for stress s^{ij} and strain γ_{ij}. Our attention focuses upon the variation:

$$\dot{v}^* = \int_{-h/2}^{h/2} \left\{\left(\frac{\partial u_0}{\partial \gamma_{ij}} - s^{ij}\right)\dot{\gamma}_{ij} - \dot{s}^{ij}\left[\gamma_{ij} - \tfrac{1}{2}(\boldsymbol{G}_i \cdot \boldsymbol{G}_j - g_{ij})\right]\right\}\mu\, d\theta^3. \tag{9.87}$$

Our approximation (9.83) differs from the Kirchhoff-Love approximation (10.1) in one essential: $\boldsymbol{A}_3 \neq \hat{\boldsymbol{N}}$. The Cauchy-Green strain at the reference surface is

$${}_0\gamma_{\alpha\beta} = \tfrac{1}{2}(A_{\alpha\beta} - a_{\alpha\beta}). \tag{9.88}$$

According to (9.82), (9.84a), and (9.88),

$$\gamma_{\alpha\beta} = {}_0\gamma_{\alpha\beta} + \theta^3 \rho_{\alpha\beta} + O[(\theta^3)^2\rho^2] + O[(\theta^3)^2\gamma^2], \tag{9.89}$$

where

$$\rho_{\alpha\beta} = -B_{\alpha\beta} + b_{\alpha\beta} + \gamma_\alpha\|_\beta + \gamma_\beta\|_\alpha, \tag{9.90a}$$

$$= \acute{\rho}_{\alpha\beta} + \gamma_\alpha\|_\beta + \gamma_\beta\|_\alpha, \tag{9.90b}$$

$$\acute{\rho}_{\alpha\beta} \equiv -B_{\alpha\beta} + b_{\alpha\beta}.$$

Here, the most prominent omissions are terms $O[(\theta^3)^2\gamma^2]$ which are negligible ($\gamma \ll 1$); these are traced to the terms produced by the transverse shear ($\gamma_\alpha\|_\beta + \gamma_\beta\|_\alpha$). Throughout, the flexural strain differs from that of the Kirchhoff-Love theory (see Chapter 10); the curvature $-B_{\alpha\beta}$ is supplanted by the term

$$-B_{\alpha\beta} + \gamma_\alpha\|_\beta + \gamma_\beta\|_\alpha,$$

attributed to the additional turning of the normal. Otherwise, the strain $\rho_{\alpha\beta}$ is the change-of-curvature. Indeed, only the transverse shear stress and strain distinguishes the present formulation.

Since our current interest is the presence of the small transverse shear strain, we turn to the term of (9.87) which provides the relation between the stress $N^{\alpha 3}$ and strain γ_α:

$$\int_{-h/2}^{h/2} \left(\frac{\partial u_0}{\partial \gamma_{\alpha 3}} - s^{\alpha 3} \right) \dot{\gamma}_{\alpha 3}\, \mu\, d\theta^3 = 0. \tag{9.91}$$

First, we note that the shear strain is given, according to (9.84a, b):

$$\gamma_{\alpha 3} = \gamma_{3\alpha} = \tfrac{1}{2}\boldsymbol{G}_3 \cdot \boldsymbol{G}_\alpha = \gamma_\alpha + O(\theta^3\gamma^2).$$

Since the strains are small, we can neglect the higher powers; this implies that $\gamma_{\alpha 3}$ is constant through the thickness. Then, by (9.85) and (9.91),

$$N^{\alpha 3} \equiv \int_{-h/2}^{h/2} s^{\alpha 3}\, \mu\, d\theta^3 \doteq 2\gamma_\beta \int_{-h/2}^{h/2} E^{\alpha 3 \beta 3}\, \mu\, d\theta^3.$$

Since the shell is thin ($h/R \ll 1$), we neglect the higher powers in the integrand, i.e., $E^{\alpha 3\beta 3} \doteq E^{\alpha 3\beta 3}(\theta^3 = 0) \equiv {}_0E^{\alpha 3\beta 3}$, $\mu \doteq 1$. Then, we have

$$N^{\alpha 3} \doteq 2h\, {}_0E^{\alpha 3\beta 3}\gamma_\beta. \tag{9.92}$$

We know that the shear stress $s^{\alpha 3} = s^{3\alpha}$ vanishes or, at most, has the small value of a traction at the surfaces s_+ and s_-. We adopt the simplest approximation consistent with our observations and the definitions (9.36a) and (9.38b):

$$s^{\alpha 3} = s^{3\alpha} \doteq \frac{3}{2h} N^{\alpha 3}\left[1 - \left(\frac{2\theta^3}{h}\right)^2\right]. \tag{9.93}$$

Then, it is reasonable to assume the transverse shear strain in the form of (9.93), i.e.,

$$\gamma_{\alpha 3} = \gamma_\alpha \left[1 - \left(\frac{2\theta^3}{h}\right)^2\right]. \tag{9.94}$$

When the forms (9.93) and (9.94) are introduced into (9.91), we obtain

$$\int_{-h/2}^{h/2} \left[\frac{3}{2h} N^{\alpha 3} - 2\, {}_0E^{\alpha 3\beta 3}\gamma_\beta\right]\left[1 - \left(\frac{2\theta^3}{h}\right)^2\right] \mu\, d\theta^3 = 0.$$

Again, neglecting the higher powers in the thickness, we obtain

$$N^{\alpha 3} \doteq \tfrac{4}{3} h\, {}_0E^{\alpha 3\beta 3}\gamma_\beta. \tag{9.95}$$

The normal component s^{33} is also absent, or very small, at the surfaces s_+ and s_-. Since our shell is presumably thin, we employ the relation

$$s^{33} = E^{33\gamma\eta}\gamma_{\gamma\eta} + E^{3333}\gamma_{33} - \alpha^{33}(T - T_0) = 0,$$

from which we obtain

$$\gamma_{33} = -\frac{E^{33\gamma\eta}}{E^{3333}}\gamma_{\gamma\eta} + \frac{\alpha^{33}}{E^{3333}}(T - T_0).$$

With the notations,

$$C^{\alpha\beta\gamma\eta} \equiv E^{\alpha\beta\gamma\eta} - \frac{E^{33\alpha\beta}E^{33\gamma\eta}}{E^{3333}}, \qquad \bar{\alpha}^{\alpha\beta} = \alpha^{\alpha\beta} - \frac{E^{33\alpha\beta}}{E^{3333}}\alpha^{33}, \tag{9.96a}$$

the free energy (9.85) takes the following form:

$$\rho_0\mathcal{F} = \tfrac{1}{2}C^{\alpha\beta\gamma\eta}\gamma_{\alpha\beta}\gamma_{\gamma\eta} + 2E^{\alpha 3\beta 3}\gamma_{\alpha 3}\gamma_{\beta 3} - \bar{\alpha}^{\alpha\beta}\gamma_{\alpha\beta}(T - T_0). \tag{9.96b}$$

Again, we neglect higher-order terms (e.g., $\theta^3\kappa \ll 1$, etc.).‡ With the aforementioned simplifications, the complete potential density (per unit area s_0) for our approximation of the *thin* Hookean shell follows:

$$\phi = \frac{h}{2}\,{}_0C^{\alpha\beta\gamma\eta}\,{}_0\gamma_{\alpha\beta}\,{}_0\gamma_{\gamma\eta} - h\,{}_0\bar{\alpha}^{\alpha\beta}\,{}_0\gamma_{\alpha\beta}\,\Delta T_0$$

$$+ \frac{h^3}{24}\,{}_0C^{\alpha\beta\gamma\eta}\,\acute{\rho}_{\alpha\beta}\acute{\rho}_{\gamma\eta} - h^2\,{}_0\bar{\alpha}^{\alpha\beta}\,\acute{\rho}_{\alpha\beta}\,\Delta T_1$$

$$+ 2h(F_s)\,{}_0E^{\alpha3\beta3}\gamma_\alpha\gamma_\beta, \tag{9.97}$$

where

$$\Delta T_0 \equiv \frac{1}{h}\int_{-h/2}^{h/2}(T - T_0)\,d\theta^3, \qquad \Delta T_1 \equiv \frac{1}{h^2}\int_{-h/2}^{h/2}(T - T_0)\theta^3\,d\theta^3.$$

The potential is thereby split into three parts; (1) membrane, (2) bending, and (3) transverse shear. The factor F_s depends upon the approximation of the shear strains (e.g., $F_s = 1$ for constant distribution through the thickness). All coefficients depend upon the actual constituency; in some instances, a shell might be composed of lamina with less stiffness in shear. If the shell is homogeneous, then

$${}_0C^{\alpha\beta\gamma\eta} = \frac{E}{2(1+\nu)}\left[a^{\alpha\gamma}a^{\beta\eta} + a^{\alpha\eta}a^{\beta\gamma} + \frac{2\nu}{1-\nu}a^{\alpha\beta}a^{\gamma\eta}\right], \tag{9.98a}$$

$${}_0E^{\alpha3\beta3} = \frac{E}{2(1+\nu)}a^{\alpha\beta}. \tag{9.98b}$$

‡Further arguments, as given in Section 10.6, support these approximations.

Chapter 10

Theories under the Kirchhoff-Love Constraint

10.1 Kinematics

In his theory of plates, G. R. Kirchhoff [152] introduced the hypothesis:

The normal line θ^3 remains normal and unstretched.

This hypothesis was used by H. Aron [151] and by A. E. H. Love [153] in their theories of shells. By the Kirchhoff hypothesis

$$\boldsymbol{R} = {}_0\boldsymbol{R} + \theta^3 \hat{\boldsymbol{N}}, \tag{10.1}$$

$$\boldsymbol{G}_\alpha = \boldsymbol{A}_\alpha - \theta^3 B_\alpha^\beta \boldsymbol{A}_\beta, \tag{10.2a}$$

$$\boldsymbol{G}_3 = \boldsymbol{A}_3 \equiv \hat{\boldsymbol{N}}. \tag{10.2b}$$

The last equation (10.2b) asserts the orthogonality of the vectors $\boldsymbol{A}_\alpha$ and $\boldsymbol{A}_3 \equiv \hat{\boldsymbol{N}}$. This means that the rotation [Figure 9.8, equation (9.8)] which carries the initial triad ($\boldsymbol{a}_1$, $\boldsymbol{a}_2$, $\hat{\boldsymbol{a}}_3 = \hat{\boldsymbol{n}}$) to the convected (but undeformed) triad ($\boldsymbol{b}_1$, $\boldsymbol{b}_2$, $\boldsymbol{b}_3 = \hat{\boldsymbol{N}}$) is also the rotation of the principal lines (see Section 9.5). In words, θ^3 is a principal direction. The only motion of the normal $\hat{\boldsymbol{N}}$ is a rotation and that is determined by the motion of the surface s_0, specifically, the motion of the tangents $\boldsymbol{A}_\alpha$. We require the relation between the spin $\dot{\boldsymbol{\Omega}}$,

$$\dot{\boldsymbol{\Omega}} = \dot{\omega}^\alpha \boldsymbol{A}_\alpha + \dot{\omega}^3 \hat{\boldsymbol{N}},$$

and the motion of the surface (${}_0\dot{\boldsymbol{R}}_{,\alpha} = \dot{\boldsymbol{A}}_\alpha$). That is dictated by the orthogonality:

$$\boldsymbol{A}_\alpha \cdot \hat{\boldsymbol{N}} = 0.$$

It follows that

$$\dot{\boldsymbol{A}}_\alpha \cdot \hat{\boldsymbol{N}} = -\boldsymbol{A}_\alpha \cdot (\hat{\boldsymbol{N}})^{\cdot} = -\boldsymbol{A}_\alpha \cdot (\dot{\boldsymbol{\Omega}} \times \hat{\boldsymbol{N}}), \tag{10.3a}$$

$$\dot{\boldsymbol{\Omega}} \times \hat{\boldsymbol{N}} = \dot{\omega}^\alpha \, \overline{E}_{\mu\alpha} \boldsymbol{A}^\mu, \tag{10.3b}$$

$$\dot{\boldsymbol{\Omega}} \times \boldsymbol{A}_\alpha = \dot{\omega}^\mu \, \overline{E}_{\mu\alpha} \hat{\boldsymbol{N}} + \dot{\omega}^3 \, \overline{E}_{\alpha\mu} \boldsymbol{A}^\mu, \tag{10.3c}$$

where

$$\overline{E}_{\alpha\beta}(\theta^1, \theta^2) = E_{\alpha\beta 3}(\theta^1, \theta^2, 0) = \sqrt{A}\, \epsilon_{\alpha\beta 3},$$

$$\overline{E}^{\alpha\beta}(\theta^1, \theta^2) = E^{\alpha\beta 3}(\theta^1, \theta^2, 0) = \epsilon^{\alpha\beta 3}/\sqrt{A},$$

$$\sqrt{A} = \hat{\boldsymbol{N}} \cdot (\boldsymbol{A}_1 \times \boldsymbol{A}_2).$$

If $\hat{\boldsymbol{t}}_\alpha$ denote unit tangent vectors along orthogonal lines (lengths c_1 and c_2), then $\hat{\boldsymbol{t}}_\alpha \cdot \hat{\boldsymbol{N}} = 0$ and

$$(\hat{\boldsymbol{t}}_\alpha)^{\cdot} \cdot \hat{\boldsymbol{N}} = -\hat{\boldsymbol{t}}_\alpha \cdot (\hat{\boldsymbol{N}})^{\cdot} = \hat{\boldsymbol{N}} \cdot \frac{\partial\, {}_0\dot{\boldsymbol{R}}}{\partial c_\alpha}. \tag{10.4a, b}$$

To examine the power of the stresses, we require $\dot{\boldsymbol{G}}_i$:

$$\begin{aligned} \dot{\boldsymbol{G}}_\alpha &= \dot{\boldsymbol{A}}_\alpha + \theta^3 (\hat{\boldsymbol{N}}_{,\alpha})^{\cdot} \\ &= \dot{\boldsymbol{A}}_\alpha - \theta^3 (B_{\alpha\beta} \boldsymbol{A}^\beta)^{\cdot}. \end{aligned} \tag{10.5}$$

Following the decomposition of Section 3.18, we can express the rates $\dot{\boldsymbol{A}}_\alpha$ and $(\hat{\boldsymbol{N}})^{\cdot}$ in terms of the strain and spin:

$$\dot{\boldsymbol{A}}_\alpha = ({}_0\dot{\gamma}_{\alpha\beta} + \dot{\omega}_{\beta\alpha}) \boldsymbol{A}^\beta + \dot{\omega}_{3\alpha} \hat{\boldsymbol{N}}, \tag{10.6a}$$

$$(\hat{\boldsymbol{N}})^{\cdot} = \dot{\omega}_{\beta 3} \boldsymbol{A}^\beta, \tag{10.6b}$$

where

$$\dot{\omega}_{\beta\alpha} \equiv \dot{\omega}^3 \, \overline{E}_{\alpha\beta}, \qquad \dot{\omega}_{\beta 3} \equiv \dot{\omega}^\phi \, \overline{E}_{\beta\phi}.$$

Since $\boldsymbol{A}^\gamma \cdot \boldsymbol{A}_\alpha = \delta^\gamma_\alpha$, it follows that

$$\dot{\boldsymbol{A}}^\alpha = -({}_0\dot{\gamma}_\gamma^{\cdot\,\alpha} + \dot{\omega}^\alpha_{\cdot\,\gamma})\boldsymbol{A}^\gamma + \dot{\omega}_3^{\cdot\,\alpha}\,\hat{\boldsymbol{N}}. \tag{10.6c}$$

Here, the mixed components are associated with the metric $(A_{\alpha\beta},\, A^{\alpha\beta})$ of the deformed surface, i.e.,

$${}_0\dot{\gamma}_\gamma^{\cdot\,\beta} = A^{\beta\alpha}\,{}_0\dot{\gamma}_{\gamma\alpha}, \qquad \dot{\omega}^\beta_{\cdot\,\gamma} = A^{\beta\alpha}\dot{\omega}_{\alpha\gamma}.$$

10.2 Stresses and Strains

The power of stresses (per unit area) is

$$\dot{u}_s = \int_{-t_-}^{t_+} \left[\boldsymbol{s}^\alpha \cdot \dot{\boldsymbol{G}}_\alpha + \boldsymbol{s}^3 \cdot \dot{\boldsymbol{G}}_3\right] \mu\, d\theta^3, \tag{10.7a}$$

$$= \int_{-t_-}^{t_+} \left[\boldsymbol{s}^\alpha \cdot \dot{\boldsymbol{A}}_\alpha + \theta^3 \boldsymbol{s}^\alpha \cdot (\hat{\boldsymbol{N}}_{,\alpha})^\cdot + \boldsymbol{s}^3 \cdot (\hat{\boldsymbol{N}})^\cdot\right] \mu\, d\theta^3. \tag{10.7b}$$

By substituting (10.5) and (10.6a–c), and using the definitions (9.36a, b) and (9.37), we obtain

$$\dot{u}_s = \boldsymbol{N}^\alpha \cdot \dot{\boldsymbol{A}}_\alpha - \boldsymbol{M}^\alpha \big(\dot{B}_{\alpha\gamma}\boldsymbol{A}^\gamma + B_{\alpha\gamma}\dot{\boldsymbol{A}}^\gamma\big) + \boldsymbol{T} \cdot \big(\dot{\omega}_{\gamma 3}\boldsymbol{A}^\gamma\big). \tag{10.8}$$

To avail ourselves of the symmetric geometrical relations of the deformed surface S_0 $(\boldsymbol{A}_\alpha \cdot \boldsymbol{A}_\beta \equiv A_{\alpha\beta},\ \boldsymbol{A}_{\alpha,\beta} \cdot \hat{\boldsymbol{N}} \equiv B_{\alpha\beta})$, we employ components associated with the triad $\boldsymbol{A}_i$:

$$N^{\alpha\beta} \equiv \boldsymbol{A}^\beta \cdot \boldsymbol{N}^\alpha, \qquad N^{\alpha 3} \equiv \hat{\boldsymbol{N}} \cdot \boldsymbol{N}^\alpha, \qquad T^\alpha \equiv \boldsymbol{A}^\alpha \cdot \boldsymbol{T}, \tag{10.9a–c}$$

$$M^{\alpha\beta} \equiv \boldsymbol{A}^\beta \cdot \boldsymbol{M}^\alpha, \qquad M^{\alpha 3} \equiv \hat{\boldsymbol{N}} \cdot \boldsymbol{M}^\alpha. \tag{10.9d, e}$$

Substituting (10.6a) and (10.6c) for the rates $\dot{\boldsymbol{A}}_\alpha$ and $\dot{\boldsymbol{A}}^\alpha$ and employing the components (10.9a–e), we obtain

$$\dot{u}_s = \big(N^{\alpha\beta} + B^\beta_\gamma M^{\gamma\alpha}\big)\,{}_0\dot{\gamma}_{\alpha\beta} - M^{\alpha\beta}\dot{B}_{\alpha\beta} + \big(N^{\alpha\beta} + B^\beta_\gamma M^{\gamma\alpha}\big)\dot{\omega}^3\,\overline{E}_{\alpha\beta}$$

$$+ \big(N^{\alpha 3} - T^\alpha - B^\alpha_\gamma M^{\gamma 3}\big)\dot{\omega}^\phi\,\overline{E}_{\phi\alpha}. \tag{10.10}$$

Since no work is done upon the (rigid) spin $\dot{\boldsymbol{\Omega}}$,

$$N^{\alpha\beta} + B^{\beta}_{\gamma} M^{\gamma\alpha} = N^{\beta\alpha} + B^{\alpha}_{\gamma} M^{\gamma\beta}, \tag{10.11a}$$

$$N^{\alpha 3} = T^{\alpha} + B^{\alpha}_{\gamma} M^{\gamma 3}. \tag{10.11b}$$

As noted previously [see Section 9.11, equation (9.64)], the final term $B^{\alpha}_{\gamma} M^{\gamma 3}$ has little consequence; indeed the stress $M^{\gamma 3}$ is itself small. Then, (10.11b) is simplified:

$$N^{\alpha 3} \doteq T^{\alpha}. \tag{10.12}$$

We note the most convenient choices for the strains and stresses (cf. [157], [159]). These are the symmetric tensors:

$${}_0\gamma_{\alpha\beta} \equiv \tfrac{1}{2}(A_{\alpha\beta} - a_{\alpha\beta}), \qquad {}_0\dot{\gamma}_{\alpha\beta} = \tfrac{1}{2}\dot{A}_{\alpha\beta}, \tag{10.13a}$$

$$\rho_{\alpha\beta} \equiv -(B_{\alpha\beta} - b_{\alpha\beta}), \qquad \dot{\rho}_{\alpha\beta} = -\dot{B}_{\alpha\beta}, \tag{10.13b}$$

$$n^{\alpha\beta} \equiv N^{\alpha\beta} + B^{\beta}_{\gamma} M^{\gamma\alpha} = n^{\beta\alpha}, \tag{10.14a}$$

$$m^{\alpha\beta} \equiv \tfrac{1}{2}(M^{\alpha\beta} + M^{\beta\alpha}) = m^{\beta\alpha}. \tag{10.14b}$$

In terms of these variables,

$$\dot{u}_s = n^{\alpha\beta}\, {}_0\dot{\gamma}_{\alpha\beta} + m^{\alpha\beta} \dot{\rho}_{\alpha\beta}. \tag{10.15}$$

The reader might compare (10.10) with the corresponding result (9.78). The latter (9.78) includes the work upon a transverse shear strain $\dot{h}_{3\beta}$. Here, the stresses of (10.9a–e) differ from those of (9.38a, b) to (9.40a, b), but only by the stretch which carries $\boldsymbol{b}_{\alpha}$ to $\boldsymbol{A}_{\alpha}$:
By (9.38a, b)

$$\boldsymbol{N}^{\alpha} \equiv N^{\alpha\beta} \boldsymbol{b}_{\beta} + N^{\alpha 3} \hat{\boldsymbol{N}};$$

by (10.9a, b)

$$\boldsymbol{N}^{\alpha} \equiv N^{\alpha\beta} \boldsymbol{A}_{\beta} + N^{\alpha 3} \hat{\boldsymbol{N}}.$$

Since (9.78) applies to small strain ($\epsilon_{ij} \ll 1$; $h_{ij} \doteq \gamma_{ij}$), the distinction is irrelevant.

10.3 Equilibrium

The total virtual work upon the "Kirchhoff-Love" shell includes the work of internal forces $\dot{u}_s$ of (10.7a, b) and the work of external forces, force $\boldsymbol{F}$ and couple $\boldsymbol{C}$ upon the surface, and force $\overline{\boldsymbol{N}}$ and couple $\overline{\boldsymbol{M}}$ on the edge:

$$\dot{W} = \iint_{s_0} \left[\boldsymbol{N}^\alpha \cdot {}_0\dot{\boldsymbol{R}}_{,\alpha} + \boldsymbol{M}^\alpha \cdot (\hat{\boldsymbol{N}}_{,\alpha})^{\cdot} + \boldsymbol{T} \cdot (\hat{\boldsymbol{N}})^{\cdot} - \boldsymbol{F} \cdot {}_0\dot{\boldsymbol{R}} - \boldsymbol{C} \cdot (\hat{\boldsymbol{N}})^{\cdot} \right] ds_0$$

$$- \int_{c_t} \left[\overline{\boldsymbol{N}} \cdot {}_0\dot{\boldsymbol{R}} + \overline{\boldsymbol{M}} \cdot (\hat{\boldsymbol{N}})^{\cdot} \right] dc. \tag{10.16}$$

Here, the force $\boldsymbol{F}$ and couple $\boldsymbol{C}$ include surface tractions on s_- and s_+ and body force $\boldsymbol{f}$, as defined by (9.46a–d) and (9.49a, b); edge force $\overline{\boldsymbol{N}}$ and couple $\overline{\boldsymbol{M}}$ are the actions per unit length as defined by (9.47a, b). Integration results in the form of (9.49c) and the differential equations (9.50a, b). There is a *fundamental difference:* In the present case $(\hat{\boldsymbol{N}})^{\cdot}$ is *not* independent of ${}_0\dot{\boldsymbol{R}}_{,\alpha}$; these are related according to equation (10.3a). Specifically, we have

$$(\hat{\boldsymbol{N}}_{,\alpha})^{\cdot} = -\dot{B}^\beta_\alpha \boldsymbol{A}_\beta - B^\beta_\alpha \dot{\boldsymbol{A}}_\beta, \tag{10.17a}$$

where

$$A_{\gamma\beta}\dot{B}^\beta_\alpha = (A_{\gamma\beta}B^\beta_\alpha)^{\cdot} - B^\beta_\alpha \dot{A}_{\gamma\beta}$$

$$= \dot{B}_{\gamma\alpha} - B^\beta_\alpha \dot{A}_{\gamma\beta}, \tag{10.17b}$$

$$\dot{B}_{\gamma\alpha} = \dot{\boldsymbol{A}}_{\alpha,\gamma} \cdot \hat{\boldsymbol{N}} + \boldsymbol{A}_{\alpha,\gamma} \cdot (\hat{\boldsymbol{N}})^{\cdot}$$

$$= \dot{\boldsymbol{A}}_{\alpha,\gamma} \cdot \hat{\boldsymbol{N}} + ({}^*\overline{\Gamma}^\mu_{\alpha\gamma} \boldsymbol{A}_\mu + B_{\alpha\gamma}\hat{\boldsymbol{N}}) \cdot (\hat{\boldsymbol{N}})^{\cdot}. \tag{10.17c}$$

Since $\hat{\boldsymbol{N}} \cdot \boldsymbol{A}_\alpha = 0$ and $\hat{\boldsymbol{N}} \cdot \hat{\boldsymbol{N}} = 1$,

$$\boldsymbol{A}_\alpha \cdot (\hat{\boldsymbol{N}})^{\cdot} = -\hat{\boldsymbol{N}} \cdot \dot{\boldsymbol{A}}_\alpha, \qquad \hat{\boldsymbol{N}} \cdot (\hat{\boldsymbol{N}})^{\cdot} = 0. \tag{10.17d}$$

Therefore,

$$\dot{B}_{\gamma\alpha} = \dot{\boldsymbol{A}}_{\alpha,\gamma} \cdot \hat{\boldsymbol{N}} - {}^*\overline{\Gamma}^\mu_{\alpha\gamma} \hat{\boldsymbol{N}} \cdot \dot{\boldsymbol{A}}_\mu. \tag{10.17e}$$

By means of (10.17a–e), every incremental term of (10.16) has an incremental factor: ${}_0\dot{\boldsymbol{R}}$, $\dot{\boldsymbol{A}}_\alpha = {}_0\dot{\boldsymbol{R}}_{,\alpha}$, or $\dot{\boldsymbol{A}}_{\alpha,\gamma} = {}_0\dot{\boldsymbol{R}}_{,\alpha\gamma}$. Additionally, we note that the following sum vanishes by (10.11b):

$$N^{\alpha 3}\hat{\boldsymbol{N}}\cdot{}_0\dot{\boldsymbol{R}}_{,\alpha} + T^\alpha \boldsymbol{A}_\alpha\cdot(\hat{\boldsymbol{N}})^{\boldsymbol{\cdot}} - B^\alpha_\beta M^{\beta 3}\hat{\boldsymbol{N}}\cdot\dot{\boldsymbol{A}}_\alpha$$

$$= (N^{\alpha 3} - T^\alpha - B^\alpha_\beta M^{\beta 3})\hat{\boldsymbol{N}}\cdot\dot{\boldsymbol{A}}_\alpha = 0.$$

Then, the work $\dot{W}$ of (10.16) assumes the form:

$$\dot{W} = \iint_{s_0}\Big[(N^{\alpha\beta} + B^\beta_\phi M^{\phi\alpha})\boldsymbol{A}_\beta\cdot{}_0\dot{\boldsymbol{R}}_{,\alpha} + C^\alpha\hat{\boldsymbol{N}}\cdot{}_0\dot{\boldsymbol{R}}_{,\alpha}$$

$$+ {}^*\overline{\Gamma}^\alpha_{\phi\beta}M^{\beta\phi}\hat{\boldsymbol{N}}\cdot{}_0\dot{\boldsymbol{R}}_{,\alpha} - M^{\alpha\beta}\hat{\boldsymbol{N}}\cdot{}_0\dot{\boldsymbol{R}}_{,\alpha\beta} - \boldsymbol{F}\cdot{}_0\dot{\boldsymbol{R}}\Big]\,ds_0$$

$$- \int_{c_t}\big[\overline{\boldsymbol{N}}\cdot{}_0\dot{\boldsymbol{R}} + \overline{\boldsymbol{M}}\cdot(\hat{\boldsymbol{N}})^{\boldsymbol{\cdot}}\big]\,dc.$$

Following the integrations by parts, the integrand of the surface integral assumes the form

$$\boldsymbol{\mathcal{L}}\left(N^{\alpha\beta}, M^{\alpha\beta}, C^\alpha, F^i\right)\cdot{}_0\dot{\boldsymbol{R}}.$$

Simply stated, we obtain one vectorial equation of equilibrium; $\boldsymbol{\mathcal{L}} = \mathbf{o}$ asserts the vanishing of force. The same result is obtained by eliminating $T^\alpha = N^{\alpha 3}$ [see equation (10.11b)] from (9.50a, b); the tensorial components (see Section 9.9) are expressed here in terms of the symmetric stresses ($n^{\alpha\beta}$, $m^{\alpha\beta}$) of (10.14a, b):

$$(n^{\alpha\beta} - B^\beta_\gamma m^{\alpha\gamma})\|_\alpha - B^\beta_\alpha m^{\gamma\alpha}\|_\gamma - C^\alpha B^\beta_\alpha + F^\beta = 0, \qquad (10.18a)$$

$$m^{\gamma\alpha}\|_{\gamma\alpha} + (n^{\alpha\beta} - B^\beta_\gamma m^{\alpha\gamma})B_{\alpha\beta} + C^\alpha\|_\alpha + F^3 = 0. \qquad (10.18b)$$

Equations (10.18a, b) are not exact, because the stress ($\boldsymbol{N}^\alpha$, $\boldsymbol{M}^\alpha$, $\boldsymbol{T}$) are based upon *undeformed* lengths and work is expressed per unit of *undeformed* area, whereas the basis vectors ($\boldsymbol{A}_\alpha$, $\hat{\boldsymbol{N}}$) are those of the *deformed* system. Mathematically, derivatives require Christoffel symbols of both metrics (${}^*\overline{\Gamma}^\alpha_{\beta\gamma}$ and $\overline{\Gamma}^\alpha_{\beta\gamma}$). Exact versions have the same form [161], wherein all quantities are based upon the *deformed* system. Our versions account for the *deformed* curvatures $B_{\alpha\beta}$, but neglect stretching; hence, the indicated differentiation ($\|$) entails the metric of the initial surface.

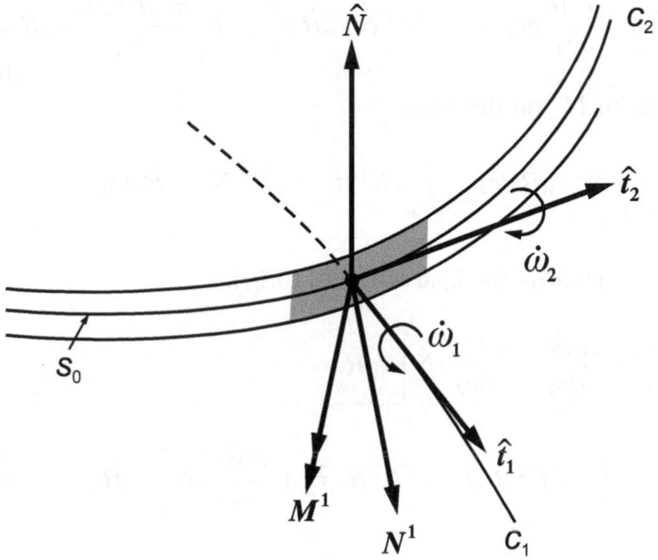

Figure 10.1 Edge conditions

The interdependence of the normal rotation and surface motion has an *essential* bearing on the edge conditions. If c_2 denotes distance along the edge and c_1 distance along a normal on the extended surface s_0, then their unit tangents are $\hat{\boldsymbol{t}}_2$ and $\hat{\boldsymbol{t}}_1$, as depicted in Figure 10.1. Vectors $\boldsymbol{N}^\alpha$ and $\boldsymbol{M}^\alpha$ are defined by (9.36a, b). Notice that the moment *vector* is

$$\boldsymbol{m}^1 = \acute{m}^{11}\,\hat{\boldsymbol{t}}_1 + \acute{m}^{12}\,\hat{\boldsymbol{t}}_2 \equiv \hat{\boldsymbol{N}} \times \boldsymbol{M}^1, \qquad \boldsymbol{M}^1 = \acute{M}^{11}\,\hat{\boldsymbol{t}}_1 + \acute{M}^{12}\,\hat{\boldsymbol{t}}_2;$$

twisting couple $\acute{m}^{11} = -\acute{M}^{12}$, bending couple $\acute{m}^{12} = \acute{M}^{11}$. Note too that all quantities are associated with the edge coordinates c_α. Clearly, the motion of $\hat{\boldsymbol{t}}_1$ and $\hat{\boldsymbol{t}}_2$ are determined by the deformation/motion of surface s_0 and, since $\hat{\boldsymbol{N}}$ remains orthogonal, the motion of $\hat{\boldsymbol{N}}$ is also determined. In accordance with (10.4a, b), the work of couple $\boldsymbol{M}^1$ is

$$\int_{c_2} \boldsymbol{M}^1 \cdot (\hat{\boldsymbol{N}})^{\cdot}\, dc_2 = \int_{c_2} \left[\acute{M}^{11}\,\hat{\boldsymbol{t}}_1 \cdot (\hat{\boldsymbol{N}})^{\cdot} - \acute{M}^{12}\hat{\boldsymbol{N}} \cdot \frac{\partial_0 \dot{\boldsymbol{R}}}{\partial c_2} \right] dc_2.$$

The second term can be integrated along the edge c_2:

$$-\int_a^b \acute{M}^{12}\hat{\boldsymbol{N}} \cdot \frac{\partial\, {}_0\dot{\boldsymbol{R}}}{\partial c_2}\, dc_2 = -\left(\acute{M}^{12}\hat{\boldsymbol{N}} \cdot {}_0\dot{\boldsymbol{R}}\right)\Big|_a^b + \int_{c_2} \frac{\partial(\acute{M}^{12}\hat{\boldsymbol{N}})}{\partial c_2} \cdot {}_0\dot{\boldsymbol{R}}\, dc_2. \tag{10.19}$$

The work of force $\boldsymbol{N}^1$ on the edge,

$$\int_{c_2} \boldsymbol{N}^1 \cdot {}_0\dot{\boldsymbol{R}}\, dc_2 = \int_{c_2} (\acute{N}^{1\alpha}\hat{\boldsymbol{t}}_\alpha + \acute{N}^{13}\hat{\boldsymbol{N}}) \cdot {}_0\dot{\boldsymbol{R}}\, dc_2,$$

is thereby augmented by the final term of (10.19):

$$\left[\acute{M}^{12}\frac{\partial \hat{\boldsymbol{N}}}{\partial c_2} + \frac{\partial \acute{M}^{12}}{\partial c_2}\hat{\boldsymbol{N}}\right] \cdot {}_0\dot{\boldsymbol{R}}$$

$$= \left[-\acute{M}^{12}\acute{B}_{12}\hat{\boldsymbol{t}}_1 - \acute{M}^{12}\acute{B}_{22}\hat{\boldsymbol{t}}_2 + \frac{\partial \acute{M}^{12}}{\partial c_2}\hat{\boldsymbol{N}}\right] \cdot {}_0\dot{\boldsymbol{R}}, \tag{10.20}$$

where $\acute{B}_{\alpha\beta}$ denote the actual curvatures and twists on c_β (see Section 8.7):

$$\acute{B}_{\alpha\beta} \equiv -\hat{\boldsymbol{t}}_\alpha \cdot \frac{\partial \hat{\boldsymbol{N}}}{\partial c_\beta}.$$

On the other hand, the work of the component $\acute{M}^{11}$ is

$$\acute{M}^{11}\hat{\boldsymbol{t}}_1 \cdot (\hat{\boldsymbol{N}})^{\cdot} = -\acute{M}^{11}\hat{\boldsymbol{N}} \cdot (\hat{\boldsymbol{t}}_1)^{\cdot} = \acute{M}^{11}\dot{\boldsymbol{\Omega}} \cdot \hat{\boldsymbol{t}}_2 = \acute{M}^{11}\dot{\omega}_2.$$

In summary, on the edge with normal $\hat{\boldsymbol{t}}_1 = n_{1\alpha}\boldsymbol{A}^\alpha$:

$$\acute{M}^{11} = M^{\alpha\beta}n_{1\alpha}n_{1\beta}, \tag{10.21a}$$

$$\acute{N}^{11} - \acute{M}^{12}B_{12} = N^{\alpha\beta}n_{1\alpha}n_{1\beta} - M^{\alpha\beta}n_{1\alpha}n_{2\beta}B_{12}, \tag{10.21b}$$

$$\acute{N}^{12} - \acute{M}^{12}B_{22} = N^{\alpha\beta}n_{1\alpha}n_{2\beta} - M^{\alpha\beta}n_{1\alpha}n_{2\beta}B_{22}, \tag{10.21c}$$

$$\acute{N}^{13} + \frac{\partial \acute{M}^{12}}{\partial c_2} = N^{\alpha 3}n_{1\alpha} + \frac{\partial}{\partial c_2}(M^{\alpha\beta}n_{1\alpha}n_{2\beta}). \tag{10.21d}$$

The reduction from five independent actions ($\acute{M}^{11}$, $\acute{M}^{12}$, $\acute{N}^{11}$, $\acute{N}^{12}$, $\acute{N}^{13}$) to these four is attributed to G. R. Kirchhoff. An account of the historical controversy is given by W. Thomson and P. G. Tait [193]. Additional details

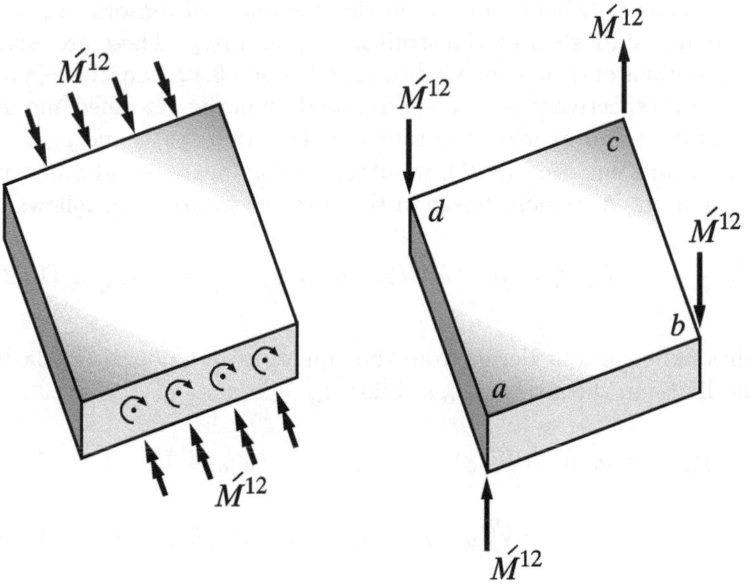

Figure 10.2 Twist via concentrated forces at the corners of a rectangular plate

and physical interpretations are contained in articles by W. T. Koiter [194] and G. A. Wempner [195].

We must also acknowledge the evaluation at end points a and b in equation (10.19): If the boundary is closed these are canceled. Otherwise, the effect of twisting couple $\acute{M}^{12}$ is a concentrated force. This is well illustrated by the action of the constant twist $\acute{M}^{12}$ upon the opposing edges of the rectangular plate in Figure 10.2; in the Kirchhoff-Love theory, this is equivalent to the concentrated forces at a and b (and c and d to maintain equilibrium).

10.4 Compatibility Equations, Stress Functions, and the Static-Geometric Analogy

In the search for mathematical solutions of the governing differential equations, one can only circumvent the displacement $({}_0\boldsymbol{R} - {}_0\boldsymbol{r})$ *if* one satisfies the necessary and sufficient conditions for existence of that continuous function. These are the compatibility equations. In the Kirchhoff-Love theory, these are basically conditions upon the fundamental tensors, $A_{\alpha\beta}$ and $B_{\alpha\beta}$, which are established by the strains, ${}_0\gamma_{\alpha\beta}$ and $\rho_{\alpha\beta}$. These are essentially the equations of Gauss and Codazzi [see Section 8.12, equations (8.91) and (8.93a, b), respectively], which are imposed upon the deformed and undeformed surfaces and expressed in terms of the strains ${}_0\gamma_{\alpha\beta}$ and $\rho_{\alpha\beta}$.

From the Gauss equation (8.91), we obtain a nonlinear second-order differential equation. A version, *linear* in the surface strains ${}_0\gamma_{\alpha\beta}$, follows:

$$\bar{e}^{\lambda\alpha}\,\bar{e}^{\mu\beta}\,{}_0\gamma_{\lambda\beta}\|_{\alpha\mu} + \widetilde{k}\;{}_0\gamma^{\alpha}_{\alpha} = \tfrac{1}{2}\bar{e}^{\lambda\alpha}\,\bar{e}^{\mu\beta}(b_{\alpha\mu}\rho_{\lambda\beta} + b_{\lambda\beta}\rho_{\alpha\mu} + \rho_{\alpha\beta}\rho_{\lambda\mu}). \quad (10.22)$$

Two additional equations derive from the equations of Codazzi (8.93a, b). A version, linear in the strains ${}_0\gamma_{\alpha\beta}$, follows:

$$\rho_{\alpha\mu}\|_{\beta} - \rho_{\alpha\beta}\|_{\mu} + b^{\eta}_{\mu}({}_0\gamma_{\alpha\eta}\|_{\beta} + {}_0\gamma_{\beta\eta}\|_{\alpha} - {}_0\gamma_{\alpha\beta}\|_{\eta})$$

$$- b^{\eta}_{\beta}({}_0\gamma_{\alpha\eta}\|_{\mu} + {}_0\gamma_{\mu\eta}\|_{\alpha} - {}_0\gamma_{\alpha\mu}\|_{\eta}) = 0. \quad (10.23)$$

As previously noted, in this and subsequent sections, the approximation of small strain has been introduced. Therefore, in deriving relation (10.23), terms of the type $(\rho^{\eta}_{\mu}\;{}_0\gamma_{\alpha\eta}\|_{\beta})$ have been neglected in comparison with terms of the type $\rho_{\alpha\beta}\|_{\mu}$. The interested reader can find further details in the works of W. T. Koiter and J. G. Simmonds ([161], [196]).

From a practical point of view, a formulation in terms of the strains $({}_0\gamma_{\alpha\beta}, \rho_{\alpha\beta})$ has some utility. In a region of small strain and rotation, the rotation and then the displacement can be expressed as integrals of the strains $({}_0\gamma_{\alpha\beta}, \rho_{\alpha\beta})$ [197]. For example, the *relative* rotations at opposing edges of a finite element are small, though global rotations may be large; then such description plays a role in the assembly of elements.

As the equations of compatibility (10.22) and (10.23) govern small strains, the increments of a finite strain $({}_0\dot{\gamma}_{\alpha\beta} = \dot{A}_{\alpha\beta}/2,\ \dot{\rho}_{\alpha\beta} = -\dot{B}_{\alpha\beta})$ must satisfy similar equations, *but* referred to the deformed surface. If an asterisk (*) or majuscule signifies the deformed version(s), then the strain increments must satisfy the following equations:

$$\overline{E}^{\lambda\alpha}\overline{E}^{\mu\beta}({}_0\dot{\gamma}_{\lambda\beta}\|^{*}_{\alpha\mu} - B_{\lambda\beta}\,\dot{\rho}_{\alpha\mu}) + \widetilde{K}\;{}_0\dot{\gamma}^{\alpha}_{\alpha} = 0, \quad (10.24)$$

$$\overline{E}^{\lambda\alpha}\overline{E}^{\beta\mu}[\dot{\rho}_{\alpha\beta}\|^{*}_{\mu} + B^{\eta}_{\beta}({}_0\dot{\gamma}_{\alpha\eta}\|^{*}_{\mu} + {}_0\dot{\gamma}_{\mu\eta}\|^{*}_{\alpha} - {}_0\dot{\gamma}_{\alpha\mu}\|^{*}_{\eta})] = 0. \quad (10.25)$$

Following W. T. Koiter [161], we consider the virtual work of the stresses [see equation (10.15)] upon the incremental strains. However, the strain variations (increments) must satisfy the compatibility conditions, (10.24) and (10.25). These are imposed, as auxiliary conditions, via Lagrangean multipliers: multipliers t_λ on (10.25) and multiplier T on (10.24). In the absence of external loads, the virtual work (with the auxiliary terms) follows:

$$\dot{W} = \iint_{s_0} \Big\{ n^{\alpha\beta}\, {}_0\dot{\gamma}_{\alpha\beta} + m^{\alpha\beta}\,\dot{\rho}_{\alpha\beta}$$

$$- t_\alpha\, \overline{E}^{\alpha\beta}\overline{E}^{\lambda\mu} \big[\dot{\rho}_{\beta\lambda}\|^*_\mu + B^\eta_\lambda ({}_0\dot{\gamma}_{\beta\eta}\|^*_\mu + {}_0\dot{\gamma}_{\mu\eta}\|^*_\beta - {}_0\dot{\gamma}_{\beta\mu}\|^*_\eta)\big]$$

$$+ T\big[\overline{E}^{\lambda\alpha}\overline{E}^{\mu\beta}({}_0\dot{\gamma}_{\lambda\beta}\|^*_{\alpha\mu} - B_{\lambda\beta}\,\dot{\rho}_{\alpha\mu}) + \widetilde{K}\,{}_0\dot{\gamma}^\alpha_\alpha\big]\Big\}\, ds_0 = 0. \quad (10.26)$$

By the principle of virtual work, the Euler equations (stationary conditions) are the equilibrium conditions in the absence of loading.

The Euler equations of the variation (10.26) are obtained in the usual way: Specifically, terms containing derivatives of variations (${}_0\dot{\gamma}_{\alpha\beta}$ and $\dot{\rho}_{\alpha\beta}$) are integrated by parts until only the variations remain in the integrand. The resulting equations are subjected to simplifications which enlist geometrical properties of the surface s_0 [e.g., Gauss equation (8.91)]. The results, as given by W. T. Koiter, follow:

$$n^{\alpha\beta} = \overline{E}^{\alpha\lambda}\overline{E}^{\beta\mu}\,\big[T\|^*_{\lambda\mu} - \widetilde{K}A_{\lambda\mu}T + B^\gamma_\lambda\|^*_\mu t_\gamma + \tfrac{1}{2}B^\gamma_\gamma(t_\lambda\|^*_\mu + t_\mu\|^*_\lambda)$$

$$+ \tfrac{1}{2}B^\gamma_\lambda(t_\gamma\|^*_\mu - t_\mu\|^*_\gamma) + \tfrac{1}{2}B^\gamma_\mu(t_\gamma\|^*_\lambda - t_\lambda\|^*_\gamma)\big], \quad (10.27)$$

$$m^{\alpha\beta} = \overline{E}^{\alpha\lambda}\overline{E}^{\beta\mu}\,\big[\tfrac{1}{2}(t_\lambda\|^*_\mu + t_\mu\|^*_\lambda) - B_{\lambda\mu}T\big]. \quad (10.28)$$

The stresses of equations (10.27) and (10.28) represent general solutions of the homogeneous equations of equilibrium, specifically, the components of (10.18a, b). The arbitrary functions, T and t_α, are the *stress functions.* In problems of extensional deformations of plates ($B^\alpha_\beta = 0$), the function T is known as the Airy stress function (see Section 7.11).

The choice of strain components is somewhat arbitrary; our choices, ${}_0\gamma_{\alpha\beta}$ and $\rho_{\alpha\beta}$ [see (10.13a, b)], are simply the changes in the two fundamental tensors from the initial to the deformed surface (s_0 to S_0). The use of an alternative "change of curvature" leads to analogous forms for the compatibility and equilibrium equations. The modified strain $\bar{\rho}_{\alpha\beta}$ is defined by the combination:

$$\bar{\rho}_{\alpha\beta} = -\rho_{\alpha\beta} - \tfrac{1}{2}B^\gamma_\alpha\,{}_0\gamma_{\gamma\beta} - \tfrac{1}{2}B^\gamma_\beta\,{}_0\gamma_{\gamma\alpha}. \quad (10.29)$$

Then the virtual work [see (10.15)] assumes the form:

$$\dot{u}_s = \bar{n}^{\alpha\beta}\,{}_0\dot{\gamma}_{\alpha\beta} + \overline{m}^{\alpha\beta}\,\dot{\bar{\rho}}_{\alpha\beta}, \tag{10.30}$$

where

$$\bar{n}^{\alpha\beta} = n^{\alpha\beta} - \tfrac{1}{2}B^{\alpha}_{\mu}m^{\mu\beta} - \tfrac{1}{2}B^{\beta}_{\mu}m^{\mu\alpha}, \tag{10.31a}$$

$$\overline{m}^{\alpha\beta} = -\,m^{\alpha\beta}. \tag{10.31b}$$

The change of sign in (10.29) and (10.31b) is a matter of deference to usage. For example, F. I. Niordson [160] and W. T. Koiter [161] adopt the convention $\rho_{\alpha\beta} = B_{\alpha\beta} - b_{\alpha\beta}$. Both present the modified strain and discuss the consequences.

Compatibility equations and equilibrium equations take different forms when expressed in terms of the modified strains $\bar{\rho}_{\alpha\beta}$ and stresses $\bar{n}^{\alpha\beta}$ and $\overline{m}^{\alpha\beta}$. These follow:

$$\overline{E}^{\lambda\alpha}\,\overline{E}^{\mu\beta}({}_0\dot{\gamma}_{\lambda\beta}\|^*_{\alpha\mu} - B_{\lambda\beta}\,\dot{\bar{\rho}}_{\alpha\mu}) = 0, \tag{10.32a}$$

$$(\overline{E}^{\alpha\beta}\,\overline{E}^{\lambda\mu}\,\dot{\bar{\rho}}_{\beta\lambda})\|^*_{\mu} + \overline{E}^{\phi\beta}\,\overline{E}^{\gamma\eta}B^{\alpha}_{\gamma}\,{}_0\dot{\gamma}_{\phi\eta}\|^*_{\beta}$$

$$-\tfrac{1}{2}(\overline{E}^{\phi\alpha}\,\overline{E}^{\mu\gamma}B^{\beta}_{\mu}\,{}_0\dot{\gamma}_{\phi\gamma} - \overline{E}^{\phi\beta}\,\overline{E}^{\mu\gamma}B^{\alpha}_{\mu}\,{}_0\dot{\gamma}_{\phi\gamma})\|^*_{\beta} = 0, \tag{10.32b}$$

$$\overline{m}^{\alpha\beta}\|^*_{\alpha\beta} - B_{\alpha\beta}\,\bar{n}^{\alpha\beta} = 0, \tag{10.33a}$$

$$\bar{n}^{\beta\alpha}\|^*_{\beta} - \tfrac{1}{2}(B^{\beta}_{\mu}\,\overline{m}^{\alpha\mu} - B^{\alpha}_{\mu}\,\overline{m}^{\beta\mu})\|^*_{\beta} + B^{\alpha}_{\lambda}\,\overline{m}^{\beta\lambda}\|^*_{\beta} = 0. \tag{10.33b}$$

The equilibrium equations (10.33a, b) are without external loads $\boldsymbol{F}$ and $\boldsymbol{C}$. These forms exhibit the remarkable static-geometric analogy: The former (10.32a, b) are obtained from the latter (10.33a, b) upon the substitutions:

$$\overline{E}^{\alpha\gamma}\overline{E}^{\lambda\beta}\,\dot{\bar{\rho}}_{\gamma\lambda} \quad \text{for } \bar{n}^{\alpha\beta}, \tag{10.34a}$$

$$\overline{E}^{\phi\alpha}\overline{E}^{\beta\gamma}\,{}_0\dot{\gamma}_{\phi\gamma} \quad \text{for } \overline{m}^{\alpha\beta}. \tag{10.34b}$$

The static-geometric analogy was discovered by A. I. Lur'e ([154], [198]) and A. L. Gol'denveizer [199]. Here, we present the equations as they apply to the strain increments and stresses associated with the deformed surface S_0 (covariant derivative $\|^*$, tensors $A_{\alpha\beta}$ and $B_{\alpha\beta}$). They hold as well for small strains, wherein the derivatives and properties are associated with the initial surface s_0.

The practical value of the analogy rests with the solution of the equations: General solutions of the compatibility equations (${}_0\dot{\gamma}_{\alpha\beta}$ and $\dot{\bar{\rho}}_{\alpha\beta}$) are provided by any displacement ${}_0\dot{\boldsymbol{V}}$ of the surface. By analogy, the equations which provide those strains from the displacement (components ${}_0\dot{V}_3$, ${}_0\dot{V}_\alpha$) also provide the stresses ($\bar{n}^{\alpha\beta}$, $\overline{m}^{\alpha\beta}$). In the latter instance, the three arbitrary functions are the *stress functions*. The appropriate equations can be obtained by the variational scheme, as the equations (10.27) and (10.28) are obtained from the variation (10.26). In the form given by W. T. Koiter,

$$\bar{n}^{\alpha\beta} = \overline{E}^{\alpha\lambda}\overline{E}^{\beta\mu}\big[T\|^*_{\lambda\mu} + B^\gamma_\lambda\|^*_\mu t_\gamma + \tfrac{1}{4}B^\gamma_\lambda(3t_\gamma\|^*_\mu - t_\mu\|^*_\gamma)$$
$$+ \tfrac{1}{4}B^\gamma_\mu(3t_\gamma\|^*_\lambda - t_\lambda\|^*_\gamma)\big], \tag{10.35a}$$

$$\overline{m}^{\alpha\beta} = -\overline{E}^{\alpha\lambda}\overline{E}^{\beta\mu}\big[\tfrac{1}{2}(t_\lambda\|^*_\mu + t_\mu\|^*_\lambda) - B_{\lambda\mu}T\big]. \tag{10.35b}$$

10.5 Constitutive Equations of the Hookean Shell

The preceding developments of the Kirchhoff-Love theory are independent of the constituency of the shell. The material need only behave as a continuous cohesive medium on a macroscopic scale. The results are applicable to inelastic or elastic material.

Under the Kirchhoff-Love hypothesis the strain distribution is determined by two symmetric tensors, either $({}_0\gamma_{\alpha\beta}, \rho_{\alpha\beta})$ or $({}_0\gamma_{\alpha\beta}, \bar{\rho}_{\alpha\beta})$. The power of the stresses has the form (10.15) or (10.30):

$$\dot{u}_s = \bar{n}^{\alpha\beta}\,{}_0\dot{\gamma}_{\alpha\beta} + \overline{m}^{\alpha\beta}\dot{\bar{\rho}}_{\alpha\beta}. \tag{10.36}$$

If the shell is elastic, then this rate/increment represents the variation of a potential ϕ (energy per unit area of surface s_0). In general, that potential depends upon the temperature T as well as the strains:

$$\dot{u}_s = \dot{\phi}({}_0\gamma_{\alpha\beta}, \bar{\rho}_{\alpha\beta}; T). \tag{10.37}$$

It follows from (10.36) and (10.37) that

$$\bar{n}^{\alpha\beta} = \frac{\partial\phi}{\partial\,{}_0\gamma_{\alpha\beta}}, \qquad \overline{m}^{\alpha\beta} = \frac{\partial\phi}{\partial\bar{\rho}_{\alpha\beta}}. \tag{10.38a, b}$$

If the behavior is Hookean, i.e., characterized by a linear stress-strain-temperature relation, then the function ϕ is a quadratic form:

$$\phi = \frac{h}{2}\left(\overline{C}^{\alpha\beta\gamma\eta}\,{}_0\gamma_{\alpha\beta}\,{}_0\gamma_{\gamma\eta} + h\overline{D}^{\alpha\beta\gamma\eta}\,{}_0\gamma_{\alpha\beta}\,\bar{\rho}_{\gamma\eta} + h^2\,\overline{E}^{\alpha\beta\gamma\eta}\,\bar{\rho}_{\alpha\beta}\,\bar{\rho}_{\gamma\eta}\right)$$

$$- h\left(\bar{\alpha}^{\alpha\beta}\,{}_0\gamma_{\alpha\beta} + h\bar{\beta}^{\alpha\beta}\bar{\rho}_{\alpha\beta}\right)(T - T_0). \tag{10.39}$$

The thickness ($h = t_+ + t_-$) is inserted to provide consistent dimensions; also, in the special case of a thin plate, $\overline{C}^{\alpha\beta\gamma\eta}$ *is* the elastic coefficient for a state of "plane" stress ($t^{33} = s^{33} = 0$). This result (10.39) applies to heterogeneous, anisotropic or homogeneous isotropic media. In all cases, the coefficients possess limited symmetry:

$$\overline{C}^{\alpha\beta\gamma\eta} = \overline{C}^{\gamma\eta\alpha\beta} = \overline{C}^{\beta\alpha\gamma\eta} = \overline{C}^{\alpha\beta\eta\gamma},$$

$$\overline{D}^{\alpha\beta\gamma\eta} = \overline{D}^{\beta\alpha\gamma\eta} = \overline{D}^{\alpha\beta\eta\gamma},$$

$$\overline{E}^{\alpha\beta\gamma\eta} = \overline{E}^{\gamma\eta\alpha\beta} = \overline{E}^{\beta\alpha\gamma\eta} = \overline{E}^{\alpha\beta\eta\gamma},$$

$$\bar{\alpha}^{\alpha\beta} = \bar{\alpha}^{\beta\alpha}, \qquad \bar{\beta}^{\alpha\beta} = \bar{\beta}^{\beta\alpha}. \tag{10.40}$$

Additionally, the properties of the shell may exhibit symmetries, orthotropy or homogeneity (see Sections 5.14 to 5.17). Further simplifications are contingent upon approximations of geometrical properties, particularly, the assessment of the relative thickness/thinness.

10.6 Constitutive Equations of the Thin Hookean Shell

At the surfaces S_- and S_+, the transverse normal stress s^{33} is that imposed by tractions; it is typically much less than the stresses $s^{\alpha\beta}$ which are caused by forces and couples on a section. Furthermore, when the shell is thin, it is unlikely that this normal stress s^{33} increases significantly in the interior. Accordingly, it is reasonable to neglect that effect in the constitutive equations. A sufficiently general form of Hooke's law is expressed by the quadratic potential (9.96b); in the Kirchhoff-Love theory, the energy of transverse shear is absent. Then

$$u = \tfrac{1}{2} C^{\alpha\beta\gamma\eta} \gamma_{\alpha\beta} \gamma_{\gamma\eta} - \bar{\alpha}^{\alpha\beta} \gamma_{\alpha\beta} (T - T_0). \tag{10.41}$$

Let us assume that the elastic properties do not vary through the thickness of the shell. In other words, the *physical* components of the stiffness tensor $C^{\alpha\beta\gamma\eta}$ are constant. These *moduli* are

$$K^{\alpha\beta\gamma\eta} = \sqrt{g_{\underline{\alpha\alpha}}\, g_{\underline{\beta\beta}}\, g_{\underline{\gamma\gamma}}\, g_{\underline{\eta\eta}}}\, C^{\alpha\beta\gamma\eta}. \tag{10.42a}$$

This follows from our identification of the physical components of strain [see Section 3.6, equation (3.35a)]. Here,

$$g_{\underline{\alpha\alpha}} \doteq a_{\underline{\alpha\alpha}} - 2\theta^3 a_{\underline{\alpha}\gamma} b^{\gamma}_{\underline{\alpha}}.$$

More specifically, *if* the surface coordinates are lengths along lines of principal curvature, then

$$g_{\underline{\alpha\alpha}} = 1 - \kappa_{\alpha} \theta^3.$$

The tensorial components of the moduli are *not* independent of coordinate θ^3, though the *material* properties are constant

$$C^{\alpha\beta\gamma\eta} = \frac{K^{\alpha\beta\gamma\eta}}{\sqrt{g_{\underline{\alpha\alpha}}\, g_{\underline{\beta\beta}}\, g_{\underline{\gamma\gamma}}\, g_{\underline{\eta\eta}}}}. \tag{10.42b}$$

Let ${}_0C^{\alpha\beta\gamma\eta}$ denote the tensorial component at the mid-surface $\theta^3 = 0$; then

$${}_0C^{\alpha\beta\gamma\eta} = \frac{K^{\alpha\beta\gamma\eta}}{\sqrt{a_{\underline{\alpha\alpha}}\, a_{\underline{\beta\beta}}\, a_{\underline{\gamma\gamma}}\, a_{\underline{\eta\eta}}}}, \tag{10.43}$$

$$C^{\alpha\beta\gamma\eta} = {}_0C^{\alpha\beta\gamma\eta} [1 + O(\kappa\theta^3)]. \tag{10.44}$$

The bracketed term contains a power series in the curvatures (and torsion, if the coordinates are not along principal lines). Similar examination of the thermal coefficients leads to a comparable result:

$$\bar{\alpha}^{\alpha\beta} = {}_0\alpha^{\alpha\beta} [1 + O(\kappa\theta^3)]. \tag{10.45}$$

The potential density of the shell [see equation (10.39)] is obtained by integrating (10.41) through the thickness; with the reference surface at the middle ($t_- = t_+ = h/2$):

$$\phi = \tfrac{1}{2}\,{}_0C^{\alpha\beta\gamma\eta} \int_{-h/2}^{h/2} \gamma_{\alpha\beta}\gamma_{\gamma\eta}\,[1 + O(\kappa\theta^3)]\,\mu\,d\theta^3$$

$$- {}_0\alpha^{\alpha\beta} \int_{-h/2}^{h/2} \gamma_{\alpha\beta}\,[1 + O(\kappa\theta^3)]\,\Delta T\mu\,d\theta^3, \tag{10.46}$$

where [see equation (9.33)]

$$\mu = \sqrt{\frac{g}{a}} = 1 - 2\widetilde{h}\theta^3 + \widetilde{k}(\theta^3)^2, \qquad \Delta T = T - T_0.$$

Under the Kirchhoff hypothesis (10.1), with the notations of (10.2a) and (10.13a, b):

$$\gamma_{\alpha\beta} \equiv \tfrac{1}{2}(\boldsymbol{G}_\alpha \cdot \boldsymbol{G}_\beta - g_{\alpha\beta})$$

$$\doteq {}_0\gamma_{\alpha\beta} + \theta^3\rho_{\alpha\beta} - (\theta^3)^2\left[\tfrac{1}{2}(b^\gamma_\alpha\rho_{\gamma\beta} + b^\gamma_\beta\rho_{\gamma\alpha}) + b^\mu_\alpha b^\gamma_\beta\,{}_0\gamma_{\gamma\mu}\right]. \tag{10.47}$$

The *approximation* (10.47) displays the terms that are *linear* in the strains ${}_0\gamma_{\alpha\beta}$ and $\rho_{\alpha\beta}$; higher-order terms [e.g., $(\theta^3)^2\,b^\mu_\alpha\,\rho^\gamma_\beta\,{}_0\gamma_{\gamma\mu}$, etc.] are not shown. Integrating (10.46) with the approximation (10.47) and considering the relation

$$A^{\alpha\beta} - a^{\alpha\beta} \doteq -2a^{\alpha\mu}a^{\beta\eta}\,{}_0\gamma_{\mu\eta},$$

we obtain

$$\phi = \frac{h}{2}\,{}_0C^{\alpha\beta\gamma\eta}\left[{}_0\gamma_{\alpha\beta}\,{}_0\gamma_{\gamma\eta} + \frac{h^2}{12}\rho_{\alpha\beta}\,\rho_{\gamma\eta}\right]$$

$$+ \frac{h^3}{24}\,{}_0C^{\alpha\beta\gamma\eta}\left[{}_0\gamma_{\alpha\beta}\,{}_0\gamma_{\gamma\eta}\,O(\kappa^2) - {}_0\gamma_{\alpha\beta}(b^\phi_\gamma\rho_{\phi\eta} + b^\phi_\eta\rho_{\phi\gamma})\right.$$

$$\left. + \rho_{\alpha\beta}\,\rho_{\gamma\eta}\,O(\kappa^2\,h^2) + \cdots\right]$$

$$- h\,{}_0\alpha^{\alpha\beta}\left[{}_0\gamma_{\alpha\beta}\,\Delta T_0 + h\rho_{\alpha\beta}\,\Delta T_1\right] + \cdots, \tag{10.48}$$

where

$$\Delta T_0 = \frac{1}{h}\int_{-h/2}^{h/2} (T - T_0)\, d\theta^3, \qquad \Delta T_1 = \frac{1}{h^2}\int_{-h/2}^{h/2} (T - T_0)\theta^3\, d\theta^3.$$

The approximation of Love consists of the two terms:

$$\phi = \phi_M + \phi_B. \tag{10.49}$$

The potential is thus split into the contributions from membrane (stretching) and bending energies:

$$\phi_M = \frac{h}{2}\, {}_0C^{\alpha\beta\gamma\eta}\, {}_0\gamma_{\alpha\beta}\, {}_0\gamma_{\gamma\eta}, \tag{10.50}$$

$$\phi_B = \frac{h^3}{24}\, {}_0C^{\alpha\beta\gamma\eta}\, \rho_{\alpha\beta}\, \rho_{\gamma\eta}. \tag{10.51}$$

If $\kappa = 1/r$ denotes the largest curvature, then the magnitudes of the neglected terms are, at most[‡]

$$\frac{h}{24}\left(\frac{h}{r}\right)^2 {}_0C^{\alpha\beta\gamma\eta}\, {}_0\gamma_{\alpha\beta}\, {}_0\gamma_{\gamma\eta} = \frac{1}{12}\left(\frac{h}{r}\right)^2 \phi_M,$$

$$\frac{h^3}{24}\, {}_0C^{\alpha\beta\gamma\eta}\, {}_0\gamma_{\alpha\beta}(b^\phi_\gamma \rho_{\phi\eta} + b^\phi_\eta \rho_{\phi\gamma}) = O\left[\left(\frac{h}{r}\right)\sqrt{\phi_M \phi_B}\right],$$

$$\frac{h^3}{24}\, {}_0C^{\alpha\beta\gamma\eta}\, \rho_{\alpha\beta}\, \rho_{\gamma\eta}\left(\frac{h}{r}\right)^2 = \left(\frac{h}{r}\right)^2 \phi_B.$$

If the shell is *thin*, then $h/r \ll 1$. In this case, the omitted terms are much less than those contained in Love's approximation (10.49). Additionally, we include the thermal contribution:

$$\phi = \phi_M + \phi_B + \phi_T, \tag{10.52}$$

where

$$\phi_T \equiv -h\, {}_0\alpha^{\alpha\beta}({}_0\gamma_{\alpha\beta}\, \Delta T_0 + h\rho_{\alpha\beta}\, \Delta T_1). \tag{10.53}$$

[‡]Note that $2\phi_M\phi_B \le \phi_M^2 + \phi_B^2$.

It is noteworthy that the omission of the term $O\left[(h/r)\sqrt{\phi_M \phi_B}\right]$ is no more or less than replacing the change-of-curvature $\rho_{\alpha\beta}$ by the modified change-of-curvature $\bar{\rho}_{\alpha\beta}$ [see equation (10.29)]. Alternative expressions for bending strains, i.e., $\rho_{\alpha\beta}$, and their implications are discussed by W. T. Koiter [158].

The stress-strain relations follow from (10.52):

$$n^{\alpha\beta} = \frac{\partial \phi}{\partial\, {}_0\gamma_{\alpha\beta}} = h({}_0C^{\alpha\beta\gamma\eta}\, {}_0\gamma_{\gamma\eta} - {}_0\alpha^{\alpha\beta}\Delta T_0), \tag{10.54a}$$

$$m^{\alpha\beta} = \frac{\partial \phi}{\partial \rho_{\alpha\beta}} = h\left(\frac{h^2}{12}\, {}_0C^{\alpha\beta\gamma\eta}\rho_{\gamma\eta} - h\, {}_0\alpha^{\alpha\beta}\Delta T_1\right). \tag{10.54b}$$

These apply to anisotropic shells with uniformity through the thickness or, at least, symmetry with respect to the middle surface. Composite shells with symmetric laminations or reinforcements are governed by *similar* equations. In general, asymmetric composition introduces coupling between the extensional and bending deformations; then the potential assumes the form of (10.39).

10.7 Intrinsic Kirchhoff-Love Theories

The term *intrinsic*‡ signifies a theory which embodies no explicit references to the displacements of particles nor to the rotations of lines, i.e., the rigid-body motion of an element. As such, the intrinsic theories consist of the equilibrium, compatibility, and constitutive equations which govern the stresses and strains. Indeed, our prior developments provide the bases; only increments ($\dot{\boldsymbol{R}}$ and $\dot{\boldsymbol{\Omega}}$) of displacements and rotations appear.

In his comprehensive treatment of thin elastic shells, W. T. Koiter ([161], Part III) provides a critical examination and a systematic simplification of the governing equations; specifically, the equilibrium and compatibility equations augmented by the linear constitutive equations of the homogeneous and isotropic Kirchhoff-Love shell [see equations (10.54a, b) and (9.98a)]. Our purpose is to set forth those equations in general, to present the essential bases for simplification, and to exhibit the simpler theories. The interested reader can follow these to their source [161].

‡This approach can be traced to the report by J. L. Synge and W. Z. Chien [155].

First, we recall the equations of equilibrium (10.18a, b) and compatibility (10.22) and (10.23). The former are valid for finite deformations, the latter are restricted to small surface strains ${}_0\gamma_{\alpha\beta}$ but do admit finite flexure $\rho_{\alpha\beta}$. Consistent with the assumption of small strain ($\gamma \ll 1$), we drop the asterisk (*) which connotes the deformed surface; in other words, the covariant derivatives embrace the initial metric $a_{\alpha\beta}$. Then, with the definition (10.13b),

$$B_{\alpha\beta} \equiv b_{\alpha\beta} - \rho_{\alpha\beta},$$

the approximation

$$B^{\beta}_{\alpha} \doteq b^{\beta}_{\alpha} - \rho^{\beta}_{\alpha},$$

and neglect of the external couple C^{α}, the equilibrium equations (10.18a, b) assume the forms:

$$(n^{\alpha\beta} - \underline{m^{\alpha\gamma} b^{\beta}_{\gamma}} + \underline{m^{\alpha\gamma}\rho^{\beta}_{\gamma}})\|_{\alpha} - \underline{(b^{\beta}_{\alpha} - \rho^{\beta}_{\alpha}) m^{\gamma\alpha}\|_{\gamma}} + F^{\beta} = 0, \tag{10.55a}$$

$$m^{\gamma\alpha}\|_{\gamma\alpha} + (b_{\alpha\beta} - \rho_{\alpha\beta}) n^{\alpha\beta} - \underline{b^{\beta}_{\gamma} b_{\alpha\beta} m^{\alpha\gamma}}$$

$$+ \underline{(b^{\beta}_{\gamma}\rho_{\alpha\beta} + b^{\beta}_{\alpha}\rho_{\gamma\beta}) m^{\alpha\gamma}} - \underline{m^{\alpha\gamma}\rho^{\beta}_{\gamma}\rho_{\alpha\beta}} + F^{3} = 0. \tag{10.55b}$$

The compatibility equations (10.22) and (10.23) follow:

$$\bar{\epsilon}^{\lambda\alpha}\bar{\epsilon}^{\mu\beta}[{}_0\gamma_{\lambda\beta}\|_{\alpha\mu} - \tfrac{1}{2}(b_{\alpha\mu}\rho_{\lambda\beta} + b_{\lambda\beta}\rho_{\alpha\mu} + \rho_{\alpha\beta}\rho_{\lambda\mu})] + \underline{\widetilde{k}\,{}_0\gamma^{\alpha}_{\alpha}} = 0, \tag{10.56a}$$

$$\bar{\epsilon}^{\alpha\beta}\bar{\epsilon}^{\lambda\mu}[\rho_{\beta\lambda}\|_{\mu} + \underline{b^{\eta}_{\lambda}({}_0\gamma_{\beta\eta}\|_{\mu} + {}_0\gamma_{\mu\eta}\|_{\beta} - {}_0\gamma_{\beta\mu}\|_{\eta})}$$

$$\underline{- \rho^{\eta}_{\lambda}({}_0\gamma_{\beta\eta}\|_{\mu} + {}_0\gamma_{\mu\eta}\|_{\beta} - {}_0\gamma_{\beta\mu}\|_{\eta})}] = 0. \tag{10.56b}$$

The constitutive equations of the homogeneous isotropic Kirchhoff-Love shell follow from (10.54a, b) and (9.98a):

$$n^{\alpha\beta} = \frac{Eh}{2(1+\nu)}\left(a^{\alpha\gamma}a^{\beta\eta} + a^{\alpha\eta}a^{\beta\gamma} + \frac{2\nu}{1-\nu}a^{\alpha\beta}a^{\gamma\eta}\right){}_0\gamma_{\gamma\eta}, \tag{10.57a}$$

$$m^{\alpha\beta} = \frac{Eh^3}{24(1+\nu)}\left(a^{\alpha\gamma}a^{\beta\eta} + a^{\alpha\eta}a^{\beta\gamma} + \frac{2\nu}{1-\nu}a^{\alpha\beta}a^{\gamma\eta}\right)\rho_{\gamma\eta}. \tag{10.57b}$$

The underscored terms of (10.55a, b) and (10.56a, b) are candidates for omission: In the equilibrium equations these terms are products of curvatures and bending couples, or their derivatives. Recall our earliest observations about the dominant role of membrane actions. In the compatibility equations, the candidates are products of curvatures and surface strains; in the Hookean shell the latter, ${}_0\gamma_{\alpha\beta}$, must remain small whereas changes-of-curvature, $\rho_{\alpha\beta}$, can be large. Deleting the underscored terms of the equilibrium equations and the final term of the compatibility equation (10.56b), we obtain the simplified system:

$$n^{\alpha\beta}\|_{\alpha} + F^{\beta} = 0, \quad (10.58a)$$

$$\underline{m^{\gamma\alpha}\|_{\gamma\alpha}} + (b_{\alpha\beta} - \underline{\rho_{\alpha\beta}})n^{\alpha\beta} + F^{3} = 0, \quad (10.58b)$$

$$\overline{e}^{\lambda\alpha}\,\overline{e}^{\mu\beta}[{}_0\gamma_{\lambda\beta}\|_{\alpha\mu} - \tfrac{1}{2}(b_{\alpha\mu}\rho_{\lambda\beta} + b_{\lambda\beta}\rho_{\alpha\mu} + \rho_{\alpha\beta}\rho_{\lambda\mu})] + \underline{\widetilde{k}\,{}_0\gamma^{\alpha}_{\alpha}} = 0, \quad (10.59a)$$

$$\overline{e}^{\alpha\beta}\,\overline{e}^{\lambda\mu}[\rho_{\beta\lambda}\|_{\mu} + \underline{b^{\eta}_{\lambda}({}_0\gamma_{\beta\eta}\|_{\mu} + {}_0\gamma_{\mu\eta}\|_{\beta} - {}_0\gamma_{\beta\mu}\|_{\eta})}] = 0. \quad (10.59b)$$

These appear to be rational simplifications: Equation (10.58a) implies that tangential loads F^{β} are resisted by membrane forces and, in turn, such loads cause the changes in membrane stresses. As anticipated, bending stresses are influenced by the normal loads F^3 and such loads are resisted also by membrane stresses as a consequence of the prevailing curvature $(b_{\alpha\beta} - \rho_{\alpha\beta})$. All experiences tell us that the nonlinear terms $(\rho_{\alpha\beta}\,n^{\alpha\beta})$ play the major role in analyses of buckling. The simplifications in the compatibility equations must be based on geometrical arguments which depend in part on the deformational pattern, e.g., the fluctuation of strains—the magnitudes of the derivatives ${}_0\gamma_{\beta\eta}\|_{\mu}$. The interested reader can find such assessments in the work of W. T. Koiter [161].

It is also interesting to consider the circumstance wherein the loads are effectively resisted *entirely* by membrane stresses and changes of curvature are small $(\rho \ll \kappa)$. This is a well designed shell! Then the underscored terms of the equilibrium equation (10.58b) are neglected and also the one nonlinear term in the compatibility equation (10.59a). The resulting system follows:

$$n^{\alpha\beta}\|_{\alpha} + F^{\beta} = 0, \qquad b_{\alpha\beta}n^{\alpha\beta} + F^{3} = 0, \quad (10.60a, b)$$

$$\overline{e}^{\lambda\alpha}\,\overline{e}^{\mu\beta}\left[{}_0\gamma_{\lambda\beta}\|_{\alpha\mu} - \tfrac{1}{2}(b_{\alpha\mu}\rho_{\lambda\beta} + b_{\lambda\beta}\rho_{\alpha\mu})\right] + \widetilde{k}\;{}_0\gamma^{\alpha}_{\alpha} = 0, \quad (10.61a)$$

$$\overline{e}^{\alpha\beta}\,\overline{e}^{\lambda\mu}\left[\rho_{\beta\lambda}\|_{\mu} + b^{\eta}_{\lambda}({}_0\gamma_{\beta\eta}\|_{\mu} + {}_0\gamma_{\mu\eta}\|_{\beta} - {}_0\gamma_{\beta\mu}\|_{\eta})\right] = 0. \quad (10.61b)$$

This simplification, as noted by W. T. Koiter, is entirely *linear*. The equilibrium equations govern the membrane theory. The constitutive equations and the compatibility equations provide the means to ascertain the flexural strains $\rho_{\alpha\beta}$, a posteriori.

Finally, we consider a further simplification of the system (10.58a, b) and (10.59a, b). This must be justified by geometrical arguments, wherein one neglects the underscored terms of (10.59a, b), viz., $\widetilde{k}\,{}_0\gamma^\alpha_\alpha$, $b^\eta_\lambda\,{}_0\gamma_{\beta\eta}\|_\mu$, etc.; these anticipate small initial curvatures, small strains, and gradients. The simplified versions of (10.59a, b) follow:

$$\bar{e}^{\lambda\alpha}\,\bar{e}^{\mu\beta}\left[{}_0\gamma_{\lambda\beta}\|_{\alpha\mu} - \tfrac{1}{2}(b_{\alpha\mu}\rho_{\lambda\beta} + b_{\lambda\beta}\rho_{\alpha\mu} + \rho_{\alpha\beta}\rho_{\lambda\mu})\right] = 0, \quad (10.62a)$$

$$\bar{e}^{\alpha\beta}\,\bar{e}^{\lambda\mu}\rho_{\beta\lambda}\|_\mu = 0. \quad (10.62b)$$

Following W. T. Koiter, we eliminate the stress $m^{\alpha\beta}$ and the strain ${}_0\gamma_{\alpha\beta}$ from the system, equations (10.58a, b) and (10.62a, b), via the stress-strain relations (10.57a, b):

$$n^{\alpha\beta}\|_\alpha + F^\beta = 0, \quad (10.63a)$$

$$\frac{Eh^3}{12(1-\nu^2)}\,\rho^\eta_\eta\|^\alpha_\alpha + (b_{\alpha\beta} - \rho_{\alpha\beta})n^{\alpha\beta} + F^3 = 0, \quad (10.63b)$$

$$n^\alpha_\alpha\|^\beta_\beta - Eh(b^\alpha_\alpha\rho^\beta_\beta - b^\alpha_\beta\rho^\beta_\alpha)$$

$$+\tfrac{1}{2}Eh(\rho^\alpha_\alpha\rho^\beta_\beta - \rho^\alpha_\beta\rho^\beta_\alpha) + (1+\nu)F^\beta\|_\beta = 0, \quad (10.64a)$$

$$\rho^\alpha_\beta\|_\gamma - \rho^\alpha_\gamma\|_\beta = 0. \quad (10.64b)$$

We recall the expressions for the covariant derivative [equation (8.77a, b)] and equations (8.88) and (8.91). Furthermore, according to (8.96) and (8.97), if the Gaussian curvature $\widetilde{k}$ is *relatively* small ($\widetilde{k}\,V^\alpha \ll 1$), the order of covariant differentiation is immaterial. Then, the general solution of equation (10.64b) can be expressed in terms of a surface invariant W:

$$\rho_{\alpha\beta} = W\|_{\alpha\beta} \doteq W\|_{\beta\alpha}. \quad (10.65)$$

W. T. Koiter [161] terms this invariant the "curvature function." In similar fashion, the homogeneous solution of equation (10.63a) is given by the Airy stress function F; if $P^{\alpha\beta}$ denotes the particular solution ($P^{\alpha\beta}\|_\alpha = -F^\beta$), then

$$n^{\alpha\beta} = P^{\alpha\beta} + \bar{e}^{\alpha\lambda}\bar{e}^{\beta\mu} F\|_{\lambda\mu}. \tag{10.66}$$

When the solutions (10.65) and (10.66) are substituted into the remaining differential equations, (10.63b) and (10.64a), we obtain the two nonlinear equations governing the functions W and F:

$$\frac{Eh^3}{12(1-\nu^2)} W\|^{\alpha\beta}_{\alpha\beta} + \bar{e}^{\alpha\lambda}\bar{e}^{\beta\mu} F\|_{\lambda\mu}(b_{\alpha\beta} - W\|_{\alpha\beta})$$

$$+ (b_{\alpha\beta} - W\|_{\alpha\beta})\, P^{\alpha\beta} + F^3 = 0, \tag{10.67}$$

$$\frac{1}{Eh} F\|^{\alpha\beta}_{\alpha\beta} - \bar{e}^{\alpha\lambda}\bar{e}^{\beta\gamma}(b_{\alpha\beta} - \frac{1}{2} W\|_{\alpha\beta})\, W\|_{\lambda\gamma}$$

$$+ \frac{1}{Eh}[\, P^{\alpha}_{\alpha}\|^{\beta}_{\beta} - (1+\nu) P^{\alpha\beta}\|_{\alpha\beta}\,] = 0. \tag{10.68}$$

The simplifications and solution expressed by equations (10.65) to (10.68) constitute Koiter's theory of "quasi-shallow shells." This is an appropriate designation for two reasons: First, his arguments and approximations are appropriate to a deformational mode which is limited to a shallow *portion* of the shell. For example, the deflections of a cylinder might be localized, within a region much less than the radius. From an analytical viewpoint, such shallowness can be characterized by the smallness of the Gaussian curvature $\widetilde{k}$; radii of curvature are large compared to the size of the deformed region. Second, these equations are essentially those employed in the usual theory of shallow shells; the function W is then the normal deflection and $W\|_{\alpha\beta}$ is an acceptable approximation of the curvature. It is also important to note that the theory does accommodate large deflections. In practice, the loading terms, $P^{\alpha\beta}$ and F^i, may pose difficulties, particularly in circumstances of large deformations, since the underlying theory refers the loads to the rotated triad ($\boldsymbol{F} = F^i \boldsymbol{A}_i$). Simple pressure does not pose such problem ($F^3 = p$).

10.8 Plasticity of the Kirchhoff-Love Shell

10.8.1 Introduction

Much of the preceding analysis is independent of the material behavior; only continuity and cohesion are implicit throughout. The kinematics

(Sections 9.5 and 10.1), stresses and strains (Section 10.2), and equations of equilibrium and compatibility (Sections 10.3 and 10.4) are drawn from geometrical and mechanical arguments; in particular, the principle of virtual work applies to conservative (elastic) systems as well as to nonconservative (inelastic) systems. Indeed, the underlying kinematic hypothesis of Kirchhoff-Love has been applied successfully to describe the behavior of thin inelastic shells. We adopt those foundations here to focus upon the central issues of elastic-plastic shells; specifically, we address the inherent features of the constitutive relations between the stresses ($\bar{n}^{\alpha\beta}$, $\overline{m}^{\alpha\beta}$) and strains (${}_0\gamma_{\alpha\beta}$, $\bar{\rho}_{\alpha\beta}$).

First, we recall the expression of internal work [equation (10.36)]; we drop the prefix ($_0$) and overbar ($^-$) from the strains ($\gamma_{\alpha\beta} \equiv {}_0\gamma_{\alpha\beta}$ and $\rho_{\alpha\beta} \equiv \bar{\rho}_{\alpha\beta}$):

$$\dot{u}_s = \bar{n}^{\alpha\beta}\dot{\gamma}_{\alpha\beta} + \overline{m}^{\alpha\beta}\dot{\rho}_{\alpha\beta}. \tag{10.69}$$

We suppose that the strains are small and confine ourselves to the usual assumptions of classical plasticity.

With our concern for the overriding features of the behavior and a view toward pragmatism, we must acknowledge the physical attributes of the stresses and strains. Essentially, the stresses ($\bar{n}^{\alpha\beta}$, $\overline{m}^{\alpha\beta}$) are forces and couples, respectively; the strains ($\gamma_{\alpha\beta}$, $\rho_{\alpha\beta}$) are extensions and curvature changes of the reference surface. If stresses are imposed which caused yielding, then, upon removal of such stresses, some permanent extensions $\gamma^P_{\alpha\beta}$ and curvature changes $\rho^P_{\alpha\beta}$ remain. If one accepts the arguments of classical plasticity (e.g., [63]), then work is dissipated:

$$\dot{w}_D = \bar{n}^{\alpha\beta}\,\dot{\gamma}^P_{\alpha\beta} + \overline{m}^{\alpha\beta}\,\dot{\rho}^P_{\alpha\beta}. \tag{10.70}$$

Also, according to the accepted concepts of elastic-plastic behavior (Section 5.24), yielding is characterized by a condition:

$$F(\bar{n}^{\alpha\beta}, \overline{m}^{\alpha\beta}) = 0. \tag{10.71}$$

That condition may be viewed as a surface in the space of the six stresses. The shell can be described as "strain-hardening" if the function F changes with the plastic deformation; F is then a functional of the plastic strains. By the usual arguments (see Section 5.29) such "strain-hardening" is characterized by the condition:

$$F = 0, \qquad dF = \frac{\partial F}{\partial \bar{n}^{\alpha\beta}}\,d\bar{n}^{\alpha\beta} + \frac{\partial F}{\partial \overline{m}^{\alpha\beta}}\,d\overline{m}^{\alpha\beta} > 0. \tag{10.72a, b}$$

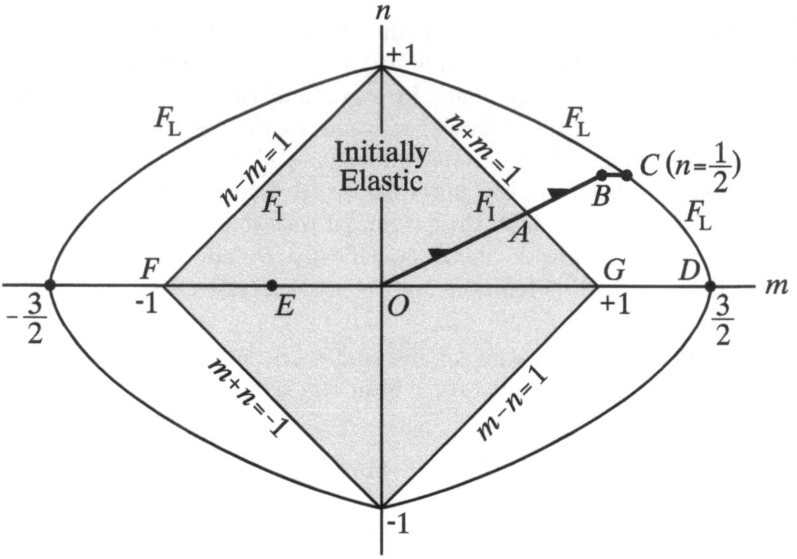

Figure 10.3 Initial yield and limit conditions for a beam

Additionally, the classical theory leads to the flow law:

$$\dot{\gamma}^{P}_{\alpha\beta} = \dot{\lambda}\frac{\partial F}{\partial \bar{n}^{\alpha\beta}}, \qquad \dot{\rho}^{P}_{\alpha\beta} = \dot{\lambda}\frac{\partial F}{\partial \overline{m}^{\alpha\beta}}. \tag{10.73a, b}$$

Here, the overdot ($\dot{}$) signifies the increment. Recall that these "stresses" are quite different physically; $\bar{n}^{\alpha\beta}$ measures a mean value and $\overline{m}^{\alpha\beta}$ a variation of the stress distribution upon a section. *Approximately*, the stress distribution in the HI-HO‡ shell is linear:

$$\sigma^{\alpha\beta} = \frac{\overline{N}^{\alpha\beta}}{h} + 12\frac{\overline{M}^{\alpha\beta}}{h^3}z, \tag{10.74}$$

where z is the distance along the normal from the reference surface and $(\overline{N}^{\alpha\beta}, \overline{M}^{\alpha\beta})$ are the physical components of $(\bar{n}^{\alpha\beta}, \overline{m}^{\alpha\beta})$.

It remains to explore the behavior of the shell as it relates to the actual material and to ascertain the validity of the foregoing concepts. In particular, the equations (10.71) and (10.73a, b) serve no purpose without useful forms of the function F. As one resorts to simple experiments (see Section 5.10) to formulate the constitutive equations of a material, so one

‡HI-HO = Homogeneous Isotropic HOokean

can turn to the simplest experiments for the shell. Our experiments here are computational and focus upon the interaction of the different stresses. We seek some insights by investigation of the loadings, n and m, upon a beam; the existence of curvature has little relevance.

10.8.2 Computational Experiments

If a HI-HO beam (plate or shell) were subjected to the simple actions of force N and couple M, then the only significant stress is the normal stress

$$\sigma = \frac{N}{bh} + z\frac{12\,M}{bh^3}, \tag{10.75}$$

where b and h are the width and depth of the rectangular section. If $\pm Y$ denotes the yield stress in simple tension/compression, we define the dimensionless quantities:

$$s = \frac{\sigma}{Y}, \qquad n = \frac{N}{bhY}, \qquad m = \frac{6M}{bh^2Y}, \qquad \theta \equiv \frac{2z}{h}. \tag{10.76a–d}$$

Then,

$$s = n + \theta m. \tag{10.77a}$$

Yielding is initiated at the upper ($\theta = +1$) or lower ($\theta = -1$) surfaces when

$$s = 1 = n \pm m \quad \text{or} \quad s = -1 = n \pm m. \tag{10.77b, c}$$

These four lines define the yield condition and enclose the region of HI-HO behavior, as depicted in Figure 10.3.

Let us pursue our experiment upon a beam of ideally plastic material, i.e., at yield $s = \pm 1$ the material exhibits unrestricted plastic deformation ϵ^P, but recovery exhibits Hookean behavior, i.e., $\Delta\epsilon = \Delta\epsilon^E = \Delta\sigma/E$. If loading ensues, yielding progresses from the top, or bottom surface; Figure 10.4a depicts a circumstance of positive force and couple. Figure 10.4b illustrates a limiting condition; i.e., the entire section has attained the yield stress in tension ($z > -p$) or compression ($z < -p$):

The limit condition (b) is described by the equation (cf. P. G. Hodge [200])

$$2m + 3n^2 = 3. \tag{10.78a}$$

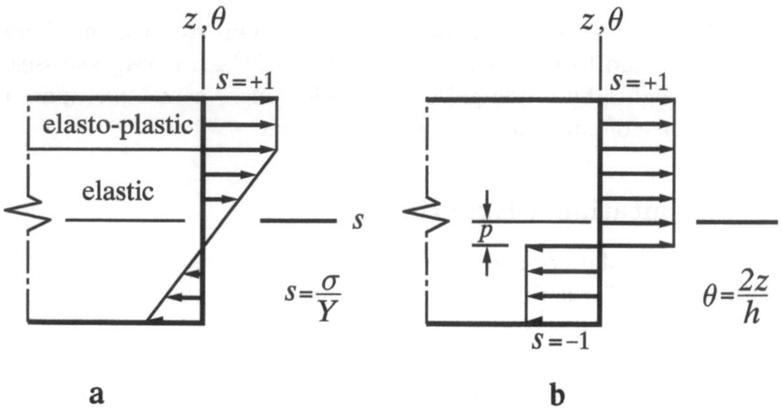

Figure 10.4 Progression of yielding in a beam (a: initial, b: limit)

Similarly, negative values of couple m produce the limit condition

$$2m - 3n^2 = -3. \tag{10.78b}$$

These parabolic curves (10.78a, b) are also shown in Figure 10.3.

The material of our shell is ideally plastic, yet the progression of yielding from initiation (curve F_I) to the limiting condition (curve F_L) exhibits the characteristics of "strain-hardening," i.e., the "yield condition" has changed from F_I to F_L. This is a consequence of the progression of yielding, from the top and bottom surfaces toward the interior, as caused by the flexure, i.e., m and ρ (curvature).

As a second experiment, let us consider the evolution of plastic flow from the onset (Figure 10.4a) to the limit (Figure 10.4b). To be specific, let us pursue a path, wherein the neutral line remains at $p = h/4$ (see Figure 10.4b). This means that strains and stresses increase/decrease monotonously at each point of the section. As an alternative to the curvature change ρ, we define a dimensionless "strain":

$$\kappa \equiv h\rho/6. \tag{10.79}$$

Also, let us employ a nondimensional expression of internal work

$$\dot{w} \equiv \frac{\dot{W}}{bhY} = \frac{N}{bhY}\dot{\gamma} + \frac{6M}{bh^2Y}\left(\frac{h\dot{\rho}}{6}\right) = n\dot{\gamma} + m\dot{\kappa}. \tag{10.80}$$

At the three stages shown in Figure 10.5a–c, our computations carry us to the states A, B, C, respectively, in Figure 10.3. Of special interest are

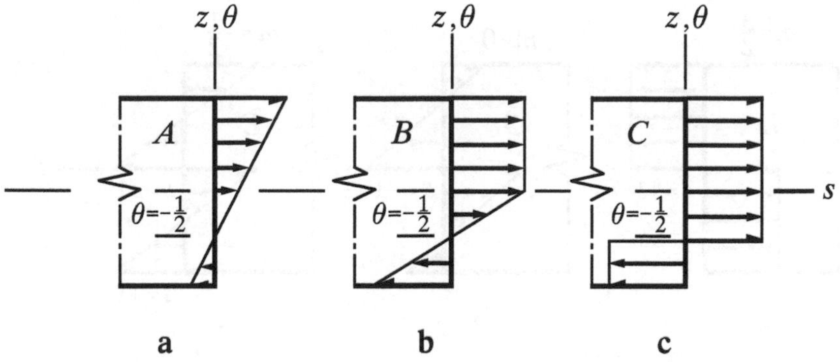

Figure 10.5 Progressive stages of yielding in a beam

the incremental strains that accompany incremental stresses. At state A, we find that the incremental stress-strain relation is the linearly elastic version:

$$\dot{N} = bhE\dot{\gamma}, \qquad \dot{M} = \frac{bh^3E}{12}\dot{\rho},$$

$$\dot{n} = \frac{E}{Y}\dot{\gamma}, \qquad \dot{m} = 3\frac{E}{Y}\dot{\kappa}. \tag{10.81a, b}$$

At state B,

$$\dot{n} = 0, \qquad \dot{m} = \frac{3}{8}\frac{E}{Y}\dot{\kappa}. \tag{10.82a, b}$$

Finally, at the limit state C, deformation, γ and κ, are unrestricted. Bear in mind that our experiment has enforced a strain path; specifically, the neutral line has remained at $z = -h/4$ ($\theta = -1/2$), so that

$$\gamma - \frac{h}{4}\rho = 0, \qquad \gamma - \frac{3}{2}\kappa = 0.$$

Only at the limit curve (where $n = 1/2$, $m = 9/8$) the incremental strain is normal to the curve; stated otherwise, if

$$F = 2m + 3n^2 - 3 = 0$$

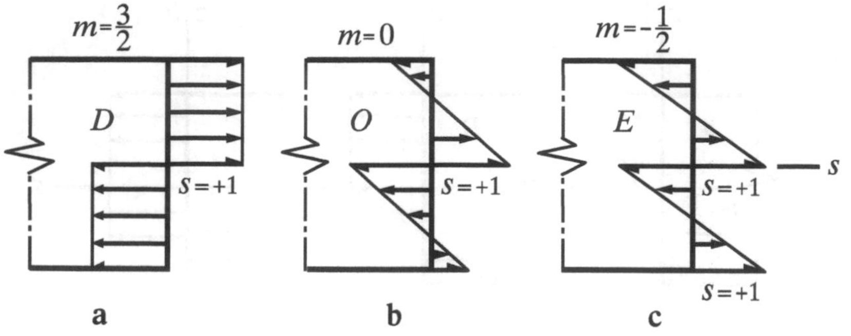

Figure 10.6 Stress distributions upon unloading (a: limit state, b: unloaded, c: reloaded)

[see equation (10.78a)], then

$$\{\dot{\gamma}, \dot{\kappa}\} = \dot{\lambda} \left\{ \frac{\partial F}{\partial n}, \frac{\partial F}{\partial m} \right\}, \tag{10.83}$$

where $\dot{\lambda}$ is some positive scalar.

It is the "stress" m that significantly effects the "hardening," i.e., the change from F_I to F_L. This suggests that we examine the consequences of unloading from a state of flexure. For simplicity, let us consider the removal of the "stress" $m = 3/2$ from the limit state D of Figure 10.3. The distribution of stress on the section, before and after, are depicted in Figure 10.6a, b; the latter is a consequence of elastic unloading, wherein

$$\Delta s = \theta \Delta m.$$

A change $\Delta s = \mp(3/2)$ occurs at top and bottom. Reloading to $m = -(1/2)$ results in the state of Figure 10.6c; this is a state of incipient yielding, i.e., $s = \mp 1$ at top and bottom. It is most significant that the condition for initial yield is now at state E (see Figure 10.3). As a consequence of the prior inelastic deformation, the yield condition has shifted from F $(m = -1)$ to E $(m = -1/2)$.

Our simple experiments, the loading sequence $O-B-C$ and the loading/unloading sequence O–D–O–E offer some physical insights. Although the model and experiments are very simple, we note the curious strain-hardening (from F_I to F_L) and the shifting from F to E. These effects can be attributed to the partial elastic/plastic behavior through the thickness, and also to the creation of residual stresses as exhibited in Figure 10.6b. The difficulties in formulating a "yield" function F and its evolution have led many researchers to revert to the three-dimensional descriptions and the related computational procedures. In other words, many utilize the two-dimensional equations which describe the *kinematics* and *statics* (or *dynamics*) of the Kirchhoff-Love theory, but revert to the three-dimensional viewpoint (introduce distributions in the distance z) to accommodate the inelasticity. Such *quasi*-shell procedures are the subject of our next section. Subsequently, we offer some formulations of yield and flow equations for the two-dimensional theory.

10.8.3 Quasi-Shell or Multi-Layer Model

We have noted the difficulties of devising a theory/approximation for the plastic behavior of shells [equations relating ($n^{\alpha\beta}$, $m^{\alpha\beta}$) and ($\dot{\gamma}^P_{\alpha\beta}$, $\dot{\rho}^P_{\alpha\beta}$)]. We also recognize the growing efficiencies and capacities of electronic computers. In view of these circumstances, many practitioners find it more expedient to revert to the three-dimensional models of elasto-plastic behavior, while retaining the kinematic-static (or dynamic) foundations of Kirchhoff-Love. Consistent with the notions of a thin shell, one presumes a state of "plane" stress and strain, i.e., $s^{33} = s^{3\alpha} = \epsilon_{\alpha 3} = 0$. One then imposes the established equations of elasto-plasticity (see Sections 5.24 through 5.34) at a number of stations through the thickness. The information regarding the state (stresses $s^{\alpha\beta}$ and requisite hardening parameters) must be stored at each site. The number of such stations is limited by computational capabilities, efficiencies, costs, and requisite precision.

As previously noted, there is no inherent difficulty in the establishment of an initial yield condition for the shell, i.e., surfaces in the ($n^{\alpha\beta} - m^{\alpha\beta}$) space. Loading to initial yielding can be treated by any of the methods for a Hookean shell. Further loading and elasto-plastic behavior is accommodated by monitoring the states (i.e., $s^{\alpha\beta}$) at each of the stations. Initial yielding at any station is signaled by the yield condition $\mathcal{Y}(s^{\alpha\beta})$ (Section 5.24). Subsequent loading at that station is governed by the incremental stress-strain relations of the elasto-plastic behavior (Section 5.28).

Now, the Kirchhoff-Love assumption governs the strain $\epsilon_{\alpha\beta}$; during loading,

$$\dot{\epsilon}_{\alpha\beta} = \dot{\epsilon}^E_{\alpha\beta} + \dot{\epsilon}^P_{\alpha\beta} = \dot{\gamma}_{\alpha\beta} + z\dot{\rho}_{\alpha\beta}. \tag{10.84}$$

To convert the consequent stress $\dot{s}^{\alpha\beta}(z)$ to the shell stresses ($\dot{n}^{\alpha\beta}$, $\dot{m}^{\alpha\beta}$) we require the stress-strain relation for the elasto-plastic increment:

$$\dot{s}^{\alpha\beta} = C_T^{\alpha\beta\gamma\eta}\dot{\epsilon}_{\gamma\eta}. \tag{10.85}$$

Note that $\dot{\epsilon}_{\alpha\beta}$ is the elastic-plastic increment of (10.84); hence, $C_T^{\alpha\beta\gamma\eta}$ is the so-called "tangent modulus" which depends upon the prevailing state.

Increments of stress $\dot{s}^{\alpha\beta}$ and elastic strain $\dot{\epsilon}^{E}_{\alpha\beta}$ are presumably related as before (Sections 10.6 and 5.22):

$$\dot{s}^{\alpha\beta} = C^{\alpha\beta\gamma\eta}\dot{\epsilon}^{E}_{\gamma\eta}. \tag{10.86}$$

In accordance with the classical theory (see Section 5.29),

$$\dot{\epsilon}^{P}_{\alpha\beta} = \dot{\lambda}\frac{\partial \mathcal{Y}}{\partial s^{\alpha\beta}}, \tag{10.87}$$

where $\dot{\lambda}$ is a positive scalar. If the material strain-hardens [see inequalities (5.120b) and (5.131)] there exists a functional G^P, such that

$$\dot{\lambda}\, G^P \equiv \frac{\partial \mathcal{Y}}{\partial s^{\alpha\beta}}\dot{s}^{\alpha\beta} > 0. \tag{10.88}$$

Note that G^P is a function of the stress $s^{\alpha\beta}$ but a functional of plastic strain $\dot{\epsilon}^{P}_{\alpha\beta}$. According to (10.84), (10.86), and (10.87), we have

$$C^{\alpha\beta\gamma\eta}\dot{\epsilon}_{\gamma\eta} = \dot{s}^{\alpha\beta} + C^{\alpha\beta\gamma\eta}\frac{\partial \mathcal{Y}}{\partial s^{\gamma\eta}}\dot{\lambda}, \tag{10.89}$$

$$\frac{\partial \mathcal{Y}}{\partial s^{\alpha\beta}}C^{\alpha\beta\gamma\eta}\dot{\epsilon}_{\gamma\eta} = \frac{\partial \mathcal{Y}}{\partial s^{\alpha\beta}}\dot{s}^{\alpha\beta} + C^{\alpha\beta\gamma\eta}\frac{\partial \mathcal{Y}}{\partial s^{\alpha\beta}}\frac{\partial \mathcal{Y}}{\partial s^{\gamma\eta}}\dot{\lambda}.$$

By means of the definition (10.88), the last equation assumes the form:

$$\frac{\partial \mathcal{Y}}{\partial s^{\alpha\beta}}C^{\alpha\beta\gamma\eta}\dot{\epsilon}_{\gamma\eta} = \left(G^P + C^{\alpha\beta\gamma\eta}\frac{\partial \mathcal{Y}}{\partial s^{\alpha\beta}}\frac{\partial \mathcal{Y}}{\partial s^{\gamma\eta}}\right)\dot{\lambda}. \tag{10.90}$$

The scalar $\dot{\lambda}$ can be eliminated from (10.89) by means of (10.90) to provide an explicit form for the "tangent modulus":

$$C_T^{\alpha\beta\gamma\eta} = C^{\alpha\beta\gamma\eta} - B^{\alpha\beta\gamma\eta}, \tag{10.91}$$

where

$$B^{\alpha\beta\gamma\eta} \equiv \left(C^{\alpha\beta\phi\mu}\frac{\partial \mathcal{Y}}{\partial s^{\phi\mu}}C^{\delta\epsilon\gamma\eta}\frac{\partial \mathcal{Y}}{\partial s^{\delta\epsilon}}\right)\Big/\left(G^P + C^{\alpha\beta\gamma\eta}\frac{\partial \mathcal{Y}}{\partial s^{\alpha\beta}}\frac{\partial \mathcal{Y}}{\partial s^{\gamma\eta}}\right). \tag{10.92}$$

The "smooth" form (i.e., no corners) of the von Mises criterion (see Section 5.27) is the most readily implemented. Then the gradients $(\partial \mathcal{Y}/\partial s^{\alpha\beta})$ are the deviatoric components of stress.

Within the context of the Kirchhoff-Love foundations, the multi-layer model admits very precise studies of elasto-plastic behavior. As such, it provides a computational tool for assessing the validity and limitations of simpler alternatives, e.g., the sandwich models or direct theories, which are described in the following.

Practically, one can only monitor the state and impose the appropriate incremental relations at a limited number of stations. There remains the question of approximation between these discrete sites. One simple approach presumes a homogeneous state in a small region about each site. In effect, that amounts to representing the shell by a finite number of layers; at any stage, the layer is presumably subject to a membrane state of stress.‡ Simple approximations by a few layers are imprecise but less costly; such models are often called "sandwich" shells.

10.8.4 Approximation by a "Sandwich" Shell

A very simple approximation of elasto-plastic behavior is provided by the ideal sandwich, wherein all resistance to extension and flexure is vested in two identical layers; the layers are separated by a core which resists transverse shear. Mathematically, the initial and limit conditions, such as F_I and F_L in Figure 10.3, are coincident. This idealization is most easily implemented. Practically, it is useful when plastic deformations predominate and/or the assessment of plastic collapse is the principal concern. Also, the model applies to certain composites which are actually fabricated of two facings and a lightweight core; much as conventional I-beams, such composites offer optimal stiffness and minimal weight.

A better, yet efficient, model for the homogeneous shell is provided by the "club sandwich" consisting of four layers (see Figure 10.7). Certain advantages are evident:

- The model exhibits an initial yield condition, followed by a transition to the limit condition. Positions and relative thicknesses of the layers can be chosen to provide a best fit.
- The model produces states of residual stress which simulate those resulting from cycles of loading and unloading, e.g., the distribution depicted in Figure 10.6b. Studies ([202], [203]) have shown that this model possesses the essential attributes of the homogeneous shell

‡An elaboration on such approximation in the two-dimensions of the surface is given in the authors' article [201].

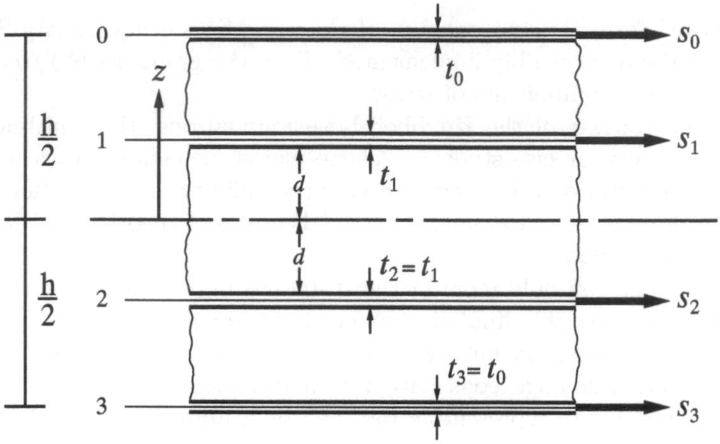

Figure 10.7 Club sandwich

and provides similar responses to various histories of loading; indeed, the "club sandwich" is a very practical alternative to more costly procedures of many layers. We sketch here only the key features:

Figure 10.7 serves to identify the elements of the four layer club sandwich. Additionally, we ascribe different moduli, E_0 and E_1, to the outer and inner layers, respectively. As before, E and $\pm Y$ denote the modulus and tensile/compressive yield stress of the actual (homogeneous) shell. The following dimensionless parameters are available for purposes of fitting the model:

$$\alpha = h/2d, \qquad \beta = t_1/t_0, \qquad k_0 = E_0/E, \qquad k_1 = E_1/E. \quad (10.93\text{a–d})$$

Here, we employ the dimensionless variables:

$$\theta = 2z/h, \qquad s^{\alpha\beta} = \sigma^{\alpha\beta}/Y, \quad (10.94\text{a, b})$$

$$_0\epsilon_{\alpha\beta} = (E/Y)\,\gamma_{\alpha\beta}, \qquad \kappa_{\alpha\beta} = (Eh/2Y)\,\rho_{\alpha\beta}. \quad (10.94\text{c, d})$$

In accord with the Kirchhoff-Love theory, the strain is expressed in the form:

$$(E/Y)\epsilon_{\alpha\beta} = {_0\epsilon_{\alpha\beta}} + \theta\kappa_{\alpha\beta}.$$

Nondimensional force and couple are defined as follows:

$$n^{\alpha\beta} = \frac{N^{\alpha\beta}}{2(1+\beta)t_0 Y}, \qquad m^{\alpha\beta} = \frac{M^{\alpha\beta}}{(1+\beta)t_0 Y h}. \quad (10.95\text{a, b})$$

Here, $N^{\alpha\beta}$ and $M^{\alpha\beta}$ denote the physical components of force and couple, respectively. Most conveniently, we can employ the yield condition of von Mises and the associated flow relations to each layer. These follow:

$$\tfrac{3}{2}\big(s^{\alpha\beta}s_{\alpha\beta} - \tfrac{1}{3}s^{\alpha}_{\alpha}s^{\beta}_{\beta}\big) - 1 = 0, \tag{10.96}$$

$$\dot{s}^{\alpha\beta} = \big(C^{\alpha\beta\gamma\eta} - B^{\alpha\beta\gamma\eta}\big)\dot{\epsilon}_{\gamma\eta}. \tag{10.97}$$

Here, $C^{\alpha\beta\gamma\eta}$ is the dimensionless version of the stiffness tensor for plane stress:

$$C^{\alpha\beta\gamma\eta} = \frac{1}{1+\nu}\big(\delta^{\alpha\gamma}\delta^{\beta\eta} + \frac{\nu}{1-\nu}\delta^{\alpha\beta}\delta^{\gamma\eta}\big). \tag{10.98}$$

In these studies, the material is presumed ideally plastic; that provides the more extreme test of our model. Then, the hardening parameter of equation (10.88) vanishes, so that

$$B^{\alpha\beta\gamma\eta} = \frac{C^{\alpha\beta\mu\phi}C^{\gamma\eta\kappa\delta}S_{\mu\phi}S_{\kappa\delta}}{C^{\alpha\beta\gamma\eta}S_{\alpha\beta}S_{\gamma\eta}}, \tag{10.99}$$

where $S^{\alpha\beta}$ denotes a component of the stress deviator:

$$S^{\alpha\beta} = s^{\alpha\beta} - \tfrac{1}{3}s^{\eta}_{\eta}\delta^{\alpha\beta}. \tag{10.100}$$

The stresses $s^{\alpha\beta}_{N}$ on the layers $(0, 1, 2, 3)$ are readily transformed to the stresses, $n^{\alpha\beta}$ and $m^{\alpha\beta}$:

$$n^{\alpha\beta} = \frac{1}{2(1+\beta)}\sum_{i=0}^{3}\left(\frac{t_i}{t_0}\right)s^{\alpha\beta}_{i}, \tag{10.101}$$

$$m^{\alpha\beta} = \frac{1}{2(1+\beta)}\sum_{i=0}^{3}\theta_i\left(\frac{t_i}{t_0}\right)s^{\alpha\beta}_{i}. \tag{10.102}$$

Two of the parameters (10.93a–d) are chosen to achieve the appropriate initial (elastic) stiffnesses:

$$\frac{dn^{11}}{d\,{}_0\epsilon_{11}} = 1, \qquad \frac{dm^{11}}{d\kappa_{11}} = \frac{1}{3}. \tag{10.103a, b}$$

Additionally, we require the correct limit moment (in bending $m^{11} = 1/2$).

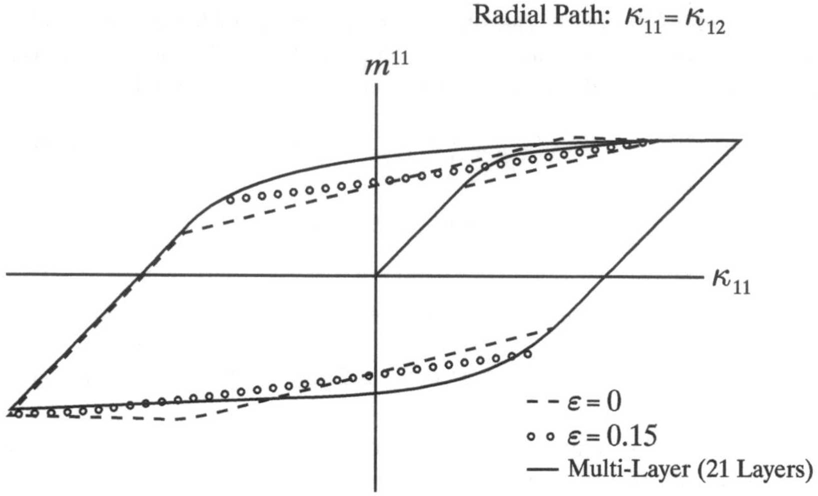

Figure 10.8 Comparison of club sandwich and multi-layer responses (radial path: $\kappa_{11} = \kappa_{12}$)

The latter imposes the condition:

$$\beta = 1 + 2\beta/\alpha. \tag{10.103c}$$

If we impose the further requirement for initial yielding in simple bending, viz., $m^{11} = 1/3$, then all four parameters (α, β, k_0, k_1) are fixed and the constitutive equations are fully prescribed. Instead, we provide for a best overall fit and admit a disparity ε, such that the initial yield moment is given by the form:

$$m_0^{11} = \frac{1+\varepsilon}{3} = \frac{1}{1+\beta}\left[1 + k_1 \frac{\beta}{\alpha^2}(1+\varepsilon)\right]. \tag{10.103d}$$

For any given ε, the four equations (10.103a–d) determine the four parameters (α, β, k_0, k_1). If membrane forces dominate, then a small value $[0 \leq \varepsilon \ll (1/2)]$ provides a good fit. If bending dominates, then a positive value $\varepsilon < (1/2)$ provides a better agreement between the piecewise linear constitutive equations and the actual equations of the homogeneous plate/shell.

Early studies were conducted with various strain histories, combinations of extension and bending (e.g., ${}_0\epsilon_{11} = \kappa_{11}$), bending and torsion (e.g., $\kappa_{11} = \kappa_{12}$), and prestrain (e.g., ${}_0\epsilon_{11} = 1.0$) followed by bending (e.g., κ_{11}). In each case, deformations extended well into the plastic range, fol-

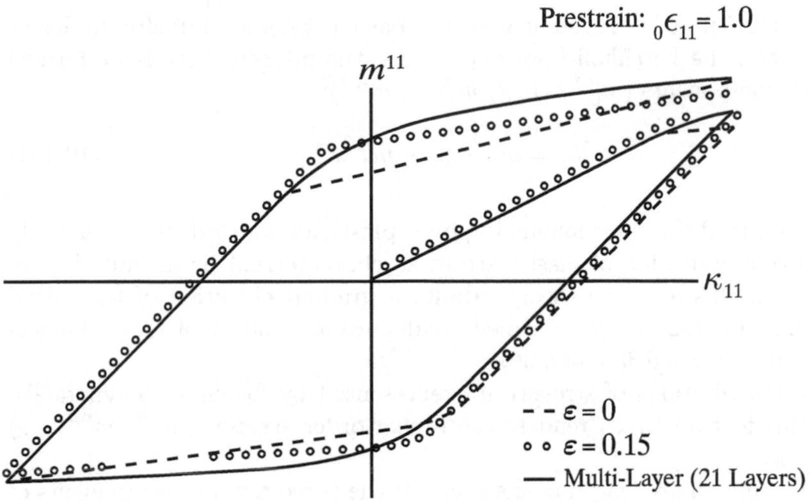

Figure 10.9 Comparison of club sandwich and multi-layer responses (prestrain: $_0\epsilon_{11} = 1.0$)

lowed by complete reversals and reloading. The two traces of Figure 10.8 (radial path: $\kappa_{11} = \kappa_{12}$) and Figure 10.9 (prestrain $_0\epsilon_{11} = 1.00$) are displayed to show the significance of adjusting, via the positive parameter $\varepsilon = 0.15$. Paths of combined extension and flexure (e.g., $_0\epsilon_{\alpha\beta} = \kappa_{\alpha\beta}$) are less sensitive to the postyielding approximation. In each example, the four-layer models ($\varepsilon = 0,\ 0.15$) are compared to the accurate response of a model with 21 layers.

A few practical remarks are appropriate: The model provides the basis of a subroutine in a computational program for the discrete approximation of an elastic-plastic plate or shell. At each node of the surface, the state of stress is characterized by 12 components ($s_i^{\alpha\beta}$, $i = 1, 2, 3, 4$). The ensuing deformations ($_0\dot{\epsilon}_{\alpha\beta}$, $\dot{\kappa}_{\alpha\beta}$) are determined by the subroutine; these are piecewise linear as the traces of Figures 10.8 and 10.9. Inherent errors are less if the material exhibits strain-hardening.

10.8.5 A Derived Theory

By "derived theory," we mean one that acknowledges a continuous distribution of stress through the thickness (z, or θ) and *derives* the two-dimensional constitutive equations [in ($n^{\alpha\beta}$, $m^{\alpha\beta}$) and ($\dot{\gamma}_{\alpha\beta}$, $\dot{\kappa}_{\alpha\beta}$)] from a three-dimensional theory (in $s^{\alpha\beta}$ and $\dot{\epsilon}_{\alpha\beta}$): The one that we recount here retains certain attributes of the simpler theories; specifically, the state of stress can be represented by a discrete number of surface tensors $m_i^{\alpha\beta}$

$(i = 0, 1, 2, 3)$. It is consistent with the basic expression (10.69): In accordance with the Kirchhoff-Love hypothesis, the internal work is performed by two components ($m_0^{\alpha\beta} \equiv n^{\alpha\beta}$, $m_1^{\alpha\beta} \equiv m^{\alpha\beta}$):

$$\dot{u}_s = m_0^{\alpha\beta}\dot{\gamma}_{\alpha\beta} + m_1^{\alpha\beta}\dot{\kappa}_{\alpha\beta}. \tag{10.104}$$

Our derived theory is founded upon a plasticity which does *not* embody the abrupt initiation of plastic strain as characterized by an initial yield condition. Instead, the theory admits a gradual evolution of the initial yielding; this follows the so-called "endochronic" theory of K. C. Valanis [89] (see Sections 5.35 and 5.36).

The distributions of stresses are represented by Legendre polynomials; it is this feature which renders the higher-order stresses ($m_2^{\alpha\beta}$, $m_3^{\alpha\beta}, \dots$) workless.

We define an arc length in the space of the three relevant components of strain:

$$\dot{\zeta}^2 = \dot{\epsilon}^{\alpha\beta}\dot{\epsilon}_{\alpha\beta} + \dot{\epsilon}_\alpha^\alpha\dot{\epsilon}_\beta^\beta. \tag{10.105}$$

We suppose that the increment of plastic strain is measured by a scalar $\dot{\lambda}$ (see Sections 5.29 and 5.35). Now, plastic strain evolves with the scalar λ, such that

$$\frac{d\lambda}{d\zeta} > 0 \qquad (\zeta > 0).$$

The rate of such plastic deformation is to depend upon stress; thus one can simulate the more abrupt transition approaching a yield condition. Specifically, we choose

$$\dot{\lambda} = \sqrt{\frac{3}{2}}\,\bar{\sigma}^n\,\dot{\zeta}, \tag{10.106}$$

where $\bar{\sigma}$ denotes the second invariant of the stress deviator ($s^{33} = 0$):

$$\bar{\sigma}^2 = \tfrac{3}{2}(s^{\alpha\beta}s_{\alpha\beta} - \tfrac{1}{3}s_\eta^\eta s_\mu^\mu). \tag{10.107}$$

Clearly, this evolution of the plastic strain $\dot{\lambda}$ is akin to the Prandtl-Reuss equations (see Section 5.33), as the strain is associated with the von Mises yield condition (see Section 5.27), i.e., $\dot{\lambda}$ depends on $\bar{\sigma}$. Accordingly, we take‡

$$\dot{\epsilon}_{\alpha\beta} = \dot{\epsilon}_{\alpha\beta}^E + \dot{\epsilon}_{\alpha\beta}^P, \tag{10.108a}$$

$$= D_{\alpha\beta\gamma\eta}\,\dot{s}^{\gamma\eta} + (s_{\alpha\beta} - \tfrac{1}{3}s_\mu^\mu\,g_{\alpha\beta})\dot{\lambda}. \tag{10.108b}$$

‡The flexibility tensor $D_{\alpha\beta\gamma\eta}$ is the inverse of the stiffness $C_{\alpha\beta\gamma\eta}$ in equation (10.86).

Our theory admits only the elastic strains during unloading as signified by the condition:

$$\dot{s}^{\alpha\beta}\dot{\epsilon}^P_{\alpha\beta} < 0, \qquad \dot{\epsilon}^P_{\alpha\beta} = 0. \tag{10.109}$$

The strain distribution is expressed in accordance with the Kirchhoff-Love hypothesis:

$$\dot{\epsilon}_{\alpha\beta} = \dot{\gamma}_{\alpha\beta} P_{(0)} + \sqrt{3}\,\dot{\kappa}_{\alpha\beta} P_{(1)}, \tag{10.110}$$

where $P_{(N)}$ denotes the Legendre polynomial of degree N. The stress distribution is approximated by a series of polynomials

$$s^{\alpha\beta} = \sqrt{1+2N}\, m^{\alpha\beta}_{(N)} P_{(N)}(\theta), \tag{10.111}$$

where, as before, $\theta = 2z/h$ and the repeated indices imply summation. Results indicate that four terms ($N = 0, 1, 2, 3$) are adequate for practical purposes; one can represent the distribution $s^{\alpha\beta}$ by more terms and achieve greater precision. The orthogonality of the polynomials admits the inverse:

$$m^{\alpha\beta}_{(N)} = \frac{\sqrt{1+2N}}{2} \int_{-1}^{1} s^{\alpha\beta} P_{(N)}\, d\theta. \tag{10.112}$$

Since the derived theory is founded upon the work/energy criteria, we derive the form (10.104) from the three-dimensional version:[‡]

$$\dot{u}_s = \tfrac{1}{2} \int_{-1}^{1} s^{\alpha\beta} \dot{\epsilon}_{\alpha\beta}\, d\theta. \tag{10.113}$$

Two versions of $\dot{u}_s$ result from the alternative expressions for the strain rate $\dot{\epsilon}_{\alpha\beta}$; equations (10.108b) and (10.110), respectively, provide

$$\dot{u}_s = m_0^{\alpha\beta} \dot{\gamma}_{\alpha\beta} + m_1^{\alpha\beta} \dot{\kappa}_{\alpha\beta}, \tag{10.114a}$$

$$= D_{\alpha\beta\gamma\eta} m^{\alpha\beta}_{(N)} \dot{m}^{\gamma\eta}_{(N)} + m^{\alpha\beta}_{(N)} (m_{\alpha\beta(M)} - \tfrac{1}{3} m^{\mu}_{\mu(M)} a_{\alpha\beta}) \dot{f}_{(NM)}, \tag{10.114b}$$

where

$$\dot{f}_{(NM)} = \frac{\sqrt{(1+2N)(1+2M)}}{2} \int_{-1}^{1} P_{(N)} P_{(M)} \dot{\lambda}\, d\theta. \tag{10.115}$$

[‡]Variations of the metric $\sqrt{g/a}$ are neglected through the thickness of our *thin* shell.

The final term of (10.114b) represents the dissipation associated with the plastic deformation. That final term has the character of the Prandtl-Reuss/von Mises plasticity, but the integral $\dot{f}_{(NM)}$ (analogous to $\dot{\lambda}$) serves to associate the two-dimensional plastic "strains"

$$(m_{\alpha\beta(M)} - \tfrac{1}{3} m^{\mu}_{\mu(M)} a_{\alpha\beta}) \dot{f}_{(NM)}$$

and "stresses" $m^{\alpha\beta}_{(N)}$.

The equality of the forms (10.114a, b) is satisfied if we equate coefficients of $m^{\alpha\beta}_{(N)}$:

$$\dot{\gamma}_{\alpha\beta} = D_{\alpha\beta\gamma\eta} \dot{m}^{\gamma\eta}_{0} + (m_{\alpha\beta(M)} - \tfrac{1}{3} m^{\mu}_{\mu(M)} a_{\alpha\beta}) \dot{f}_{(0M)}, \tag{10.116a}$$

$$\dot{\kappa}_{\alpha\beta} = D_{\alpha\beta\gamma\eta} \dot{m}^{\gamma\eta}_{1} + (m_{\alpha\beta(M)} - \tfrac{1}{3} m^{\mu}_{\mu(M)} a_{\alpha\beta}) \dot{f}_{(1M)}, \tag{10.116b}$$

$$0 = D_{\alpha\beta\gamma\eta} \dot{m}^{\gamma\eta}_{(N)} + (m_{\alpha\beta(M)} - \tfrac{1}{3} m^{\mu}_{\mu(M)} a_{\alpha\beta}) \dot{f}_{(NM)}. \tag{10.116c}$$

The integer N in equation (10.116c) takes values $N = 2, 3, \cdots$. The first term on the right sides of equations (10.116a–c) are elastic strains $\dot{e}^{(N)}_{\alpha\beta}$; the second is the plastic strain $\dot{p}^{(N)}_{\alpha\beta}$:

$$\dot{e}^{(N)}_{\alpha\beta} = D_{\alpha\beta\gamma\eta} \dot{m}^{\gamma\eta}_{(N)}, \tag{10.117}$$

$$\dot{p}^{(N)}_{\alpha\beta} = (m_{\alpha\beta(M)} - \tfrac{1}{3} m^{\mu}_{\mu(M)} a_{\alpha\beta}) \dot{f}_{(NM)}. \tag{10.118}$$

We follow the model of three-dimensional plasticity and identify the two-dimensional counterpart of the inequality (10.109): Elastic unloading occurs ($\dot{f}_{(NM)} = 0$), if

$$\dot{m}^{\alpha\beta}_{(N)} \dot{p}^{(N)}_{\alpha\beta} < 0. \tag{10.119}$$

The foregoing theory results in six incremental equations (10.116a, b) relating the six incremental stresses ($\dot{m}^{\alpha\beta}_{0}$, $\dot{m}^{\alpha\beta}_{1}$) to the six incremental strains ($\dot{\gamma}_{\alpha\beta}$, $\dot{\kappa}_{\alpha\beta}$) and, additionally $3(N-1)$ equations which determine the $3(N-1)$ incremental stresses ($\dot{m}^{\alpha\beta}_{2}, \ldots, \dot{m}^{\alpha\beta}_{(N)}$). All evolve with the plastic strains as expressed by the scalars $\dot{f}_{(NM)}$ and the prevailing state of stress $m^{\alpha\beta}_{(N)}$. At each step, the scalar $\dot{f}_{(NM)}$ must be evaluated via equation (10.115); the latter is determined by equations (10.106) and (10.107), in

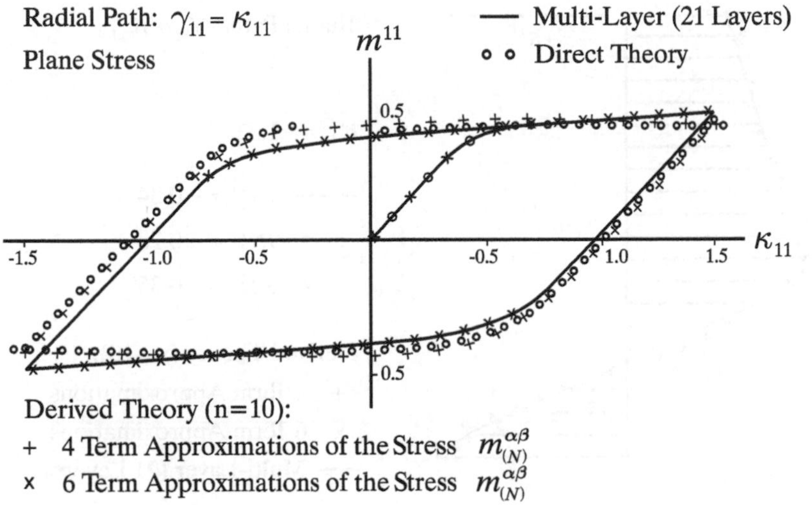

Figure 10.10 Comparison of derived, direct, and multi-layer models

which scalars $\bar{\sigma}$ and $\dot{\zeta}$ are determined in accordance with (10.105), (10.110), and (10.111). In short, the implementation requires the storage of the stresses $m^{\alpha\beta}_{(N)}$ and the evaluation of the integrals $\dot{f}_{(NM)}$ at each step.

Studies (cf. [95]) indicate that an approximation of stress by four terms $[(P_0, P_1, P_2, P_3)$ in equation (10.111)] provides very good agreement with results using 21 stations (see Section 10.8.3). Such comparisons were made for various loading paths which included radial paths (e.g., $\gamma_{11} = \kappa_{11}$), prestrain (e.g., γ_{11} and κ_{12}) followed by flexure (κ_{11}), complete unloadings and reversals. The interested reader can find details and results in the previous publication [95].

The derived theory offers one distinct advantage: Like the multi-layer approach, an approximation of the distribution $s^{\alpha\beta}(\theta)$ can be recovered at any stage. Specifically, following an inelastic deformation and subsequent unloading, the approximate state of residual stress is available.

Figure 10.10 shows the results obtained through a complete cycle of loading-unloading-reloading along the radial path $\gamma_{11} = \kappa_{11}$. Results are shown for computations employing four and six term approximations of the stress $m^{\alpha\beta}_{(N)}$; also shown are the more precise results of a multi-layer (21 layers) approach and the trace of Bieniek's direct theory (see the next Section 10.8.6). Figure 10.11 exhibits the approximation of stress $s^{11}(\theta)$ following the path of Figure 10.10 to the state of unloading *and* reversal $(m^{11}_1 = -0.34,\ \kappa_{11} = -1.31)$.

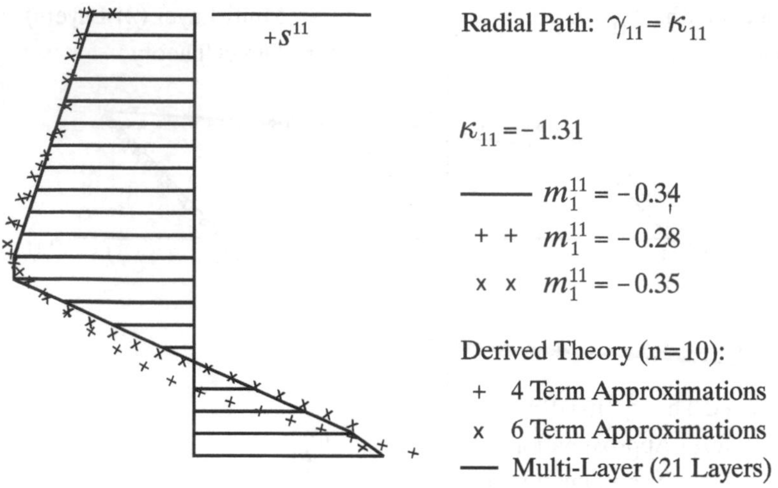

Figure 10.11 Stress approximation following loading, unloading, and reversal

10.8.6 A Direct Theory

A "direct theory" is one that does *not* derive from considerations of the distribution of stress $s^{\alpha\beta}(z)$ through the thickness, but postulates forms and criteria for the evolution of a yield condition in terms of the shell stresses ($n^{\alpha\beta}$, $m^{\alpha\beta}$). The descriptions of plastic deformation, criteria for loading and unloading, are typically drawn from the classical concepts of three-dimensional plasticity (see Section 5.29). The interested reader can find an extensive treatment, various alternatives and references in the text by Y. Başar and W. B. Krätzig [177]. Here we recount the general features of such theory.

As previously noted, the initial yield condition poses no inherent difficulties. The state of stress at a surface ($z = \pm h/2$) of a *thin* shell follows from equation (10.74):

$$\sigma^{\alpha\beta} = \frac{\overline{N}^{\alpha\beta}}{h} \pm \frac{6\,\overline{M}^{\alpha\beta}}{h^2} \quad \text{or} \quad s^{\alpha\beta} = n^{\alpha\beta} \pm m^{\alpha\beta}, \quad (10.120\text{a, b})$$

where

$$s^{\alpha\beta} \equiv \frac{\sigma^{\alpha\beta}}{Y}, \qquad n^{\alpha\beta} \equiv \frac{\overline{N}^{\alpha\beta}}{hY}, \qquad m^{\alpha\beta} \equiv \frac{6\,\overline{M}^{\alpha\beta}}{h^2 Y}. \quad (10.121\text{a–c})$$

Enforcing the yield condition at the top/bottom surfaces provides the initial yield condition in terms of the shell stresses ($n^{\alpha\beta}$, $m^{\alpha\beta}$). The von Mises criterion provides the condition in terms of the dimensionless stresses:

$$F_I = I_{NN} \pm I_{NM} + I_{MM} = 1, \tag{10.122}$$

where

$$I_{NN} = \tfrac{3}{2}(n^{\alpha\beta} n_{\alpha\beta} - \tfrac{1}{3} n^{\mu}_{\mu} n^{\eta}_{\eta}), \tag{10.123a}$$

$$I_{NM} = 3n^{\alpha\beta} m_{\alpha\beta} - n^{\mu}_{\mu} m^{\eta}_{\eta}, \tag{10.123b}$$

$$I_{MM} = \tfrac{3}{2}(m^{\alpha\beta} m_{\alpha\beta} - \tfrac{1}{3} m^{\mu}_{\mu} m^{\eta}_{\eta}). \tag{10.123c}$$

Yielding is initiated at the positive value $F_I = 1$; hence, the condition (10.122) has the alternative form:

$$I_{NN} + |I_{NM}| + I_{MM} = 1. \tag{10.124}$$

In the absence of bending, i.e., $m^{\alpha\beta} = 0$, the initial and limit conditions for ideal plasticity are the same, viz., $I_{NN} = 1$. Any bending introduces changes akin to hardening (see Subsection 10.8.2). One might conjecture that the general form of equation (10.122) might serve with modifications which account for such hardening. Such yield condition was given by A. A. Ilyushin [204] in the form:

$$F = I_{NN} + A\,|I_{NM}| + B\,I_{MM} = 1. \tag{10.125}$$

Ilyushin's condition provided for an evolution of the invariants (I_{NM}, I_{MM}), while the coefficients, A and B, are constant. V. I. Rozenblyum [205] developed a form wherein $A = 0$. A useful comparison of these forms was given by A. Robinson [206], [207]. Modifications of Ilyushin's theory were developed by M. A. Crisfield [208].

Our observations of simple extension/bending (Subsection 10.8.2) indicate the differences between the initial yield curve F_I and limit curve F_L. The changes are associated with bending. Specifically, we note the shifting of the yield condition (from -1 at F to $-1/2$ at E in Figure 10.3) as a consequence of prior plastic bending (from O to D). It is also significant that the elastic *range* (-1 to $+1$) is unchanged ($-1/2$ to $+3/2$). This suggests that the general yield "surface" in the stress-space ($n^{\alpha\beta}$, $m^{\alpha\beta}$) possesses certain attributes: Beginning at the surface F_I [equation (10.124)], the surface $F(n^{\alpha\beta}, m^{\alpha\beta})$ evolves with plastic strain and, specifically, translates with respect to bending stresses $m^{\alpha\beta}$; additionally, the surface for ideal plasticity has a limit F_L. The direct theory of M. P. Bieniek and J. R. Funaro [209] provides an effective mathematical description of such elasto-plastic behavior, the yield function F, the plastic deformations and "hardening." We recount the essentials here:

The limit condition F_L has the form (10.125). Coefficient $B = 4/9$ provides the correct limit in bending. Coefficient A is chosen to provide the correct value for the state of maximum I_{NM}, viz.,

$$n^{11} = n^{22}, \qquad m^{11} = m^{22}, \qquad n^{12} = m^{12} = 0.$$

That maximum occurs for a distribution as in Figure 10.4 with $p = 1/(2\sqrt{3})$; the corresponding values of the stresses are $n^{11} = n^{22} = 1/\sqrt{3}$, $m^{11} = m^{22} = 1$, and $I_{NN} = 1/3$, $I_{MM} = 1$ and $I_{NM} = 2/\sqrt{3}$. It follows that $A = 1/(3\sqrt{3})$.‡ Accordingly, the limit condition assumes the form:

$$F_L = I_{NN} + \frac{1}{3\sqrt{3}}\,|I_{NM}| + \frac{4}{9}\,I_{MM} = 1. \tag{10.126}$$

There remains the crucial development of the general condition F as elasto-plastic deformations progress from the initial F_I to the limit condition F_L.

The Bieniek theory adopts a form like (10.126),†

$$F = I_{NN} + \alpha\,|I_{NM}| + I^*_{MM} = 1. \tag{10.127}$$

An invariant I^*_{MM} accounts for the translation with respect to moment by introducing "hardening parameters" $m^{\alpha\beta}_*$; these have the character of residual moment stress. The invariant I^*_{MM} has the form of I_{MM} with $(m^{\alpha\beta} - m^{\alpha\beta}_*)$ replacing $m^{\alpha\beta}$:

$$I^*_{MM} = \tfrac{3}{2}[(m^{\alpha\beta} - m^{\alpha\beta}_*)(m_{\alpha\beta} - m^*_{\alpha\beta}) - \tfrac{1}{3}(m^{\mu}_{\mu} - m^{\mu *}_{\mu})(m^{\eta}_{\eta} - m^{\eta *}_{\eta})]. \tag{10.128}$$

One anticipates a coefficient α [see equation (10.124)] which has value between 1 (as in F_I) and $1/(3\sqrt{3})$ (as in F_L). The Bieniek theory adopts the latter $[\alpha = 1/(3\sqrt{3})]$ accepting the error in the initial condition F_I.

‡It should be noted that Bieniek's I_{NM} differs by the factor (1/2), hence, the coefficient by the factor 2.

†Note that F in (10.127) differs from F in (10.71). The latter is equal to F of (10.127) minus 1.

The conditions for plastic deformation follow the classical criteria (see Section 5.29):

$$F = 1, \qquad \dot{F} \equiv \frac{\partial F}{\partial n^{\alpha\beta}} \dot{n}^{\alpha\beta} + \frac{\partial F}{\partial m^{\alpha\beta}} \dot{m}^{\alpha\beta} > 0. \tag{10.129}$$

Only elastic behavior ensues if

$$F < 1 \quad \text{or} \quad \dot{F} \leq 0. \tag{10.130}$$

The "hardening parameters" $m_*^{\alpha\beta}$ evolve with plastic strain $\dot{\kappa}^P_{\alpha\beta}$, the inelastic changes-of-curvature:

$$\dot{m}^*_{\alpha\beta} = 3\frac{E}{Y}\beta(1 - F_L)\frac{F_S^2}{F_M^2}\dot{\kappa}^P_{\alpha\beta}, \tag{10.131}$$

where

$$F_S^2 \equiv \frac{\partial F}{\partial n^{\alpha\beta}} \frac{\partial F}{\partial n^{\alpha\beta}} + \frac{\partial F}{\partial m^{\alpha\beta}} \frac{\partial F}{\partial m^{\alpha\beta}}, \tag{10.132}$$

$$F_M^2 \equiv \frac{\partial F}{\partial m^{\alpha\beta}} \frac{\partial F}{\partial m^{\alpha\beta}}, \tag{10.133}$$

and $\kappa^P_{\alpha\beta}$ denotes a dimensionless version of the plastic strain [$\kappa_{\alpha\beta} = h\rho_{\alpha\beta}/6$ as in equation (10.79)].

Note: F_S is the absolute value of grad F; F_M is the part attributed to bending. As the state approaches the limit condition ($F_L \to 1$), the hardening diminishes ($m_*^{\alpha\beta} \to 0$).

The remainder of the Bieniek-Funaro theory follows the concepts of classical plasticity; however, to complete the formulation, we adopt the authors' notations: Firstly, we express the six stresses ($n^{\alpha\beta} = n^{\beta\alpha}$ and $m^{\alpha\beta} = m^{\beta\alpha}$) in the form s^i ($i = 1, \ldots, 6$) and the six strains ($\gamma_{\alpha\beta} = \gamma_{\beta\alpha}$ and $\kappa_{\alpha\beta} = \kappa_{\beta\alpha}$) in the form e_i ($i = 1, \ldots, 6$). Then, the coefficients of elasticity [see (10.54a, b)] are represented by coefficients E^{ij} such that

$$\dot{s}^i = E^{ij}\dot{e}^E_j = E^{ij}(\dot{e}_j - \dot{e}^P_j). \tag{10.134}$$

Specifically, if

$$\{\, s^1 \;\; s^2 \;\; s^3 \,\} \equiv \{\, n^{11} \;\; n^{12} \;\; n^{22} \,\}, \tag{10.135a}$$

$$\{\, s^4 \;\; s^5 \;\; s^6 \,\} \equiv \{\, m^{11} \;\; m^{12} \;\; m^{22} \,\}, \tag{10.135b}$$

$$\{\, e_1 \;\; e_2 \;\; e_3 \,\} \equiv \{\, \gamma_{11} \;\; \gamma_{12} \;\; \gamma_{22} \,\}, \tag{10.136a}$$

$$\{\, e_4 \;\; e_5 \;\; e_6 \,\} \equiv \{\, \kappa_{11} \;\; \kappa_{12} \;\; \kappa_{22} \,\}, \tag{10.136b}$$

then

$$\begin{Bmatrix} E^{11} & E^{12} & E^{13} \\ E^{21} & E^{22} & E^{23} \\ E^{31} & E^{32} & E^{33} \end{Bmatrix} \equiv h \begin{Bmatrix} {}_0C^{1111} & 0 & {}_0C^{1122} \\ 0 & {}_0C^{1212} & 0 \\ {}_0C^{2211} & 0 & {}_0C^{2222} \end{Bmatrix}, \tag{10.137a}$$

$$\begin{Bmatrix} E^{33} & E^{34} & E^{35} \\ E^{43} & E^{44} & E^{45} \\ E^{53} & E^{54} & E^{55} \end{Bmatrix} \equiv \frac{h^3}{12} \begin{Bmatrix} {}_0C^{1111} & 0 & {}_0C^{1122} \\ 0 & {}_0C^{1212} & 0 \\ {}_0C^{2211} & 0 & {}_0C^{2222} \end{Bmatrix}. \tag{10.137b}$$

By the associated "flow" rule,

$$\dot{e}_i^P = \dot{\lambda}\frac{\partial F}{\partial s^i}, \quad \text{if} \quad F = 1 \quad \text{and} \quad \frac{\partial F}{\partial s^i}\dot{s}^i > 0. \tag{10.138}$$

On the other hand,

$$\dot{e}_i^P = 0, \quad \text{if} \quad F < 1 \quad \text{or} \quad \frac{\partial F}{\partial s^i}\dot{s}^i \leq 0. \tag{10.139}$$

Also, at the yield condition,

$$\dot{F} = \frac{\partial F}{\partial s^i}\dot{s}^i + \frac{\partial F}{\partial s_*^i}\dot{s}_*^i = 0. \tag{10.140}$$

The parameter $\dot{\lambda}$ is eliminated as before [see equations (10.89) to (10.92) of Subsection 10.8.3]. Then

$$\dot{s}^i = D^{ij}\dot{e}_j. \tag{10.141}$$

Now, the tangent modulus has a form [similar to $C_T^{\alpha\beta\gamma\eta}$ in equation (10.91)]:

$$D^{ij} = E^{ij} - \left(E^{ik}\frac{\partial F}{\partial s^k}E^{jl}\frac{\partial F}{\partial s^l}\right) \Big/ \left(E^{ij}\frac{\partial F}{\partial s^i}\frac{\partial F}{\partial s^j} - A\frac{\partial F}{\partial s^i}\frac{\partial F}{\partial s_*^i}\right), \tag{10.142}$$

where

$$A = 3\beta\frac{E}{Y}(1 - F_L)\frac{F_S^2}{F_M^2}. \tag{10.143}$$

A very practical theory is completed with the assignment of the constant β. We quote from the original work: "...a constant value $\beta = 2$ has been found reasonably satisfactory for solid shells." A key feature of the theory is the "hardening law" represented by equation (10.131) and again in equation (10.143). Again, we quote the authors: "It is motivated solely by the fact that it represents fairly closely the actual behavior of a solid shell in the plastic range. It reproduces also the lowered yield point ("Bauschinger effect") which manifests itself if the bending moment is reversed, the shell is unloaded and then loaded in the opposite direction."

One result of the Bieniek-Funaro theory is included in Figure 10.10 for comparisons with those of the multi-layer and derived theory.

10.9 Strain-Displacement Equations

To this point, the displacements and rotations have not entered our development of shell theory. Some parts, notably those invoking Hooke's law, are limited to small strains. Thin shells are particularly prone to moderately large rotations and displacements, though strains are usually small. Now, we turn to this final aspect, the relation of strains, ${}_0\gamma_{\alpha\beta}$ and $\rho_{\alpha\beta}$, and displacement:

$${}_0\boldsymbol{V} = {}_0\boldsymbol{R} - {}_0\boldsymbol{r}. \tag{10.144}$$

In the Kirchhoff-Love theory only the displacement ${}_0\boldsymbol{V}$ of the surface is required. By hypothesis, particles elsewhere lie on the normal, before and after deformation:

$$\boldsymbol{r} = {}_0\boldsymbol{r} + \theta^3\hat{\boldsymbol{n}}, \qquad \boldsymbol{R} = {}_0\boldsymbol{R} + \theta^3\hat{\boldsymbol{N}}. \tag{10.145a, b}$$

Realistically, transverse shear is accommodated by the slight rotation [see Section 9.5, equation (9.5)], which carries $\hat{\boldsymbol{N}}$ to $\boldsymbol{A}_3$. Then,

$$\boldsymbol{R} = {}_0\boldsymbol{R} + \theta^3\boldsymbol{G}_3 = {}_0\boldsymbol{R} + \theta^3\boldsymbol{A}_3. \tag{10.146a, b}$$

As before [see Section 9.13, equation (9.84b)],

$$\boldsymbol{G}_3 = \boldsymbol{A}_3 = \hat{\boldsymbol{N}} + 2\gamma_\mu\boldsymbol{A}^\mu. \tag{10.147}$$

The strain-displacement relations follow from the definition:

$$\gamma_{\alpha\beta} = \tfrac{1}{2}(\boldsymbol{G}_\alpha \cdot \boldsymbol{G}_\beta - \boldsymbol{g}_\alpha \cdot \boldsymbol{g}_\beta).$$

Using the relations

$$\boldsymbol{G}_\alpha = \boldsymbol{R}_{,\alpha} = {}_0\boldsymbol{R}_{,\alpha} + \theta^3 \boldsymbol{A}_{3,\alpha}, \tag{10.148a}$$

$$= {}_0\boldsymbol{r}_{,\alpha} + {}_0\boldsymbol{V}_{,\alpha} + \theta^3(-B^\mu_\alpha \boldsymbol{A}_\mu + 2\gamma_\mu\|^*_\alpha \boldsymbol{A}^\mu + 2\gamma_\mu B^\mu_\alpha \hat{\boldsymbol{N}}), \tag{10.148b}$$

$$\boldsymbol{g}_\alpha = {}_0\boldsymbol{r}_{,\alpha} - \theta^3 b^\mu_\alpha \boldsymbol{a}_\mu = \boldsymbol{a}_\alpha - \theta^3 b^\mu_\alpha \boldsymbol{a}_\mu, \tag{10.149}$$

we obtain

$$\gamma_{\alpha\beta} = {}_0\gamma_{\alpha\beta} + \theta^3 \acute{\rho}_{\alpha\beta} + (\theta^3)^2\, \psi_{\alpha\beta}, \tag{10.150}$$

$${}_0\gamma_{\alpha\beta} = \tfrac{1}{2}(\boldsymbol{a}_\alpha \cdot {}_0\boldsymbol{V}_{,\beta} + \boldsymbol{a}_\beta \cdot {}_0\boldsymbol{V}_{,\alpha} + {}_0\boldsymbol{V}_{,\alpha} \cdot {}_0\boldsymbol{V}_{,\beta}), \tag{10.151}$$

$$\acute{\rho}_{\alpha\beta} = -(K_{\alpha\beta} - b_{\alpha\beta}), \tag{10.152a}$$

$$K_{\alpha\beta} = B_{\alpha\beta} - \gamma_\alpha\|^*_\beta - \gamma_\beta\|^*_\alpha, \tag{10.152b}$$

$$\psi_{\alpha\beta} = \tfrac{1}{2}(B^\mu_\alpha B_{\mu\beta} - b^\mu_\alpha b_{\mu\beta})$$

$$+ (-B^\mu_\alpha \gamma_\mu\|^*_\beta - B^\mu_\beta \gamma_\mu\|^*_\alpha + 2\gamma_\mu\|^*_\alpha \gamma_\nu\|^*_\alpha A^{\mu\nu} + 2\gamma_\mu\gamma_\nu B^\mu_\alpha B^\nu_\beta). \tag{10.153}$$

Here, only the strain of the surface ${}_0\gamma_{\alpha\beta}$ is given explicitly in terms of the displacement of the surface ${}_0\boldsymbol{V}$. In the Kirchhoff-Love theory that is all that is needed.

In the Kirchhoff-Love theory only,

$$K_{\alpha\beta} = B_{\alpha\beta}, \tag{10.154a}$$

$$= \hat{\boldsymbol{N}} \cdot \boldsymbol{A}_{\alpha,\beta}, \tag{10.154b}$$

$$= \hat{\boldsymbol{N}} \cdot (\overline{\Gamma}^\mu_{\alpha\beta}\, \boldsymbol{a}_\mu + b_{\alpha\beta}\, \hat{\boldsymbol{n}} + {}_0\boldsymbol{V}_{,\alpha\beta}). \tag{10.154c}$$

We recall that the *finite* rotation which carries the initial triad ($\boldsymbol{a}_1$, $\boldsymbol{a}_2$, $\hat{\boldsymbol{n}}$) to the convected triad ($\boldsymbol{b}_1$, $\boldsymbol{b}_2$, $\boldsymbol{b}_3 \equiv \hat{\boldsymbol{N}}$) is expressed by the transformation (9.8):

$$\boldsymbol{b}_i = \bar{r}_i^{\bullet j} \boldsymbol{a}_j. \tag{10.155}$$

Therefore,

$$K_{\alpha\beta} = \overline{\Gamma}_{\alpha\beta\mu} \bar{r}_3^{\bullet\mu} + b_{\alpha\beta} \bar{r}_3^{\bullet 3} + \hat{\boldsymbol{N}} \cdot {}_0\boldsymbol{V}_{,\alpha\beta},$$

$$\bar{r}_3^{\bullet 3} = r^3_{\bullet 3} = \boldsymbol{b}_3 \cdot \boldsymbol{a}^3 = \hat{\boldsymbol{N}} \cdot \hat{\boldsymbol{n}}.$$

The flexural strain of the Kirchhoff-Love theory follows:

$$\begin{aligned} \rho_{\alpha\beta} &= -(B_{\alpha\beta} - b_{\alpha\beta}) \\ &= -\overline{\Gamma}^{\mu}_{\alpha\beta} \hat{\boldsymbol{N}} \cdot \boldsymbol{a}_\mu - b_{\alpha\beta} \hat{\boldsymbol{n}} \cdot \hat{\boldsymbol{N}} - \hat{\boldsymbol{N}} \cdot {}_0\boldsymbol{V}_{,\alpha\beta} + b_{\alpha\beta}. \end{aligned} \tag{10.156}$$

The form of (10.156) is revealing: Firstly, note that $\hat{\boldsymbol{N}} \cdot \boldsymbol{a}_\alpha$ is typically small; it is significant only if rotations are very large. Secondly, note the result *if* there is *no* rotation of the normal ($\hat{\boldsymbol{n}} \cdot \hat{\boldsymbol{N}} = 1$, $\boldsymbol{a}_\alpha \cdot \hat{\boldsymbol{N}} = 0$) and then

$$\rho_{\alpha\beta} = -\hat{\boldsymbol{N}} \cdot {}_0\boldsymbol{V}_{,\alpha\beta}$$

[see remarks in Section 10.7 following equations (10.67) and (10.68)]. Of course, equation (10.156) applies to finite rotations as well.

Any alterations of the general forms for the flexural strain, $\acute{\rho}_{\alpha\beta}$ or $\rho_{\alpha\beta}$, are largely academic. In practical contemporary problems of finite deformation, one is most likely to resort to discrete approximation and electronic computation for the algebraic/numerical system. A general method for the treatment of the nonlinear systems is one of successive linear increments. Implementation entails the successive revision of the variables: Some variables (small strains and stresses) are additive; *rotation* is *not*. To pursue such strategy, we require the relation between the rotation, tensor $\bar{r}_i^{\bullet j}$, and the increment $\dot{\bar{r}}_i^{\bullet j}$; the *small* increment can be approximated by the vector $\dot{\boldsymbol{\omega}} = \dot{\bar{\omega}}^i \boldsymbol{b}_i$ [see Subsection 3.14.3, equations (3.105) and (3.108) and also Section 9.5, equations (9.24a–c) and (9.25a–c)]:

$$\begin{aligned} \dot{\bar{r}}_i^{\bullet j} &= \dot{\bar{\omega}}^k e_{kil}\, r^{lj} \\ &= \tfrac{1}{2}(\dot{\boldsymbol{b}}_i \cdot \boldsymbol{b}_l - \boldsymbol{b}_i \cdot \dot{\boldsymbol{b}}_l) r^{lj}. \end{aligned} \tag{10.157}$$

10.10 Approximation of Small Strains and Moderate Rotations

Our notion of a moderate rotation is one that admits the approximation by a vector [see Section 3.22]. The deformed vector $\boldsymbol{A}_\alpha$ is represented as a linear combination:

$$\boldsymbol{A}_\alpha = \boldsymbol{a}_\alpha + {}_0\boldsymbol{V}_{,\alpha} = \boldsymbol{a}_\alpha + (e_{i\alpha} + \Omega_{i\alpha})\boldsymbol{a}^i, \tag{10.158a}$$

$$\boldsymbol{A}_3 = \hat{\boldsymbol{n}} + (e_{\alpha 3} + \Omega_{\alpha 3})\boldsymbol{a}^\alpha. \tag{10.158b}$$

Here e_{ij} and Ω_{ij} are the symmetric and antisymmetric tensors defined by (3.159) and (3.160), respectively. Once again, we preclude stretching of the normal:

$$\boldsymbol{A}_3 \cdot \hat{\boldsymbol{N}} \doteq 1.$$

Following the earlier arguments [see Section 3.22], $\Omega_{ij} = -\Omega_{ji}$ is associated with the rotation of the surface:

$$\Omega^i = \boldsymbol{a}^i \cdot \boldsymbol{\omega} = \tfrac{1}{2} e^{imn}\Omega_{nm}.$$

Here, we admit also a small transverse shear ${}_0\gamma_{\alpha 3} \doteq e_{\alpha 3}$. Then rotation of the normal can be represented by a vector $\boldsymbol{\beta}$; but this differs from $\boldsymbol{\omega}$ by virtue of the transverse shear:

$$\boldsymbol{A}_3 = \hat{\boldsymbol{n}} + \boldsymbol{\beta} \times \hat{\boldsymbol{n}}, \tag{10.158c}$$

$$\beta^\alpha = \bar{e}^{\gamma\alpha}(e_{\gamma 3} + \Omega_{\gamma 3}), \qquad \beta^3 = \Omega^3. \tag{10.159a, b}$$

In terms of displacement:

$$\Omega_{\alpha\beta} = \tfrac{1}{2}(\boldsymbol{a}_\alpha \cdot {}_0\boldsymbol{V}_{,\beta} - \boldsymbol{a}_\beta \cdot {}_0\boldsymbol{V}_{,\alpha}), \tag{10.160a}$$

$$\Omega_{3\alpha} = \tfrac{1}{2}(\boldsymbol{a}_3 \cdot {}_0\boldsymbol{V}_{,\alpha} - \boldsymbol{a}_\alpha \cdot \hat{\boldsymbol{N}}), \tag{10.160b}$$

$$\beta^\mu \bar{e}_{\gamma\mu} = e_{\gamma 3} + \Omega_{\gamma 3} = \boldsymbol{a}_\gamma \cdot \boldsymbol{A}_3. \tag{10.161}$$

In the Kirchhoff-Love theory, $e_{\gamma 3} = 0$, $\boldsymbol{A}_3 = \hat{\boldsymbol{N}}$, and $\boldsymbol{\beta} = \boldsymbol{\omega}$.

Here, as before (see Section 3.22) the components $e_{ij} = e_{ji}$ have the magnitude of the strain. We neglect their products in the surface strain:

$$ {}_0\gamma_{\alpha\beta} = \tfrac{1}{2}(\boldsymbol{A}_\alpha \cdot \boldsymbol{A}_\beta - a_{\alpha\beta}), \tag{10.162a} $$

$$ \doteq e_{\alpha\beta} + \tfrac{1}{2}\Omega_{i\alpha}\Omega^{i}_{\cdot\beta}. \tag{10.162b} $$

Additionally, we note that rotation about the normal ($\boldsymbol{\beta}\cdot\hat{\boldsymbol{n}} \doteq \boldsymbol{\omega}\cdot\hat{\boldsymbol{n}}$) is usually small of the order of the strain $[\Omega_{\alpha\beta} = O({}_0\gamma_{\alpha\beta})]$. Accordingly, we neglect that component in the product $(\Omega_{\mu\alpha}\Omega^{\mu}_{\cdot\beta})$ of (10.162b). However, the reader must be alert to exceptional circumstances: A simple example is a thin elastic tube that has a slot along the entire length of a generator. When subjected to axial torque, the tube exhibits pronounced rotation about a normal; one can readily perceive the movement at edges of the slot and also the warping of a cross-section. In the remaining strains, ${}_0\gamma_{\alpha 3}$ and $\acute{\rho}_{\alpha\beta}$, we neglect the nonlinear terms. Then,

$$ {}_0\gamma_{\alpha 3} = e_{\alpha 3} = \tfrac{1}{2}(\boldsymbol{a}_\alpha \cdot \boldsymbol{A}_3 + \boldsymbol{a}_3 \cdot \boldsymbol{A}_\alpha), \tag{10.163a} $$

$$ = \tfrac{1}{2}[\boldsymbol{a}_\alpha \cdot (\boldsymbol{\beta} \times \hat{\boldsymbol{n}}) + \boldsymbol{a}_3 \cdot {}_0\boldsymbol{V}_{,\alpha}], \tag{10.163b} $$

$$ = \tfrac{1}{2}[\boldsymbol{\beta} \cdot (\hat{\boldsymbol{n}} \times \boldsymbol{a}_\alpha) + \boldsymbol{a}_3 \cdot {}_0\boldsymbol{V}_{,\alpha}]. \tag{10.163c} $$

Since we include the strain ${}_0\gamma_{\alpha 3}$, the normal $\hat{\boldsymbol{n}}$ turns to $\boldsymbol{A}_3$ according to (10.158b, c). Then, the strain $\rho_{\alpha\beta}$ of (10.13b) is altered accordingly:

$$ \acute{\rho}_{\alpha\beta} \equiv \boldsymbol{A}_\alpha \cdot \boldsymbol{A}_{3,\beta} + b_{\alpha\beta}. \tag{10.164a} $$

By means of (10.158a), (10.15c), and (10.159a, b), we obtain

$$ \acute{\rho}_{\alpha\beta} = \tfrac{1}{2}(\bar{e}_{\alpha\mu}\boldsymbol{a}^{\mu} \cdot \boldsymbol{\beta}_{,\beta} + \bar{e}_{\beta\mu}\boldsymbol{a}^{\mu} \cdot \boldsymbol{\beta}_{,\alpha} - \underline{b^{\mu}_{\alpha}\, {}_0\gamma_{\mu\beta}} - \underline{b^{\mu}_{\beta}\, {}_0\gamma_{\mu\alpha}}). \tag{10.164b} $$

Note that the latter is a symmetric version, the only part that contributes to the internal work of the symmetric stress tensor $m^{\alpha\beta}$.

Once again, we remind the reader that the membrane stretching and membrane forces play the dominant role in thin shells. It remains to establish the equilibrium equations *consistent* with the foregoing approximations. These must follow from the principle of virtual work; $\dot{W} = 0$, where

$$\dot{W} = \iint_{s_0} [n^{\alpha\beta}\,{}_0\dot{\gamma}_{\alpha\beta} + m^{\alpha\beta}(\acute{\rho}_{\alpha\beta})^{\cdot} + 2\,T^{\alpha}\,{}_0\dot{\gamma}_{\alpha 3}$$

$$- \boldsymbol{F} \cdot {}_0\dot{\boldsymbol{V}} - \boldsymbol{C} \cdot (\dot{\boldsymbol{\beta}} \times \hat{\boldsymbol{N}})]\, ds_0$$

$$- \int_{c_2} [\boldsymbol{N} \cdot {}_0\dot{\boldsymbol{V}} + \boldsymbol{M} \cdot (\dot{\boldsymbol{\beta}} \times \hat{\boldsymbol{N}})]\, dc_2. \qquad (10.165a)$$

The reader is referred to Figure 10.1 (see Section 10.3), where c_1 and c_2 denote distance along the normal and edge, respectively. The integrand of (10.165a) contains terms of the type $[\ \cdots\ ]^{\alpha} \cdot {}_0\dot{\boldsymbol{V}}_{,\alpha}$ and $[\ \cdots\ ]^{\alpha} \cdot \dot{\boldsymbol{\beta}}_{,\alpha}$. Both require an integration by parts. The final result has the form

$$\dot{W} = - \iint_{s_0} \{[\ \textbf{Force}\ ]^{i}\, \boldsymbol{a}_i \cdot {}_0\dot{\boldsymbol{V}} + [\ \textbf{Moment}\ ]^{\alpha}\, \boldsymbol{a}_{\alpha} \cdot \dot{\boldsymbol{\beta}}\}\, ds_0$$

$$- \int_{c_2} \{[\ \textbf{Force}\ ] \cdot {}_0\dot{\boldsymbol{V}} + [\ \textbf{Moment}\ ] \cdot \dot{\boldsymbol{\beta}}\}\, dc_2. \qquad (10.165b)$$

Each bracketed term must vanish for equilibrium.

The differential equations governing components of force follow:

$$(n^{\alpha\beta} - \underline{b^{\beta}_{\mu} m^{\alpha\mu}})\|_{\beta} - (n^{\phi\beta} - \underline{b^{\beta}_{\mu} m^{\phi\mu}}) b^{\alpha}_{\beta} \Omega_{3\phi} - T^{\phi} b^{\alpha}_{\phi} + F^{\alpha} = 0, \quad (10.166a)$$

$$(n^{\alpha\beta} - \underline{b^{\beta}_{\mu} m^{\alpha\mu}}) b_{\alpha\beta} + [(n^{\alpha\beta} - \underline{b^{\beta}_{\mu} m^{\alpha\mu}})\Omega_{3\alpha}]\|_{\beta} + T^{\alpha}\|_{\alpha} + F^{3} = 0. \quad (10.166b)$$

The differential equations governing components of moment follow:

$$m^{\alpha\beta}\|_{\beta} - T^{\alpha} + C^{\alpha} = 0, \qquad (10.166c)$$

$$m^{\alpha\beta} b^{\gamma}_{\beta} = m^{\gamma\beta} b^{\alpha}_{\beta}. \qquad (10.166d)$$

At the boundary c_2 we have the following conditions on components of force:

$$(n^{\alpha\beta} - \underline{b^{\beta}_{\mu} m^{\alpha\mu}}) n_{\beta} + (n^{\alpha\beta} - \underline{b^{\beta}_{\mu} m^{\alpha\mu}}) \Omega_{3\alpha} n_{1\beta} = \boldsymbol{N} \cdot \boldsymbol{A}^{\alpha}, \qquad (10.167a)$$

$$(n^{\alpha\beta} - \underline{b^{\beta}_{\mu} m^{\alpha\mu}}) \Omega_{3\alpha} n_{\beta} + T^{\beta} n_{1\beta} = \boldsymbol{N} \cdot \boldsymbol{A}_3. \qquad (10.167b)$$

The conditions on moment follow:

$$\bar{e}_{\alpha\phi} m^{\alpha\beta} n_{1\beta} \boldsymbol{a}^{\phi} = \boldsymbol{m}^1 \equiv \boldsymbol{n} \times \boldsymbol{M}^1. \tag{10.167c}$$

Note that the moderate rotation enters the conditions on force; in particular,

$$\boldsymbol{A}^{\alpha} \doteq \boldsymbol{a}^{\alpha} + \boldsymbol{\omega} \times \boldsymbol{a}^{\alpha}.$$

Here (see Figure 10.1)

$$n_{1\alpha} = \boldsymbol{a}_{\alpha} \cdot \hat{\boldsymbol{t}}_1 = \frac{\partial c_1}{\partial \theta^{\alpha}}.$$

Since the rotation $\boldsymbol{\beta}$ is not dictated by the surface deformation, i.e., $\boldsymbol{A}_3 \neq \hat{\boldsymbol{N}}$, we obtain the three components of force (10.166a, b). By means of (10.166c), we can readily eliminate the component T^{α}; then, (10.166a, b) assume the forms:

$$(n^{\alpha\beta} - \underline{b^{\beta}_{\mu} m^{\alpha\mu}})\|_{\beta} - (n^{\phi\beta} - \underline{b^{\beta}_{\mu} m^{\phi\mu}}) b^{\alpha}_{\beta} \Omega_{3\phi}$$
$$- b^{\alpha}_{\phi} m^{\phi\beta}\|_{\beta} - b^{\alpha}_{\phi} C^{\phi} + F^{\alpha} = 0, \tag{10.168a}$$

$$(n^{\alpha\beta} - \underline{b^{\beta}_{\mu} m^{\alpha\mu}}) b_{\alpha\beta} + [(n^{\alpha\beta} - \underline{b^{\beta}_{\mu} m^{\alpha\mu}}) \Omega_{3\alpha}]\|_{\beta}$$
$$+ m^{\alpha\beta}\|_{\beta\alpha} + C^{\alpha}\|_{\alpha} + F^3 = 0. \tag{10.168b}$$

The linear versions of (10.168a, b) ($\Omega_{3\alpha} = 0$) and the linear versions of (10.18a, b) ($B^{\alpha}_{\beta} \doteq b^{\alpha}_{\beta}$) are fully consistent. Note too that the occurrence of a *small* transverse shear strain is inconsequential.

We recall that the moments of (10.167c) are related to the moment *vector* (see Section 10.3 and Figure 10.1). From (10.167c) we have the two conditions:

$$n_{1\alpha} n_{1\beta} m^{\alpha\beta} = \acute{M}^{11}, \tag{10.169a}$$

$$n_{1\alpha} n_{2\beta} m^{\alpha\beta} = \acute{M}^{12}. \tag{10.169b}$$

A *small* transverse shear strain effects only the edge conditions [see Section 10.3, equation (10.20)]. In the Kirchhoff-Love theory, the rotation of normal $\hat{\boldsymbol{n}}$ about the normal $\hat{\boldsymbol{t}}_1$ to curve c_2 is determined by the displacement, viz.,

$$\dot{\boldsymbol{\omega}} \cdot \hat{\boldsymbol{t}}_1 = (\hat{\boldsymbol{n}})^{\cdot} \cdot \hat{\boldsymbol{t}}_2 = -\hat{\boldsymbol{n}} \cdot \frac{\partial_0 \dot{\boldsymbol{V}}}{\partial c_2}.$$

In the manner of condition (10.20), the normal component of (10.167b) takes the form [see (10.21d)]:

$$T^\beta n_{1\beta} + \left[(n^{\alpha\beta} - b^\beta_\mu m^{\alpha\mu})\Omega_{3\alpha}\right] n_{1\beta} + \frac{\partial(n_{1\alpha} n_{2\beta} m^{\alpha\beta})}{\partial c_2} = T^1 + \frac{\partial \acute{M}^{12}}{\partial c_2}. \quad (10.170)$$

Only the tangential component of the moment $\acute{M}^{11}$ of equation (10.169a) is prescribed on c_2. Note that the moment *vector* is

$$\boldsymbol{m}^1 = \hat{\boldsymbol{n}} \times \boldsymbol{M}^1;$$

twisting couple $\acute{m}^{11} = -\acute{M}^{12}$, bending couple $\acute{m}^{12} = \acute{M}^{11}$ (see also Section 10.3).

In most practical applications, a shell does not undergo significant rotations about the normal; $\beta^3 = \Omega^3$ has the order-of-magnitude of the strains. Exceptions occur: A helicoidal shell ([210] to [212]), like a spring, or a cylindrical tube [213] opened with a longitudinal slit are extremely flexible; they flex and twist readily *because* they offer little resistance to membrane forces. With the understanding that the results must be employed cautiously, we have simplified the preceding equations by neglecting products with the normal rotation.

Various authors, cf. W. T. Koiter ([158], [161]) have noted the negligible role of the terms $(b^\mu_\beta \, {}_0\gamma_{\mu\alpha})$, underlined in equation (10.164b); see also Section 10.4 and the modified strain $\bar{\rho}_{\alpha\beta}$ of equation (10.29). Note too that those terms in the strain of (10.164b) account for the terms $(b^\beta_\mu m^{\alpha\mu})$, underlined in the equilibrium equations, (10.166a, b) and (10.167a, b). The latter also play a negligible role ($b^\beta_\mu m^{\alpha\mu} \ll n^{\alpha\beta}$). These terms are a curiosity, which deserve some scrutiny. In particular, we note their entrance via the virtual work of moment, viz.,

$$\boldsymbol{N}^\alpha \cdot \dot{\boldsymbol{A}}_3 \equiv \int_{t_-}^{t_+} \boldsymbol{s}^\alpha \cdot \dot{\boldsymbol{A}}_3 \, \mu \, d\theta^3 = N^{\alpha\beta} \boldsymbol{A}_\beta \cdot \dot{\boldsymbol{A}}_3.$$

The stresses, $s^{\alpha\beta}$ and $N^{\alpha\beta}$, are based upon the *deformed* vectors, $\boldsymbol{G}_\alpha$ and $\boldsymbol{A}_\alpha$. The latter is stretched in accordance with (10.158a) ($e_{\alpha\beta} \doteq {}_0\gamma_{\alpha\beta}$). This is the origin of the small terms in question. They reflect a change in the stress vector. An alternative definition, e.g., components associated with the triad $\boldsymbol{b}_\alpha$ [$N^{\alpha\beta} \equiv \boldsymbol{b}^\beta \cdot \boldsymbol{N}^\alpha$, see Section 9.6 and equation (9.38a)] circumvents the dilemma.

10.11 Theory of Shallow Shells

We consider now a shell in which lengths L on the reference surface are small compared to the radii of curvature ($|L\kappa| \ll 1$); strains are small enough to interchange the order of covariant differentiation [see Section 10.7, equations (10.65)]. Additionally, we presume that rotations about the normal are small, as the strains, and that tangential displacements are small compared to normal displacement [$(a_{\underline{\alpha\alpha}})^{1/2}\, V^\alpha \ll V^3$]. Of course, these are strictly valid for plates; our shallow shell exhibits similar attributes.

Since transverse shear strain is absent, rotation of the normal is determined by the rotation of the surface:

$$\Omega_{3\alpha} = \tfrac{1}{2}(\hat{\boldsymbol{n}} \cdot {}_0\boldsymbol{V}_{,\alpha} - \boldsymbol{a}_\alpha \cdot \hat{\boldsymbol{N}}) \doteq \hat{\boldsymbol{n}} \cdot {}_0\boldsymbol{V}_{,\alpha} \doteq {}_0V_{3,\alpha}. \tag{10.171}$$

Since rotations about the normal are small and the order of covariant differentiation is immaterial, we have the approximations [see equations (10.162a, b) and (10.164a, b)]:

$$\begin{aligned} {}_0\gamma_{\alpha\beta} &\doteq {}_0e_{\alpha\beta} + \tfrac{1}{2}\Omega_{3\alpha}\Omega_{3\beta} \\ &\doteq \tfrac{1}{2}({}_0V_\alpha\|_\beta + {}_0V_\beta\|_\alpha - 2b_{\alpha\beta}\,{}_0V_3) + \tfrac{1}{2}\,{}_0V_{3,\alpha}\,{}_0V_{3,\beta}, \end{aligned} \tag{10.172}$$

$$\rho_{\alpha\beta} \doteq \tfrac{1}{2}(\Omega_{\alpha 3}\|_\beta + \Omega_{\beta 3}\|_\alpha) = -{}_0V_3\|_{\alpha\beta}. \tag{10.173}$$

Our consistent approximation of the compatibility equation (10.56a) follows:

$$\bar{\epsilon}^{\lambda\alpha}\bar{\epsilon}^{\mu\beta}\left({}_0\gamma_{\lambda\beta}\|_{\alpha\mu} + b_{\alpha\mu}\,{}_0V_3\|_{\lambda\beta} + \tfrac{1}{2}\,{}_0V_3\|_{\alpha\mu}\,{}_0V_3\|_{\lambda\beta}\right) = 0. \tag{10.174}$$

Consistent approximations of the equations of equilibrium follow from the principle of virtual work:

$$n^{\alpha\beta}\|_\alpha + F^\beta = 0, \tag{10.175a}$$

$$m^{\alpha\beta}\|_{\beta\alpha} + (b_{\alpha\beta} + {}_0V_3\|_{\alpha\beta})n^{\alpha\beta} + F^3 = 0. \tag{10.175b}$$

These are simply the equations (10.58a, b) with the approximation (10.173).

The stress $m^{\alpha\beta}$ can be eliminated by the constitutive equations; equations (10.57b) apply for the isotropic shell. Then, (10.175b) assumes the form:

$$-\frac{Eh^3}{12(1-\nu^2)}\,{}_0V_3\|^{\alpha\beta}_{\alpha\beta} + (b_{\alpha\beta} + {}_0V_3\|_{\alpha\beta})n^{\alpha\beta} + F^3 = 0. \tag{10.176}$$

Likewise, the compatibility equation (10.174) is expressed in terms of the stress $n^{\alpha\beta}$ via the constitutive equation (10.57a) for the isotropic shell:

$$n^{\alpha}_{\alpha}\|^{\beta}_{\beta} + Eh(b^{\alpha}_{\alpha}\,{}_0V_3\|^{\beta}_{\beta} - b^{\alpha}_{\beta}\,{}_0V_3\|^{\beta}_{\alpha})$$

$$+\frac{Eh}{2}({}_0V_3\|^{\alpha}_{\alpha}\,{}_0V_3\|^{\beta}_{\beta} - {}_0V_3\|^{\alpha}_{\beta}\,{}_0V_3\|^{\beta}_{\alpha}) + (1+\nu)F^{\beta}\|_{\beta} = 0. \tag{10.177}$$

Our complete system of differential equations consist of (10.175a), (10.176), and (10.177) which govern the variables $n^{\alpha\beta}$ and ${}_0V_3$. If $P^{\alpha\beta}$ is the particular solution of (10.175a), then the solution is given in the form:

$$n^{\alpha\beta} = P^{\alpha\beta} + \overline{e}^{\alpha\lambda}\,\overline{e}^{\beta\mu}F\|_{\lambda\mu}, \tag{10.178a}$$

where

$$P^{\alpha\beta}\|_{\alpha} = -F^{\beta}. \tag{10.178b}$$

When this solution is substituted into (10.176) and (10.177), we obtain the two equations governing the two invariants F and ${}_0V_3$:

$$\frac{1}{Eh}F\|^{\alpha\beta}_{\alpha\beta} + \overline{e}^{\alpha\lambda}\,\overline{e}^{\beta\mu}(b_{\alpha\beta} + \frac{1}{2}\,{}_0V_3\|_{\alpha\beta})\,{}_0V_3\|_{\lambda\mu}$$

$$+\frac{1}{Eh}[P^{\alpha}_{\alpha}\|^{\beta}_{\beta} - (1+\nu)P^{\alpha\beta}\|_{\alpha\beta}] = 0, \tag{10.179}$$

$$\frac{Eh^3}{12(1-\nu^2)}\,{}_0V_3\|^{\alpha\beta}_{\alpha\beta} - \overline{e}^{\alpha\lambda}\,\overline{e}^{\beta\mu}(b_{\alpha\beta} + {}_0V_3\|_{\alpha\beta})F\|_{\lambda\mu}$$

$$-(b_{\alpha\beta} + {}_0V_3\|_{\alpha\beta})P^{\alpha\beta} - F^3 = 0. \tag{10.180}$$

These equations are the same as those obtained by W. T. Koiter [see equations (10.67) and (10.68) for "quasi-shallow shells"]. As previously noted, they are applicable if only the pattern of deflection occupies a relatively small portion, viz., breadth L is small compared to the radii of curvature r.

10.12 Buckling of Thin Elastic Shells

10.12.1 Introduction

The reader must be forewarned that the stability of a thin shell is often very different than the stability of other structural forms, for example, rods, frames, or plates. A thin shell can sustain great loads by virtue of its curvature and the attendant membrane actions. However, slight imperfections in form or an external disturbance can cause the shell to snap abruptly to a severely deformed state. This happens because the bending actions of adjacent configurations cannot sustain the extreme loads supported by the membrane actions of the ideal form. For example, a thin spherical shell of radius r sustains an external pressure p by the action of a uniform membrane force $\mathcal{N} = pr/2$. The spherical form is unstable if the pressure exceeds a critical value

$$p_C = \left(\frac{h}{r}\right)^2 \frac{2E}{\sqrt{3(1-\nu^2)}},$$

but a shell with a small dimple may crumple under a much lower pressure, for example, $0.1p_C$. In general, a well-designed shell, like the sphere under uniform pressure, supports loads primarily by membrane action, but ironically the efficient shell is often most susceptible to snap-through buckling. Such sensitivity to imperfections and to disturbances is the price for the inherent strength and stiffness of thin shells.

A thorough study of stability, particularly, effects of imperfections, dynamic loading, and postbuckling states are beyond the scope of our present effort. However, an illustration may serve to indicate the precarious nature of nearly critical equilibrium states.

Suppose that a perfect cylindrical shell is subjected to an axial load P as shown in Figure 10.12. With the freedom to expand radially at the edge, the shell is in equilibrium under the action of the simple membrane force $\mathcal{N}^{22} = P/2\pi r$, $\mathcal{N}^{12} = \mathcal{N}^{21} = \mathcal{N}^{11} = 0$. The Hookean shell experiences a small axial strain proportional to the load. The plot of load P versus axial displacement W traces the straight line OA of Figure 10.12. The line extended beyond A represents equilibrium states which are unstable. Hence, the load of point A is the critical load, or the so-called classical

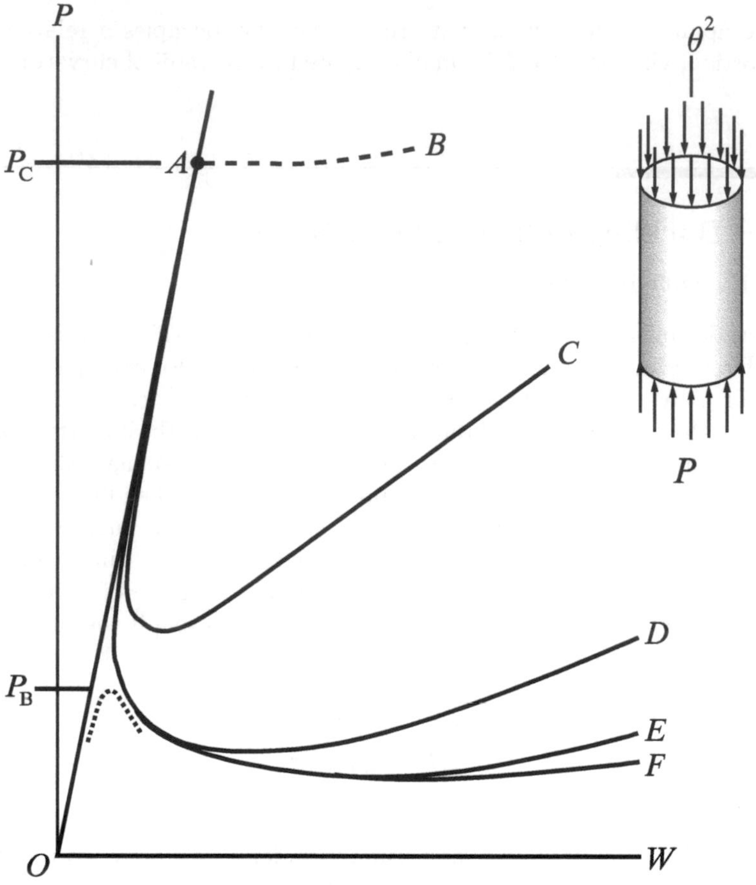

Figure 10.12 Buckling and postbuckling of a thin cylindrical shell

buckling load P_C. In the present case (see [214], p. 462)

$$P_C \doteq \frac{2\pi E h^2}{\sqrt{3(1-\nu^2)}}.$$

But, numerous experiments have placed the actual buckling load as low as $0.1P_C$ [215].

The point A of Figure 10.12 is termed a *branching*, or *bifurcation*, point, because another path emanates from point A; the other path represents postbuckled configurations, which are not axially symmetric. To illustrate the phenomena graphically, we can plot the load P versus a *mean* displacement W. Recall that the postbuckled states of a beam or a *plate* trace a gradually ascending path like AB. By contrast, the postbuckled path of the cylindrical shell drops precipitously as shown [216].

In 1941, Th. von Kármán and H. S. Tsien [217] employed an approximation of the buckled form and obtained a curve like AC. Subsequently, refinements have led to lower and lower curves, like D, E, and F [218]. It appears now that the minimum postbuckling load is so small that it provides little basis for estimating the actual buckling load.

The results of experiments show widely scattered plots of the actual buckling loads upon cylindrical shells, but always much less than P_C. These discrepancies have been attributed to the great influence of initial imperfections [219]. Such sensitivity to imperfections and to disturbances is demonstrated by the simple experiment upon common beverage cans (see Figure 9.2) described in Section 9.1 (cf. [106], p. 578). The imperfect shell undergoes bending at the onset of loading. A plot of load versus deflection follows the dotted path of Figure 10.12, which indicates buckling at a load P_B, much less than the critical value P_C.

The sensitivity to imperfections and disturbances is closely related to the character of equilibrium states at the critical point A. In the case of the perfect cylinder, the critical and postcritical states are unstable. The instability is indicated by the descending curve and by the character of the potential energy of the postbuckled states [107]. Consequently, the shell snap-buckles toward a state of stable equilibrium, presumably on the ascending portion of the postbuckled curve AF. The location or existence of such stable states are academic questions, for the actual shell is irreparably damaged before such configurations can be realized.

Clearly, the important questions concern the actual buckling load P_B and the effects of imperfections. A complete answer requires the solution of the nonlinear equations that govern equilibrium states of the imperfect shell. However, important answers are provided by studies of the *initial* stages of postbuckling: in particular, a study of the initial stages indicates whether the shell is stable at the buckling load or unstable, that is, whether the shell can sustain the load or collapses. The theory of Koiter [220] sets forth criteria based upon the potential energy of the initial stages. By examining the potential energy of the imperfect shell near the critical load, W. T. Koiter also gives a means for estimating the effects of imperfections. The underlying concepts, methodology, and consequences are described briefly in Sections 6.11 to 6.13.

In the following, we approximate the energy potential of a thin elastic shell under conservative loads and develop the equations of neutral equilibrium according to the criteria of E. Trefftz. To that extent, our development follows W. T. Koiter, but stops short of further considerations concerning postbuckling and the effects of imperfections. The interested reader can follow the works of J. Kempner [216], B. O. Almroth [218], W. T. Koiter [220], B. Budiansky and J. W. Hutchinson [221], [222], and J. W. Hutchinson [223].

10.12.2 Equations of a Critical State

Let us recall some notions about the stability of an equilibrium state: a state is stable if the potential energy is a minimum. Therefore, stability is assured if, in a variation of displacement, the second-degree terms of the potential variation assume a positive definite form. A critical state exists when the second variation is positive semidefinite. The semidefinite form vanishes for one (or more) displacements $\boldsymbol{v}$ which constitute a buckling mode. The buckling mode, at the critical load, is governed by linear differential equations (linear in a small perturbation of displacement $\boldsymbol{v}$); these are the Euler equations of the Trefftz stationary condition which implies that a minimum (zero value) of the second variation occurs for a nonzero displacement $\boldsymbol{v}$. In other words, the shell is in a neutral state wherein a small perturbation $\boldsymbol{v}$ to an adjacent state requires no change (zero value) in the second-degree variation of the potential. To establish the equations of the critical state, we require an approximation of the potential variation through the terms of second degree.

We are concerned about the stability of an arbitrary configuration which we call the reference state. Therefore, we employ the notations of an arbitrarily deformed shell to denote quantities of the reference state: $\boldsymbol{A}_\alpha$ denotes the tangent vector, $A_{\alpha\beta}$ the metric component, $B_{\alpha\beta}$ the curvature, etc. We mark the corresponding quantities of an adjacent state by a bar $(-)$: $\bar{A}_{\alpha\beta}$, $\bar{B}_{\alpha\beta}$, etc. A displacement $\boldsymbol{v} = v^i \boldsymbol{A}_i$ carries the shell from the reference state to an adjacent state and produces incremental changes in the strains, ${}_0\gamma_{\alpha\beta}$ and $\rho_{\alpha\beta}$:

$$ {}_0\acute{\gamma}_{\alpha\beta} = \tfrac{1}{2}(\bar{A}_{\alpha\beta} - A_{\alpha\beta}), \tag{10.181a} $$

$$ \acute{\rho}_{\alpha\beta} = -(\bar{B}_{\alpha\beta} - B_{\alpha\beta}). \tag{10.181b} $$

In accordance with the Kirchhoff-Love theory, the total strains are the sums: ${}_0\bar{\gamma}_{\alpha\beta} + {}_0\acute{\gamma}_{\alpha\beta}$ and $\bar{\rho}_{\alpha\beta} + \acute{\rho}_{\alpha\beta}$. In an isothermal Hookean deformation, the incremental *change* of internal energy follows in accordance with equations (10.49) to (10.51):

$$ '\phi = $$

$$ h\,{}_0C^{\alpha\beta\gamma\eta}\Big({}_0\bar{\gamma}_{\gamma\eta}\,{}_0\acute{\gamma}_{\alpha\beta} + \frac{h^2}{12}\bar{\rho}_{\gamma\eta}\acute{\rho}_{\alpha\beta}\Big) + \frac{h}{2}{}_0C^{\alpha\beta\gamma\eta}\Big({}_0\acute{\gamma}_{\alpha\beta}\,{}_0\acute{\gamma}_{\gamma\eta} + \frac{h^2}{12}\acute{\rho}_{\alpha\beta}\acute{\rho}_{\gamma\eta}\Big). $$

In accordance with (10.54a, b), this change has the form:

$$'\phi \doteq \bar{n}^{\alpha\beta}\,{}_0\acute{\gamma}_{\alpha\beta} + \overline{m}^{\alpha\beta}\acute{\rho}_{\alpha\beta} + \frac{h}{2}\,{}_0C^{\alpha\beta\gamma\eta}\left({}_0\acute{\gamma}_{\alpha\beta}\,{}_0\acute{\gamma}_{\gamma\eta} + \frac{h^2}{12}\,\acute{\rho}_{\alpha\beta}\acute{\rho}_{\gamma\eta}\right), \quad (10.182)$$

where $\bar{n}^{\alpha\beta}$ and $\overline{m}^{\alpha\beta}$ are the prevailing stresses of the prebuckled reference state.

As previously noted, the well-designed shell supports loading primarily by membrane action; membrane stresses $\bar{n}^{\alpha\beta}$ play the dominant role while flexural stresses are usually less consequential. Therefore, in most practical circumstances, the latter can be neglected in the increment $'\phi$. The approximation of (10.182) follows:

$$'\phi \doteq \bar{n}^{\alpha\beta}\,{}_0\acute{\gamma}_{\alpha\beta} + \frac{h}{2}\,{}_0C^{\alpha\beta\gamma\eta}\left({}_0\acute{\gamma}_{\alpha\beta}\,{}_0\acute{\gamma}_{\gamma\eta} + \frac{h^2}{12}\acute{\rho}_{\alpha\beta}\acute{\rho}_{\gamma\eta}\right). \quad (10.183)$$

Since our analysis is concerned with *small* excursions from the fundamental state ($A_{\alpha\beta}$, $B_{\alpha\beta}$), we assume that these are accompanied by *small* strains and moderate rotations; specifically [see (10.162a, b) and (10.164a, b)]

$${}_0\acute{\gamma}_{\alpha\beta} = \acute{e}_{\alpha\beta} + \tfrac{1}{2}\acute{\omega}_{\alpha}^{\bullet\, i}\,\acute{\omega}_{\beta i}, \quad (10.184)$$

$$\acute{\rho}_{\alpha\beta} = \tfrac{1}{2}(\acute{\omega}_{\alpha 3}\|_{\beta} + \acute{\omega}_{\beta 3}\|_{\alpha} - b^{\mu}_{\alpha}\acute{\omega}_{\mu\beta} - b^{\mu}_{\beta}\acute{\omega}_{\mu\alpha}). \quad (10.185)$$

Note: In some instances, the approximation (10.162a, b) and equation (10.184) may be inadequate. W. T. Koiter mentions the buckling of a cylinder, wherein (10.184) must be augmented by the products:

$$\tfrac{1}{2}(\acute{e}_{\mu\alpha}\acute{\omega}^{\mu}_{\bullet\,\beta} + \acute{e}_{\mu\beta}\acute{\omega}^{\mu}_{\bullet\,\alpha}).$$

Strictly speaking, the covariant derivatives are based upon the metric tensor $A_{\alpha\beta}$ of the reference state. The components, $\acute{e}_{\alpha\beta}$, $\acute{\omega}_{\alpha i}$, are obtained in the manner of (10.160a, b); here, we employ the triad $\boldsymbol{A}_i$ and the small displacement $\boldsymbol{v}$ from the prebuckled state:

$$\acute{e}_{\alpha\beta} = \tfrac{1}{2}(\boldsymbol{A}_{\alpha}\cdot\boldsymbol{v}_{,\beta} + \boldsymbol{A}_{\beta}\cdot\boldsymbol{v}_{,\alpha}), \quad (10.186)$$

$$\acute{\omega}_{\alpha\beta} = \tfrac{1}{2}(\boldsymbol{A}_{\alpha}\cdot\boldsymbol{v}_{,\beta} - \boldsymbol{A}_{\beta}\cdot\boldsymbol{v}_{,\alpha}), \qquad \acute{\omega}_{\alpha 3} = -(\boldsymbol{A}_3\cdot\boldsymbol{v}_{,\alpha}). \quad (10.187\text{a, b})$$

The last holds because $\acute{e}_{\alpha 3} \doteq 0$. By (10.185) and (10.187b),

$$\acute{\rho}_{\alpha\beta} = {}^*\overline{\Gamma}^{\mu}_{\alpha\beta}\, \boldsymbol{A}_3 \cdot \boldsymbol{v}_{,\mu} - \boldsymbol{A}_3 \cdot \boldsymbol{v}_{,\alpha\beta}. \tag{10.188}$$

The reference state is presumed to be a state of equilibrium; therefore, the first-order variation of the potential vanishes. If the external forces are constant, so-called "dead loads," the potential of those loads is entirely first-order in the displacement. Then, the total variation of potential is the integral of the higher-order terms in $'\phi$ [equation (10.183)]; those higher order terms follow:

$$\acute{\Phi} \equiv \iint_{S_0} \left[\frac{1}{2} \bar{n}^{\alpha\beta} (\acute{\omega}_{3\alpha}\, \acute{\omega}_{3\beta} + \acute{\omega}^{\mu}_{\cdot\,\alpha}\, \acute{\omega}_{\mu\beta}) \right.$$

$$\left. + \frac{h}{2}\, {}_0C^{\alpha\beta\gamma\eta} \left({}_0\acute{\gamma}_{\alpha\beta}\, {}_0\acute{\gamma}_{\gamma\eta} + \frac{h^2}{12} \acute{\rho}_{\alpha\beta} \acute{\rho}_{\gamma\eta} \right) \right] dS_0. \tag{10.189}$$

In view of equations (10.184) to (10.187a, b), the variation $\acute{\Phi}$ is a functional of the displacement $\boldsymbol{v}$ with terms of degree two and four. If $\acute{\Phi}_2$ denotes the aggregate of all terms of second degree, then the Trefftz criterion is the stationary condition:

$$\delta \acute{\Phi}_2 = 0. \tag{10.190}$$

Following W. T. Koiter, we *define*

$$\overset{*}{n}^{\alpha\beta} \equiv h\, {}_0C^{\alpha\beta\gamma\eta}\, {}_0\acute{e}_{\gamma\eta}, \qquad \acute{m}^{\alpha\beta} \equiv \frac{h^3}{12}\, {}_0C^{\alpha\beta\gamma\eta}\, {}_0\acute{\rho}_{\gamma\eta}.$$

It follows that

$$\delta \acute{\Phi}_2 = \iint_{S_0} \left[\left(\overset{*}{n}^{\alpha\beta} + \tfrac{1}{2} \bar{n}^{\beta\mu} \acute{\omega}^{\alpha}_{\cdot\,\mu} - \tfrac{1}{2} \bar{n}^{\alpha\mu} \acute{\omega}^{\beta}_{\cdot\,\mu} \right) (\boldsymbol{A}_\alpha \cdot \delta \boldsymbol{v}_{,\beta}) \right.$$

$$+ \left(\bar{n}^{\alpha\beta} \acute{\omega}_{3\alpha} + \acute{m}^{\alpha\mu}\, {}^*\overline{\Gamma}^{\beta}_{\alpha\mu} \right) (\boldsymbol{A}_3 \cdot \delta \boldsymbol{v}_{,\beta})$$

$$\left. - \acute{m}^{\alpha\mu} (\boldsymbol{A}_3 \cdot \delta \boldsymbol{v}_{,\alpha\beta}) \right] dS_0 = 0. \tag{10.191a}$$

Again, we suppose that the shell has an edge defined by a curve C_2 with arc length C_2 on the surface S_0; a curve C_1 on S_0, normal to C_2, has arc length C_1 [see Figure 10.1]. After the integration of (10.191a) by parts, the variation consists of an integral on surface S_0 and another along curve C_2:

$$\delta\acute{\Phi}_2 = -\iint_{S_0}\left\{\frac{1}{\sqrt{A}}\left[\left(\overset{*}{n}^{\alpha\beta} + \tfrac{1}{2}\bar{n}^{\beta\mu}\acute{\omega}^{\alpha}_{\cdot\mu} - \tfrac{1}{2}\bar{n}^{\alpha\mu}\acute{\omega}^{\beta}_{\cdot\mu}\right)\sqrt{A}\,\boldsymbol{A}_\alpha\right.\right.$$

$$\left.+\left(\bar{n}^{\alpha\beta}\acute{\omega}_{3\alpha} + \acute{m}^{\alpha\mu}\,{}^{*}\bar{\Gamma}^{\beta}_{\alpha\mu}\right)\sqrt{A}\,\boldsymbol{A}_3\right]_{,\beta}$$

$$\left.+\frac{1}{\sqrt{A}}\left(\acute{m}^{\alpha\mu}\sqrt{A}\,\boldsymbol{A}_3\right)_{,\alpha\beta}\right\}\cdot\delta\boldsymbol{v}\;dS_0$$

$$+\int_{C_2}\left[\left(\overset{*}{n}^{\alpha\beta} + \tfrac{1}{2}\bar{n}^{\beta\mu}\acute{\omega}^{\alpha}_{\cdot\mu} - \tfrac{1}{2}\bar{n}^{\alpha\mu}\acute{\omega}^{\beta}_{\cdot\mu}\right)\boldsymbol{A}_\alpha N^1_\beta\right.$$

$$+\left(\bar{n}^{\alpha\beta}\acute{\omega}_{3\alpha} + \acute{m}^{\alpha\mu}\,{}^{*}\bar{\Gamma}^{\beta}_{\alpha\mu}\right)\boldsymbol{A}_3 N^1_\beta$$

$$\left.+\frac{1}{\sqrt{A}}\left(\acute{m}^{\alpha\mu}\sqrt{A}\,\boldsymbol{A}_3\right)_{,\beta}N^1_\alpha\right]\cdot\delta\boldsymbol{v}\;dC_2$$

$$-\int_{C_2}\acute{m}^{\alpha\beta}N^1_\beta\boldsymbol{A}_3\cdot\delta\boldsymbol{v}_{,\alpha}\;dC_2 = 0, \tag{10.191b}$$

where

$$N^\alpha_\beta = \frac{\partial C_\alpha}{\partial\theta^\beta}.$$

Here, as before [see Section 10.2, equations (10.19) and (10.20), and Figure 10.1] the twist M^{12} is not independent, but contributes to the transverse shear, as a consequence of the Kirchhoff hypothesis. Since the variation must vanish for arbitrary $\delta\boldsymbol{v}$ on S_0, we have the equilibrium equation:‡

$$\left[\sqrt{A}\left(\overset{*}{n}^{\alpha\beta} + \tfrac{1}{2}\bar{n}^{\beta\mu}\acute{\omega}^{\alpha}_{\cdot\mu} - \tfrac{1}{2}\bar{n}^{\alpha\mu}\acute{\omega}^{\beta}_{\cdot\mu} - \acute{m}^{\mu\beta}B^{\alpha}_{\mu}\right)\boldsymbol{A}_\alpha\right.$$

$$\left.+\sqrt{A}\left(\acute{m}^{\mu\beta}\|^{*}_{\mu} + \bar{n}^{\mu\beta}\acute{\omega}_{3\mu}\right)\boldsymbol{A}_3\right]_{,\beta} = \mathbf{o}. \tag{10.192}$$

‡Here the double bar with the asterisk ($\|^*$) signifies the covariant derivative with respect to the metric $A_{\alpha\beta}$ of the reference state.

If the edge is fixed, then the variations $\delta \boldsymbol{v}$ and $\partial \delta \boldsymbol{v} / \partial C_1$ vanish on C_2. If the edge is free, then the corresponding force and bending moment vanish:

$$
\begin{aligned}
&\left(\overset{*}{n}^{\alpha\beta} + \tfrac{1}{2}\bar{n}^{\beta\mu}\acute{\omega}^{\alpha}_{\bullet\,\mu} - \tfrac{1}{2}\bar{n}^{\alpha\mu}\acute{\omega}^{\beta}_{\bullet\,\mu} - \acute{m}^{\mu\beta}B^{\alpha}_{\mu}\right)\boldsymbol{A}_{\alpha}N^{1}_{\beta} \\
&+ \left(\acute{m}^{\mu\beta}\|^{*}_{\mu} + \bar{n}^{\mu\beta}\acute{\omega}_{3\mu}\right)N^{1}_{\beta}\,\boldsymbol{A}_3 + \frac{\partial}{\partial C_2}\left(\acute{m}^{\alpha\beta}N^{1}_{\beta}\frac{\partial C_2}{\partial \theta^{\alpha}}\boldsymbol{A}_3\right) = \mathbf{o}, \qquad (10.193)
\end{aligned}
$$

$$
\acute{m}^{\alpha\beta}N^{1}_{\alpha}N^{1}_{\beta} = 0. \qquad (10.194)
$$

These edge conditions are comparable to the equations (10.169a, b) and (10.170), but apply here to the actions which accompany buckling.

On a practical note, the prebuckled deformations of the Hookean shell are typically so small that the distinction between the metrics $A_{\alpha\beta}$ and $a_{\alpha\beta}$ is inconsequential in the evaluation of the covariant derivative $[\,(.)\|^{*} \doteq (.)\|\,]$.

10.13 Refinements-Limitations-References

Most theories are intended to describe the behavior of thin shells; most are founded on the Kirchhoff hypothesis and Hookean elasticity. Our treatment sets forth the basics and provides an access to the vast literature, the formulations and solutions, based on those foundations. The underlying kinematics and statics might be applied, but cautiously, to other circumstances, even inelastic behavior.

For practical considerations, to reduce weight, to enhance surface properties, to gain stiffness and/or strength, shells are composed of layers. In such circumstances, one might adapt all or part of the Kirchhoff-Love theory to each layer. For example, a shell might be composed of two different layers with dissimilar thicknesses and materials. The hypothesis of Kirchhoff might hold for each layer; the interface (not the mid-surface) would be the appropriate reference surface. Such treatment of sandwich shells was employed in the previous work of the first author ([224], [225]). Extensive formulations for composite shells are contained in the treatise of L. Librescu [185].

Many theories have been advanced to accommodate the behavior of thick shells. Such theories are typically founded upon alternative/better approximations of displacement, strain and/or stress distributions through the thickness. Progressively higher-order theories can be achieved by expand-

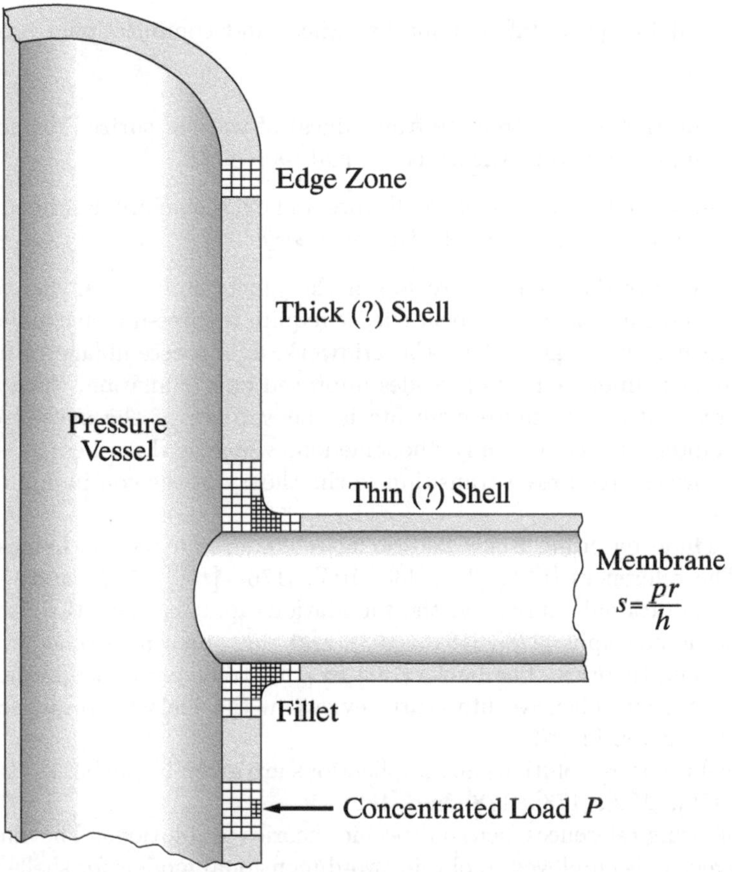

Figure 10.13 Diverse conditions in shell-like structures

ing in powers of the variable θ^3, or other functions; these lead to multi-stress theories, e.g., $n^{\alpha\beta}$, $m^{\alpha\beta}$, $m_2^{\alpha\beta}$,..., $m_n^{\alpha\beta}$ [146]. Orthogonal functions (e.g., Legendre polynomials in the interval $-1 \leq 2\theta^3/h \leq +1$) are an alternative; they have been used to devise a theory for inelastic shells [95]. Inevitably, these refinements tend toward more complicated mathematical theories. The interested reader may gain insights from the article by E. Reissner [226]. The recent work of E. Ramm (see, e.g., [227]) offers refinements to accommodate various material properties.

Certain practical situations defy description by the refined theories: Localized deformations at supports, sites of loading or junctures, call for a transition to alternatives, essentially three-dimensional analysis and approximation. As an example, we depict the intersection of two cylindrical shells in Figure 10.13. We can perceive very different physical conditions and the need for quite different mathematical and computational procedures:

1. In interior regions (remote from edges) of the pressurized cylinder, a simple-membrane theory is adequate $s = pr/h$.

2. Near the edges and juncture, flexure and extension are described by a shell theory, e.g., Kirchhoff-Love theory.

3. Very near the reentrant region of the juncture and near the very concentrated load, the local effects require a three-dimensional approach. As suggested by the gridwork, a fine assemblage of ever smaller finite elements provides improved approximation. Such approximation via finite elements is the subject of the subsequent chapter. Of course, any phenomenon, which is described by singularities, requires a transition to the theory of the continuum.

We conclude by citing a few additional references on the mechanics of shells: The references [142], [3], [143], [105], [176], [160], [177], and [180] constitute additional sources on the foundations and mathematical analysis. The monographs [228], [181], [163], and [182] present geometrically nonlinear shell theories. The books [189] to [191] treat plasticity problems of shells. A comprehensive literature review on the inelastic response of thin shells is given in [229].

Classical theories, solutions and applications are given in the books [140], [168] to [170], [172], [173], [178], and [179].

The following references focus on specific theories or solutions: The mathematical methods employed to obtain two-dimensional models for shells are presented in the monograph [183]. Asymptotic theories for spherical and cylindrical shells are given in [149] and [150]. Reference [230] presents shell theories based on the notion of the Cosserat surface. The book [171] deals with shells on elastic foundations. The work [184] treats anisotropic shells. Contact problems are studied in [231]. The monograph [175] emphasizes the effect of localized loads. The mechanics of shell vibrations and analytical solutions can be found in a number of publications, e.g., in the monographs [186] to [188]. Numerous books are devoted to shell buckling: [232], [174], [233], and [234]. Finally, the references [185], [235], and [236] deal with composite shells.

Our listing of references is very limited; most are monographs which are available in English. The interested reader can locate a multitude of additional texts and articles in the journals and proceedings.

Chapter 11

Concepts of Approximation

11.1 Introduction

Most analyses and designs of structural and mechanical elements are based upon the foregoing theories of continua, yet we are seldom able to obtain the mathematical solutions for the actual configurations. Even a cursory review of our subject shows that solutions of the boundary-value problems exist only for the most regular forms. Specifically, we note solutions for bodies with rectangular, cylindrical, spherical, and other regular geometrical boundaries. To be sure, very important practical results have been gleaned from exact solutions: The bending and torsion of prismatic members (see Sections 7.14 and 7.15) are basic to the design of simple structures and machines. Stress concentrations are revealed by the solutions for plates with holes and shafts with peripheral flutes (see Section 7.12 and Subsection 7.15.4). Still such solutions for the multitude of actual structural forms are unobtainable. Consequently, engineers must accept approximations. In general, such approximations entail some a priori choices of N functions and then a means to determine the N algebraic quantities (e.g., coefficients) which together provide the best approximations within the *computational limitations*.

Computational limitations have been the greatest obstacle to the approximation of the solutions. Specifically, prior to the advent of solid-state electronics and microcircuits (i.e., electronic digital computers), the numerical solution of large systems of equations was not a practical option. Mathematicians, scientists, and engineers resorted to a variety of procedures: Orthogonal functions in the manner of Rayleigh-Ritz ([237] to [239]) have been widely used, but essentially limited to simple geometrical boundaries. The relaxation method [240] was a means to avoid the solution of large algebraic systems. The "method of weighted residuals" [241] (point/subdomain collocation, Galerkin method [242], etc.) has also been employed to obtain approximations, but all were limited by computational requirements.

11.2 Alternative Means of Approximation

Our attention is directed to the approximations of solids and shells, which are founded upon theories governing the continuous bodies. An approximation is a mechanical model with finite degrees-of-freedom; N discrete variables supplant the continuous functions. The transition from the continuous description to the discrete approximation can be termed "discretization." In the following, we address the alternative means to achieve such discretization.

Our discussion of the mathematical means to achieve an approximation must acknowledge three distinct facets:

I. The theory of the continuous body can be expressed in alternative, but equivalent, forms:

   1. The solution is given by the function, or functions, which renders a stationary (or minimum) condition of a *functional* (or potential) and satisfies prescribed constraints.
   2. The solution satisfies the governing *differential equation* (or equations) and prescribed boundary conditions.

   Here, we avoid mathematical intricacies, but note that the differential equations of (2) are the stationary conditions of the functional (1); hence, the alternative bases are fully equivalent.

II. The approximation of a solution, the model described by N discrete variables, can be achieved via either of two routes:

   1. We can introduce the approximation (functions defined by N discrete variables) into the functional of (I.1), obtain the discrete version (an algebraic form in N variables) and, then apply the stationary criteria to obtain the algebraic equations governing the N variables.
   2. We can apply the stationary criteria to the functional of (I.1), obtain the differential equations of (I.2) and, then introduce the discrete approximation to obtain the algebraic equations governing the N variables.

   The former (1) is the route which is usually taken to obtain the algebraic equations governing the discrete model when viewed as an assembly of finite elements. The latter (2) is the route which is customarily taken to arrive at so-called "difference" equations. Indeed, the term *difference* is descriptive of the equations which supplant the *differential* equations. The viewpoints are quite different, as we reverse the order of the two procedures, approximation and variation. Nevertheless, the same approximations lead to the same discrete models, the same algebraic/difference equations. This is illustrated by our example of Sections 11.4, 11.6, and 11.7.

III. The representation of an approximation can assume either of two distinctly different forms:

1. The approximation can be represented by a linear combination of N functions, each defined (and usually continuous) throughout the region occupied by the body. Typically, their coefficients constitute the discrete variables which then determine the approximation. An important characteristic of such approximation is evident when viewed from a variational standpoint: A variation of the discrete variable (i.e., coefficient) constitutes a variation *throughout* the body.‡ This attribute is at odds with the fundamental lemma of the variational calculus. From a physical viewpoint, one can imagine that such functions might be inappropriate for an approximation of localized effects.

2. Another form of approximation is constituted of N values of the approximation at N prescribed sites or nodes, together with a preselected interpolation between those nodal values. Here the important attribute is again evident from the variational standpoint: A variation of the discrete nodal value produces a variation which is confined to the region delimited by adjoining nodes. This is consistent with the variational concept and, from the physical viewpoint, admits the more general and localized effects.

The latter form of approximation is inherent in the *difference* approximation of the differential equations, and also in nodal approximations via the stationarity of a functional; it is the usual basis of "finite element" approximations. Either form of representation can be based upon the differential equations or the functional. Functions of global support, typically orthogonal functions (e.g., vibrational modes), were widely used prior to the advent of electronic digital computers. Now, our computational capabilities and also the more immediate physical interpretations (e.g., finite elements) favors the nodal representation. In the subsequent treatment, we limit our attention to the latter and to the most basic mathematical and physical foundations. Our intent is a lucid correlation of such methods of discrete approximation and their analytical foundations. Such correlations provide the assurance that a given approximation is an

‡Such approximating functions are said to have *global* "support."

adequate description, that it *would* converge to the solution. However, it cannot represent any manner of singularity and associated phenomena (e.g., certain theories of cracks).

In our short introduction to the topic, we provide a perspective and methodology which is couched in the mechanics of the continua as presented in the preceding chapters.

11.3 Brief Retrospection

Contemporary successes in the approximation and computation of solids and shells can be traced to the works of disparate schools: The first is constituted of the early scholars of classical mechanics (Fourier, Lagrange, and Hamilton are but a few of many). Their concepts of work and energy, specifically, the notions of virtual work, stationary and minimum potential provide the foundations for most of our approximations. A second school is comprised of those physicists and engineers who explored the properties of semiconductors and devised the means to produce the microcircuitry of contemporary digital computers. Of course, the marriage of classical mechanics and solid-state physics embodies the mathematical logic of the binary system. Likewise, the implementation of the mechanical concepts requires precise mathematical tools, tensors, vectors, invariants, and the variational methods. Together these disparate works of mechanics, mathematics, and physics have wrought a revolution in the treatment of solids and shells. While the concepts of the continua remain as the foundation, computational procedures have largely supplanted analysis as a means to solutions and numerical results.

The development of the modern electronic digital computer coalesced with a recognition of the finite element form of approximation. Although the essential features of the method were evident in the early works of A. Hrenikoff ([243], 1941), R. Courant ([244], 1943), and D. McHenry ([245], 1943), the rapid evolution of electronic computers (circa 1950) gave impetus to the method. Engineers were quick to seize these tools and apply them to the problems of structural/solid mechanics. Computational strategies were developed to adapt the methodology and the computers to treat large systems. Pioneering work was done by J. H. Argyris ([246], 1954) and M. J. Turner et al. ([247], 1956). R. W. Clough ([248], 1960) introduced the term "finite element."

The early work was not without mathematical foundations, but was much influenced by the prior engineering experiences. Most engineers viewed such discrete approximation as a mechanical assembly of elements as opposed to a piecewise approximation of a continuous body. Attempts to fit some traditional models into the new framework led to conflicts. For example, a preconceived description of plates and shells, viz., the hypothesis of Kirchhoff-Love, is not readily amenable to approximation by simple elements. We address the inherent features and attendant problems in Sections 11.10 and 11.13.

The early developments followed various paths which led to alternative methods of implementation. These included the introduction of special coordinates to describe elements of particular shapes; specific systems were devised for quadrilaterals ([249], [250]) and for triangles [251]. Eventually, elements were derived via the many alternative criteria of work and energy (see Chapter 6); modifications were incorporated to accommodate discontinuities (see Sections 11.12.3 and 11.12.4). Engineers have adapted the finite element methodology to all forms of solids and structures, to simple parts and complicated shell-like structures. Today the engineer can frequently turn to existing computational programs to obtain a requisite approximation. An early (mid 1960's) program with broad capabilities was presented by J. H. Argyris [252]. Still, we are reluctant to proclaim "general-purpose" programs, since technological advances always pose unanticipated challenges.

Together, the methodologies of finite elements and the electronic computer have provided the engineer with powerful tools for the treatment of the many practical problems of solids and shells. Many diverse schemes, elemental approximations, computational procedures, and eventually, mathematical analyses have emerged to utilize these tools. The evolution of the methodology has been accompanied by a plethora of special devices, or elements, to accommodate specific problems; often these are assigned descriptive names or acronyms. That terminology can be a barrier to the uninitiated. Here, we strive to couch the concepts and formulations in the traditional terms of mechanics and mathematics, to view the method as a means of approximating the solutions and to reveal the analogies between the discrete formulations and their continuous counterparts.

The reader is reminded that we intend no discourse on the finite element method, which has widespread applicability to diverse topics of mathematics, science, and engineering. Our focus is the correlation with the mechanics of continuous solid bodies. Broader aspects and historical accounts are contained in books which specifically address the finite element method (see, e.g., [253] to [265]).

11.4 Concept of Finite Differences

We recall the most basic concepts of finite differences as a precursor to our subsequent correlation with the similar approximations which are drawn from the stationarity of a functional. The knowledgeable reader can surely skip our abbreviated account.

The approximation of a differential equation by a finite difference equation is natural. One merely reverts to the elementary notion of the derivative of a continuous function, but precludes the limit, i.e.,

$$f' \equiv \frac{df}{dx} \equiv \lim_{\Delta x \to 0} \frac{\Delta f}{\Delta x} \doteq \frac{\Delta f}{\Delta x},$$

where

$$\Delta f = f_n - f_{n-1}, \qquad \Delta x = x_n - x_{n-1}.$$

The approximation is the derivative of a function at some intermediate point, $(x_{n-1} < x < x_n)$. Again, we signify the approximation of an equality $(=)$ with the overdot $(\doteq)$. The solution is approximated by the nodal values $f_n \equiv f(x_n)$, wherein a linear interpolation is implied. If we consider a typical point n and its immediate neighbors $n-1$ and $n+1$, then the *central difference approximation* for the derivative of f at the point x_n, f'_n, assumes the form:

$$f'_n \doteq \frac{f_{n+1} - f_{n-1}}{x_{n+1} - x_{n-1}} = \frac{f_{n+1} - f_{n-1}}{2\Delta x}.$$

Here, the tangent of f at x_n is replaced by the chord between the values of the function f at the points x_{n+1} and x_{n-1}. The distance between two successive grid points has been assumed equal:

$$x_{n+1} - x_n = x_n - x_{n-1} = \Delta x.$$

In like manner, the second derivative has the approximation at x_n:

$$f'' \equiv \frac{df'}{dx} \doteq 2\left(\frac{f_{n+1} - f_n}{x_{n+1} - x_n} - \frac{f_n - f_{n-1}}{x_n - x_{n-1}}\right) \Big/ (x_{n+1} - x_{n-1}).$$

Assuming equal distances Δx, we obtain

$$f''_n \doteq \frac{1}{\Delta x^2}(f_{n+1} - 2f_n + f_{n-1}).$$

To place this in context, consider a simple, but relevant, example: The axial displacement $u(x)$ of a <u>h</u>omogeneous, <u>i</u>sotropic and <u>Ho</u>okean (HI-HO) rod under axially symmetric load $P(x)$ has a continuous model governed by the linear equation:

$$\frac{d^2u}{dx^2} + \frac{P(x)}{EA} = 0. \tag{11.1a}$$

The difference approximation follows

$$\frac{1}{\Delta x^2}(u_{n+1} - 2u_n + u_{n-1}) + \frac{1}{EA}\left(\frac{1}{6}P_{n+1} + \frac{2}{3}P_n + \frac{1}{6}P_{n-1}\right) = 0. \tag{11.1b}$$

The final term of (11.1b) is the mean value of P in the interval (x_{n-1}, x_{n+1}) when P, like u, is approximated by linear interpolation. If the end $x = 0$ is fixed, then

$$u(0) = u_0 = 0. \tag{11.2}$$

If the end $x = l$ is free, then

$$\sigma(l) = E\frac{du}{dx}\bigg]_l = 0. \tag{11.3}$$

The consistent approximation of (11.3) is not immediately apparent (see Section 11.6).

11.5 Stationarity of Functionals; Solutions and Forms of Approximation

The preceding section provides a brief and limited account of difference equations, as drawn directly from the *differential equations*. The thrust of our presentation addresses such equations as drawn from the *stationary* criteria for a *functional*. To fully appreciate the very basic and distinctive character of alternative forms of approximation, e.g., finite elements, let us recall the fundamental logic of the variational method.

As employed repeatedly in Chapter 6, the procedure requires successive integration upon a functional $V(u)$ until one arrives at an integral of the form:

$$\delta V = \int_{x_0}^{x_l} \mathcal{L}[u(x)]\,\delta u\,dx + \cdots. \tag{11.4}$$

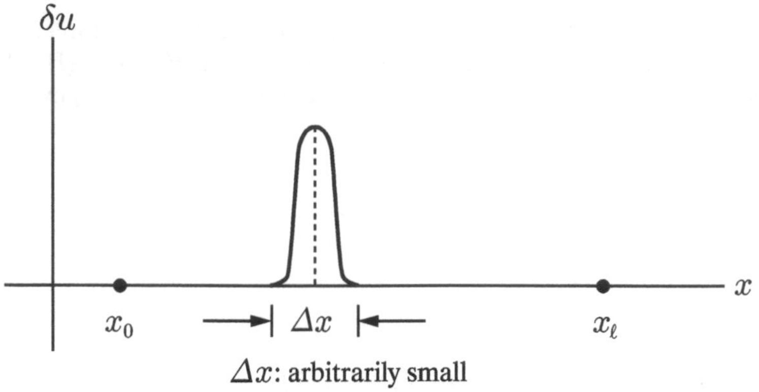

Figure 11.1 An arbitrary variation

Here $\mathcal{L}$ signifies a differential operator, linear *or* nonlinear, u the required function, i.e., the solution, and δu the variation. Of course, depending upon the underlying principle, the functional might depend on more than one function; a form such as (11.4) emerges if a potential $V(u)$ depends upon one scalar u, e.g., the lateral displacement of a beam. The logic applies as well to the general case, wherein $\boldsymbol{u}(x_i)$ is a vector function in the three-dimensional space, or $V(\boldsymbol{u}, \boldsymbol{\sigma})$ is a functional of two (or more) vectors $(\boldsymbol{u}, \boldsymbol{\sigma})$.

The fundamental lemma of the variational calculus requires that the function $u(x)$ satisfies the differential equation *everywhere* in (x_0, x_l):

$$\mathcal{L}(u) = 0. \tag{11.5}$$

Additionally, u must satisfy the requisite end conditions, inferred in (11.4). The argument rests on the arbitrariness of the variation $\delta u(x)$. This is illustrated by Figure 11.1, wherein the variation can be nonvanishing in an *arbitrarily small* interval Δx. Moreover, the mean value (the value as $\Delta x \to 0$) can be positive or negative. By this logic, $\mathcal{L}(u)$ must vanish everywhere. Since the end values of u (or derivatives) are also arbitrary, if unconstrained, we obtain the "natural" end conditions as well. It is important to note that the form of $\delta u(x)$ is *irrelevant*, only that it can be nonzero in an *arbitrarily* small segment Δx. It is this feature that provides the crucial link with the analogous criterion of a finite element formulation.

The practitioner seeks an approximation of the solution by functions (or a function) as prescribed by discrete parameters. The *functional* is thereby approximated by a *function* of those parameters. The best approximation by the preselected functions renders the function stationary with respect to the discrete parameters. The crucial step is the choice of the functions. Two types have very different and distinctive properties, and also very different implications: These are functions of *global* versus *local* support. In either case, we can express our approximation in a form:

$$u(x) \doteq \sum_{n=1}^{m} A_n g_n(x). \tag{11.6}$$

Here, $g_n(x)$ are preselected functions and A_n the discrete parameters which determine the approximation via the stationarity conditions, viz.,

$$\frac{\partial V}{\partial A_n} = 0. \tag{11.7}$$

For many years engineers, following the methods of Rayleigh-Ritz ([237], [239]), used functions of global support, i.e., $g_n(x)$ is defined and nonzero *throughout* the region (x_0, x_l). An example is the choice $g_n(x) = \sin(n\pi x/l)$ to describe the deflection of a beam $(0, l)$. Observe that the *variation* δu is then accomplished by a *discrete* variation δA_n; *but*, such variation, unlike the arbitrary variation of Figure 11.1, alters u *throughout* the interval. We note that such form of approximation is contrary to the fundamental lemma. Of course, experience tells us that such forms of approximation can work, but depend very crucially on a good choice of the functions $g_n(x)$. Indeed, one might obtain a very good approximation by a judicious choice; e.g., $A_1 \sin(\pi x/l)$ provides the *exact* solution for a simply-supported HI-HO (*h*omogeneous, *i*sotropic and *Ho*okean) column under axial thrust.

It is the essence of finite element approximations to approximate the solution by discrete nodal values and a preselected form of interpolation. Then the parameters A_n of equation (11.6) become the nodal values u_n of the function and its derivatives, depending on the form of interpolation. Specifically, the approximation may take the forms:[‡]

$$u(x) \doteq \sum_{n=1}^{m} u_n \widetilde{w}_n(x), \tag{11.8a}$$

or

$$u(x) \doteq \sum_{n=1}^{m} [u_n w_n(x) + u'_n \overline{w}_n(x)]. \tag{11.8b}$$

[‡]Higher-order approximations provide continuity for derivatives of any order.

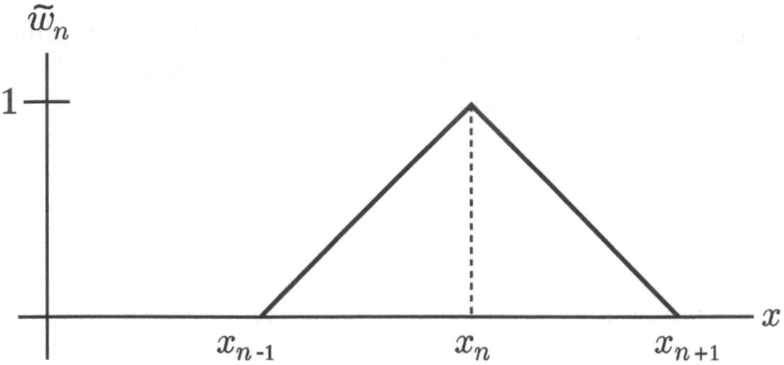

Figure 11.2 Shape function of Lagrangean (linear) interpolation

Relationship (11.8a) admits a linear (Lagrangean) interpolation [u_n are the nodal values, $u_n = u(x_n)$]. The latter (11.8b) admits cubic (Hermitian) interpolation, wherein the approximation provides continuity of the function and its first derivative [$u_n' = u'(x_n)$]. It is especially noteworthy that the functions $\widetilde{w}_n(x)$ in (11.8a), $w_n(x)$ and $\overline{w}_n(x)$ of (11.8b) have *local* support; these are the so-called "shape" functions depicted in Figures 11.2 and 11.3. As before, our approximation of the functional is again a function of discrete parameters. However, now the parameters are the nodal values; more *important*, a *discrete* variation, say δu_n, alters the approximation only in the finite region adjoining the node (x_{n-1}, x_{n+1}). That region of support can be taken ever smaller, much as the region about the variation $\delta u(x)$ in Figure 11.1. Each of the m equations, e.g., $\partial V/\partial u_n$, is a difference equation which involves only the nodal values adjoining the nth node; each entails an integral over the *small* region of the local support (the

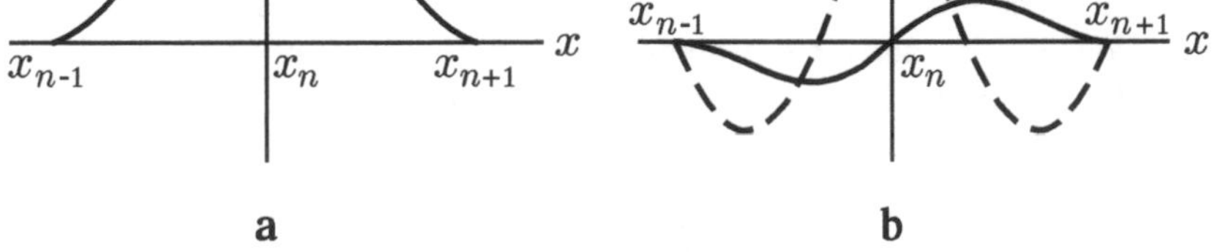

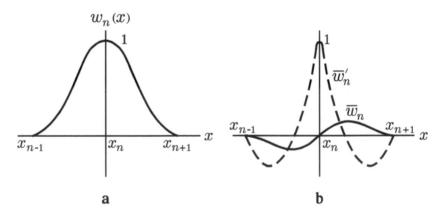

Figure 11.3 Shape functions of Hermitian interpolation

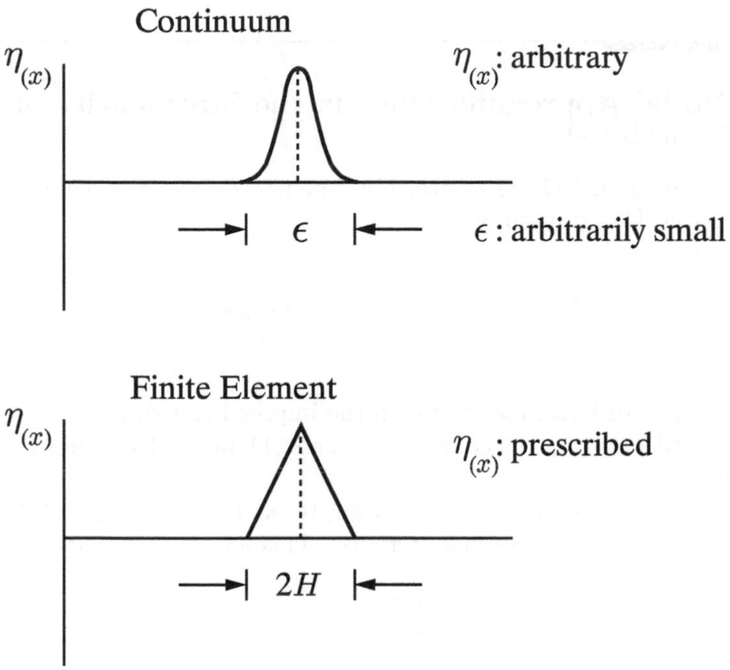

Figure 11.4 Analogous concepts of variation

region of contiguous elements). By contrast with approximations/functions of global support, the approximations of the form (11.8a, b) govern events in the immediate neighborhood, much as the differential equation governs at each point. Indeed, we see analogous logic and anticipate the convergence; as the region (x_{n-1}, x_{n+1}) shrinks, the difference equation(s) approach(es) the differential equation. The arbitrarily small support of the variation (see Figure 11.4) is essential to the argument that the differential equation(s) (Euler-Lagrange) vanish *everywhere* in the region. In Figure 11.4, H denotes the width of an element.

Approximations defined by nodal values and polynomial interpolation are termed *spline* functions. Traditionally, engineers have employed "splines" (thin pliable strips) to fit smooth curves through plotted points, as we now seek to fit spline functions through nodal values. The Lagrangean and Hermitian interpolations of (11.8a, b) provide the simplest spline functions. In the subsequent sections, we use these examples to illustrate the implications of alternative spline and associated shape functions.

11.6 Nodal Approximations via the Stationarity of a Functional

The continuous HI-HO rod governed by equation (11.1a) is in equilibrium if the potential V is stationary; viz.,

$$V = \int_0^l \left[\frac{EA}{2} \left(\frac{du}{dx} \right)^2 - Pu \right] dx. \tag{11.9}$$

The stationarity of $V(u)$, together with the imposed constraint $[u(0) = 0]$, is fully equivalent to the differential equation (11.1a) and the end condition (11.2).

Now, let us represent our approximation by nodal values u_n and linear interpolation as in the preceding sections. This can be expressed in the form:

$$u(x) \doteq \sum_n u_n \widetilde{w}_n(x),$$

where $\widetilde{w}_n(x)$ is the "shape function" defined as follows:

$$\widetilde{w}_n(x) = \begin{cases} 0, & x \leq x_{n-1} \\ \left(\dfrac{x - x_{n-1}}{x_n - x_{n-1}} \right), & x_{n-1} \leq x \leq x_n \\ \left(\dfrac{x - x_{n+1}}{x_n - x_{n+1}} \right), & x_n \leq x \leq x_{n+1} \\ 0. & x \geq x_{n+1} \end{cases}$$

The function $\widetilde{w}_n(x)$ is depicted in Figure 11.2.

A variation of the nodal value u_n and the requirement of stationarity follows:

$$\frac{\partial V}{\partial u_n} = \int_{x_{n-1}}^{x_{n+1}} \left[EA \sum_m (u_m \widetilde{w}'_m(x) \widetilde{w}'_n(x) - P_m \widetilde{w}_m(x) \widetilde{w}_n(x) \right] dx = 0. \tag{11.10}$$

Here, to be consistent, we approximate $P(x)$ as we approximate $u(x)$. Then, we obtain

$$\frac{EA}{\Delta x}(-u_{n-1} + 2u_n - u_{n+1}) - \left(\frac{1}{6} P_{n-1} + \frac{2}{3} P_n + \frac{1}{6} P_{n+1} \right) \Delta x = 0.$$

Apart from an irrelevant constant $(-\Delta x)$ this is equation (11.1b). Note that (11.1b) holds only at *intermediate* nodes, but *not* at the end node $(x = l)$. There,

$$\frac{\partial V}{\partial u_l} = \int_{x_l - \Delta x}^{x_l} \left[EA \sum_m (u_m \widetilde{w}'_m(x) \widetilde{w}'_l(x) - P_m \widetilde{w}_m(x) \widetilde{w}_l(x) \right] dx = 0. \quad (11.11)$$

Our result follows:

$$E \left[\frac{u(l) - u(l - \Delta x)}{\Delta x} \right] - \frac{1}{A} \left[\frac{P(l)}{3} + \frac{P(l - \Delta x)}{6} \right] \Delta x = 0. \qquad (11.12)$$

The meaning of this approximation deserves physical interpretation. The initial term yields:

$$\lim_{\Delta x \to 0} E \left[\frac{u(l) - u(l - \Delta x)}{\Delta x} \right] = E \left. \frac{du}{dx} \right]_l = \sigma]_l .$$

The second term of (11.12) vanishes in the limit:

$$\lim_{\Delta x \to 0} \frac{1}{A} \left[\frac{P(l)}{3} + \frac{P(l - \Delta x)}{6} \right] \Delta x = 0.$$

It follows that the approximation (11.12) approaches $\sigma]_l = 0$. The *consistent* approximation (11.12) of this end condition (11.3) is hardly obvious, but emerges from the variational criterion (11.11). It should come as no surprise that the same difference equations result from *consistent* approximations of the differential equations and the functional whence they derive.

11.7 Higher-Order Approximations with Continuous Derivatives

The foregoing treatment (Section 11.6) provides for continuity of the approximation; the derivative is discontinuous at each node. One can adopt higher-order forms of interpolation which require also continuity of derivatives (of any order). *Note:* Throughout we denote the derivative with a prime ($'$). The implications can be demonstrated by the foregoing model, wherein we now require continuity of the first derivative of the approximation. Stated otherwise, we employ Hermitian interpolation rather than the

Lagrangean (linear) interpolation of the preceding examples (Sections 11.4 and 11.6). Then, the function is expressed in terms of nodal values and derivatives:

$$u(x) \doteq \sum_n [u_n w_n(x) + u'_n \overline{w}_n(x)]. \tag{11.13}$$

The *shape* functions, $w_n(x)$ and $\overline{w}_n(x)$ are the cubics as depicted in Figure 11.3.

Our approximation of the stationary criterion on the functional of (11.9) is achieved by the discrete variations of nodal values, u_n and u'_n. Corresponding to the variation of u_n, we obtain

$$\begin{aligned} -\frac{6EA}{5\Delta x}(u_{n-1} - 2u_n + u_{n+1}) + \frac{EA}{10}(u'_{n+1} - u'_{n-1}) \\ -\frac{1}{70}(9P_{n-1} + 52P_n + 9P_{n+1})\Delta x \\ -\frac{13}{420}(P'_{n-1} - P'_{n+1})\Delta x^2 = 0. \end{aligned} \tag{11.14}$$

In the limit

$$\lim_{\Delta x \to 0} \frac{1}{10\Delta x}(u'_{n+1} - u'_{n-1}) = \frac{d^2u}{5dx^2},$$

$$\lim_{\Delta x \to 0} \frac{6}{5\Delta x^2}(-u_{n-1} + 2u_n - u_{n+1}) = -\frac{6}{5}\frac{d^2u}{dx^2},$$

$$\lim_{\Delta x \to 0} -\frac{1}{70}(9P_{n-1} + 52P_n + 9P_{n+1}) = -P_n,$$

$$\lim_{\Delta x \to 0} -\frac{13}{420}(P'_{n-1} - P'_{n+1})\Delta x = 0.$$

It follows that the *difference* equation (11.14) approaches ($\Delta x \to 0$)

$$\frac{d^2u}{dx^2} + \frac{P}{EA} = 0.$$

The approximation (11.13) is not determined by the system (11.14); it contains but one equation for each node n corresponding to the variation δu_n. Additionally, we have the equations corresponding to the variation

$\delta u'_n$, viz.,

$$\frac{\partial}{\partial u'_n}\int_{x_0}^{x_l}\left[\frac{EA}{2}(u')^2 - Pu\right]dx.$$

That is a difference equation in the following form:

$$\frac{EA}{15}\left[\left(-u'_{n-1} + 2u'_n - u'_{n+1}\right)\frac{1}{\Delta x^2}\right]\Delta x^3$$

$$+\frac{EA}{30}\left[\left(u'_{n-1} + 4u'_n + u'_{n+1}\right) - 6\left(\frac{u_{n+1} - u_{n-1}}{2\Delta x}\right)\right]\Delta x$$

$$-\frac{\Delta x^3}{420}\left[26\left(\frac{P_{n+1} - P_{n-1}}{2\Delta x}\right) + \left(-3P'_{n-1} + 8P'_n - 3P'_{n+1}\right)\right] = 0. \quad (11.15)$$

Since our approximation insures continuity of the derivative, we have the following limits ($\Delta x \to 0$):

$$\frac{u_{n+1} - u_{n-1}}{2\Delta x} \to u'_n, \qquad \frac{P_{n+1} - P_{n-1}}{2\Delta x} \to P'_n.$$

Accordingly, the second bracketed term of (11.15) vanishes *in the limit* ($\Delta x \to 0$); the final term approaches $(-\Delta x^3/15)P'_n$ and the difference equation approaches the following:

$$\left[-EA\frac{\left(u'_{n-1} - 2u'_n + u'_{n+1}\right)}{\Delta x^2} - P'_n\right]\frac{\Delta x^3}{15} = 0.$$

This is the counterpart of the differential equation

$$-EA\frac{d^3u}{dx^3} - \frac{dP}{dx} = 0.$$

In effect, the use of the higher-order approximation (continuous first derivative) imposes yet the additional condition, analogous to the derivative of the differential equation.

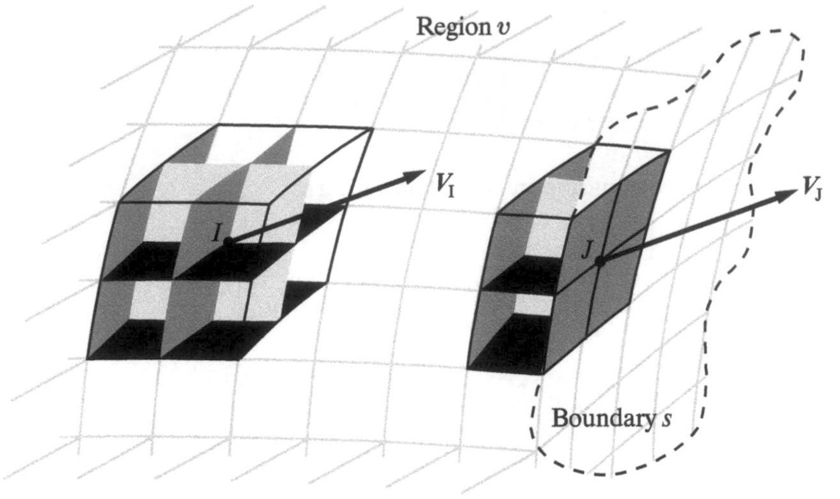

Figure 11.5 Finite elements of a solid body

11.8 Approximation by Finite Elements; Physical and Mathematical Implications

In the preceding text we focused upon the alternative forms and bases of approximations: We cited some mathematical advantages of using nodal values and interpolation (shape functions with local support) and deriving such approximations by the stationary criteria on the functional. That is a mathematical description of *a* finite element method. Physical interpretations provide additional insights and suggest means which enable the practitioner to extend and enhance the method. In the present context, these interpretations are drawn from a view of the solid body subdivided into the finite elements by surfaces (e.g., coordinate surfaces) which intersect at the nodes. Such subdivision into quadrilateral elements is depicted in Figure 11.5. A typical node I and the adjoining elements are isolated along coordinate surfaces θ^i in Figure 11.6. Our use of coordinate surfaces enables us to correlate certain quantities in the discrete approximation with their counterparts in the continuum, e.g., difference equations versus differential equations.

To admit the most meaningful and general interpretations, let us focus on a basic variational expression, the virtual work. The appropriate terms are given in Chapter 6: The virtual work of the stresses (per unit volume) takes the various forms of (6.101a–c), wherein the stress can assume the forms of (6.105a, b). The virtual work of body forces and surface tractions

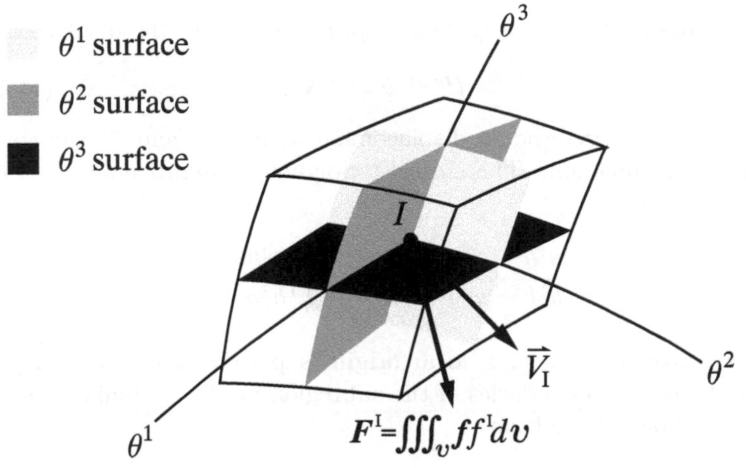

Figure 11.6 Region adjoining node *I*

takes the forms of (6.102) and (6.103), respectively (per units of volume and area). Together these contribute to the total virtual work of (6.110a, b), viz.,

$$\delta W = -\iiint_v \left[\frac{1}{\sqrt{g}} (\sqrt{g}\, \boldsymbol{s}^i)_{,i} + \boldsymbol{f} \right] \cdot \delta \boldsymbol{R}\, dv$$

$$- \iint_{s_t} (\overline{\boldsymbol{T}} - \boldsymbol{s}^i n_i) \cdot \delta \boldsymbol{R}\, ds + \iint_{s_v} \boldsymbol{s}^i n_i \cdot \delta \boldsymbol{R}\, ds. \qquad (11.16)$$

Here, as before, the *variation* of position (the *virtual* displacement) is signified by $\delta \boldsymbol{R}$, whereas dv and ds signify the *differentials* of volume and surface, respectively. It is *important* to note that equation (11.16) and the principle ($\delta W = 0$) applies to any body; that is, the body could be inelastic. In mathematical parlance, the variation δW need not be the Fréchet differential of a functional (or potential). On the other hand, the virtual work could be a differential of a potential [$\delta W = -\delta \mathcal{V}$, as in (6.121)]; it could be a differential of the modified potential [$\delta \mathcal{V}^*$ of (6.126)]. Of course in the theory of the continuum, the displacement $\boldsymbol{V}$ is continuous and $\delta \boldsymbol{R}$ is arbitrary so that the bracketed term and the term in parenthesis must vanish at each point (the first in v, the second on s_t). Now, we consider the implications of spline approximations of displacement $\boldsymbol{V}$ and the corresponding variations of $\delta \boldsymbol{R} \equiv \delta \boldsymbol{V}$.

The typical interior node I in Figure 11.5 is contiguous to the eight hexahedral elements isolated in Figure 11.6. The region occupied by these eight elements is designated v_I. Any spline approximation of the displacement has the form:

$$\boldsymbol{V} = f^I(\theta^1, \theta^2, \theta^3)\boldsymbol{V}_I, \tag{11.17}$$

where $\boldsymbol{V}_I$ is the discrete nodal displacement at node I and f^I the corresponding shape function. The simplest polynomial approximation is the trilinear expression:

$$f^I = 1 - \frac{\theta^1}{h^1} - \frac{\theta^2}{h^2} - \frac{\theta^3}{h^3} + \frac{\theta^1\theta^2}{h^1h^2} + \frac{\theta^2\theta^3}{h^2h^3} + \frac{\theta^1\theta^3}{h^1h^3} - \frac{\theta^1\theta^2\theta^3}{h^1h^2h^3} \quad 0 \le \theta^i \le h^i. \tag{11.18}$$

Here, for convenience only, a local origin is placed at the node I and $\theta^i = \pm h^i$ defines the boundaries of the subregion v_I. With similar interpolation throughout v_I, we have

$$\iiint_{v_I} f^I \, d\theta^1 \, d\theta^2 \, d\theta^3 = 1.$$

Furthermore, if s_i denotes the interface along the θ^i surface through node I,

$$\iint_{s_i} f^I \, d\theta^j \, d\theta^k = 1 \quad (i \ne j \ne k \ne i).$$

We reemphasize the key feature of such approximation: A *nodal variation* $\delta\boldsymbol{V}_I$ produces a *variation* $\delta\boldsymbol{V}$ only within the subregion v_I (the contiguous elements).

Let us return now to the primitive form of δW, viz., equation (6.110a); as applied to our variation $\delta\boldsymbol{V}_I$:

$$\delta\boldsymbol{V}_I \cdot \left[\iiint_{v_I} (\boldsymbol{s}^i f^I_{,i} - \boldsymbol{f} f^I) \, dv - \iint_{s_t} \overline{\boldsymbol{T}} f^I \, ds \right]. \tag{11.19}$$

By the principle of virtual work, the bracketed term must vanish for every node I of our discrete model. The final integral appears only if the region of nonvanishing f^I includes a portion of boundary s_t. First, let us consider the condition for an interior node as shown in Figure 11.6; v_I includes only the contiguous elements, wherein $f^I \ne 0$. Integration by parts in the region v_I yields

$$- \iiint_{v_I} \left[\frac{1}{\sqrt{g}} (\sqrt{g}\, \boldsymbol{s}^i)_{,i} + \boldsymbol{f} \right] f^I \, dv$$

$$+ \sum_n \iint_{s_n} (\boldsymbol{s}^i_+ - \boldsymbol{s}^i_-) \, n_i f^I \, ds_n + \iint_{s_I} (\boldsymbol{s}^i n_i - \overline{\boldsymbol{T}}) \, f^I \, ds = 0. \tag{11.20}$$

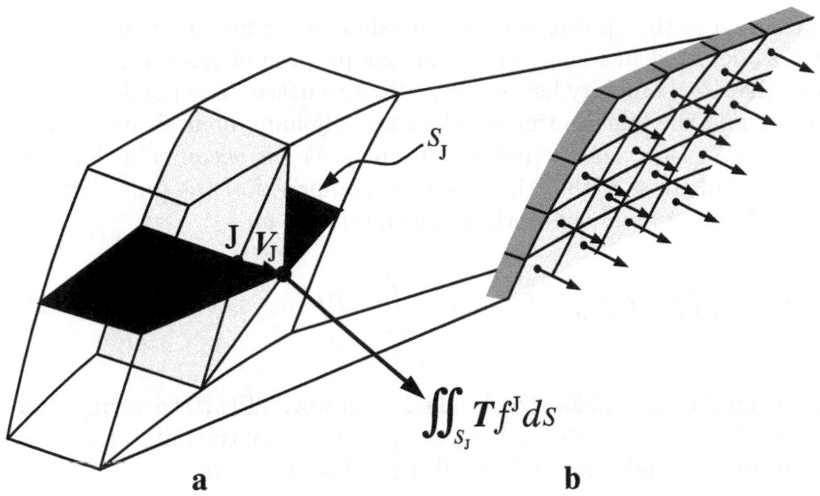

Figure 11.7 Region adjoining boundary node J

Here, s_I signifies the boundary of region v_I. Furthermore, we employ the symbol s_n to signify an *interface*, i.e., one of the *three* coordinate surfaces intersecting the node. This admits the possibility of discontinuous stress $\boldsymbol{s}^i$ upon the interior faces. Since the shape function vanishes upon the exterior surface s_I, the final term in (11.20) vanishes. If the stress (or approximation of the stress) is continuous, then the equilibrium condition for our *discrete model is* a weighted average of the differential equation governing the continuous body. This is akin to the Galerkin method of approximation, as drawn from the differential equation. Note too that diminution of region v_I (refinement of the mesh) is fully analogous to the limiting argument ($\Delta x \to 0$ in Figure 11.1) of the variational calculus. If the stress has discontinuities, then equation (11.20) includes weighted averages of those jumps. These differences contribute as the differential in the continuum.

If the approximation of strain is compatible with the spline approximation of displacements, and if the material is elastic, then each equilibrium equation for the discrete model, e.g., (11.20), contains only the nodal displacements in the neighborhood of I, i.e., at node I and the neighboring nodes; for the approximation (11.18), each equation contains no more than 81 components of displacement. That number multiplies with the order of the spline approximation; the spline which preserves continuity of derivatives includes also the nodal value of the derivatives (see the example of Section 11.7). From the computational perspective the bandwidth of matrices are thereby expanded.

If the approximation of stress is discontinuous at interfaces, then a stress does not exist in the usual sense (as $v_I \to 0$). To establish a meaningful interpretation of the approximation, we examine the action at an exterior node J as depicted in Figure 11.5. For our purpose of interpretation, we suppose that the boundary lies on a coordinate surface. The portion of that surface s_J and the four contiguous elements adjoining node J are isolated in Figure 11.7. Now, the virtual displacement $\delta \boldsymbol{V}_J$ is accompanied by the virtual work of stress within the region v_J comprised of the four elements [as in v_I of (11.19)] and the work of the traction $\boldsymbol{T}$ on s_J:

$$\delta \boldsymbol{V}_J \cdot \left[\iiint_{v_J} (\boldsymbol{s}^i f^J{}_{,i} - \boldsymbol{f} f^J)\, dv - \iint_{s_J} \boldsymbol{T} f^J\, ds \right] \equiv \delta W_J. \qquad (11.21)$$

Here, the final term, unlike that in (11.19), is work of the stress upon the surface s_J through the node J. Now, the integration of the initial term and enforcement of equilibrium ($\delta W_J = 0$), provides the condition:

$$\iint_{s_J} \boldsymbol{T} f^J\, ds = \iint_{s_J} \boldsymbol{s}^i n_i f^J\, ds - \iiint_{v_J} \left[\frac{1}{\sqrt{g}} (\sqrt{g}\, \boldsymbol{s}^i)_{,i} + \boldsymbol{f} \right] f^J\, dv$$

$$+ \sum_{\alpha} \iint_{s_\alpha} (\boldsymbol{s}^i_+ - \boldsymbol{s}^i_-) n_i f^J\, ds_\alpha. \qquad (11.22)$$

Again, s_α ($\alpha = 1, 2$) signifies an *interface* through node J. The final term vanishes if $\boldsymbol{s}^i$ has a continuous approximation in v_J. Then,

$$\iint_{s_J} \boldsymbol{T} f^J\, ds = \iint_{s_J} \boldsymbol{s}^i n_i f^J\, ds - \iiint_{v_J} \left[\frac{1}{\sqrt{g}} (\sqrt{g}\, \boldsymbol{s}^i)_{,i} + \boldsymbol{f} \right] f^J\, dv. \qquad (11.23)$$

The left side of (11.23) is the "generalized" force associated with node J, a weighted integral of tractions upon s_J. If we divide by the area s_J and pass to the limit, we obtain:

$$\lim_{s_J \to 0} \frac{1}{s_J} \iint_{s_J} \boldsymbol{T} f^J\, ds \equiv \boldsymbol{T}.$$

Likewise from (11.23), we have

$$\boldsymbol{T} = \boldsymbol{s}^i n_i - \lim_{s_J \to 0} \frac{1}{s_J} \iiint_{v_J} \left[\frac{1}{\sqrt{g}} (\sqrt{g}\, \boldsymbol{s}^i)_{,i} + \boldsymbol{f} \right] f^J\, dv.$$

If only the derivatives $\boldsymbol{s}^i{}_{,i}$ are *bounded*, then the final term *vanishes* and we recover the boundary condition of the continuum, viz.,

$$\boldsymbol{T} = \boldsymbol{s}^i n_i. \tag{11.24}$$

11.9 Approximation via the Potential; Convergence

A stable equilibrated state of an elastic body is characterized by the minimum of the potential $\mathcal{V}$. It follows that the optimum parameters for a *fully compatible* discrete model are likewise determined by that minimal criterion. Such compatibility requires that the internal energy be derived from a continuous displacement vector $\boldsymbol{V}$ with components V_i in full accordance with the continuum theory. Then, for any a priori choice of functions f_i^n, e.g., $V_i \doteq f_i^n A_{n\underline{i}}$, the best consistent choice of parameters A_{ni} is governed by the stationary criteria: $\partial\mathcal{V}/\partial A_{ni} = 0$. If stability is not an issue, the approach is always valid though not necessarily efficient.

A compatible discrete model, which is based on the stationarity of potential, and spline approximation (compatible finite elements), offers the assurance of converge. The proof was presented by M. W. Johnson and R. W. McLay [266]. We reiterate their logic:

Let $\boldsymbol{V}_E$ denote the displacement of the exact solution; this renders the minimum potential $\mathcal{V}_E$ among *all* admissible fields. Let $\boldsymbol{V}_A \doteq f^n \boldsymbol{V}_n$ provide the minimum $\mathcal{V}_A$ for the model defined by the nodal approximations $\boldsymbol{V}_n$. It is assumed that the chosen approximation (spline/shape functions) does not violate geometrical or mechanical requirements. Stated otherwise, the solution $\boldsymbol{V}_E$ can be approximated in the form $\boldsymbol{V}_E \doteq f^n \widetilde{\boldsymbol{V}}_n$. Indeed, we define a field

$$\boldsymbol{V}_I \equiv f^n \overline{\boldsymbol{V}_n}, \tag{11.25}$$

where

$$\overline{\boldsymbol{V}_n} \equiv \boldsymbol{V}_E(\theta_n^1, \theta_n^2, \theta_n^3), \tag{11.26}$$

i.e., $\overline{\boldsymbol{V}_n}$ are the values of the exact solution at the nodes. Let $\mathcal{V}_B$ denote the potential provided by the approximation (11.25). This is greater than *the* minimum $\mathcal{V}_E$:

$$\mathcal{V}_E < \mathcal{V}_B.$$

However, our approximation $\boldsymbol{V}_A$ was chosen to provide the minimum potential $\mathcal{V}_A$ among *all* fields of form (11.25). It follows that

$$\mathcal{V}_E < \mathcal{V}_A \leq \mathcal{V}_B. \tag{11.27}$$

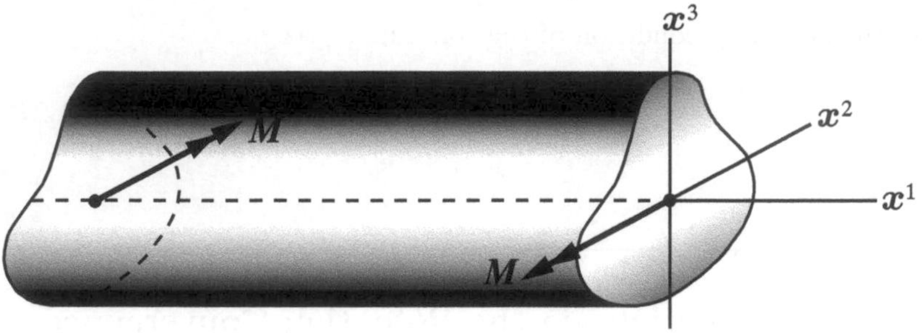

Figure 11.8 Simple bending of a symmetrical beam

By our definition of $\boldsymbol{V}_n$ and $\mathcal{V}_B$, the potential $\mathcal{V}_B$ must approach $\mathcal{V}_E$ as the mesh is refined ($h_n \to 0$). It follows from (11.27) that $\mathcal{V}_A$ approaches the exact minimum $\mathcal{V}_E$. — Quod Erat Demonstrandum.

11.10 Valid Approximations, Excessive Stiffness, and Some Cures

Having demonstrated the mathematical validity of the direct approximation via the minimal criterion, we must now confront certain practical failings. As stated before, any approximation of the configuration produces an excessive energy in the elastic body. Approximating the deformation is analogous to imposing physical constraints which stiffen the body and enforce additional internal energy. In certain circumstances, the effect is so severe that the direct approach via the minimal criterion and a compatible approximation is impractical; although theoretically correct, mesh refinement and computational costs are prohibitive. The difficulties are especially manifested in thin bodies (beams, plates, and shells). Extraordinary behavior is to be anticipated in these circumstances: As previously observed (see Chapter 9), certain strains (and stresses), specifically the transverse shear and normal stresses, contribute little of consequence to the internal energy while tangential components predominate, both mean values (membrane actions) *and* gradients (flexural actions). This is evident in the simplest example, "pure" bending of the beam in Figure 11.8 [267].

For illustrative purposes, let us adopt the reduced expression of internal energy (assume $s^{22} = s^{33} = s^{32} = s^{21} = 0$); these are the customary assumptions in a theory of beams. Each element of the beam in Figure 11.9 is deformed according to the simplest bilinear approximation of the displace-

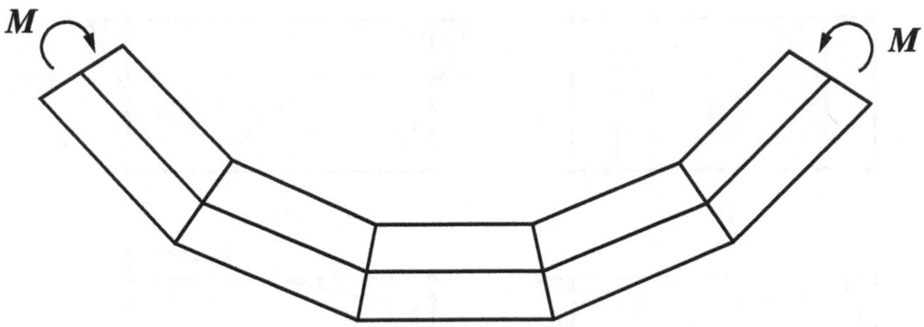

Figure 11.9 Approximation of simple bending

ment:

$$V_1 = \overline{V}_1 + \bar{\epsilon}_{11}x^1 + \bar{\epsilon}_{13}x^3 + \bar{\kappa}_{11}x^1x^3, \tag{11.28a}$$

$$V_3 = \overline{V}_3 + \bar{\epsilon}_{31}x^1 + e_{31}x^3x^1. \tag{11.28b}$$

The coordinates originate at the center of the element as depicted in Figure 11.10. In relations (11.28a, b), the notations for each coefficient is chosen in anticipation of their roles, e.g., constants $\overline{V}_1$ and $\overline{V}_3$ correspond to rigid translation.

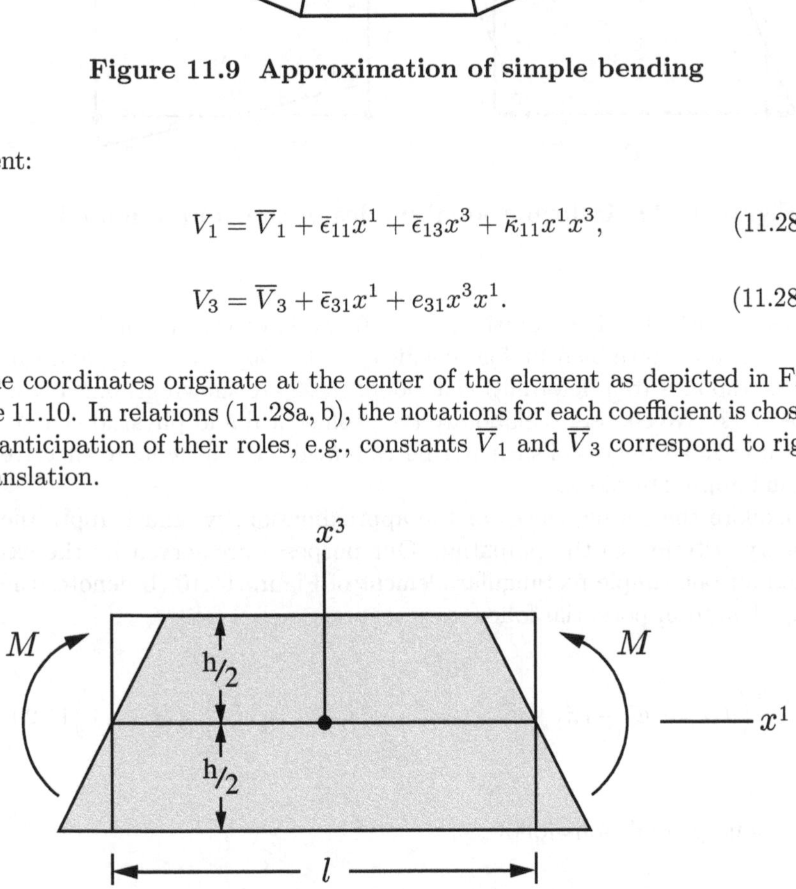

Figure 11.10 Rectangular element under pure bending

The compatible approximations of small strains and rotation follow:

$$\epsilon_{11} = \bar{\epsilon}_{11} + \bar{\kappa}_{11}x^3, \qquad \epsilon_{13} = \tfrac{1}{2}(\bar{\epsilon}_{13} + \bar{\epsilon}_{31}) + \tfrac{1}{2}(e_{31}x^3 + \bar{\kappa}_{11}x^1),$$

$$\Omega_2 = \tfrac{1}{2}(\bar{\epsilon}_{13} - \bar{\epsilon}_{31}).$$

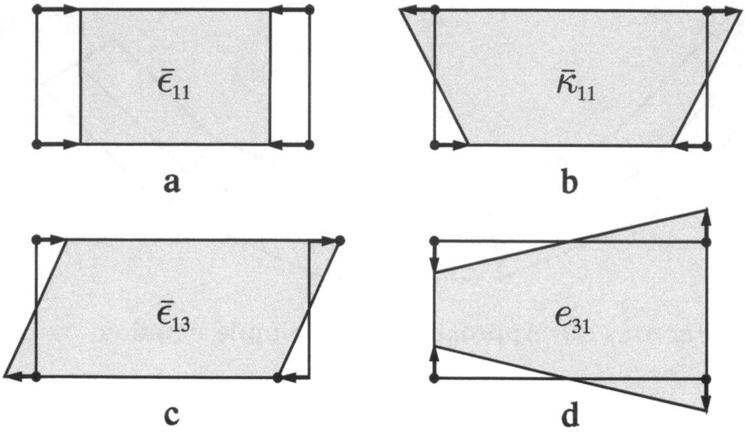

Figure 11.11 Deformational modes of the simple model

Dispensing with the rigid rotation, $\Omega_2 = 0$, we have the relevant components of strain determined by four coefficients: $\bar{\epsilon}_{11}$, $\bar{\kappa}_{11}$, $\bar{\epsilon}_{13} = \bar{\epsilon}_{31}$, and e_{31}. Each corresponds to a deformational mode; these are shown graphically in Figure 11.11. We choose to elaborate the geometrical and physical character because an awareness of such attributes provides the insights to effective *and* valid approximations.

To explore the consequences of the approximation, we must apply the stationary criterion to the potential. Our purposes are served by the expression for one simple rectangular element of Figure 11.10 (b denotes the width). The total potential follows:

$$\mathcal{V} = \frac{lbh}{2}\left(E\,\bar{\epsilon}_{11}^2 + E\,\frac{h^2}{12}\,\bar{\kappa}_{11}^2 + 4G\,\bar{\epsilon}_{13}^2 + G\,\frac{l^2}{12}\,\bar{\kappa}_{11}^2 + G\,\frac{h^2}{12}\,e_{31}^2\right) - Ml\bar{\kappa}_{11}. \quad (11.29)$$

The stationary conditions follow:

$$\frac{\partial V}{\partial \bar{\epsilon}_{11}} = 0, \qquad \Longrightarrow \qquad \bar{\epsilon}_{11} = 0,$$

$$\frac{\partial V}{\partial \bar{\epsilon}_{13}} = 0, \qquad \Longrightarrow \qquad \bar{\epsilon}_{13} = 0,$$

$$\frac{\partial V}{\partial e_{31}} = 0, \qquad \Longrightarrow \qquad e_{31} = 0,$$

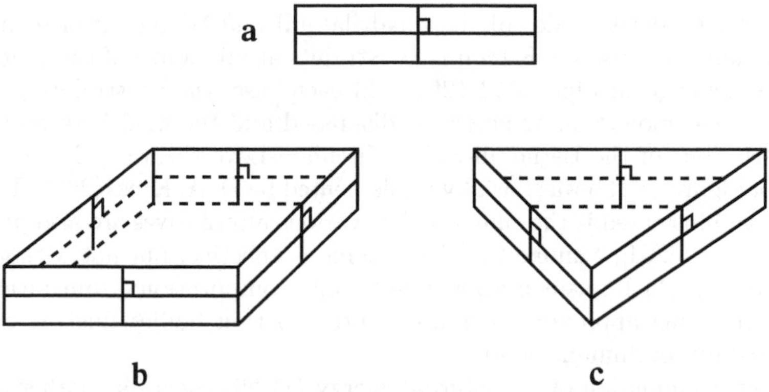

Figure 11.12 Discrete Kirchhoff constraints

$$\frac{\partial V}{\partial \bar{\kappa}_{11}} = 0, \qquad \Longrightarrow \qquad M = \frac{bh}{12}(Eh^2 + Gl^2)\bar{\kappa}_{11}.$$

Of course, we know the exact solution and indeed our approximation (as expected) provides that answer *in the limit* $l \to 0$. Unless the length "l" of our finite element is much less than the depth "h" of the beam, the approximation (due to the term $Gbhl^2\bar{\kappa}_{11}/12$) is much too stiff; in short, the direct approach via the approximation (11.28a, b) is *practically worthless*. In the words of B. Irons [255], this is a "legal" element, but useless in practice.

Many students of finite elements have confronted the particular stiffness illustrated above. The distain for such stiffness and impracticality, led to the term "locking." Since this particular stiffness is traceable to the shear energy, it has been termed "shear locking." Various schemes have been devised to overcome the difficulty (see, e.g., R. H. MacNeal [261]). An early means was proposed by the first author: The origins of the stiffness were noted. It was also recognized that most practical problems of thin elastic bodies (beams, plates, and shells) were quite adequately treated by the theories of Bernoulli-Euler and Kirchhoff-Love. In short, the Bernoulli/Kirchhoff/Love hypothesis is usually adequate: Normals to the midsurface are assumed to remain normal. These observations by G. A. Wempner led him to the discrete counterpart ([116], [268]): At an intermediate point of the element one introduces the *discrete* constraint; for example, at the midpoint of the beam element the normal is constrained to remain normal. The constraint is depicted in Figure 11.12a. Similar excessive stiffness occurs in the simple quadrilateral or triangular elements of a plate or shell (see the early works presented in [116], [268] to [271]). The deformational modes of the shell elements are described in Section 11.13.

The transverse shear modes of the quadrilateral and triangular elements are constrained by *discrete* Kirchhoff constraints at midpoints of the edges; these are depicted in Figures 11.12b, c. In each case, the constraints prohibit the shear modes; shear energy is dismissed and the model converges to the solutions of the Bernoulli-Euler/Kirchhoff-Love theories. The constraint of Figure 11.12b was initially implemented by D. A. Kross [268]. The mechanism illustrated is the simplest, but various alternatives are also possible (see, e.g., [264], Volume 2). The scheme is effective, but not without shortcomings: (1) Implementation poses some computational difficulties. (2) The resulting approximations are limited to thin bodies, just as the corresponding continuum theories.

A closer examination of the internal energy (11.29) suggests another effective means to avoid the shear stiffness and also the inherent shortcomings of the shear constraint [267]. We observe that the deformational mode $\bar{\kappa}_{11}$ (mode (b) in Figure 11.11) produces internal energy via *two* terms in the potential (11.29). What is more important, the contribution to the shear energy is a *higher-order* term; in the limit $l \to 0$, whereas h remains finite. As we explore the matter further (see Section 11.11), we have frequent cause to reiterate: "higher-order" term; consequently, we beg the reader's indulgence with a second acronym: (H.O.). In any model of finite elements, we must only take care that the mean values of each strain are present ($\bar{\epsilon}_{11}$, $\bar{\epsilon}_{13}$, etc.) and *also* that every higher-order deformational mode (e.g., $\bar{\kappa}_{11}$) is inhibited by one (or more) contribution(s) to the internal energy, i.e., there is no need for a second H.O. term [e.g., $Gl^3bh\bar{\kappa}_{11}^2/24$ in equation (11.29)]. These features of the deformation and associated energy are crucial to effective approximation, as exemplified by the simple beam. Omission of the nonessential H.O. term does not compromise, but vastly improves the model. In general, the attributes (e.g., H.O. terms) are not so evident as those perceived in our example of simple bending. An effective means to suppress such excessive energy and stiffness is achieved via the modified potential, as described in the following (Section 11.11).

We must acknowledge other attempts to circumvent the excessive stiffness ("locking"). Some have devised computational schemes which replace the actual integrals of strain energies by evaluations based on isolated values. Such evaluations, called reduced/selective integration, identify the number and sites of the values used ([272] to [274]). For example, if we choose *only* the midpoint value of the transverse shear γ_{13} (one-point integration), then the unwanted contribution ($Gl^2\bar{\kappa}_{11}^2/12$) is absent from the energy (11.29) and the desired result emerges. If we are to apply such procedures with confidence, we must appreciate the underlying mechanics and logic: To wit, the omitted H.O. term is nonessential, *but only* because the associated mode is inhibited by another source of strain energy.

11.11 Approximation via the Modified Potential; Convergence and Efficiency

A generalization of the principle of stationary potential (see Section 6.16) provides an alternative and effective basis for a discrete model via spline approximations (finite elements). We recall that this generalization employs the modified potential of Hu-Washizu ([114], [115]), a functional of displacement, strain, and stress; variations of the three fields provide equations of equilibrium, stress-strain and strain-displacement, respectively. These governing equations are thereby rendered in their primitive forms (e.g., equilibrium conditions are expressed in terms of stresses). We emphatically employ the term modified *potential* since the functional has the value of the potential if the fields are fully compatible; this feature is vital to the convergence of an approximation.

As the stationary criteria provide the primitive forms of the differential and algebraic equations governing the continuous body, the corresponding criteria (i.e., variations of discrete parameters) provide the algebraic versions governing the discrete model. Here, we introduce discrete approximations of each field; these are interrelated through the stationary conditions. Accordingly, the scheme admits independent forms of approximation for the variables, i.e., full compatibility of displacement, strain, and stress is not a prerequisite. The *requisite* "compatibility" is provided by the stationary criteria. The relaxation of compatibility between the fields enables the suppression of unnecessary H.O. terms, avoidance of unwarranted stiffness and, consequently, *more rapid convergence.*

To illustrate, we employ the trilinear approximation of displacement within a three-dimensional element, as described by rectangular Cartesian coordinates (x^1, x^2, x^3 or x^i, $i = 1, 2, 3$). Because we intend subsequently to explore the approximation of a plate (or shell), we identify the two surface coordinates (x^1, x^2 or x^α, $\alpha = 1, 2$); then the third (x^3) denotes distance through the thickness. With such notations, components of displacement within an element have similar trilinear approximations:

$$V_1 = \overline{V}_1 + \bar{\epsilon}_{1i}x^i + e_{12}x^1x^2 + \bar{\kappa}_{11}x^1x^3 + \bar{\kappa}_{12}x^2x^3 + \kappa_1 x^1x^2x^3, \tag{11.30a}$$

$$V_2 = \overline{V}_2 + \bar{\epsilon}_{2i}x^i + e_{21}x^2x^1 + \bar{\kappa}_{22}x^2x^3 + \bar{\kappa}_{21}x^1x^3 + \kappa_2 x^1x^2x^3, \tag{11.30b}$$

$$V_3 = \overline{V}_3 + \bar{\epsilon}_{3i}x^i + e_{31}x^3x^1 + e_{32}x^3x^2 + \gamma_3 x^1x^2 + \kappa_3 x^1x^2x^3. \tag{11.30c}$$

As usual, the repeated indices imply the summation ($i = 1, 2, 3$). The coefficients in (11.30a–c) are labeled to distinguish them, physically and

Deformational Modes

$$
\begin{array}{rcl}
 & \textit{Extensional} & \textit{Flexural} \\
\epsilon_{11} = & \bar{\epsilon}_{11} + e_{12}x^2 & + \quad \underline{\bar{\kappa}_{11}}x^3 + \underline{\kappa_1}x^2x^3 \\
\epsilon_{22} = & \bar{\epsilon}_{22} + e_{21}x^1 & + \quad \underline{\bar{\kappa}_{22}}x^3 + \underline{\kappa_2}x^1x^3 \\
2\epsilon_{12} = & {}_m\bar{\epsilon}_{12} + (e_{12}x^1 + e_{21}x^2) & + \quad (\bar{\kappa}_{12} + \bar{\kappa}_{21})x^3 + (\underline{\kappa_1}x^1 + \underline{\kappa_2}x^2)x^3 \\
\epsilon_{33} = & \bar{\epsilon}_{33} + e_{31}x^1 & + \quad \epsilon_{32}x^2 + \kappa_3x^2x^1
\end{array}
$$

$$
\begin{array}{rcl}
 & \textit{Transverse Shear} & \textit{Higher-Order Transverse Shear} \\
2\epsilon_{13} = & {}_m\bar{\epsilon}_{13} + (\bar{\kappa}_{12} + \gamma_3)x^2 & + \quad (\underline{\bar{\kappa}_{11}}x^1 + e_{31}x^3) + (\underline{\kappa_1}x^1 + \kappa_3x^3)x^2 \\
2\epsilon_{23} = & {}_m\bar{\epsilon}_{23} + (\bar{\kappa}_{21} + \gamma_3)x^1 & + \quad (\underline{\bar{\kappa}_{22}}x^2 + e_{32}x^3) + (\underline{\kappa_2}x^2 + \kappa_3x^3)x^1
\end{array}
$$

$$
{}_m\bar{\epsilon}_{12} = \bar{\epsilon}_{12} + \bar{\epsilon}_{21}, \qquad {}_m\bar{\epsilon}_{13} = \bar{\epsilon}_{13} + \bar{\epsilon}_{31}, \qquad {}_m\bar{\epsilon}_{23} = \bar{\epsilon}_{23} + \bar{\epsilon}_{32}
$$

Table 11.1 Approximation of strains in trilinear displacement

mathematically. First, we note that the leading terms $\overline{V_i}$ represent translation. Rigid rotation is also included; in the linear theory (small rotations) these are, approximately,

$$
\Omega_1 \doteq \tfrac{1}{2}(\bar{\epsilon}_{32} - \bar{\epsilon}_{23}), \qquad \Omega_2 \doteq \tfrac{1}{2}(\bar{\epsilon}_{13} - \bar{\epsilon}_{31}), \qquad \Omega_3 \doteq \tfrac{1}{2}(\bar{\epsilon}_{21} - \bar{\epsilon}_{12}). \quad (11.31\text{a–c})
$$

The remaining 18 degrees-of-freedom comprise the deformational modes, those that impart strain energy. Since *relative* rotations and strains are typically small *within* an element, we confine our attention to the linear approximation of the strains as displayed in Table 11.1. The format and groupings are arranged for the subsequent approximations in two-dimensions, specifically "plane stress" ($s^{33} = 0$).

The keys to our approximation are:

1. The identification of the higher-order (H.O.) terms
2. The realization that these *reappear* in different components

The terms of higher order vanish from the energy density in the *limit*. Still, each must be present *somewhere* in the *finite* element to inhibit that mode; however, it is sufficient to retain such term in but one of the components. If a higher-order term is omitted, then the element cannot provide the necessary resistance against the corresponding deformational mode. Such deformational mode occurs without expending work/energy; it is known as a "spurious" or "zero energy" mode.

Strictly speaking, all terms in Table 11.1 are of higher order except the constants $\bar{\epsilon}_{ij}$. Stated otherwise, in the limit the strain energy *density* of three dimensions is

$$u = \tfrac{1}{2}E^{ijkl}\bar{\epsilon}_{ij}\bar{\epsilon}_{kl}.$$

In the two-dimensional case (plate or shell) the *one* dimension (thickness x^3) remains *finite*; then the limit of the density (per unit area) has the form:

$$u = \frac{h}{2}E^{ijkl}\bar{\epsilon}_{ij}\bar{\epsilon}_{kl} + \frac{h^3}{24}D^{\alpha\beta\gamma\delta}\bar{\kappa}_{\alpha\beta}\bar{\kappa}_{\gamma\delta}, \tag{11.32}$$

where $(\alpha, \beta, \gamma, \delta = 1, 2)$. In the latter case, the flexural terms $\bar{\kappa}_{\alpha\beta}$ *cannot* be treated as H.O. terms and, indeed, may dominate the membrane terms $\bar{\epsilon}_{\alpha\beta}$. Transverse shears $[(\bar{\epsilon}_{13} + \bar{\epsilon}_{31})$ and $(\bar{\epsilon}_{23} + \bar{\epsilon}_{32})]$ remain in any case; this is a distinct advantage of our present formulation and *without adverse consequences.*

Now, we recall that the functional of Hu-Washizu admits independent variations of strain and stress components; hence, we can admit independent approximations in our finite element. Specifically, our approximations of the strains need not retain the H.O. terms in two different components. Suppressing such terms in one, or the other, serves to reduce the internal energy and improve convergence. For example, our approximations of the strain must contain a term of the form $e_{12}x^2$ in the component ϵ_{11} *or* one of the form $(e_{12}x^1)/2$ in component ϵ_{12}. By omitting one or the other, we anticipate a reduction in strain energy and stiffness, and therefore we also expect improved, though not monotonic, convergence. Of course, the best choice depends upon the nature of the problem. Note that the *only* requirement is the retention of one H.O. term to inhibit each mode.

In the case of a thin plate or shell, the simplification is evident. Then the transverse stress s^{33} and strain ϵ_{33} are eliminated; four deformational modes $(\bar{\epsilon}_{33}$, e_{31}, e_{32}, and $\kappa_3)$ are absent. The remaining H.O. modes (underscored) in the transverse shears (ϵ_{13} and ϵ_{23}) are already inhibited by flexural terms (viz., $\bar{\kappa}_{11}$, $\bar{\kappa}_{22}$, κ_1, and κ_2 in ϵ_{11} and ϵ_{22}). These H.O. terms are the offensive ("locking") terms; their presence is unacceptable as previously illustrated by the beam of Section 11.10. By our previous arguments, such terms can be safely omitted in the transverse shears (ϵ_{13} and ϵ_{23}); moreover, their omission *improves convergence.*

In the formulation of a three-dimensional element, one can also simplify the approximations *and* improve convergence by suppressing such nonessential terms. One must only inhibit the mode via one strain and the consequent energy. Examples are given in an earlier article [275].

The foregoing example in rectangular coordinates is admittedly, and intentionally, simple. Curvilinear and/or nonorthogonal systems certainly introduce additional coupling of modes, but do not invalidate the foregoing arguments. We note too that the interior of any body can be discretized by rectangular elements (or elements with edges parallel to the curvilinear coordinates) and that a shell can be approximated by sufficiently small flat elements (coupling is incorporated in assembly). Additionally, we observe that any smooth surface is described most simply by the orthogonal lines of principal curvature. Only at intersections and edges must we confront triangular elements; however, the behavior in such regions is seldom described by theories of two dimensions.

The reliability of the foregoing approach stems from the fact that the functional of Hu-Washizu is a modification of the potential. If the approximations of strains (and stresses) are compatible with displacements, then that functional *is* equal to the potential. Since the incompatibilities are only in the H.O. terms, the convergence is not compromised. The extent and origins of the improvements require detailed examination (see C. D. Pionke [276]).

We can safely introduce those approximations of stresses which are consistent with the strains. Still, alternative approximations (e.g., different forms for stresses and strains) are effective in some circumstances ([201], [275], [277], [278]). The functional does admit different approximations, but caution is in order, since the approximations of stress also modify the relations between the discrete strains (deformational modes) and displacements. Such possibilities offer yet another avenue for further development of efficient elemental approximations.

11.12 Nonconforming Elements; Approximations with Discontinuous Displacements

11.12.1 Introduction

In most instances, displacement approximations, as determined by nodal values and shape functions, preserve interelement continuity; the elements are said to conform. However, some nonconformity need not invalid an approximation. One can perceive nodal connection with nonconforming interfaces. For example, the upper surface a-b-c-d of element A in Figure 11.13 might be deformed according to the approximation:

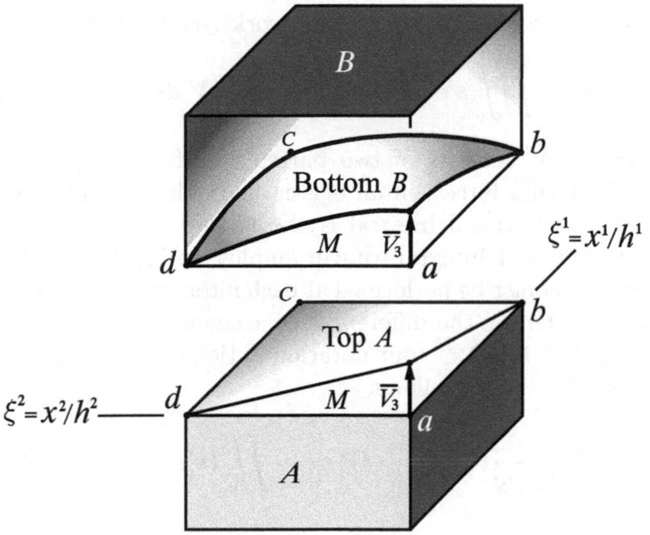

Figure 11.13 Nonconforming surfaces

$$V_3^A = (1 - \xi^1 - \xi^2 + \xi^1\xi^2)\overline{V}_3. \tag{11.33a}$$

With origin ($x^i = 0$) at corner a, $\xi^i = x^i/h^{\underline{i}}$; h^i denotes the length of the respective edge. The lower surface of the adjoining element B might be deformed according to the approximation:

$$V_3^B = (1 - \xi^1 - \xi^2 + \xi^1\xi^2)\overline{V}_3 + [\xi^1 - (\xi^1)^2]\alpha_1 + [\xi^2 - (\xi^2)^2]\alpha_2. \tag{11.33b}$$

Note that the nodes remain joined; all other particles are separated. However, as the *finite* element shrinks ($h_i \to 0$), the surfaces *approach* congruence. In other words, *any smooth* surface can be approximated by quadrilateral platelets as simple as the surface (11.33a); these converge ($h_i \to 0$). Practically, this means that some nonconformity need not preclude the convergence of an approximation. Indeed such nonconforming elements are admissible, though precautions are required. The renowned "patch test" ([279], [255], [253]) provides a necessary condition.

11.12.2 Patch Test

Consider a body bounded by a surface s_t (applied traction $\overline{\boldsymbol{T}}$) and subdivided by interfaces s_n (interelement surfaces, $n = 1, \ldots N$). In the absence of a body force $\boldsymbol{f}$ the principle of virtual work (see Section 6.14) asserts that

$$\delta W = \iiint_v \boldsymbol{s}^i \cdot \delta \boldsymbol{V}_{,i}\, dv - \iint_{s_t} \overline{\boldsymbol{T}} \cdot \delta \boldsymbol{V}\, ds = 0. \tag{11.34}$$

The virtual work δW consists of two parts: The first integral accounts for the work of internal forces; in an elastic body that is the variation of internal energy. The second integral is the virtual work of the external tractions $\overline{\boldsymbol{T}}$. To the first integral we can employ an integration by parts: Now the evaluations must be performed at each interface s_n, since $\boldsymbol{V}$ can be discontinuous. We denote the different displacements on the positive and negative sides by $\boldsymbol{V}_+^N$ and $\boldsymbol{V}_-^N$ (our notation anticipates directions relative to a coordinate). The result follows:

$$\delta W = -\iiint_v \frac{1}{\sqrt{g}}(\sqrt{g}\, \boldsymbol{s}^i)_{,i} \cdot \delta \boldsymbol{V}\, dv + \iint_{s_t} (\boldsymbol{s}^i n_i - \overline{\boldsymbol{T}}) \cdot \delta \boldsymbol{V}\, ds$$

$$+ \sum_n \iint_{s_n} \boldsymbol{s}^i \cdot (\delta \boldsymbol{V}_+^N - \delta \boldsymbol{V}_-^N)\, ds. \tag{11.35}$$

We now invoke a patch test: We suppose that our body is an assembly of elements (a patch) subjected to a *homogeneous equilibrated* state of stress. Then the first and second integrals vanish; both are *necessary* conditions for equilibrium. The condition ($\delta W = 0$) is reduced to the work performed by the stress upon the discontinuity of displacements ($\delta \boldsymbol{V}_+^N - \delta \boldsymbol{V}_-^N$) at all interfaces. Recall however that the virtual displacement in a continuum is an arbitrary function. The counterpart in our approximation is the arbitrary variation of each discrete value, i.e., nodal displacement $\boldsymbol{V}_I$ and such other generalized coordinate, say α_J, which collectively determine the approximation.

Finally, we reiterate the condition: For *arbitrary* variations of the discrete coordinates, the work performed by a homogeneous state of stress upon the discontinuities ($\delta \boldsymbol{V}_+^N - \delta \boldsymbol{V}_-^N$) must vanish:

$$\sum_n \iint_{s_n} \boldsymbol{s}^i \cdot (\delta \boldsymbol{V}_+^N - \delta \boldsymbol{V}_-^N)\, ds = 0. \tag{11.36}$$

Note: The summation remains, since we cannot preclude the circumstance, wherein a variation effects displacements along numerous interfaces.

Example:

Consider a problem of plane stress ($\boldsymbol{s}^3 = 0$), wherein the body is subdivided into rectangular elements. Suppose that we test an approximation

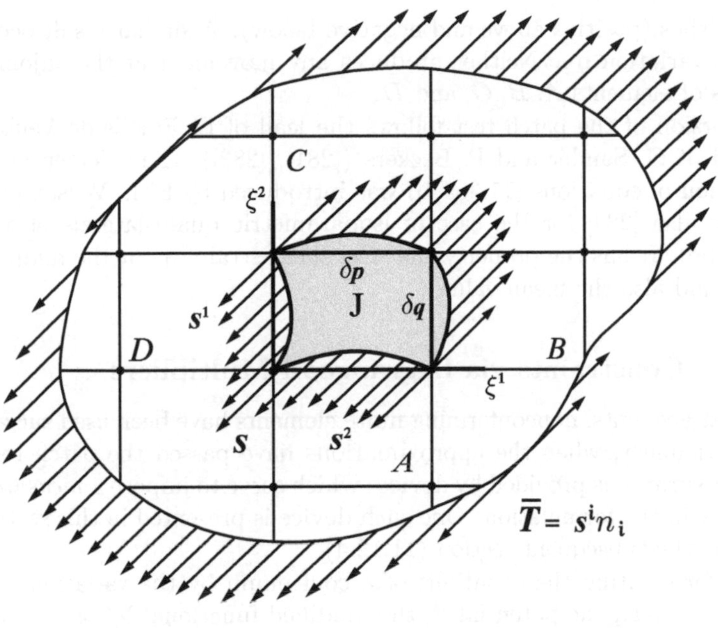

Figure 11.14 Homogeneous stress on a "patch"

$\boldsymbol{V}$ comprised of a conforming part $\boldsymbol{V}_C$ and a nonconforming part $\boldsymbol{V}_D$:

$$\boldsymbol{V} = \boldsymbol{V}_C + \boldsymbol{V}_D,$$

where

$$\boldsymbol{V}_D = g^J(\xi^1)\,\boldsymbol{p}_J + h^J(\xi^2)\,\boldsymbol{q}_J. \tag{11.37}$$

The conforming part might be the bilinear form, the two-dimensional counterpart of equations (11.17) and (11.18). That part is irrelevant as it contributes nothing to our test. For our example, we take

$$g^J = 4[\xi^1 - (\xi^1)^2], \tag{11.38a}$$

$$h^J = 4[\xi^2 - (\xi^2)^2]. \tag{11.38b}$$

The form applies to each element with the origin at the lower left node. The arbitrary patch of Figure 11.14 is subjected to a homogeneous stress $\overline{\boldsymbol{T}}$. This acts on the surfaces of a typical interior element J. The two modes of equations (11.38a, b) are depicted, much exaggerated: The first, $\delta\boldsymbol{p}_J$, produces the bulge at the top (convex) and a similar depression (concave) at the bottom. Without evaluation, it is evident that the consequent virtual work vanishes (positive above and negative below). A similar result occurs upon the variation $\boldsymbol{q}_J$. Neither produces any movement at the adjoining interfaces of elements A, B, C, and D.

Our version of the patch test follows the lead of B. Fraeijs de Veubeke [280] and of G. Sander and P. Beckers ([281], [282]). The elemental approximation of equations (11.38a, b) was introduced by E. L. Wilson et al. [283] (see also [284] for the case of isoparametric quadrilaterals of arbitrary shape). It has the property that the shear strain γ_{12} at the midpoint vanishes and also the mean value.

11.12.3 Constraints via Lagrangean Multipliers

By most accounts, nonconforming finite elements have been used successfully, particularly, when the approximations have passed the patch tests. Further assurance is provided by devices which serve to impose a *measure* of conformity in the formulation. One such device is presented in this section, another in the subsequent section (11.12.4).

When formulating the equations of a continuum by the variations of a functional F, e.g., the potential $\mathcal{V}$, the modified functional $\mathcal{V}_c^*$ of Reissner, or the modified potential $\mathcal{V}^*$ of Hu-Washizu , the variations of the displacement $\boldsymbol{V}$ are confined (a priori) to a class of *continuous* functions. Viewed from that perspective, any additional condition, such as

$$\boldsymbol{V}_+^n = \boldsymbol{V}_-^n$$

on an intermediate surface s_n, alters nothing. Indeed, any of the functionals $F(\boldsymbol{V},\ldots)$ is unchanged by the addition of terms of the form:

$$\iint_{s_n} \boldsymbol{\lambda}_n \cdot (\boldsymbol{V}_+^n - \boldsymbol{V}_-^n)\, ds.$$

However, in anticipation of our intentions (to formulate nonconforming elements), we augment the functional F (either $\mathcal{V}$, $\mathcal{V}^*$ or $\mathcal{V}_c^*$) to form the modified functional:

$$\overline{F}(\boldsymbol{V},\ldots,\boldsymbol{\lambda}_n) = F(\boldsymbol{V},\ldots,\boldsymbol{\lambda}_n) - \sum_n \iint_{s_n} \boldsymbol{\lambda}_n \cdot (\boldsymbol{V}_+^n - \boldsymbol{V}_-^n)\, ds. \qquad (11.38)$$

We note that the modified potential $\mathcal{V}^*$ is also a functional of stress $\boldsymbol{s}^i$ and strain γ_{ij}, and the modified functional $\mathcal{V}_c^*$ is also dependent on stress. Such dependence is not relevant to our immediate concern, viz., the role of displacement $\boldsymbol{V}$ and its continuity in an approximation. In each instance, the variations of functions $\boldsymbol{V}$ and $\boldsymbol{\lambda}$ produce the same result:

$$\delta\overline{F} = \iiint_v \boldsymbol{s}^i \cdot \delta\boldsymbol{V}_{,i}\, dv - \iint_{s_t} \overline{\boldsymbol{T}} \cdot \delta\boldsymbol{V}^t\, ds$$

$$+ \sum_n \iint_{s_n} (\boldsymbol{s}^i n_i - \boldsymbol{\lambda}_n) \cdot (\delta\boldsymbol{V}^n_+ - \delta\boldsymbol{V}^n_-)\, ds$$

$$- \sum_n \delta\boldsymbol{\lambda}_n \cdot (\boldsymbol{V}^n_+ - \boldsymbol{V}^n_-) = 0. \tag{11.39}$$

The variation $\delta\overline{F}$ must vanish for *arbitrary* variations $\delta\boldsymbol{V}$ in the interior and $(\delta\boldsymbol{V}^t,\, \delta\boldsymbol{V}^n_+,\, \delta\boldsymbol{V}^n_-,\, \delta\boldsymbol{\lambda}_n)$ on the respective surfaces: Equilibrium insures the vanishing of the first and second terms. The arbitrariness of $\boldsymbol{\lambda}_n$ in the final term requires that

$$\boldsymbol{V}^n_+ = \boldsymbol{V}^n_- \quad \text{at each } s_n. \tag{11.40}$$

The third term serves to identify the multiplier:

$$\boldsymbol{\lambda}_n = n_i \boldsymbol{s}^i \big]_{s_n}. \tag{11.41}$$

Again, (11.40) and (11.41) add nothing to our theory of the *continuous* body but serve a purpose in the approximation, wherein $\boldsymbol{V}^n_+ \neq \boldsymbol{V}^n_-$.

Let us examine the implications of the modified functional $\overline{F}$ as applied to approximations by nonconforming elements ($\boldsymbol{V}^n_+ \neq \boldsymbol{V}^n_-$). Typically, the approximation of displacement can be expressed in the form:

$$\boldsymbol{V} \doteq \widetilde{\boldsymbol{V}} \equiv f^I(\theta^1, \theta^2, \theta^3)\boldsymbol{V}_I + \alpha^J(\theta^1, \theta^2, \theta^3)\, \boldsymbol{q}_J. \tag{11.42}$$

The vector $\boldsymbol{V}_I$ signifies a nodal displacement; $\boldsymbol{q}_J$ can be any generalized coordinate. The latter are typically associated with an element. One might consider indices I and J to be associated with nodes and elements, respectively. In either case, summation is implied by the repeated indices. In mathematical parlance, the approximation has *local support*; variations of the discrete variables ($\boldsymbol{V}_I$ and $\boldsymbol{q}_J$) confine the variation $\widetilde{\boldsymbol{V}}$ to the adjoining elements. The functions $\boldsymbol{\lambda}_n$ are defined *only on interfaces*; their approximation can have the form

$$\boldsymbol{\lambda}_n \doteq \tilde{\boldsymbol{\lambda}}_n = g^M \boldsymbol{\lambda}_M. \tag{11.43}$$

The number of the discrete variables may be greater than n, since one or more discrete values $\boldsymbol{\lambda}_M$ are required at each interelement surface. In general, g^M is a function of the surface coordinates. Again, the approximation (11.43) has local support.

Let us focus on the circumstance in which our functional F is the potential $\mathcal{V}$; then, $\overline{F}$ does not entail variables other than displacement $\boldsymbol{V}$ and multipliers $\boldsymbol{\lambda}_n$. Then, the substitution of our approximation into that functional and the integration of the known functions (f^I, α^J, g^M) reduces the functional to a *function* $\widetilde{F}$ of those discrete variables. If the body in question is a Hookean body, and if the rotations are small, then the function $\widetilde{F}$ is a quadratic form in those variables ($\boldsymbol{V}_I$, $\boldsymbol{q}_J$, and $\boldsymbol{\lambda}_M$). Then, the result of the variational procedure is a linear system; the solution determines those discrete variables and the approximation (11.42).

Of particular interest is the consequence of the constraint; i.e., the discrete counterpart of the continuity condition ($\boldsymbol{V}_+^n = \boldsymbol{V}_-^n$). To that end, let us examine the condition upon an interface between two adjoining elements, such as A and B in Figure 11.13. Here, the adjoining surfaces are nonconforming; however, on the interface, say s_M, we have approximated a multiplier $[\boldsymbol{\lambda}_n \doteq g^M(\theta^1, \theta^2)\,\boldsymbol{\lambda}_M]$. The nonconforming interfaces are also approximated in a form (11.42). As before, the nonconformity is the difference, $(\boldsymbol{V}_+ - \boldsymbol{V}_-)\,]_{s_M}$. For simplicity, let us assume that the approximation of $\boldsymbol{\lambda}_n$ on s_M is defined by a single discrete value $\boldsymbol{\lambda}_M$. Then, the discrete constraint results from the variation of that one variable:

$$\delta\boldsymbol{\lambda}_M \cdot \iint_{s_M} g^M(\boldsymbol{V}_+ - \boldsymbol{V}_-)\,ds = 0,$$

or

$$\iint_{s_M} g^M(\boldsymbol{V}_+ - \boldsymbol{V}_-)\,ds = 0. \tag{11.44}$$

In short, the discrete constraint (11.44) (one of the linear equations in our system) requires that a weighted average of the discontinuity vanish. Such condition is said to provide a "weak" conformity. If $\boldsymbol{\lambda}_n$ is assumed constant at each interface (g^M = constant), then (11.44) asserts that the mean values of $\boldsymbol{V}_+]_M$ and $\boldsymbol{V}_-]_M$ are equal. Since the functions (i.e., surfaces) are smooth, convergence (conformity in the limit) is insured.

We recall that the proof of convergence (Section 11.9) presumes continuity of the approximation $\boldsymbol{V}_A$. Although our approximation exhibits interelement discontinuity, the condition (11.44) assures conformity in the limit. The preceding proof need only be amended: Specifically, we now recognize that the exact solution $\boldsymbol{V}_E$ can also be approximated in the form (11.42) with the "weak" conformity of (11.44). The potential $\mathcal{V}_B$ of inequality (11.27) is then the potential of that configuration.

Approximations of the type described here are called "hybrid." These entail approximations of two types: Those defined *within* elements [equation (11.42)] and also those defined *upon* the surfaces of elements [equation (11.43)]. As previously noted, the functional $\overline{F}$ subject to the constraints [see equation (11.38a, b)] could be the modified potential $\mathcal{V}^*(\boldsymbol{V}, s^{ij}, \gamma_{ij})$ or $\mathcal{V}_c^*(\boldsymbol{V}, s^{ij})$ (see Section 6.16). With consistent approximations of the functions, the former $\mathcal{V}^*$ converges to the potential $\mathcal{V}$; likewise, the approximation via the weakly conforming assembly converges to the solution. Hybrid elements were pioneered by T. H. H. Pian [285]; the interested reader can consult the writings of T. H. H. Pian and P. Tong (see, e.g., [286] to [289]). A very lucid description of hybrid elements is given by R. H. Gallagher [254].

11.12.4 Constraints via Penalty Functions

The imposition of the constraints via Lagrangean multipliers (previous section) has one obvious disadvantage. It introduces additional unknowns $\boldsymbol{\lambda}_M$. These can be eliminated, but additional computation is required; one solves for $\boldsymbol{\lambda}_M$ in terms of displacements $\boldsymbol{V}_I$ and $\boldsymbol{q}_J$. With any strategy, the method presents additional computational difficulty. An effective alternative is the imposition of penalty functions.

With a view toward imposing a measure of conformity, the functional $F(\boldsymbol{V}, \ldots)$ can be amended by adding positive-definite functionals, as follows:

$$\overline{F}(\boldsymbol{V}, \ldots) = F(\boldsymbol{V}, \ldots) + \sum_n \iint_{s_n} \tfrac{1}{2} K_n (\boldsymbol{V}_+^n - \boldsymbol{V}_-^n) \cdot (\boldsymbol{V}_+^n - \boldsymbol{V}_-^n)\, ds. \quad (11.45)$$

The factor (1/2) has *no* mathematical relevance, but provides a physical interpretation: The product is the square of distance between the nonconforming surfaces. The engineer can imagine an omnidirectional, linear, and continuous spring along the interface; it is installed unstretched at the conforming state. The extension (or contraction) is the distance e_n

$$e_n = |\boldsymbol{V}_+^n - \boldsymbol{V}_-^n|.$$

The energy of such spring (per unit area) is the integral of (11.45). The function K_n is the stiffness of said spring; it can be variable on s_n, i.e., a function of the surface coordinates.

We note two prominent differences between the present modification (11.45) and the previous form (11.38a, b): First, the functions K_n are not treated as unknowns; they are to be prescribed. Second, when these functions are added to the potential $\mathcal{V}$, the modified functional retains the positive-definite character. Indeed, the addition of the energy can only increase the potential; the system is *penalized*. The "penalties" increase with the nonconformity (distances e_n). Presumably, the best approximation, the minimum of the modified potential, is achieved with the best conformity.

The present method does pose a certain dilemma: On the one hand, a large K_n (a strong spring) tends to enforce a greater measure of conformity. On the other hand, if the spring is *too* stiff (K_n too large), it constraints a desirable degree of nonconformity. One must recognize that nonconforming elements (some discontinuities) are introduced *only if* such approximation offers advantages, e.g., reduction in the degrees-of-freedom. Moreover, the nonconforming elements can only converge to the continuous solution in the limit. An excessive penalty function K_n upon the nonconforming assembly of *finite* elements serves only to increase the potential and inhibit convergence. To date, the choices of the penalty factor are usually matters of experience and judgements based on computational efficiency. We cannot but wonder whether B. Irons would contrive a universal test: Imagine a patch of these constrained elements subjected to a state of known consequences and criteria to determine the best stiffness K_n.

11.13 Finite Elements of Shells; Basic Features

11.13.1 Inherent Characteristics

The continuum theories of three dimensions and two, i.e., shell theories, have been traditionally separated at birth. The classical theories of shells are typically characterized by basic assumptions/hypotheses which immediately reduce them to two dimensions. The reduction is achieved by assumptions about the distributions of displacement, strain and/or stress through the thickness. The most widely used and effective basis is the Kirchhoff-Love hypothesis (see Chapter 10). Such theories inevitably possess limitations imposed by the relative thickness and, to some extent, the loading and properties of the materials.

Any theory of shells can be viewed as a first step in an approximation of the thin three-dimensional body via finite elements; subsequent, or simultaneous, subdivision in the remaining surface provides the fully discrete model. However, as demonstrated by the examples in Sections 11.10 and 11.11, thin bodies pose unique difficulties. Consequently, traditional theories are not necessarily well suited to such discrete models. Specifically, the theory of Kirchhoff-Love requires a higher order of interpolation (continuity of derivatives ${}_0\boldsymbol{R}_{,\alpha}$) so that normals remain continuous at the contiguous edges of adjoining elements. Stated otherwise, that theory precludes any kinks in the reference surface. To use the Kirchhoff-Love theory and achieve the requisite continuity entails complicated polynomials as shape functions and, in general, all second derivatives appear as degrees of freedom at each node of a quadrilateral element (cf. [255], p. 265 and [264], Vol. 2, p. 122). Since the effective model is intended to accommodate unknown configurations, many elements are still needed; then efficiency calls for simpler elements. Such simplifications can only be achieved by forsaking Kirchhoff's hypothesis and admitting the relative rotation of normals; in other words, such simpler elements must admit transverse shear strain.

Aside from the aforementioned arguments, there are practical advantages in the inclusion of transverse components of the strains and stresses. Indeed, if all six components are present in our basic element, it possesses the essentials of a three-dimensional element. In engineering practice, we invariably encounter regions (e.g., at junctures or supports) of shell-like structures which are not amenable to the simpler theory. Then, we are compelled to accommodate more complex behavior by refinements of our model. In the spirit of discrete elements the refined shell theories can be supplanted by a progression of layers (i.e., a refined mesh). The vessel depicted in Figure 10.13 illustrates such realities. Portions of the thin cylindrical pipe are described as a simple membrane; the only significant stress is the so-called "hoop" stress, $s = pr/h$. Near the juncture, these shells exhibit bending which requires no less than the gradients of the Kirchhoff-Love theory. At the juncture, particularly at the reentrant corner, the behavior can only be described by the three-dimensional theory, a refined mesh and, perhaps, a transition to the continuum theory.

In summary, we can achieve certain simplifications (simpler interpolation in the surface coordinates), computational advantages, and enhanced capabilities by means of a simple conforming three-dimensional element.

11.13.2 Some Consequences of Thinness

The distinctive feature of a shell is thinness. Apart from isolated regions (junctures, supports, etc.), as noted above, the deformations of thin bodies typically exhibit distinctive attributes which have a bearing upon their discrete approximation: Elements can undergo large rotations, though strains and *relative* rotations (e.g., within a finite element) are small. This means that we can often focus on small strains when modeling the *element*, though the shell (i.e., assembly) exhibits finite rotations ([117], [290]). Such finite rotations appear in the assembly and account for geometrical nonlinearities.

By definition, the thickness of a shell is small compared to a radius of curvature. Also, the dimensions of an element must be small compared to the deformational pattern; otherwise, such pattern must be anticipated by a preselected shape function. It follows that the concepts and theories of shallow shells are often applicable to the individual element; this is the notion that supports Koiter's theory of "quasi-shallow" shells (see

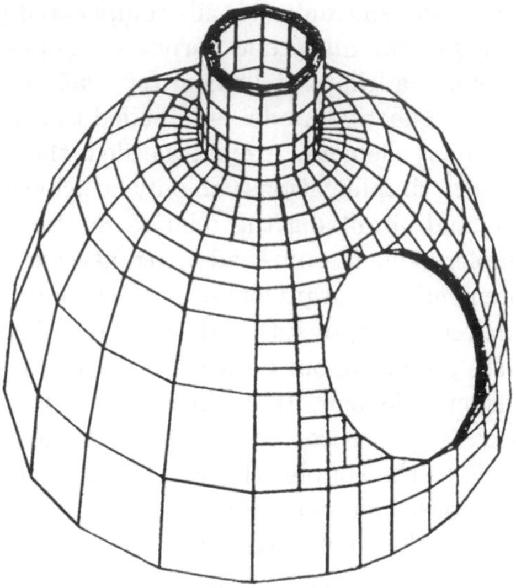

Figure 11.15 Shell-like structure

Section 10.11). We can carry the argument further: Any surface, hence any shell, can be approximated by an assembly of plane elements, plates (triangular or quadrilateral elements delineated along lines of curvatures). In the latter case, both initial and deformed curvatures must be manifested in the assembly.

11.13.3 Simple Conforming Elements

The foregoing remarks provide some guidelines that acknowledge certain peculiarities and difficulties. We turn now to basic attributes of simple elements. Figure 11.15 depicts a shell-like structure with typical features. The juncture of two different geometrical forms and an opening. Some triangular elements are required, but only at edges. The interior of the shell can be subdivided along orthogonal lines of curvature into quadrilateral elements. The most rudimentary form, rectangular, exhibits the essentials of the simplest conforming element. The three-dimensional approximation of displacement is expressed by equations (11.30a–c) and the strains are displayed in Table 11.1. The typical thin shell is adequately described by the assumption of "plane" stress, i.e., $s^{33} = 0$. Then no work is done via stretching of the normal, i.e., ϵ_{33} is irrelevant. Our attention is focused on the remaining strains and the fourteen deformational modes which contribute to the internal energy:

- *Extensional Strain* Dominant terms: $\bar{\epsilon}_{11}, \quad \bar{\epsilon}_{12} = \bar{\epsilon}_{21}, \quad \bar{\epsilon}_{22}$
  Higher-order terms: $e_{12}, \quad e_{21}$
- *Transverse Shear* Dominant terms: $\bar{\epsilon}_{31} = \bar{\epsilon}_{13}, \quad \bar{\epsilon}_{23} = \bar{\epsilon}_{32}$
  Higher-order terms: $\bar{\kappa}^*_{12} = (\bar{\kappa}_{21} - \bar{\kappa}_{12})/2, \quad \gamma_3$
- *Flexural Strain* Dominant terms: $\bar{\kappa}_{11}, \quad \bar{\kappa}_{22},$
  $\bar{\kappa}'_{12} = (\bar{\kappa}_{12} + \bar{\kappa}_{21})/2$
  Higher-order terms: $\kappa_1, \quad \kappa_2$

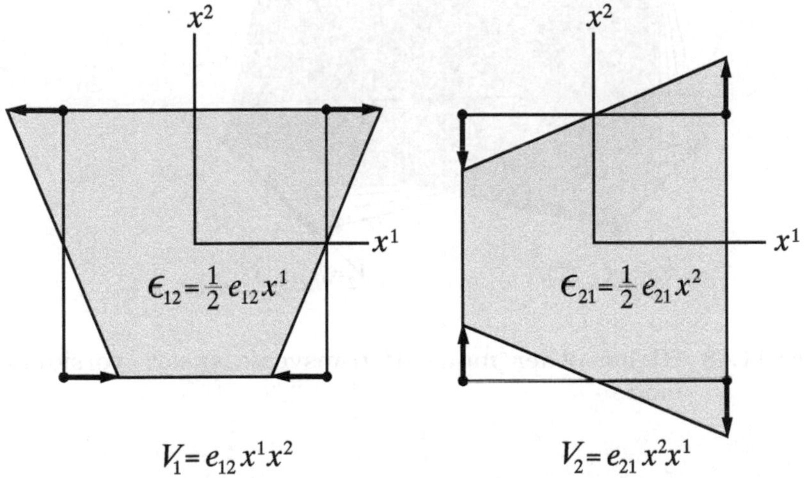

a **b**

Figure 11.16 Higher-order extensional modes

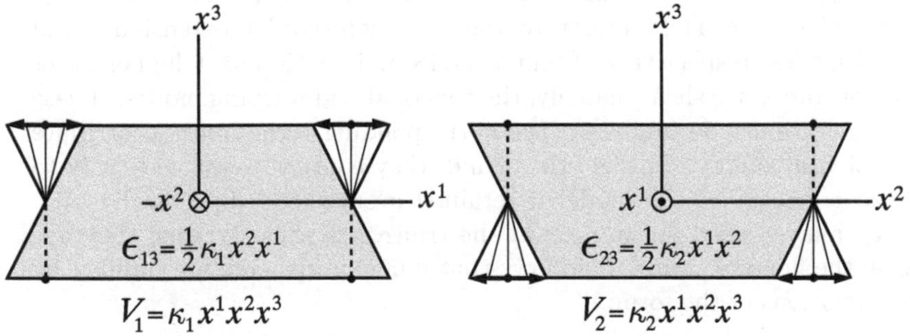

a **b**

Figure 11.17 Higher-order flexural modes

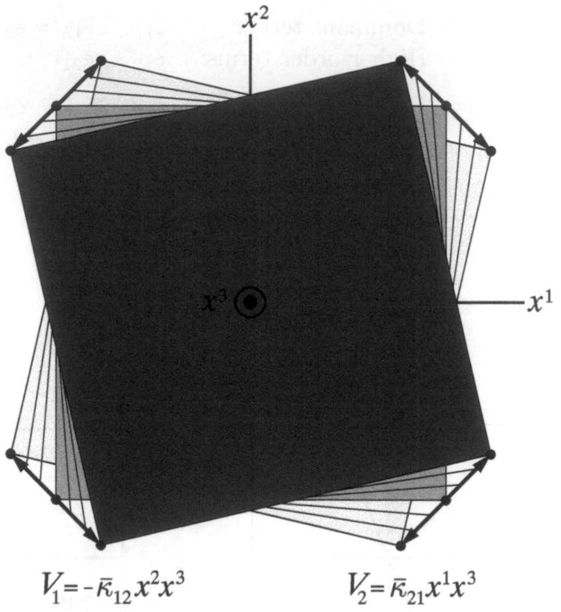

Figure 11.18 Higher-order mode of transverse shear: torsional mode

The eight dominant modes are essential to the continuum theory of the shell, a so-called "shear deformable" shell. The remaining six modes, i.e., $(e_{12}, e_{21}, \tilde{\kappa}^*_{12}, \gamma_3, \kappa_1, \kappa_2)$ constitute higher-order terms. The terms of ϵ_{33} in Table 11.1 are absent as a consequence of the "plane stress" assumption. As always, an appreciation of the mechanical/kinematical behavior is helpful; accordingly the six H.O. modes are illustrated in Figures 11.16 to 11.19. Figures 11.16 and 11.17 illustrate the two higher-order extensional and flexural modes, respectively. Figures 11.18 and 11.19 depict higher-order modes of transverse shear, namely, the torsional and warping modes. These H.O. modes must be inhibited; the corresponding terms must contribute some internal energy. On the other hand, they contribute unnecessarily to the excess energy of the model if retained everywhere. Specifically, $\bar{\kappa}_{11}$, $\bar{\kappa}_{22}$, κ_1, and κ_2 need not appear in the transverse shear strains; they appear in the flexural terms. It follows that our requirements are fulfilled by approximations of the form:

$$\epsilon_{13} = \bar{\epsilon}_{13} + \gamma_{13} x^2, \tag{11.46a}$$

$$\epsilon_{23} = \bar{\epsilon}_{23} + \gamma_{23} x^1. \tag{11.46b}$$

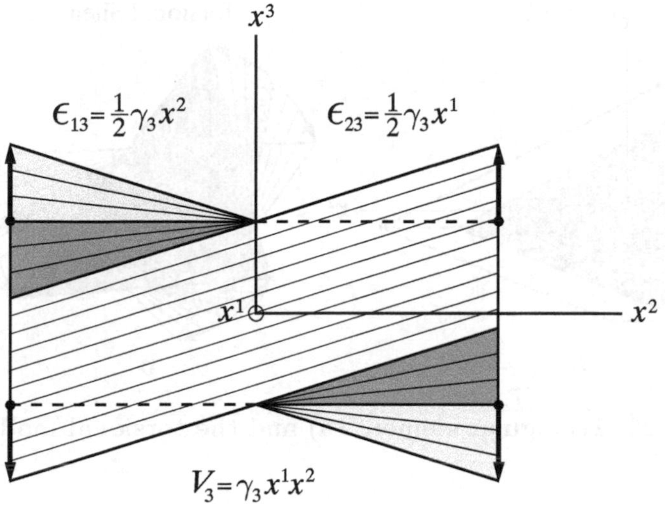

Figure 11.19 Higher-order mode of transverse shear: warping mode

We note too that the modes e_{12} and e_{21} are present in the extensional strains ϵ_{11} and ϵ_{22} as well as the shear strain ϵ_{12}. These terms can be deleted from one, but not both. The omission of such repeated H.O. terms reduces internal energy and improves convergence, when the model is based upon the modified potential.

A simple, yet consistent, element derives from the modified potential of Hu-Washizu. The element of the shell requires a simplified version of approximation (11.30a–c): Specifically, the transverse displacement is simpler as a consequence of the "plane stress" assumption, viz.,

$$V_3 = \overline{V}_3 + \bar{\epsilon}_{3\alpha} x^\alpha + \gamma_3 x^1 x^2 \qquad (\alpha = 1, 2). \tag{11.47}$$

Basically, this simple conforming element possesses 20 degrees of freedom. The transverse shear strains are adequately approximated in the forms (11.46a, b) and the tangential strains (membrane and flexure) in the following forms (see Table 11.1):

$$\epsilon_{11} = \bar{\epsilon}_{11} + \underline{e_{12}} x^2 + \bar{\kappa}_{11} x^3 + \underline{\underline{\kappa_1}} x^2 x^3, \tag{11.48a}$$

$$\epsilon_{22} = \bar{\epsilon}_{22} + \underline{e_{21}} x^1 + \bar{\kappa}_{22} x^3 + \underline{\underline{\kappa_2}} x^1 x^3, \tag{11.48b}$$

$$\epsilon_{12} = \bar{\epsilon}_{12} + \tfrac{1}{2}(\underline{e_{12}} x^1 + \underline{e_{21}} x^2) + \bar{\kappa}'_{12} x^3 + \tfrac{1}{2}(\underline{\underline{\kappa_1}} x^1 + \underline{\underline{\kappa_2}} x^2) x^3. \tag{11.48c}$$

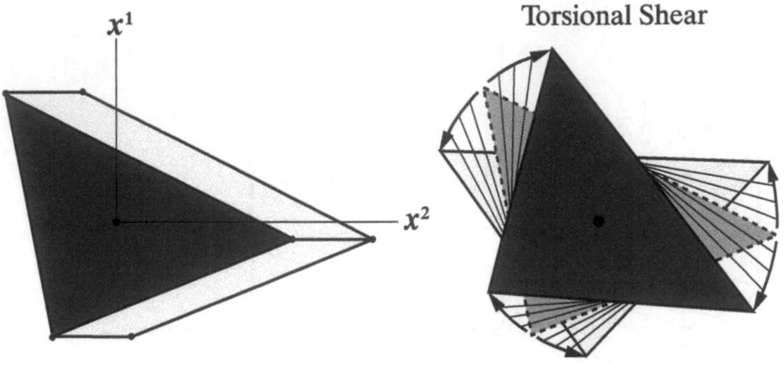

Figure 11.20 Triangular element (a) and the torsional mode (b)

Again, only excessive energy emerges if the underlined terms are retained in both places. These are H.O. terms; they are needed only to inhibit a mode. Convergence is assured if only these are retained in one strain. Here the best choice is not evident, but retention of both can only cause unwarranted "extensional" and "flexural" stiffness. The simple quadrilateral element, as described above, has been employed effectively for certain benchmark problems [267]. An alternative elemental formulation [201] incorporates piecewise constant approximations for stresses and strains which are just sufficient to accommodate the various deformational modes. The formulation leads to further simplifications and also provides a mechanism for the progressive yielding of elasto-plastic elements.

Triangular elements arise, if only at edges as seen in Figure 11.15. A simple conforming element has the displacement components:

$$V_1 = \overline{V}_1 + \bar{\epsilon}_{11}x^1 + \bar{\epsilon}_{12}x^2 + \bar{\epsilon}_{13}x^3 + \bar{\kappa}_{11}x^1x^3 + \bar{\kappa}_{12}x^2x^3, \tag{11.49a}$$

$$V_2 = \overline{V}_2 + \bar{\epsilon}_{21}x^1 + \bar{\epsilon}_{22}x^2 + \bar{\epsilon}_{23}x^3 + \bar{\kappa}_{21}x^1x^3 + \bar{\kappa}_{22}x^2x^3, \tag{11.49b}$$

$$V_3 = \overline{V}_3 + \bar{\epsilon}_{31}x^1 + \bar{\epsilon}_{32}x^2 + \underline{\bar{\epsilon}_{33}x^3 + \gamma_{31}x^1x^3 + \gamma_{32}x^2x^3}. \tag{11.49c}$$

The reader may compare (11.49a–c) with the corresponding components in (11.30a–c). We note that the element has 18 degrees of freedom, embodied in the 18 coefficients. Note too that the triangular element of Figure 11.20a has six corners, hence 18 components of nodal displacement. The coefficients are expressible in terms of those nodal displacements.

The three constants $\overline{V}_i$ represent the rigid translation. A small rigid rotation is expressed by the following constants:

$$\Omega_1 \doteq \tfrac{1}{2}(\bar{\epsilon}_{32} - \bar{\epsilon}_{23}), \qquad \Omega_2 \doteq \tfrac{1}{2}(\bar{\epsilon}_{13} - \bar{\epsilon}_{31}), \qquad \Omega_3 \doteq \tfrac{1}{2}(\bar{\epsilon}_{21} - \bar{\epsilon}_{12}). \tag{11.50a–c}$$

The remaining 12 degrees of freedom comprise the deformational modes, embodied in the strains:

$$\epsilon_{11} = \bar{\epsilon}_{11} + \bar{\kappa}_{11}x^3, \tag{11.51a}$$

$$\epsilon_{22} = \bar{\epsilon}_{22} + \bar{\kappa}_{22}x^3, \tag{11.51b}$$

$$\epsilon_{33} = \underline{\bar{\epsilon}_{33}} + \underline{\gamma_{31}x^1} + \underline{\gamma_{32}x^2}, \tag{11.51c}$$

$$\epsilon_{12} = \tfrac{1}{2}(\bar{\epsilon}_{12} + \bar{\epsilon}_{21}) + \tfrac{1}{2}(\bar{\kappa}_{12} + \bar{\kappa}_{21})x^3, \tag{11.51d}$$

$$\epsilon_{13} = \tfrac{1}{2}(\bar{\epsilon}_{13} + \bar{\epsilon}_{31}) + \tfrac{1}{2}(\bar{\kappa}_{11}x^1 + \bar{\kappa}_{12}x^2) + \underline{\tfrac{1}{2}\gamma_{31}x^3}, \tag{11.51e}$$

$$\epsilon_{23} = \tfrac{1}{2}(\bar{\epsilon}_{23} + \bar{\epsilon}_{32}) + \tfrac{1}{2}(\bar{\kappa}_{21}x^1 + \bar{\kappa}_{22}x^2) + \underline{\tfrac{1}{2}\gamma_{32}x^3}. \tag{11.51f}$$

For the thin shell, we again neglect the transverse extension; $\epsilon_{33} \doteq 0$. Then, the underlined terms in equations (11.51c, e, f) are not present. Additionally, we define

$$\kappa_{12} \equiv \tfrac{1}{2}(\bar{\kappa}_{12} + \bar{\kappa}_{21}), \qquad \kappa \equiv \tfrac{1}{2}(\bar{\kappa}_{12} - \bar{\kappa}_{21}). \tag{11.52a, b}$$

Then, according to (11.52a, b),

$$\bar{\kappa}_{12} = \kappa_{12} + \kappa, \qquad \bar{\kappa}_{21} = \kappa_{12} - \kappa. \tag{11.53a, b}$$

In these notations the shear strains of (11.51d–f) follow:

$$\epsilon_{12} = \tfrac{1}{2}(\bar{\epsilon}_{12} + \bar{\epsilon}_{21}) + \kappa_{12}x^3, \tag{11.54a}$$

$$\epsilon_{13} = \tfrac{1}{2}(\bar{\epsilon}_{13} + \bar{\epsilon}_{31}) + \tfrac{1}{2}(\underline{\bar{\kappa}_{11}x^1} + \underline{\kappa_{12}x^2} + \kappa x^2), \tag{11.54b}$$

$$\epsilon_{23} = \tfrac{1}{2}(\bar{\epsilon}_{23} + \bar{\epsilon}_{32}) + \tfrac{1}{2}(\underline{\kappa_{12}x^1} - \kappa x^1 + \underline{\bar{\kappa}_{22}x^2}). \tag{11.54c}$$

Finally, we observe that the underlined H.O. terms of (11.54b, c) are not needed, as these reflect the flexural modes of (11.51a, b) and (11.51d) or

(11.54a). The appropriate approximation follows:

$$\epsilon_{13} = \tfrac{1}{2}(\bar{\epsilon}_{13} + \bar{\epsilon}_{31}) + \tfrac{1}{2}\kappa x^2, \qquad (11.55a)$$

$$\epsilon_{23} = \tfrac{1}{2}(\bar{\epsilon}_{23} + \bar{\epsilon}_{32}) - \tfrac{1}{2}\kappa x^1. \qquad (11.55b)$$

In summary, the simple element possesses eight essential (homogeneous) deformational modes, i.e.,

$$\bar{\epsilon}_{11}, \quad \bar{\epsilon}_{22}, \quad \tfrac{1}{2}(\bar{\epsilon}_{12} + \bar{\epsilon}_{21}), \quad \bar{\kappa}_{11}, \quad \bar{\kappa}_{22}, \quad \kappa_{12}, \quad \tfrac{1}{2}(\bar{\epsilon}_{13} + \bar{\epsilon}_{31}), \quad \tfrac{1}{2}(\bar{\epsilon}_{23} + \bar{\epsilon}_{32}),$$

and one higher-order mode, κ. The latter is the torsional mode depicted in Figure 11.20b. The deletion of unnecessary terms of (11.54b, c) eliminates unwarranted internal energy, the attendant excessive stiffness, and uncouples that mode from the bending modes $(\bar{\kappa}_{11}, \bar{\kappa}_{22})$. The reader is reminded that the use of the modified principle offers the possibilities of different approximations of stress which in turn enables one to develop an efficient element. The interested reader will find further treatment of the triangular element in ([277], [278]).

In Section 11.10, we described the excessive stiffness which stems from the intrusion of flexural modes in the internal shear energy. Then, we cited the discrete Kirchhoff constraints which inhibit transverse shear. Now, we observe that the four constraints upon the rectangular element of Figure 11.12b suppress the homogeneous modes $\frac{1}{2}(\bar{\epsilon}_{13} + \bar{\epsilon}_{31})$ and $\frac{1}{2}(\bar{\epsilon}_{23} + \bar{\epsilon}_{32})$, and also the higher modes $\bar{\kappa}^*_{12}$ and γ_3 of Figures 11.18 and 11.19. Likewise, the three constraints upon the triangular element of Figure 11.12c are needed to suppress the homogeneous modes $\frac{1}{2}(\bar{\epsilon}_{13} + \bar{\epsilon}_{31})$ and $\frac{1}{2}(\bar{\epsilon}_{23} + \bar{\epsilon}_{32})$ and the one higher mode κ of Figure 11.20b.

Previously, in Section 11.10, we described the discrete Kirchhoff constraints as a means to avoid the excessive shear energy. Experience shows that the present method is equally effective, yet admits transverse shear deformations, hence it applies as well to thicker plates and shells.

11.13.4 Summary

The foregoing commentary is but a synopsis of those mechanical features that distinguish the finite elements of thin bodies. Much as the continuum theories of shells, the discrete approximations are fraught with difficulties which do not arise in their three-dimensional counterparts. The interested reader will find a profusion of literature, many alternative methods, and computational devices. Our intent is to set forth the basic concepts, couched in the context of continuum theory and illustrated by the simplest models. The potential and modified potential provide rational bases for effective models, but also offer insights to the underlying features of these thin elements.

11.14 Supplementary Remarks on Elemental Approximations

We have attempted to reveal the mechanical basis for the approximation of solid bodies via finite elements. Our emphasis is entirely mechanical, yet in practice the solution of the resulting algebraic equations is a mathematical and computational problem. The later aspects require additional tools. Still the physical and mathematical aspects are often analogous. Indeed, one finds in the treatment of the mathematical systems the questions of "consistency" and "stability." The former alludes to the convergence of the discrete equation(s) to their differential counterpart(s). In the examples of Sections 11.6 to 11.8, we see that a rational and consistent treatment of the continuous body and the assembly of elements achieves such consistency, i.e., the difference (algebraic) equations replicate those achieved via the elements and converge to the differential equations. The mathematical "instability" can be traced to a form of mechanical instability: In the examples of Sections 11.10 and 11.11, we identify modes which constitute rigid motions and deformational modes. As noted, from a mechanical perspective each must be inhibited by constraint or stiffness; the latter must produce internal energy. Otherwise, the deformable *body* (or some part) is reduced to a mechanism (unstable); the resulting mathematical system is likewise unstable. *Mathematical* questions of convergence, consistency, and stability are examined and treated in texts on algebraic systems, computational methods and finite elements (see, e.g., [253], [257] [259], [264], [291]).

11.15 Approximation of Nonlinear Paths

The prevailing method of approximating the deformation of bodies is founded on *interpolation*: The determination of nodal values establishes the spline approximation or, stated otherwise, describes the assembly of finite elements. The concepts and procedures, as described in the preceding text, apply as well to linear and nonlinear behaviors. These produce large systems of algebraic equations, linear or nonlinear. With modern computational tools, the solutions of the linear systems are straightforward; time and costs pose the only practical limitations. On the other hand, no straightforward and general means are available for nonlinear systems. Fortunately, our insights to the physical behavior of deformable bodies offer a powerful tool: We know that the responses of *most* mechanical systems exhibit a continuity; a notable exception is the bifurcation of equilibrated paths which often characterize buckling. Stated otherwise, the nonlinear response usually follows a smooth path; physically, a small step (motion plus deformation) alters, but slightly, the response to a subsequent step. Therefore, our basic tool is *extrapolation.*

A discrete model of a structural/mechanical system may be governed by a nonlinear system of algebraic equations. If N^Q denotes one of n nonlinear operators, V_Q denotes one of n variables (e.g., nodal displacements) and λ a loading parameter (or time), then an equation of our system has the form:

$$N^Q(V_Q;\lambda) = 0. \tag{11.56}$$

For a stable state $(\overline{V}_Q;\bar{\lambda})$, an incremental step is approximated by the *linear* equation:

$$\overline{\frac{\partial N^Q}{\partial V_R}}\,\Delta V_R = -\overline{\frac{\partial N^Q}{\partial \lambda}}\,\Delta\lambda. \tag{11.57}$$

Here, the repeated index (e.g., R) implies the summation ($R = 1\dots n$). A succession of such linear steps generates an approximation of the nonlinear solution. However, the approximation does stray from the actual solution. This is illustrated by Figure 11.21; this shows a simple plot of load λ versus displacement V. A linear extrapolation from state M places the approximation at N which is displaced from the correct nonlinear path OMQ. Various schemes can be employed to improve the approximation (see, e.g., [260], [262]): These include, e.g., better estimates of the mean derivative ([292], [293]) and the introduction of higher derivatives [294]. One simple approach follows [252]:

Let the barred coefficients of (11.57) be the values of the reference state $(\overline{V}_Q\,;\,\lambda)$ (e.g., state M in Figure 11.21). Then the solution of (11.57) provides an approximation $(\overline{V}_Q + \Delta V_Q\,;\;\lambda + \Delta\lambda)$ of the nearby equilibrium state (e.g., point N in Figure 11.21). Substituting these values into the left side of (11.56) and denoting the error by R^Q, we obtain for the erroneous approximation of the step (M to N in Figure 11.21)

$$R^Q \equiv N^Q(\overline{V}_Q + \Delta V_Q\,;\,\bar{\lambda} + \Delta\lambda). \tag{11.58}$$

A correction $\Delta\widetilde{V}_Q$ can be obtained by the Newton-Raphson method: If a double bar ($\overline{\overline{}}$) signifies the valuation at the current state (N in Fig-

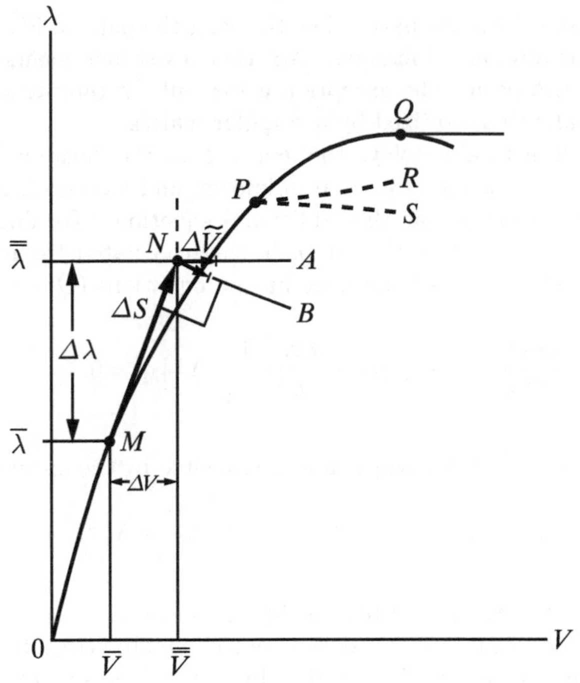

Figure 11.21 Stepwise approximation of a nonlinear path

ure 11.21), then

$$\overline{\overline{\frac{\partial N^Q}{\partial V_R}}} \, \Delta\widetilde{V}_R = -R^Q. \tag{11.59}$$

The procedure can be repeated until the error is reduced to an acceptable amount.

Since the stiffness (slope) can change significantly, uniform steps in loading $\Delta\lambda$ can produce vastly different increments in variables ΔV_Q. Moreover, at a bifurcation point P or a limit point Q in Figure 11.21 there exist neighboring values ΔV_Q at those loads, i.e., $\Delta\lambda = 0$. In short, the process of loading increments *fails*! Then, it is necessary to identify such states and to prescribe an alternative parameter other than the load. An effective alternative was introduced by G. A. Wempner.‡ The concept has been pursued and modified by other authors in subsequent works (cf. [295] to [297]). The essential feature of Wempner's method [192] is the "generalized arc-length," S, defined by:

$$\Delta S^2 \equiv \Delta V_Q \, \Delta V_Q + \Delta\lambda \, \Delta\lambda. \tag{11.60}$$

‡Presented at the ASCE Annual Meeting, Chicago, 1969, published in [192].

The two-dimensional counterpart is length along the path of Figure 11.21. Corresponding to any small increment ΔS, there exist increments ΔV_Q *and* $\Delta\lambda$ along a *smooth* path. The exception arises only at bifurcation points, which are typically characterized by a singular matrix.

The arc-length method employs the length S as the "loading" parameter, treats the load λ as an additional unknown, and incorporates a linear version of equation (11.60) as the additional equation. To illustrate, let us suppose that we seek the Mth step, having calculated the increments $\Delta V_Q]_{M-1}$ and $\Delta\lambda]_{M-1}$. As before, we have n equations ($Q = 1, \ldots n$):

$$\left.\frac{\partial N^Q}{\partial V_R}\right]_M \Delta V_R]_M + \left.\frac{\partial N^Q}{\partial \lambda}\right]_M \Delta\lambda]_M = 0. \tag{11.61}$$

That system with $n+1$ unknowns is now augmented by the linear equation:

$$\Delta V]_{M-1}\,\Delta V]_M + \Delta\lambda]_{M-1}\,\Delta\lambda]_M = \Delta S^2. \tag{11.62}$$

The increment ΔS is the prescribed loading parameter.

Of course, the arc-length method does not eliminate error. It does provide a more uniform progression, and reduces the excessive errors which otherwise occur near limit points such as Q in Figure 11.21.

G. A. Wempner also offered a consistent scheme for corrections. Instead of equation (11.59), which determines the increment $\Delta\widetilde{V}_R$ (see A in Figure 11.21), he proposed a correction $(\Delta\widetilde{V}_R]_M,\ \Delta\tilde{\lambda}_R]_M)$ "orthogonal" to the previous increment (see B in Figure 11.21):

$$\left.\frac{\partial N^Q}{\partial V_R}\right]_{M+1} \Delta\widetilde{V}_R]_M + \left.\frac{\partial N^Q}{\partial \lambda}\right]_{M+1} \Delta\tilde{\lambda}]_M = -N^Q\Big]_{M+1}, \tag{11.63a}$$

$$\Delta V_R]_M\,\Delta\widetilde{V}_R]_M + \Delta\lambda]_M\,\Delta\tilde{\lambda}]_M = 0. \tag{11.63b}$$

Again, the procedure holds on a smooth path. A bifurcation point P is characterized by the vanishing determinant

$$\left|\frac{\partial N^Q}{\partial V_R}\right| = 0. \tag{11.64}$$

At such point, the ensuing step follows the eigenvector of the homogeneous system:

$$\frac{\partial N^Q}{\partial V_R}\Delta V_R = 0. \tag{11.65}$$

Having turned the corner (at P in Figure 11.21), we proceed as before. In many cases, we require only the computations to that critical state and, perhaps, the determination of the stability (versus instability) at that state. The latter question can be resolved by the further examination of postbuckled variations in energy. The all-important question regarding the stability of bifurcation was addressed by W. T. Koiter [107]. In essence, W. T. Koiter pursued the postbuckled path in the direction of the eigenvector. The stability hinges upon the negative/positive variation of potential. His work provided invaluable insight to the snap-buckling of thin shells. The underlying concepts are described in Section 6.11.

References

[1] A. E. H. Love. *A Treatise on the Mathematical Theory of Elasticity.* Dover Publications, New York, fourth edition, 1944. Unabridged and unaltered republication of the fourth (1927) edition, Cambridge University Press.

[2] V. V. Novozhilov. *Foundations of the Nonlinear Theory of Elasticity.* Graylock Press, Rochester, N. Y., 1953. Translated from the first (1948) Russian edition.

[3] A. E. Green and W. Zerna. *Theoretical Elasticity.* Dover Publications, New York, second edition, 1992. Originally published: Oxford University Press, London, first (second) edition, 1954 (1968).

[4] C. Truesdell and R. A. Toupin. The classical field theories. In S. Flügge, editor, *Principles of Classical Mechanics and Field Theory*, volume III of *Encyclopedia of Physics*, part 1, pages 226–793. Springer-Verlag, Berlin, 1960.

[5] C. Truesdell and W. Noll. *The Non-Linear Field Theories of Mechanics.* Springer-Verlag, Berlin, second edition, 1992. Corrected edition of *The Non-Linear Field Theories of Mechanics*, which first appeared as volume 3, part 3 of the *Encyclopedia of Physics*, editor: S. Flügge, 1965.

[6] J. C. H. Gerretsen. *Lectures on Tensor Calculus and Differential Geometry.* P. Noordhoff, Groningen, The Netherlands, first edition, 1962.

[7] A. L. Cauchy. Sur la condesation et la dilatation des corps solides. *Exercices de Mathématique*, 2:60–69, 1827.

[8] G. Green. On the propagation of light in crystallized media. *Trans. Cambridge Phil. Soc.*, 7:121–140, 1841.

[9] A. Nadai. Plastic behavior of metals in the strain-hardening range. *J. Appl. Phys.*, 8:205, 1937.

[10] H. L. Langhaar. *Energy Methods in Applied Mechanics.* John Wiley & Sons, Inc., New York, first edition, 1962.

[11] B. Noble and J. W. Daniel. *Applied Linear Algebra.* Prentice-Hall, Inc., Englewood Cliffs, New Jersey, second edition, 1977.

[12] A. Tucker. *Linear Algebra—An Introduction to the Theory and Use of Vectors and Matrices.* Macmillan Publishing Company, New York, 1993.

[13] J. L. Synge and B. A. Griffith. *Principles of Mechanics.* McGraw-Hill, New York, third edition, 1959.

[14] L. A. Pars. *A Treatise on Analytical Dynamics.* Heinemann Educational Books Ltd., London, 1965.

[15] M. A. Biot. *Mechanics of Incremental Deformations.* John Wiley & Sons, Inc., New York, London, 1965.

[16] R. A. Toupin. The elastic dielectric. *J. Rational Mech. Anal.*, 5: 849–914, 1956. Reprinted in: *Foundations of Elasticity Theory.* Internat. Sci. Rev. Ser., Gordon & Breach, New York, 1965.

[17] B. Fraeijs de Veubeke. *A Course in Elasticity.* Springer-Verlag, New York, 1979. Translation from French.

[18] R. D. Mindlin. Microstructure in linear elasticity. *Arch. Rational Mech. Anal.*, 16:51–78, 1964.

[19] A. E. Green. Micromaterial and multipolar continuum mechanics. *Internat. J. Engrg. Sci.*, 3:533–537, 1965.

[20] R. A. Toupin. Theories of elasticity with couple-stress. *Arch. Rational Mech. Anal.*, 17:85–112, 1964.

[21] A. L. Cauchy. Recherches sur l' équilibre et le mouvement intérieur des corps solides ou fluides, élastiques ou non élastiques. *Bull. Soc. Philomath.*, pages 9–13, 1823.

[22] A. L. Cauchy. Sur les pressions ou tensions supportées en un point donné d' un corps solide par trois plans perpendiculaires entre eux. *Exercices de Mathématique*, 4:30–40, 1829.

[23] G. Piola. La meccanica de' corpi naturalmente estesi trattata col calcolo delle variazioni. In *Opusc. mat. fis di diversi autori*, volume 1,

pages 201–236. Giusti, Milano, 1833.

[24] G. Kirchhoff. Über die Gleichungen des Gleichgewichts eines elastischen Körpers bei nicht unendlich kleinen Verschiebungen seiner Theile. *Sitzungsber. Akad. Wiss. Wien*, 9:762–773, 1852.

[25] E. Trefftz. Zur Theorie der Stabilität des elastischen Gleichgewichts. *Z. Angew. Math. Mech.*, 13:160, 1933.

[26] G. Jaumann. Physik der kontinuierlichen Medien. *Denkschr. Akad. Wiss. Wien*, 95:461–562, 1918.

[27] Barré de Saint-Venant. Mémoire sur la torsion des prismes. *Mémoire des Savants Étrangers*, T. XIV:297–299, 1855.

[28] I. I. Gol'denblat. *Some Problems of the Mechanics of Deformable Media.* P. Noordhoff N. V., Groningen, The Netherlands, 1962. Translation from the Russian by Z. Mróz.

[29] C. Lanczos. *The Variational Principles of Mechanics.* University of Toronto Press, fourth edition, 1970. First edition: 1949, Reprinted: 1974.

[30] M. Mooney. A theory of large elastic deformations. *J. Appl. Phys.*, 11:582–592, 1940.

[31] L. Onsager. Reciprocal relations in irreversible processes, Part I. *Phys. Rev.*, 37(4):405–426, 1931.

[32] L. Onsager. Reciprocal relations in irreversible processes, Part II. *Phys. Rev.*, 38(12):2265–2279, 1931.

[33] J. Meixner. Zur Thermodynamik der irreversiblen Prozesse in Gasen mit chemisch reagierenden, dissoziierenden und anregbaren Komponenten. *Ann. Physik*, 43(4):244–270, 1943.

[34] M. A. Biot. Theory of stress-strain relations in anisotropic viscoelasticity and relaxation phenomena. *J. Appl. Phys.*, 25(11):1385–1391, 1954.

[35] J. Kratochvil and O. W. Dilon Jr. Thermodynamics of elastic-plastic materials as a theory with internal state variables. *J. Appl. Phys.*, 40, 1968.

[36] J. M. Kelly and P. P. Gillis. Thermodynamics and dislocation mechanics. *J. Franklin Inst.*, 297, 1974.

[37] G. A. Wempner and J. Aberson. A formulation of inelasticity from

thermal and mechanical concepts. *Internat. J. Solids Structures*, 12:705–721, 1976.

[38] A. E. Green and J. E. Adkins. *Large Elastic Deformations and Non-Linear Continuum Mechanics.* Oxford University Press, London, first edition, 1960.

[39] G. A. Wempner and G. Clodfelter. Mechanical behavior of filament-wound composite tubes. *J. Composite Materials*, 14:260–268, 1980.

[40] C. L. M. H. Navier. Mémoire sur les lois de l équilibre et du movement des corps solides élastiques. *Mém. Inst. Natl.*, 1824. The paper was presented to the French Academy, May 14, 1821. An excerpt from it was printed in the *Bull. Soc. Philomath. Paris*, 1823, p. 177.

[41] A. L. Cauchy. De la pression ou tension dans un system de points matériels. *Exercices de Mathématique*, 3:213, 1828.

[42] S. D. Poisson. Mémoire sur l' équilibre et le mouvement des corps élastiques. *Mém. de l' Acad., Paris*, 8, 1829.

[43] B. F. E. Clayperon. Mémoire sur l' équilibre intérieur des corps solides homogénes. *J. f. Math.*, 7, 1831.

[44] G. Lamé. *Leçons sur la Théorie Mathématique de l' Élasticité des Corps Solides.* Gauthier-Villars, Paris, second edition, 1866.

[45] I. Todhunter and K. Pearson. *A History of the Theory of Elasticity and of the Strength of Materials from Galilei to the Present Time*, volume 1 (1886), volume 2 (1893). University Press, Cambridge, first edition, 1886–1893. Reprinted New York: Dover Publications, 1960.

[46] S. P. Timoshenko. *History of Strength of Materials.* McGraw-Hill Book Co., New York, first edition, 1953.

[47] J. N. Goodier. A general proof of Saint-Venant's principle. *Phil. Mag. Ser.*, 7(23):607, 1937. Supplementary note 24:325, 1937.

[48] O. Zanaboni. Dimonstrazione generale del principio del de Saint-Venant. *Atti. Accad. Lincei, Roma*, 25:117, 1937.

[49] O. Zanaboni. Valutazione dell errore massimo cui da luogo l' applicazione de principio del de Saint-Venant in un solids isotropo. *Atti. Accad. Lincei, Roma*, 25:595, 1937.

[50] P. Locatelli. Estensione del principio di St. Venant a corpi non perfettamente elastici. *Atti. Accad. Sci. Torino*, 75:502, 1940.

[51] P. Locatelli. Estensione del principio di St. Venant a corpi non perfettamente elastici. *Atti. Accad. Sci. Torino*, 76:125, 1941.

[52] R. von Mises. On Saint-Venant's principle. *Bull. Amer. Math. Soc.*, 51:555, 1945.

[53] E. Sternberg. On Saint-Venant's principle. *Quart. Appl. Math.*, 11:393–402, 1954.

[54] Y. C. Fung. *Foundations of Solid Mechanics*. Prentice-Hall, Inc., Englewood Cliffs, New Jersey, first edition, 1965.

[55] P. W. Bridgman. *The Physics of High Pressure*. International Textbooks of Exact Science. Macmillan, New York, 1931. Peprinted London: Bell, 1952.

[56] P. W. Bridgman. The effect of hydrostatic pressure on the fracture of brittle substances. *J. Appl. Phys.*, 18:246, 1947.

[57] P. W. Bridgman. *Studies in Large Plastic Flow and Fracture with Special Emphasis on the Effects of Hydrostatic Pressure*. McGraw-Hill, New York, 1952.

[58] B. P. Haigh. The strain energy function and the elastic limit. *Engrg.*, 109:158–160, 1920.

[59] H. M. Westergaard. On the resistance of ductile materials to combined stresses in two or three directions perpendicular to one another. *J. Franklin Inst.*, 189:627–640, 1920.

[60] H. Tresca. Mémoire sur l' écoulement des corps solides soumis à les fortes pressions. *Compt. Rend. Acad. Sci. Paris*, 59:754, 1864.

[61] R. von Mises. Mechanik der festen Körper im plastisch-deformablen Zustand. *Nachr. Akad. Wiss. Göttingen, Math.-Phys. Kl. I*, pages 582–592, 1913.

[62] J. Lubliner. *Plasticity Theory*. Macmillan Publishing Company, New York, first edition, 1990.

[63] R. Hill. *The Mathematical Theory of Plasticity*. The Oxford Engineering Science Series. Oxford University Press, London, first edition, 1950. Reprinted 1986.

[64] A. Nadai. *Theory of Flow and Fracture of Solids*, volume 1 (1950) and volume 2 (1963) of *Engineering Societies Monographs*. McGraw-Hill, New York, second (vol. 1) and first (vol. 2) edition, 1950 and 1963.

[65] L. M. Kachanov. *Fundamentals of the Theory of Plasticity.* North-Holland and Mir Publishers, Amsterdam and Moscow, first edition, 1971 and 1974. Translated from the Russian edition (1969).

[66] J. B. Martin. *Plasticity: Fundamentals and General Results.* The MIT Press, Cambridge, Massachusetts, first edition, 1975.

[67] J. Lemaitre and J.-L. Chaboche. *Mechanics of Solid Materials.* Cambridge University Press, Cambridge, 1990. Originally published by Dunod, Paris, 1985.

[68] N. Cristescu. *Dynamic Plasticity.* North-Holland, Amsterdam, 1967.

[69] W. J. Stronge and T. X. Yu. *Dynamic models for structural plasticity.* Springer-Verlag, London, first edition, 1993.

[70] E. H. Lee and R. L. Mallett, editors. *Plasticity of Metals at Finite Strain: Theory, Computation and Experiment; Research Workshop (1981)*, Stanford, 1982. Stanford University, Div. Applied Mechanics.

[71] J. C. Simo and T. J. R. Hughes. *Computational Inelasticity.* Springer-Verlag, New York, Berlin, Heidelberg, first edition, 1998.

[72] I. Doltsinis. *Elements of Plasticity: Theory and Computation.* High Performance Stuctures and Materials. WIT Press, Southampton, first edition, 2000.

[73] H. Hencky. Zur Theorie plastischer Deformationen und der hierdurch im Material hervorgerufenen Nachspannungen. *Z. Angew. Math. Mech.*, 4:323–334, 1924.

[74] F. Schleicher. Die Energiegrenze der Elastizität. *Z. Angew. Math. Mech.*, 5(6):478–479, 1925.

[75] F. Schleicher. Der Spannungszustand an der Fliessgrenze (Plastizitätsbedingung). *Z. Angew. Math. Mech.*, 6(3):199–216, 1926.

[76] R. von Mises. Discussion of the paper by F. Schleicher in Z. Angew. Math. Mech., 6(3):199–216, 1926. *Z. Angew. Math. Mech.*, 6(3):199, 1926.

[77] D. C. Drucker. A more fundamental approach to plastic stress-strain relations. In *Proc. 1st. US Natl. Congr. Appl. Mech., Chicago*, pages 487–491. Edwards Brothers Inc., 1951.

[78] R. von Mises. Mechanik der plastischen Formänderung von Kristallen. *Z. Angew. Math. Mech.*, 8(3):161–185, 1928.

[79] W. Prager. Recent developments in the mathematical theory of plasticity. *J. Appl. Phys.*, 20:235, 1949.

[80] W. T. Koiter. Stress-strain relations, uniqueness and variational theorems for elastic-plastic materials with a singular yield surface. *Quart. Appl. Math.*, 11(3):350–354, 1953.

[81] J. L. Sanders. Plastic stress-strain relations based on linear loading functions. In *Proc. 2nd. U. S. Nat. Congr. Appl. Mech. (Ann Arbor, Mich., 1954)*, pages 455–460, New York, 1955.

[82] J. J. Moreau. Application of convex analysis to the treatment of elastoplastic systems. In P. Germain and B. Nayroles, editors, *Applications of Methods of Functional Analysis to Problems in Mechanics*, page 56. Springer-Verlag, Berlin, 1976.

[83] B. de Saint-Venant. Mémoire sur l' établissement des équations différentielles des mouvements intérieurs opérés dans les corps solides ductiles au delà des limites où l' élasticité pourrait les ramener à leur premier état. *Compt. Rend. Acad. Sci. Paris*, 70:473–480, 1870.

[84] M. Lévy. Mémoire sur les équations générales des mouvements intérieurs des corps solides ductiles au delà des limites où l' élasticité pourrait les ramener à leur premier état. *Compt. Rend. Acad. Sci. Paris*, 70:1323–1325, 1870.

[85] M. Lévy. Mémoire sur équations générales des mouvements intérieurs des corps solides ductiles au delà des limites où l' élasticité pourrait les ramener à leur premier état. *J. Math. Pures Appl.*, 16:369–372, 1871.

[86] L. Prandtl. Spannungsverteilung in plastischen Körpern. In *Proc. 1st. Int. Congr. Appl. Mech., Delft*, pages 43–54, Delft, 1924.

[87] A. Reuss. Berücksichtigung der elastischen Formänderungen in der Plastizitätstheorie. *Z. Angew. Math. Mech.*, 10:266–274, 1930.

[88] A. C. Pipkin and R. S. Rivlin. Mechanics of rate independent materials. *Z. Angew. Math. Phys.*, 16:313–326, 1965.

[89] K. C. Valanis. A theory of plasticity without a yield surface. *Arch. Mech. Stos.*, 23(4):517–551, 1971.

[90] K. C. Valanis. *Irreversible Thermodynamics of Continuous Media: Internal Variable Theory*, volume CISM 77 of *Courses and Lectures/International Centre for Mechanical Sciences (CISM)*. Springer Verlag, Wien, New York, 1972.

[91] K. C. Valanis. Effect of prior deformation on cyclic response of metals. *J. Appl. Mech.*, 1974.

[92] K. C. Valanis. On the foundations of the endochronic theory of viscoplasticity. In *Proc. N. S. F. Workshop: Inelastic Constitutive Equations for Metals*, pages 172–187, 1975.

[93] Z. P. Bažant and P. Bhat. Endochronic theory of inelasticity and failure of concrete. *J. Engrg. Mech. Div., Proc., ASCE*, 102:701–721, 1976.

[94] Z. P. Bažant. *Inelastic Analysis of Structures in Civil Engineering.* John Wiley & Sons, Chichester, 1999.

[95] G. A. Wempner and C.-M. Hwang. A derived theory of elastic-plastic shells. *Internat. J. Solids Structures*, 13:1123–1132, 1977.

[96] R. M. Cristensen. *Theory of Viscoelasticity.* Academic Press, New York, first edition, 1982.

[97] W. Flügge. *Viscoelasticity.* Springer-Verlag, Berlin, second revised edition, 1975. First edition: Blaisdell Publishing Company, Waltham, MA, 1967.

[98] Y. M. Haddad. *Viscoelasticity of Engineering Materials.* Chapman & Hall, London, first edition, 1995.

[99] R. S. Lakes. *Viscoelastic Solids.* CRC Press LLC, Boca Raton, FL, first edition, 1999.

[100] J. C. Maxwell. On the dynamical theory of gases. *Philos. Mag.*, 35:129–145 and 185–217, 1868.

[101] H. Jeffreys. *The Earth.* Cambridge, London, second edition, 1929.

[102] J. L. Lagrange. *Mécanique Analytique.* Mallet-Bachelier, Paris, many later editions, e.g., third (revised by J. Bertrand) edition, 1811–1815. First edition: 1788.

[103] G. Æ. Oravas and L. McLean. Historical development of energetical principles of mechanics, Part I: From Heraclitos to Maxwell. *Appl. Mech. Rev.*, 19(8):647–658, 1966.

[104] G. Æ. Oravas and L. McLean. Historical development of energetical principles of mechanics, Part II: From Cotterill to Prange. *Appl. Mech. Rev.*, 19(11):919–933, 1966.

[105] G. A. Wempner. *Mechanics of Solids with Applications to Thin Bod-*

ies. Sijthoff & Noordhoff, Alphen aan den Rijn, The Netherlands, first edition, 1981. Originally published: McGraw-Hill, Inc., New York, 1973.

[106] G. A. Wempner. *Mechanics of Solids.* PWS Publishing Company, Boston, MA, first edition, 1995.

[107] W. T. Koiter. *Over de Stabiliteit van het Elastisch Evenwicht.* Thesis, Polytechnic Institute Delft, H. J. Paris Publisher, Amsterdam, 1945. English Translation: "On the Stability of Elastic Equilibrium," NASA TT F-10, 833 (1967) and AFFDL TR 70-25 (1970).

[108] J. W. Hutchinson and W. T. Koiter. Postbuckling theory. *Appl. Mech. Revs.*, 23:1353–1366, 1970.

[109] Z. P. Bažant and L. Cedolin. *Stability of Structures: Elastic, Inelastic, Fracture and Damage Theories.* The Oxford Engineering Science Series; 26. Oxford University Press, Inc., New York, 1991.

[110] H. Leipholz. *Stability of Elastic Systems.* Mechanics of elastic stability. Sijthoff & Noordhoff, Alphen aan den Rijn, The Netherlands, 1980.

[111] J. Roorda. Stability of structures with small imperfections. *J. Engrg. Mech. Div. ASCE*, 91(1):87–106, 1965.

[112] W. T. Koiter. Post-buckling analysis of a simple two-bar frame. Report 312, Lab. Engrg. Mech., University of Delft, 1965.

[113] W. T. Koiter. Post-buckling analysis of a simple two-bar frame. In B. Broberg et al., editor, *Recent Progress in Applied Mechanics (Folke Odqvist Volume)*, Sweden, 1967. Almqvist and Wiksell.

[114] H. C. Hu. On some variational principles in the theory of elasticity and plasticity. *Scientia Sinica*, 4(1):33–54, March 1955.

[115] K. Washizu. On the variational principles of elasticity and plasticity. Technical Report 25-18, Aeroelastic and Structures Research Laboratory, Massachusetts Institute of Technology, 1955.

[116] G. A. Wempner. New concepts for finite elements of shells. *Z. Angew. Math. Mech.*, 48:T174–T176, 1968.

[117] G. A. Wempner. Finite elements, finite rotations and small strains of flexible shells. *Int. J. Solids Structures*, 5:117–153, 1969.

[118] R. Weinstock. *Calculus of Variations.* Dover Publications, Inc., New York, 1974. Originally published by the McGraw-Hill Book Company

in 1952.

[119] B. Fraeijs de Veubeke. Displacement and equilibrium models in the finite element method. In O. C. Zienkiewicz and G. S. Holister, editors, *Stress Analysis*, chapter 9, pages 145–197. John Wiley & Sons, London, 1965.

[120] R. Courant and D. Hilbert. *Methods of Mathematical Physics*. John Wiley & Sons, New York, 1989. Translation from German.

[121] B. Fraeijs de Veubeke. A new variational principle for finite elastic displacements. *Int. J. Engrg. Sci.*, 10:745–763, 1972.

[122] G. A. Wempner. The complementary potentials of elasticity, extremal properties, and associated functionals. *J. Appl. Mech.*, 59:568–571, 1992.

[123] E. Hellinger. Die allgemeinen Ansätze der Mechanik der Kontinua. In F. Klein and C. Müller, editors, *Enz. Math. Wiss.*, volume IV, pages 602–694. Teubner Verlagsgesellschaft, Leipzig, 1914.

[124] E. Reissner. On a variational theorem in elasticity. *J. Math. Phys.*, 29(2):90–95, 1950.

[125] A. Castigliano. *Théorie de l' équilibre des systèmes élastiques et ses applications*. A. F. Negro, Turin, 1879.

[126] M. M. Vainberg. *Variational Methods for the Study of Nonlinear Operators*. Holden-Day Series in Mathematical Physics. Holden-Day, San Francisco, first edition, 1964.

[127] W. R. Hamilton. *The Mathematical Papers of Sir W. R. Hamilton, II, Dynamics*. Cambridge Press, 1940.

[128] N. I. Muskhelishvili. *Some Basic Problems of the Mathematical Theory of Elasticity*. P. Noordhoff Ltd., Groningen-The Netherlands, third, revised and augmented edition, 1953. Translated from the Russian by J. R. M. Radok.

[129] I. S. Sokolnikoff. *Mathematical Theory of Elasticity*. McGraw-Hill, Inc., New York, second edition, 1956. Reprint edition: 1983 by Krieger Publishing Company-Malabar, Florida (Original edition: 1946 by McGraw-Hill, New York).

[130] C. E. Pearson. *Theoretical Elasticity*. Harvard University Press, Cambridge, Mass., 1959.

[131] V. V. Novozhilov. *Theory of Elasticity*. Pergamon Press Ltd., Oxford,

London, first edition, 1961. Translated from the Russian by J. K. Lusher. Original volume: 1958 Leningrad, Sudpromgiz.

[132] M. Filonenko-Borodich. *Theory of Elasticity*. P. Noordhoff N. V., Groningen-The Netherlands, 1965. Translated from the Russian by M. Konayeva.

[133] S. P. Timoshenko and J. N. Goodier. *Theory of Elasticity*. McGraw-Hill Book Company, New York, third edition, 1970.

[134] H. Schäfer. Die Spannungsfunktionen des drei-dimensionalen Kontinuums und des elastischen Körpers. *Z. Angew. Math. Mech.*, 33:356–362, 1953.

[135] G. V. Kolosov. *On an Application of Complex Function Theory to a Plane Problem of the Mathematical Theory of Elasticity*. Doctoral thesis, Yuriev, 1909.

[136] I. S. Sokolnikoff and R. M. Redheffer. *Mathematics of Physics and Modern Engineering*. McGraw-Hill Book Company, New York, second edition, 1966.

[137] W. Kaplan. *Advanced Calculus*. Addison-Wesley Publishing Company, Inc., Cambridge, first edition, 1952. second printing 1963.

[138] A. J. McConnell. *Applications of Tensor Calculus*. Dover Publications, New York, first edition, 1957.

[139] I. S. Sokolnikoff. *Tensor Analysis: Theory and Applications to Geometry and Mechanics of Continua*. Applied Mathematics Series. John Wiley & Sons, Inc., New York, second edition, 1964.

[140] V. V. Novozhilov. *Thin Shell Theory*. P. Noordhoff Ltd., Groningen, The Netherlands, second edition, 1964. Translated from the second Russian edition by P. G. Lowe (editor: J. R. M. Radok).

[141] A. L. Cauchy. Sur l' équilibre et le mouvement d' une plaque solide. *Exercices de Mathématique*, 3, 1828.

[142] P. M. Naghdi. Foundations of Elastic Shell Theory. In I. N. Sneddon and R. Hill, editors, *Progress in Solid Mechanics*, volume IV. North-Holland Publishing Company, Amsterdam, 1963.

[143] P. M. Naghdi. The Theory of Shells and Plates. In S. Flügge, editor, *Principles of Classical Mechanics and Field Theory*, volume VI of *Encyclopedia of Physics*, part A2, pages 425–640. Springer-Verlag, Berlin, 1972.

[144] E. R. A. Oliveira. A theory of shells involving moments of arbitrary order. Technical report, Nat. Lab. Civil Eng., Technical University Lisbon, Lisbon, 1967.

[145] E. R. A. Oliveira. A further research on the analysis of shells using moments of arbitrary order. Technical report, Nat. Lab. Civil Eng., Technical University Lisbon, Lisbon, 1967.

[146] G. A. Wempner. Invariant multi-couple theory of shells. *J. Engrg. Mech. Div. ASCE*, 98:1397–1415, 1972.

[147] A. I. Soler. Higher-order theories for structural analysis using Legendre polynomial expansions. *J. Appl. Mech.*, 36(4):757–762, 1969.

[148] M. W. Johnson and E. Reissner. On the foundations of the theory of thin elastic shells. *J. Math. and Phys.*, 37:375–392, 1958.

[149] F. I. Niordson. An asymptotic theory for circular cylindrical shells. Report, The Danish Center for Applied Mathematics and Mechanics 575, Techn. Univ., Dep. of Solid Mechanics, Lyngby, 1998.

[150] F. I. Niordson. An asymptotic theory for spherical shells. Report, The Danish Center for Applied Mathematics and Mechanics 638, Techn. Univ., Dep. of Solid Mechanics, Lyngby, 2000.

[151] H. Aron. Das Gleichgewicht und die Bewegung einer unendlich dünnen, beliebig gekrümmten, elastischen Schale. *Z. Reine Angew. Math.*, 78:136, 1874.

[152] G. R. Kirchhoff. *Vorlesungen über mathematische Physik: Mechanik.* Leipzig, third edition, 1883. First edition: 1876.

[153] A. E. H. Love. The small free vibrations and deformations of a thin elastic shell. *Phil. Trans. Roy. Soc. London*, ser. A 179:491–546, 1888.

[154] A. I. Lur'e. The general theory of thin elastic shells (in Russian). *Prikl. Mat. Mekh.*, IV(2):7–34, 1940.

[155] J. L. Synge and W. Z. Chien. The intrinsic theory of elastic plates and shells. In *Th. von Kármán Anniversary Album*, pages 103–120. California Inst. of Technology, 1941.

[156] J. L. Sanders. An improved first-approximation theory for thin shells. NASA-Report 24, National Aeronautics and Space Administration, Washington, D.C., 1959.

[157] J. L. Sanders. Nonlinear theories for thin shells. *Quart. Appl. Math.*, 21:21–36, 1963.

[158] W. T. Koiter. A consistent first approximation in the general theory of thin elastic shells. In *Proc. IUTAM-Symposium on the Theory of Thin Elastic Shells (Delft, 1959)*, pages 12–33, Amsterdam, 1960. North-Holland Publishing Company.

[159] R. W. Leonard. *Nonlinear First-Approximation Thin Shell and Membrane Theory.* Thesis, Virginia Polytechnic Institute, Virginia, 1961. National Aeronautics and Space Administration, Langley Field, Va.

[160] F. I. Niordson. *Introduction to Shell Theory.* Technical University of Denmark, Lyngby, first edition, 1980.

[161] W. T. Koiter. On the nonlinear theory of thin elastic shells, Part I: Introductory sections, Part II: Basic shell equations, Part III: Simplified shell equations. In *Proc. Konink. Nederl. Akad. Wetensch.*, volume Ser. B, 69, No. 1, pages 1–54, Amsterdam, 1966.

[162] J. G. Simmonds and D. A. Danielson. Nonlinear theory of shells with a finite rotation vector. In *Proc. Konink. Nederl. Akad. Wetensch.*, volume 73, pages 460–478, Amsterdam, 1970.

[163] W. Pietraszkiewicz. *Geometrically Nonlinear Theories of Thin Elastic Shells*, volume 12 (number 1) of *Advances in Mechanics.* 1989. pages 52–130.

[164] E. Reissner. The effect of transverse shear deformations on the bending of elastic plates. *J. Appl. Mech.*, 12:A69–A77, 1945.

[165] R. D. Mindlin. Influence of rotary inertia and shear on flexural motions of isotropic, elastic plates. *J. Appl. Mech.*, 18:31–38, 1951.

[166] G. A. Wempner. Mechanics of shells in the age of computation. In K. S. Pister, editor, *Structural Engineering and Structural Mechanics*, pages 92–120. Prentice-Hall, Inc., Englewood Cliffs, N. J., 1980.

[167] G. A. Wempner. Mechanics and finite elements of shells. *Appl. Mech. Rev.*, 42(5):129–142, 1989. ASME Book No. AMR055.

[168] S. P. Timoshenko and S. Woinowsky-Krieger. *Theory of Plates and Shells.* McGraw-Hill Book Company, Inc., New York, second edition, 1959.

[169] A. L. Gol'denveizer. *Theory of Elastic Thin Shells.* International Series of Monographs on Aeronautics and Astronautics. Pergamon Press, New York, Oxford, 1961. Translation from the 1953 Russian edition (edited by G. Herrmann).

[170] K. Girkmann. *Flächentragwerke.* Springer-Verlag, Wien, sixth edi-

tion, 1963. First edition: Springer-Verlag, 1946.

[171] V. Z. Vlasov and N. N. Leontev. *Beams, Plates and Shells on Elastic Foundations.* Israel Program for Scientific Translations, Jerusalem, 1966. Translation from Russian.

[172] H. Kraus. *Thin Elastic Shells: An Introduction to the Theoretical Foundations and the Analysis of their Static and Dynamic Behavior.* John Wiley & Sons, Inc., New York, first edition, 1967.

[173] W. Flügge. *Stresses in Shells.* Springer-Verlag, New York, second edition, 1973.

[174] L. H. Donnell. *Beams, Plates and Shells.* Engineering Society Monographs. McGraw-Hill, New York, 1976.

[175] S. Lukasiewicz. *Local Loads in Plates and Shells*, volume 4 of *Monographs and Textbooks on Mechanics of Solids and Fluids.* Sijthoff & Noordhoff, Alphen aan den Rijn, first edition, 1979.

[176] J. Mason. *Variational, Incremental and Energy Methods in Solid Mechanics and Shell Theory*, volume 4 of *Studies in Applied Mechanics.* Elsevier Scientific Publishing Company, Amsterdam, first edition, 1980.

[177] Y. Başar and W. B. Krätzig. *Mechanik der Flächentragwerke.* Friedr. Vieweg & Sohn Verlagsgesellschaft mbH, Braunschweig, first edition, 1985.

[178] E. L. Axelrad. *Theory of Flexible Shells*, volume 28 of *North-Holland Series in Applied Mathematics and Mechanics.* North-Holland, Amsterdam, first edition, 1987.

[179] P. L. Gould. *Analysis of Shells and Plates.* Prentice Hall, Upper Saddle River, NJ, 1999. Originally published: Springer Verlag, New York, 1988.

[180] Y. Başar and W. B. Krätzig. *Theory of Shell Structures*, volume 258 of *Fortschritt-Berichte, VDI Reihe 18 (Mechanik/Bruchmechanik).* VDI Verlag GmbH, Düsseldorf, first edition, 2000.

[181] A. Libai and J. G. Simmonds. *The Nonlinear Theory of Elastic Shells: One Spatial Dimension.* Academic Press, Inc., Boston, first edition, 1988.

[182] R. Valid. *The Nonlinear Theory of Shells Through Variational Principles.* John Wiley & Sons, Chichester, New York, first edition, 1995.

[183] P. G. Ciarlet. *Theory of Shells*, volume 29 of *Studies in Mathematics and its Applications*. Elsevier, Amsterdam, first edition, 2000.

[184] S. A. Ambartsumyan. *Theory of Anisotropic Shells*, volume F-118 of *NASA Technical Translations*. National Aeronautics and Space Administration, Washington, D. C., 1964.

[185] L. Librescu. *Elastostatics and Kinetics of Anisotropic and Heterogeneous Shell-Type Structures*. Noordhoff International Publishing, Leyden, The Netherlands, first edition, 1975.

[186] A. W. Leissa. *Vibration of Shells*, volume SP 288 of *NASA Special Publications*. Scientific and Technical Information Office, National Aeronautics and Space Administration, Washington, D. C., 1973.

[187] W. Soedel. *Vibrations of Shells and Plates*, volume 86 of *Mechanical Engineering*. Dekker, New York, second edition, 1993.

[188] K. C. Le. *Vibrations of Shells and Rods*. Springer-Verlag, Berlin, 1999.

[189] W. Olszak and A. Sawczuk. *Inelastic Behaviour in Shells*. Noordhoff Series of Monographs and Textbooks on Pure and Applied Mathematics. Noordhoff, Groningen, first edition, 1967.

[190] A. Sawczuk. *Mechanics and Plasticity of Structures*. English edition first published in coedition between Ellis Horwood, Ltd. (Chichester) and PWN-Polish Scientific Publishers (Warsaw), 1989. Editor J. Sokół-Supel, Translation Editor J. M. Alexander.

[191] M. A. Save, C. E. Massonnet, and G. de Saxce. *Plastic Analysis and Design of Plates, Shells, and Disks*, volume 43 of *North-Holland Series in Applied Mathematics and Mechanics*. Elsevier, Amsterdam, 1997.

[192] G. A. Wempner. Discrete approximations related to nonlinear theories of solids. *Internat. J. Solids Structures*, 7:1581–1599, 1971.

[193] W. Thomson and P. G. Tait. *Treatise on Natural Philosophy*, volume 1, part 2, pages 645–648. Cambridge, first edition, 1867.

[194] W. T. Koiter. On the dynamic boundary conditions in the theory of thin shells. In *Proc. Konink. Nederl. Akad. Wetensch.*, volume 67, Amsterdam, 1964.

[195] G. A. Wempner. The boundary conditions for thin shells and their physical meaning. *Z. Angew. Math. Mech.*, 47(2):136, 1967.

[196] W. T. Koiter and J. G. Simmonds. Foundations of shell theory. In E. Becker and G. M. Mikhailov, editors, *Proc. 13th Internat. Congr. Theoret. and Appl. Mech., Moscow (1972)*, pages 150–176, Berlin, 1973. Springer Verlag.

[197] G. A. Wempner. The deformation of thin shells. In *Developments in Theoretical and Applied Mechanics, Proc. of the Third Southeastern Conference held in Columbia, S. Columbia (1966)*, volume 3, pages 245–254, Oxford & New York, 1967. Pergamon Press.

[198] A. I. Lur'e. On the static geometric analogue of shell theory. In *Problems of Continuum Mechanics (N. I. Muskhelishvili Anniversary Volume)*, pages 267–274. S.I.A.M., 1961.

[199] A. L. Gol'denveizer. The equations of the theory of shells (in Russian). *Prikl. Mat. Mekh.*, IV(2):35–42, 1940.

[200] P. G. Hodge. *Plastic Analysis of Structures*. McGraw-Hill, New York, 1959.

[201] D. Talaslidis and G. A. Wempner. A simple finite element for elastic-plastic deformations of shells. *Comput. Methods Appl. Mech. Engnrg.*, 34:1051–1064, 1982.

[202] G. A. Wempner and C.-M. Hwang. Elastoplasticity of the club sandwich. *Internat. J. Solids Structures*, 16:161–165, 1980.

[203] G. A. Wempner and C.-M. Hwang. A simple model of elastic-plastic plates. *Internat. J. Solids Structures*, 20(1):77–80, 1984.

[204] A. A. Ilyushin. *Plasticité*. Eyrolles, Paris, 1956. Translated from the Russian edition: *Plastichnost*, Gostekhizdat, Moscow, 1969.

[205] V. I. Rozenblyum. An approximate theory of the equilibrium of plastic shells (in Russian). *Prikl. Mat. Mekh.*, 18:289–302, 1954.

[206] A. Robinson. A comparison of yield surfaces for thin shells. *Internat. J. Mech. Sci.*, 15:345–354, 1971.

[207] A. Robinson. The effect of transverse shear stresses on the yield surface for thin shells. *Internat. J. Solids Structures*, 9:819–828, 1973.

[208] M. A. Crisfield. Approximate yield criterion for thin steel shells. Technical Report 658, Transport and Road Research Laboratory, Crowthorne, Berkshire, G. Britain, 1974.

[209] M. P. Bieniek and J. R. Funaro. Elasto-plastic behavior of plates and shells. Technical Report DNA 3954T, Defense Nuclear Agency,

Weidlinger Associates, New York, 1976.

[210] J. W. Cohen. The inadequacy of the classical stress-strain relations for the right helicoidal shell. In *Proc. IUTAM Symp. on Thin Shell Theory (1959)*, page 415, 1960.

[211] E. Reissner. On twisting and stretching of helicoidal shells. In *Proc. IUTAM Symp. on Thin Shell Theory (1959)*, pages 434–466, 1960.

[212] E. Reissner and F. Y. M. Wang. On axial extension and torsion of helicoidal shells. *J. Math. Phys.*, 47(1):1–31, 1968.

[213] E. Reissner. On the form of variationally derived shell equations. *J. Appl. Mech.*, 31:233–238, 1964.

[214] S. P. Timoshenko and J. M. Gere. *Theory of Elastic Stability*. Engineering Societies Monographs. McGraw-Hill, New York, second edition, 1961.

[215] B. O. Almroth, A. M. C. Holmes, and D. O. Brush. An experimental study of the buckling of cylinders under axial compression. *J. Exp. Mech.*, 4:263, 1964.

[216] J. Kempner. Postbuckling behavior of axially compressed circular cylindrical shells. *J. Aeron. Sci.*, 21:329, 1954.

[217] Th. von Kármán and H. S. Tsien. The buckling of thin cylindrical shells under axial compression. *J. Aero. Sci.*, 8(6):303–312, 1941.

[218] B. O. Almroth. Postbuckling behavior of axially compressed circular cylinders. *AIAA J.*, 1:630, 1963.

[219] L. H. Donnell. Effect of imperfections on buckling of thin cylinders under external pressure. *J. Appl. Mech. (ASME)*, 23(4):569, 1956.

[220] W. T. Koiter. General equations of elastic stability for shells. In *Donnell Anniversary Volume*, pages 185–225. Houston, Texas, 1967.

[221] B. Budiansky and J. W. Hutchinson. A survey of some buckling problems. Report SM-8, Harvard University, 1966.

[222] B. Budiansky and J. W. Hutchinson. A survey of some buckling problems. *AIAA J.*, 4:1505–1510, 1966.

[223] J. W. Hutchinson. Imperfection sensitivity of externally pressurized spherical shells. *J. Appl. Mech. (ASME)*, 34:49–55, 1967.

[224] G. A. Wempner and J. L. Baylor. General theory for sandwich shells with dissimilar facings. *Internat. J. Solids Structures*, 1(2):157–177,

1965.

[225] G. A. Wempner. Theory of moderately large deflections of sandwich shells with dissimilar facings. *Internat. J. Solids Structures*, 3: 367–382, 1967.

[226] E. Reissner. Reflections on the theory of elastic plates. *Appl. Mech. Rev.*, 38:1453–1464, 1985.

[227] E. Ramm, M. Braun, and M. Bischoff. Higher order nonlinear shell formulations: Theory and application. *IASS Bulletin*, 36(119):145–152, 1995.

[228] Kh. M. Mushtari and K. Z. Galimov. *Non-Linear Theory of Thin Elastic Shells*. Academy of Sciences, USSR, Kazan Branch. Tatknigoizdat, Kazan, 1957. Translated by J. Morgenstern et al. Also available as NASA-Report (No: N19980231031; NASA-TT-F-62), National Aeronautics and Space Administration, Washington, D.C.

[229] W. Olszak and A. Sawczuk. Inelastic response of thin shells. In W. Olszak, editor, *Thin Shell Theory—New Trends and Applications*, volume 240 of *CISM Courses and Lectures*, chapter 5, pages 211–241. Springer-Verlag, Berlin, 1980.

[230] M. B. Rubin. *Cosserat Theories: Shells, Rods and Points*, volume 79 of *Solid Mechanics and its Applications*. Kluwer Academic Publ., Dordrecht, first edition, 2000.

[231] E. I. Grigolyuk and V. M. Tolkachev. *Contact Problems in the Theory of Plates and Shells*. Mir Publishers, Moskow, 1987.

[232] D. O. Brush and B. O. Almroth. *Buckling of Bars, Plates and Shells*. McGraw-Hill, New York, first edition, 1975.

[233] C. R. Calladine. *Theory of Shell Structures*. Cambridge University Press, London, 1983.

[234] L. Kollár and E. Dulácska. *Buckling of Shells for Engineers*. John Wiley & Sons, Chichester, England, first edition, 1984.

[235] J. R. Vinson. *The Behavior of Shells Composed of Isotropic and Composite Materials*, volume 18 of *Solid Mechanics and its Applications*. Kluwer, Dordrecht, 1993.

[236] A. Bogdanovich. *Non-Linear Dynamic Problems for Composite Cylindrical Shells*. Elsevier Applied Science, London, New York, 1993. Translated from the Russian, Translation Editor C. W. Bert.

[237] Lord Rayleigh (J. W. Strutt). On the theory of resonance. In *Trans. Roy. Soc. (London)*, vol. A161, pages 77–118, 1870.

[238] Lord Rayleigh (J. W. Strutt). Some general theorems relating to vibrations. In *Proc. London Math. Soc.*, vol. 4, pages 357–368, 1873.

[239] W. Ritz. Über eine neue Methode zur Lösung gewisser Variationsprobleme der mathematischen Physik. *J. Reine Amgew. Math.*, 135:1–61, 1909.

[240] R. V. Southwell. *Relaxation Methods in Theoretical Physics.* The Clarendon Press, Oxford, 1946.

[241] B. A. Finlayson. *The Method of Weighted Residuals and Variational Principles*, volume 87 of *Mathematics in Science and Engineering.* Academic Press, Inc., New York, first edition, 1972.

[242] B. G. Galerkin. Series solution of some problems of elastic equilibrium of rods and plates (in Russian). *Vestn. Inzh. Tech.*, 19:897–905, 1915.

[243] A. Hrenikoff. Solution of problems in elasticity by the framework method. *J. Appl. Mech.*, 8:169–175, 1941.

[244] R. Courant. Variational methods for the solution of problems of equilibrium and vibration. *Bull. Amer. Math. Soc.*, 49:1–23, 1943.

[245] D. McHenry. A lattice analogy for the solution of plane stress problems. *J. Inst. Civil Engrg.*, 21:59–82, 1943.

[246] J. H. Argyris. Energy theorems and structural analysis. *Aircraft Engrg.*, Vol. 26 (Oct.–Nov.) and Vol. 27 (Feb.–May), 1954 and 1955.

[247] M. J. Turner, R. W. Clough, H. C. Martin, and L. J. Topp. Stiffness and deflection analysis of complex structures. *J. Aeronautical Sci.*, 23:803–823, 1956.

[248] R. W. Clough. The finite element in plane stress analysis. In *Proc. 2nd ASCE Conf. on Electronic Computation*, Pittsburgh, Pa., 1960.

[249] I. C. Taig. Structural analysis by the matrix displacement method. Report No. S017, Engl. Electric Aviation, 1961.

[250] B. M. Irons. Engineering application of numerical integration in stiffness methods. *AIAA J.*, 14:2035–2037, 1968.

[251] J. H. Argyris, I. Fried, and D. W. Scharpf. The TET 20 and the TEA 8 elements for the matrix displacement method. *Aero. J.*, 72:618–625, 1968.

[252] J. H. Argyris. Continuua and discontinua. In J. S. Przemieniecki et al., editor, *Proc. 1st Conf. Matrix Methods in Structural Analysis (1965)*, pages 11–189, Ohio, 1966. Wright-Patterson AFB, AFFDL-TR-66-80, Air Force Flight Dynamics Laboratory.

[253] G. Strang and G. J. Fix. *An Analysis of the Finite Element Method.* Prentice-Hall, Inc., Englewood Cliffs, N. J., 1973.

[254] R. H. Gallagher. *Finite Element Analysis: Fundamentals.* Prentice Hall, Inc., Englewood Cliffs, New Jersey, 1975.

[255] B. Irons and S. Ahmad. *Techniques of Finite Elements.* Ellis Horwood Limited, Chichester, 1980.

[256] J. H. Argyris and H.-P. Mlejnek. *Die Methode der Finiten Elemente in der elementaren Strukturmechanik, Band 1: Verschiebungsmethode, Band 2: Kraft- und gemischte Methoden, Nichtlinearitäten, Band 3: Einführung in die Dynamik.* Friedr. Vieweg & Son, Braunschweig, Wiesbaden, 1987.

[257] T. J. R. Hughes. *The Finite Element Method—Linear Static and Dynamic Analysis.* Prentice-Hall, Inc., Englewood Cliffs, New Jersey, 1987.

[258] R. D. Cook, D. S. Malkus, and M. E. Plesha. *Concepts and Applications of Finite Element Analysis.* John Wiley & Sons, New York, third edition, 1989.

[259] B. Szabó and I. Babuška. *Finite Element Analysis.* John Wiley & Sons, Inc., New York, 1991.

[260] E. Hinton (Editor). *Introduction to Nonlinear Finite Element Analysis.* NAFEMS, Birniehill, East Kilbride, Glasgow, 1992.

[261] R. H. MacNeal. *Finite Elements: Their Design and Performance.* Marcel Dekker, Inc., New York, first edition, 1994.

[262] M. A. Crisfield. *Non-linear Finite Element Analysis of Solids and Structures, Volume 1: Essentials, Volume 2: Advanced Topics.* John Wiley & Sons Ltd, Chichester, first edition, 1991 and 1997.

[263] K.-J. Bathe. *Finite Element Procedures.* Prentice-Hall, Inc., Englewood Cliffs, New Jersey, second edition, 1996.

[264] O. C. Zienkiewicz and R. L. Taylor. *The Finite Element Method, Volume 1: The Basis, Volume 2: Solid Mechanics.* Butterworth-Heinemann, Oxford, fifth edition, 2000. First published in 1967 by McGraw-Hill.

[265] T. Belytschko, W.-K. Liu, and B. Moran. *Nonlinear Finite Elements for Continua and Structures.* John Wiley & Sons, Chichester, 2000.

[266] M. W. Johnson and R. W. McLay. Convergence of the finite element method in the theory of elasticity. *J. Appl. Mech.*, 35:274–278, 1968.

[267] G. A. Wempner, D. Talaslidis, and C.-M. Hwang. A simple and efficient approximation of shells via finite quadrilateral elements. *J. Appl. Mech., ASME*, 49:115–120, 1982.

[268] G. A. Wempner, J. T. Oden, and D. A. Kross. Finite element analysis of thin shells. *J. Engrg. Mech. Div., ASCE*, 94(EM6):1273–1294, 1968.

[269] J. A. Stricklin et al. A rapidly converging triangular plate element. *J. AIAA*, 7:180–181, 1969.

[270] G. S. Dhatt. Numerical analysis of thin shells by curved triangular elements based on diskrete-Kirchhoff hypothesis. In W. R. Rowan and R. M. Hackett, editors, *Proc. Symp. Appl. of F. E. M. in Civil Engrg., Nashville*, pages 255–277. Vanderbilt Univ., Springer-Verlag, 1969.

[271] B. M. Irons. The semiloof shell element. In D. G. Ashwell and R. H. Gallagher, editors, *Finite Elements for Thin and Curved Members*, pages 197–222, London, N. York, 1976. John Wiley & Sons.

[272] O. C. Zienkiewicz, J. Too, and R. L. Taylor. Reduced integration technique in general analysis of plates and shells. *Internat. J. Numer. Methods Engrg.*, 3:275–290, 1971.

[273] S. F. Pawsey and R. W. Clough. Improved numerical integration of thick slab finite elements. *Internat. J. Numer. Methods Engrg.*, 3:575–586, 1971.

[274] D. S. Malkus and T. J. R. Hughes. Mixed finite element methods in reduced and selective integration techniques: A unification of concepts. *Comput. Methods Appl. Mech. Engnrg.*, 15:63–81, 1978.

[275] G. A. Wempner. Finite-element modeling of solids via the Hu-Washizu functional. *Comput. Mech. Engrg.*, 2:67–75, 1983.

[276] C. D. Pionke. *Convergence of Finite Elements Based on the Hu-Washizu Variational Theorem with Minimal Compatibility.* Doctoral thesis, Georgia Institute of Technology, Georgia, Atlanta, 1993.

[277] D. Talaslidis and G. A. Wempner. The linear isoparametric triangular element: Theory and application. *Comput. Methods Appl. Mech.*

Engnrg., 103:375–397, 1993.

[278] C. Karakostas, D. Talaslidis, and G. A. Wempner. Triangular C^0 bending elements based on the Hu-Washizu principle and orthogonality conditions. *Internat. J. Numer. Methods Engrg.*, 36:181–200, 1993.

[279] B. M. Irons and A. Razzaque. Experience with the patch test for convergence of finite elements. In A. K. Aziz, editor, *Mathematical Foundations of the Finite Element Method with Applications to Partial Differential Equations*, pages 557–587, New York, 1972. Academic Press.

[280] B. Fraeijs de Veubeke. Variational principles and the patch test. *Internat. J. Numer. Methods Engrg.*, 8:783–801, 1974.

[281] G. Sander and P. Beckers. Delinquent finite elements for shells. In J. Robinson, editor, *Proc. World Congress on Finite Element Methods in Structural Mechanics*, Bournemouth, 1975.

[282] G. Sander and P. Beckers. The influence of the choice of connectors in the finite element method. *Internat. J. Numer. Methods Engrg.*, 11:1491–1505, 1977.

[283] E. L. Wilson, R. L. Taylor, W. P. Doherty, and J. Ghaboussi. Incompatible displacement models. In S. T. Fenves et al., editor, *Numerical and Computational Methods in Structural Mechanics*, pages 43–57, New York, 1973. Academic Press, Inc.

[284] R. L. Taylor, P. J. Beresford, and E. L. Wilson. A non-conforming element for stress analysis. *Internat. J. Numer. Methods Engrg.*, 10:1211–1220, 1976.

[285] T. H. H. Pian. Derivation of element stiffness matrices by assumed stress distributions. *AIAA J.*, 2:1332–1336, 1964.

[286] T. H. H. Pian and P. Tong. Basis of finite element methods for solid continua. *Internat. J. Numer. Methods Engrg.*, 1(1):3–28, 1969.

[287] P. Tong. New displacement hybrid finite element models for solid continua. *Internat. J. Numer. Methods Engrg.*, 2:73–83, 1970.

[288] T. H. H. Pian. Hybrid methods. In S. T. Fenves et al., editor, *Numerical and Computational Methods in Structural Mechanics*, pages 59–78, New York, 1973. Academic Press, Inc.

[289] T. H. H. Pian and K. Sumihara. Rational approach for assumed stress finite elements. *Internat. J. Numer. Methods Engrg.*,

20(9):1685–1695, 1985.

[290] J. H. Argyris et al. Finite element method—the natural approach. *Comput. Methods Appl. Mech. Engrg.*, 17/18:1–106, 1979.

[291] H. Kardestuncer (Editor). *Finite Element Handbook.* McGraw-Hill Book Company, New York, 1987.

[292] R. W. Hamming. *Numerical Methods for Scientists and Engineers.* McGraw-Hill, New York, 1962. p. 185.

[293] R. Beckett and J. Hurt. *Numerical Calculations and Algorithms.* McGraw-Hill, New York, 1967. p. 180.

[294] J. M. T. Thompson and A. C. Walker. A general theory for the branching analysis of discrete structural systems. *Internat. J. Solids Structures*, 5:281–288, 1969.

[295] E. Riks. The application of Newton's method to the problem of elastic stability. *J. Appl. Mech.*, 39:1060–1066, 1972.

[296] E. Riks. An incremental approach to the solution of snapping and buckling problems. *Int. J. Solids Structures*, 15:529–551, 1979.

[297] E. Ramm. Strategies for tracing the nonlinear response near limit points. In W. Wunderlich, E. Stein, and K.-J. Bathe, editors, *Nonlinear Finite Element Analysis in Structural Mechanics*, pages 63–89, Berlin, 1981. Springer-Verlag.

Index

F

R

V

W

Y

Milton Keynes UK
Ingram Content Group UK Ltd.
UKHW021927071024
449327UK00022B/1712

9 780367 395698